The IMA Volumes
in Mathematics
and its Applications

Volume 45

Series Editors
Avner Friedman   Willard Miller, Jr.

Institute for Mathematics and
its Applications
IMA

The **Institute for Mathematics and its Applications** was established by a grant from the National Science Foundation to the University of Minnesota in 1982. The IMA seeks to encourage the development and study of fresh mathematical concepts and questions of concern to the other sciences by bringing together mathematicians and scientists from diverse fields in an atmosphere that will stimulate discussion and collaboration.

The IMA Volumes are intended to involve the broader scientific community in this process.

Avner Friedman, Director
Willard Miller, Jr., Associate Director

* * * * * * * * * *

IMA ANNUAL PROGRAMS

1982–1983	Statistical and Continuum Approaches to Phase Transition
1983–1984	Mathematical Models for the Economics of Decentralized Resource Allocation
1984–1985	Continuum Physics and Partial Differential Equations
1985–1986	Stochastic Differential Equations and Their Applications
1986–1987	Scientific Computation
1987–1988	Applied Combinatorics
1988–1989	Nonlinear Waves
1989–1990	Dynamical Systems and Their Applications
1990–1991	Phase Transitions and Free Boundaries
1991–1992	Applied Linear Algebra
1992–1993	Control Theory and its Applications
1993–1994	Emerging Applications of Probability

IMA SUMMER PROGRAMS

1987	Robotics
1988	Signal Processing
1989	Robustness, Diagnostics, Computing and Graphics in Statistics
1990	Radar and Sonar
1990	Time Series
1991	Semiconductors
1992	Environmental Studies: Mathematical, Computational, and Statistical Analysis

* * * * * * * * * *

SPRINGER LECTURE NOTES FROM THE IMA:

The Mathematics and Physics of Disordered Media

Editors: Barry Hughes and Barry Ninham
(Lecture Notes in Math., Volume 1035, 1983)

Orienting Polymers

Editor: J.L. Ericksen
(Lecture Notes in Math., Volume 1063, 1984)

New Perspectives in Thermodynamics

Editor: James Serrin
(Springer-Verlag, 1986)

Models of Economic Dynamics

Editor: Hugo Sonnenschein
(Lecture Notes in Econ., Volume 264, 1986)

David Brillinger   Peter Caines   John Geweke
Emanuel Parzen   Murray Rosenblatt   Murad S. Taqqu

Editors

New Directions in
Time Series Analysis

Part I

With 55 Illustrations

Springer-Verlag
New York Berlin Heidelberg London Paris
Tokyo Hong Kong Barcelona Budapest

David Brillinger
Department of Statistics
University of California
Berkeley, CA 94720
USA

Peter Caines
Department of Electrical Engineering
McGill University
Montreal, Quebec H3A 2A7
Canada

John Geweke
Department of Economics
University of Minnesota
Minneapolis, MN 55455
USA

Emanuel Parzen
Department of Statistics
Texas A&M University
College Station, TX 77843-3143
USA

Murray Rosenblatt
Mathematics Department
University of California, San Diego
La Jolla, CA 92093
USA

Murad S. Taqqu
Department of Mathematics
Boston University
Boston, MA 02215
USA

Mathematics Subject Classifications (1991): 62M10, 90A20, 62-06

Library of Congress Cataloging-in-Publication Data
New directions in time series analysis, part I \ [edited by] David Brillinger . . . [et al.]
        p.   cm. — (The IMA volumes in mathematics and its
    applications ; v. 45)
    "Based on the proceedings of the IMA summer program . . . "—Foreword.
    Includes bibliographical references.
    ISBN 0-387-97896-8 (U.S.)
    1. Time-series analysis.   I. Brillinger, David.   II. University of Minnesota.
    Institute for Mathematics and Its Applications.   III. Series.
    QA280.N47   1992   92-22697

Printed on acid-free paper.

Production managed by Hal Henglein; manufacturing supervised by Vincent R. Scelta.
Camera-ready copy prepared by the IMA.
Printed and bound by Edwards Brothers, Inc., Ann Arbor, MI.
Printed in the United States of America.

9 8 7 6 5 4 3 2 1

ISBN 0-387-97896-8 Springer-Verlag New York Berlin Heidelberg
ISBN 3-540-97896-8 Springer-Verlag Berlin Heidelberg New York

Math

The IMA Volumes
in Mathematics and its Applications

Current Volumes:

Volume 1: Homogenization and Effective Moduli of Materials and Media
Editors: Jerry Ericksen, David Kinderlehrer, Robert Kohn, J.-L. Lions

Volume 2: Oscillation Theory, Computation, and Methods of Compensated Compactness
Editors: Constantine Dafermos, Jerry Ericksen,
David Kinderlehrer, Marshall Slemrod

Volume 3: Metastability and Incompletely Posed Problems
Editors: Stuart Antman, Jerry Ericksen, David Kinderlehrer, Ingo Müller

Volume 4: Dynamical Problems in Continuum Physics
Editors: Jerry Bona, Constantine Dafermos, Jerry Ericksen,
David Kinderlehrer

Volume 5: Theory and Applications of Liquid Crystals
Editors: Jerry Ericksen and David Kinderlehrer

Volume 6: Amorphous Polymers and Non-Newtonian Fluids
Editors: Constantine Dafermos, Jerry Ericksen, David Kinderlehrer

Volume 7: Random Media
Editor: George Papanicolaou

Volume 8: Percolation Theory and Ergodic Theory of Infinite Particle Systems
Editor: Harry Kesten

Volume 9: Hydrodynamic Behavior and Interacting Particle Systems
Editor: George Papanicolaou

Volume 10: Stochastic Differential Systems, Stochastic Control Theory and Applications
Editors: Wendell Fleming and Pierre-Louis Lions

Volume 11: Numerical Simulation in Oil Recovery
Editor: Mary Fanett Wheeler

Volume 12: Computational Fluid Dynamics and Reacting Gas Flows
Editors: Bjorn Engquist, M. Luskin, Andrew Majda

FOREWORD

This IMA Volume in Mathematics and its Applications

NEW DIRECTIONS IN TIME SERIES ANALYSIS, PART I

is based on the proceedings of the IMA summer program "New Directions in Time Series Analysis." We are grateful to David Brillinger, Peter Caines, John Geweke, Emanuel Parzen, Murray Rosenblatt, and Murad Taqqu for organizing the program and we hope that the remarkable excitement and enthusiasm of the participants in this interdisciplinary effort are communicated to the reader.

Avner Friedman

Willard Miller, Jr.

PREFACE

Time Series Analysis is truly an interdisciplinary field because development of its theory and methods requires interaction between the diverse disciplines in which it is applied. To harness its great potential, strong interaction must be encouraged among the diverse community of statisticians and other scientists whose research involves the analysis of time series data. This was the goal of the IMA Workshop on "New Directions in Time Series Analysis."

The workshop was held July 2–July 27, 1990 and was organized by a committee consisting of Emanuel Parzen (chair), David Brillinger, Murray Rosenblatt, Murad S. Taqqu, John Geweke, and Peter Caines. Constant guidance and encouragement was provided by Avner Friedman, Director of the IMA, and his very helpful and efficient staff.

The workshops were organized by weeks. It may be of interest to record the themes that were announced in the IMA newsletter describing the workshop: 1. Non-linear and non-Gaussian models and processes (higher order moments and spectra, nonlinear systems, applications in astronomy, geophysics, engineering, simulation); 2. Self-similar processes and long-range dependence (time series with long memory, fractals, 1/f noise, stable noise); 3. Interactions of time series analysis and statistics (topics include information, model identification, categorical valued time series, nonparametric and semiparametric methods); 4. Time series research common to engineers and economists (topics include modeling of multivariate (possibly non-stationary) time series, especially by state space and adaptive methods).

The office and computing facilities provided by the IMA helped stimulate an extremely fruitful and enthusiastic meeting. Participants formed many new friendships and research collaborations. The goal of these volumes (which are a record of the papers presented at the workshop) is to enable a worldwide community dispersed in space and time to share the excitement experienced by their colleagues who were present at IMA in July 1990.

This volume is dedicated to the promotion of interdisciplinary and international collaboration, to the worldwide community of researchers who develop and apply methods of statistical analysis of time series, and to the many colleagues and staff members whose cheerful help made the workshop a successful happening which was enjoyed socially and intellectually by all participants.

We would like to express our appreciation to the agencies that provided support for the workshop: The Air Force Office of Scientific Research, the Army Research Office, the National Security Agency, the National Science Foundation and the Office of Naval Research.

David Brillinger
Peter Caines
John Geweke
Emanuel Parzen
Murray Rosenblatt
Murad S. Taqqu

CONTENTS

CONTENTS

NEW DIRECTIONS IN TIME SERIES ANALYSIS, PART II

COMPUTATION OF LOCAL POWER SPECTRA
BY THE WINDOWED FOURIER TRANSFORM

M. ASCH*, W. KOHLER†, G. PAPANICOLAOU*,
M. POSTEL*, AND B. WHITE‡

Abstract. We show how to use the windowed Fourier transform to estimate efficiently the local power spectral density of reflected signals from randomly layered media.

1. Introduction. A pulse incident on a randomly layered half space produces a reflected signal that is a stochastic process with relatively complicated structure. In [1-13] we have initiated a study of reflected signals starting from the underlying stochastic scattering problem and using ideas and methods from the asymptotic analysis of stochastic equations. Our interest in pulse reflection by randomly layered media was stimulated by the numerical simulations of Richards and Menke [15,16] who found a number of interesting qualitative properties of the reflected signals.

Here we show how to estimate the local power spectral density of reflected signals in a robust and efficient way. This is particularly important in inverse problems [4,9] where the reflected signals are used to extract information about the large-scale properties of the scattering medium.

2. Brief review of the theory. We will summarize briefly from [2] the necessary facts about reflected signals and their properties that we need. The two main hypotheses in [2] that allow detailed mathematical treatment of pulse reflection are the following:

- The medium is layered.

- The incident pulse is broad compared to the typical width of the layers but narrow compared to the large scale variations in the properties of the scattering half space.

If in dimensionless variables macroscopic variations of the medium properties occur on an order one scale then the pulse width is of order ϵ, which is a small positive parameter, and the size of the layers is of order ϵ^2 which is much smaller than ϵ.

Let $R_f^\epsilon(t)$ be the reflected signal on the surface with $f = f(t)$ denoting the shape of the incident pulse. In [2], and in full detail in [13], we show that as ϵ tends to zero $R_f^\epsilon(t)$ tends to a locally stationary process. This means in particular the following. Let

$$(1) \qquad C_f^\epsilon(t,\sigma) = E\{R_f^\epsilon(t - \frac{\epsilon\sigma}{2})R_f^\epsilon(t + \frac{\epsilon\sigma}{2})\}$$

* Courant Institute of Mathematical Sciences, 251 Mercer Street, New York, NY 10012

† Department of Mathematics, Virginia Polytechnic Institute and State University, Blacksburg, VA 24061

‡ Exxon Research and Engineering Company, Route 22 East, Annandale, NJ 08801

be the local or windowed correlation function of the signal at time t, the window center, for a window of size ϵ when σ runs over an interval that is of order one as ϵ tends to zero. Then

$$(2) \qquad \lim_{\epsilon \downarrow 0} C_f^\epsilon(t, \sigma) = \frac{1}{2\pi} \int_{-\infty}^{\infty} e^{i\omega\sigma} |\hat{f}(\omega)|^2 \Lambda(\omega, t) d\omega$$

where $\hat{f}(\omega)$ is the Fourier transform of the incident pulse and $\Lambda(\omega, t)$ is the local power spectral density. The convergence in (2) is pointwise in σ but only weakly in t which means, [13], that (2) holds when both sides are multiplied by a smooth function of compact support and integrated over t. The local power spectral density $\Lambda(\omega, t)$ is in general difficult to calculate explicitly because we must solve a system of partial differential equations for its determination [2]. However we know that $\Lambda(\omega, t)$ is nonnegative, as should be for a power spectral density, and when the medium properties have no large-scale variations then it can be computed explicitly

$$(3) \qquad \Lambda(\omega, t) = \frac{\alpha\omega^2}{(1 + \alpha\omega^2 t)^2}$$

Here α is a positive constant that depends on the statistics of the fluctuations in the medium properties. Both heuristic arguments [2] and numerical simulations [3] indicate that the law of the locally stationary process that is the weak limit of the reflected signals is Gaussian, but we have not been able to prove this so far.

From (1) and (2) we are led to consider the expression

$$(4) \qquad \int_{-\infty}^{\infty} e^{-i\omega\sigma} C_f^\epsilon(t, \sigma) d\sigma$$

as an approximation for $|\hat{f}(\omega)|^2 \Lambda(\omega, t)$ for ϵ small. We note first that although (4) is well defined for $\epsilon > 0$ it is not a nonnegative function. This is well known in signal analysis [18]. Second, if we try to use (4) as an estimate for $|\hat{f}(\omega)|^2 \Lambda(\omega, t)$ it will depend sensitively on how we truncate the infinite Fourier integral. With enough experimenting with the data we can make a suitable version of (4) provide useful estimates for $\Lambda(\omega, t)$. This was done in [3] as well as in [6] and [7] with data generated by direct numerical simulation of the scattering problem. It was actually done in a somewhat different way which is more efficient computationally and is described in detail in [3]. The question is then: what is a good way to estimate $\Lambda(\omega, t)$ from $R_f^\epsilon(t)$ when we only know the parameter ϵ? By good we mean primarily robust, that is, by an algorithm that does not need fine tuning each time we look at a different set of data. Of course the estimation must also be unbiased. We describe next a method for estimating $\Lambda(\omega, t)$ based on the local or windowed Fourier transform.

3. The windowed Fourier transform. Define the family of window functions $g^\delta(t)$ whose support is of order δ by

$$(5a) \qquad \text{support of } g^\delta = \frac{2\pi\delta}{\omega_0}$$

$$(5b) \qquad \pi < \omega_0 t_0 < 2\pi$$

(5c)
$$\sum_{n=-\infty}^{\infty} |g^\delta(t - n\delta t_0)|^2 = \frac{1}{\delta t_0}$$

(5d)
$$\int_{-\infty}^{\infty} |g^\delta(t)|^2 dt = 1$$

Note here that t_0 and ω_0 are of order 1. An example of such a function is given in [19,20] defined on $[-\frac{\delta\pi}{\omega_0}, \frac{\delta\pi}{\omega_0}]$ by

(6) $\quad g^\delta(t) = \begin{cases} (\delta t_0)^{-1/2} \sin\left[\frac{\pi}{2}\nu\left(\lambda^{-1}(\frac{\delta\pi}{\omega_0} + t)\right)\right] & \text{for } \frac{-\delta\pi}{\omega_0} \le t \le \frac{-\delta\pi}{\omega_0} + \lambda \\ (\delta t_0)^{-1/2} & \text{for } \frac{-\delta\pi}{\omega_0} + \lambda \le t \le \frac{\delta\pi}{\omega_0} - \lambda \\ (\delta t_0)^{-1/2} \cos\left[\frac{\pi}{2}\nu\left(\lambda^{-1}(t - \frac{\delta\pi}{\omega_0} + \lambda)\right)\right] & \text{for } \frac{\delta\pi}{\omega_0} - \lambda \le t \le \frac{\delta\pi}{\omega_0} \end{cases}$

where $\nu(t) = 0$ if $t \le 0$ and $\nu(t) = 1$ if $t \ge 1$ and $\lambda = \frac{2\delta\pi}{\omega_0} - \delta t_0$. Define the doubly indexed family g^δ_{mn} by

(7)
$$g^\delta_{mn}(t) = e^{-i\omega_0 mt/\delta} g^\delta(t - n\delta t_0)$$

Then [19] if $R(t) \in C_o^\infty(\mathbf{R})$ we have the representation

(8)
$$R(t) = \frac{\omega_0 t_0}{2\pi} \sum_{mn=-\infty}^{\infty} R_{mn} g^\delta_{mn}(t)$$

where

(9)
$$R_{mn} = <R, g^\delta_{mn}> = \int_{-\infty}^{+\infty} R(s)\overline{g^\delta_{mn}(s)}ds$$

We also have Parseval's identity

(10)
$$\int_{-\infty}^{\infty} |R(t)|^2 dt = \frac{\omega_0 t_0}{2\pi} \sum_{nm=-\infty}^{\infty} |R_{mn}|^2$$

from which we can extend (8) to all of $L^2(\mathbf{R})$. The meaning of the conditions (5a) and (5b) is clear from the expansion (8): the function $R(t)$ is replaced by its periodic extension outside an interval $\frac{2\pi\delta}{\omega_0}$ which is larger than the spacing of the windows δt_0 in which the function is represented by a Fourier series. Condition (5c) tells us that the translated cutoff functions form a partition of unity.

Let $R^\epsilon_f(t)$ be the reflected signal and let $R^{\delta\epsilon}_{mn} = <R^\epsilon_f, g^\delta_{mn}>$ be the local Fourier coefficients of the reflected signal, which are now random variables. We will show that if the windowed correlation function satisfies the properties

(11)
$$\lim_{\epsilon\downarrow 0} \iint \phi(\tau, \sigma)C^\epsilon_f(\tau, \sigma)d\tau d\sigma = \iint \phi(\tau, \sigma)C_f(\tau, \sigma)d\tau d\sigma$$

for each smooth function ϕ of compact support,

(12)
$$\lim_{M\uparrow\infty} \overline{\lim_{\epsilon\downarrow 0}} \sup_{|\tau|\le K} \int_{|\sigma|\ge M} |C^\epsilon_f(\tau, \sigma)|d\sigma = 0$$

for each $K < \infty$, and

(13)
$$\lim_{\delta\downarrow 0} \int_{|\tau-t|\le\delta} d\tau \int d\sigma |C_f(\tau, \sigma) - C_f(t, \sigma)| = 0$$

for each t, then

(14) $$|\hat{f}(\omega)|^2 \Lambda(\omega, t) = \lim_{\delta \downarrow 0} \lim_{\epsilon \downarrow 0} \sum_{mn=-\infty}^{\infty} E\{|R_{mn}^{\delta\epsilon}|^2\} G_{mn}^{\delta\epsilon}(t, \omega)$$

where

(15) $$G_{mn}^{\delta\epsilon}(t, \omega) = \frac{\omega_0 t_0}{2\pi} \int_{-\infty}^{\infty} e^{-i(\omega - \omega_0 m \epsilon \delta^{-1})\sigma} g^\delta(t - n\delta t_0 - \frac{\epsilon\sigma}{2}) g^\delta(t - n\delta t_0 + \frac{\epsilon\sigma}{2}) d\sigma$$

The meaning of (14) is that given the reflected signal $R_f^\epsilon(t)$, we compute its local Fourier coefficients $R_{mn}^{\delta\epsilon}$ relative to the cutoff function g^δ with δ a multiple of ϵ, say $\delta = 4\epsilon$. Then with t of the form $t = N\delta t_0$ and ω of the form $\omega_0 M \epsilon \delta^{-1}$ we estimate the local power spectral density by

(16) $$|\hat{f}(\omega)|^2 \Lambda(\omega, t) \approx |R_{MN}^{\delta\epsilon}|^2 G_0 + |R_{M+1N}^{\delta\epsilon}|^2 G_1 + |R_{M-1N}^{\delta\epsilon}|^2 G_{-1}$$

where

$$G_M = \frac{\omega_0 t_0}{2\pi} \int_{-\infty}^{\infty} e^{-iM\omega_0 \epsilon \delta^{-1}\sigma} g^\delta(-\frac{\epsilon\sigma}{2}) g^\delta(\frac{\epsilon\sigma}{2}) d\sigma$$

We only take three terms in the sum over m in (14) because G_M decays very rapidly with M. Also, when t and ω take the above special values then only one term in the sum over n enters. A naive estimate for $|\hat{f}(\omega)|^2 \Lambda(\omega, t)$ would be a multiple of $|R_{MN}^{\delta\epsilon}|^2$, which is comparable to the periodogram for stationary processes. Our estimate is a very significant improvement in practice. We note that G_1 and G_{-1} are small negative numbers which means that the additional terms in (16) are a small but significant correction to the main term $|R_{MN}^{\delta\epsilon}|^2 G_0$. Of the three hypotheses (11), (12) and (13) needed for the validity of (14), (12) has not been shown yet for reflected signals. The other two are shown in [13].

The above interpretation that we have given to (14) is not the only one possible, especially when δ is chosen to be a multiple of ϵ. When δ is not linked to ϵ, the variance of the estimate (16) is small [13], although we do not examine it in detail here. When, however, $\delta = \gamma\epsilon$, with γ a constant, and ϵ goes to zero, the situation is somewhat different. With t fixed and of the form $t = N\gamma\epsilon t_0$, where $N = 1, 2, 3, \ldots$ and ϵ goes to zero along the subsequence $\epsilon = t/N\gamma t_0$, we have

(17) $$|\hat{f}(\omega)|^2 \Lambda(\omega, t) = \lim_{\gamma \uparrow \infty} \lim_{\epsilon \downarrow 0} \sum_{mn=-\infty}^{\infty} E\{|R_{mn}^{\gamma\epsilon \, \epsilon}|^2\} G_{mn}^{\gamma\epsilon \, \epsilon}(t, \omega)$$

provided that in addition to (12), the local correlation function C_f^ϵ satisfies the uniform continuity condition

(18) $$\lim_{\delta \downarrow 0} \lim_{\epsilon \downarrow 0} \sup_{|\tau - t| \le \delta} \sup_{|\sigma| \le M} |C_f^\epsilon(\tau, \sigma) - C_f^\epsilon(t, \sigma)| = 0$$

for each t fixed and $M < \infty$, and

(19) $$\lim_{\epsilon \downarrow 0} C_f^\epsilon(t, \sigma) = C_f(t, \sigma)$$

pointwise in t and σ. Conditions (18) and (19) on the local covariance $C_f^\epsilon(\tau, \sigma)$ are much stronger than (11) and (13), needed for (14), and they have not been proven yet for reflected signals. We will consider the proof of (17) after the one for (14).

To prove (14) we proceed as follows. Note first that

$$E\{|R_{mn}^{\delta\epsilon}|^2\} = \iint E\{R_f^\epsilon(t)R_f^\epsilon(s)\}g_{mn}^\delta(t)g_{mn}^\delta(s)dtds$$

$$= \epsilon \iint C_f^\epsilon(\tau,\sigma)e^{-iw_0 m\epsilon\sigma/\delta}g^\delta(\tau - n\delta t_0 - \frac{\epsilon\sigma}{2})g^\delta(\tau - n\delta t_0 + \frac{\epsilon\sigma}{2})d\tau d\sigma$$

where we have made the change of variables

$$t = \tau - \epsilon\frac{\sigma}{2} \qquad \text{and} \qquad s = \tau + \epsilon\frac{\sigma}{2}$$

Thus

$$\sum_{mn=-\infty}^{\infty} E\{|R_{mn}^{\delta\epsilon}|^2\}G_{mn}^{\delta\epsilon}(t,\omega) =$$

$$\frac{\epsilon\omega_0 t_0}{2\pi} \sum_{mn=-\infty}^{\infty} \iiint d\tau d\sigma d\tilde\sigma C_f^\epsilon(\tau,\sigma)e^{-i\omega\tilde\sigma}e^{i(\omega_0\tilde\sigma - \omega_0\sigma)m\epsilon/\delta} \cdot$$

$$g^\delta(\tau - n\delta t_0 - \frac{\epsilon\sigma}{2})g^\delta(\tau - n\delta t_0 + \frac{\epsilon\sigma}{2})g^\delta(t - n\delta t_0 - \frac{\epsilon\tilde\sigma}{2})g^\delta(t - n\delta t_0 + \frac{\epsilon\tilde\sigma}{2})$$

We now use the Poisson summation formula [17] in the form

(20) $$\frac{1}{2\pi} \sum_{m=-\infty}^{\infty} e^{i(\omega_0\tilde\sigma - \omega_0\sigma)m\epsilon/\delta} = \frac{\delta}{\epsilon\omega_0} \sum_{m=-\infty}^{\infty} \delta(\tilde\sigma - \sigma + \frac{2\pi m\delta}{\epsilon\omega_0})$$

to get

$$\sum_{mn=-\infty}^{\infty} E\{|R_{mn}^{\delta\epsilon}|^2\}G_{mn}^{\delta\epsilon}(t,\omega) = \iint d\tau d\sigma C_f^\epsilon(\tau,\sigma)e^{-i\omega\sigma} \cdot$$

$$\delta t_0 \sum_{m=-1}^{1} e^{2\pi im\delta/\omega_0\epsilon} \sum_{n=-\infty}^{\infty} g^\delta(\tau - n\delta t_0 - \frac{\epsilon\sigma}{2})g^\delta(\tau - n\delta t_0 + \frac{\epsilon\sigma}{2}) \cdot$$

$$g^\delta(t - n\delta t_0 + \frac{\epsilon\sigma}{2} + \frac{m\delta\pi}{\omega_0})g^\delta(t - n\delta t_0 - \frac{\epsilon\sigma}{2} - \frac{m\delta\pi}{\omega_0})$$

When the Poisson summation formula is used, only the terms $m = 0, +1, -1$ in the sum contribute because of the support properties (5a) and (5b) of g^δ and this gives the above equality. Again because of the support properties of g^δ, the τ integration is only over the interval $|t - \tau| \leq 3\pi\delta\omega_0^{-1}$ and the σ integration over $|\sigma| \leq 3\pi\delta/\epsilon\omega_0$. Using hypotheses (11) and (12) we can pass to the limit $\epsilon \to 0$ and get

$$\lim_{\epsilon\downarrow 0} \sum_{mn=-\infty}^{\infty} E\{|R_{mn}^{\delta\epsilon}|^2\}G_{mn}^{\delta\epsilon}(t,\omega) =$$

$$\iint d\tau d\sigma C_f(\tau,\sigma)e^{-i\omega\sigma}\delta t_0 \sum_{n=-\infty}^{\infty} |g^\delta(\tau - n\delta t_0)|^2|g^\delta(t - n\delta t_0)|^2$$

The terms $m = \pm 1$ in the sum vanish because of the support properties (5a) and (5b) of g^δ and only the term $m = 0$ survives. Since $C_f(\tau,\sigma)$ is continuous in τ in the sense of (13) and in view of (5c) and (5d)

$$\lim_{\delta\downarrow 0}\lim_{\epsilon\downarrow 0} \sum_{mn=-\infty}^{\infty} E\{|R_{mn}^{\delta\epsilon}|^2\}G_{mn}^{\delta\epsilon}(t,\omega) = \int d\sigma C_f(\tau,\sigma)e^{-i\omega\sigma}$$

which is what we wanted to prove.

To prove (17) we note that with $\delta = \gamma\epsilon$ and ϵ going to zero so that $t = N\gamma\epsilon t_0$ with t fixed and $N = 1, 2, 3, ...$, we have

$$\sum_{mn=-\infty}^{\infty} E\{|R_{mn}^{\gamma\epsilon\ \epsilon}|^2\}G_{mn}^{\gamma\epsilon\ \epsilon}(t,\omega) = \iint d\tau d\sigma C_f^\epsilon(\tau,\sigma)e^{-i\omega\sigma} \ .$$

$$\gamma\epsilon t_0 \sum_{m=-1}^{1} e^{2\pi i m\gamma/\omega_0} \sum_{n=-\infty}^{\infty} g^{\gamma\epsilon}(\tau - n\epsilon\gamma t_0 - \frac{\epsilon\sigma}{2})g^{\gamma\epsilon}(\tau - n\epsilon\gamma t_0 + \frac{\epsilon\sigma}{2}) \ .$$

$$g^{\gamma\epsilon}((N-n)\epsilon\gamma t_0 + \frac{\epsilon\sigma}{2} + \frac{m\epsilon\gamma\pi}{\omega_0})g^{\gamma\epsilon}((N-n)\epsilon\gamma t_0 - \frac{\epsilon\sigma}{2} - \frac{m\epsilon\gamma\pi}{\omega_0})$$

The σ integral is decomposed into two parts, one over $|\sigma| \leq M$ and one over $|\sigma| > M$. The part over $|\sigma| > M$ is arbitrarily small uniformly in ϵ and γ if M is large enough because of (12) and also because $g^\delta(t) = \delta^{-1/2}g(t\delta^{-1})$ where $g(t)$ is a nonnegative function of compact support given explicitly by (6) with $\delta = 1$. In the integral over $|\sigma| \leq M$ we add and subtract the term $C_f^\epsilon(t,\sigma)$. By the uniform continuity condition (18) and the convergence (19), we see that we have

$$\lim_{\epsilon\downarrow 0} \sum_{mn=-\infty}^{\infty} E\{|R_{mn}^{\gamma\epsilon\ \epsilon}|^2\}G_{mn}^{\gamma\epsilon\ \epsilon}(t,\omega) = \lim_{M\to\infty}\lim_{\gamma\to\infty} \int_{|\sigma|\leq M} d\sigma C_f(t,\sigma)e^{-i\omega\sigma} \ .$$

$$t_0 \sum_{m=-1}^{1} e^{2\pi i m\gamma/\omega_0} \sum_{\tilde{n}=-\infty}^{\infty} \int ds\, g(s - \tilde{n}t_0 - \frac{\sigma}{2\gamma})g(s - \tilde{n}t_0 + \frac{\sigma}{2\gamma}) \ .$$

$$g(\tilde{n}t_0 + \frac{\sigma}{2\gamma} + \frac{m\pi}{\omega_0})g(\tilde{n}t_0 - \frac{\sigma}{2\gamma} - \frac{m\pi}{\omega_0})$$

which is equal to

$$\lim_{M\to\infty} \int_{|\sigma|\leq M} d\sigma C_f(t,\sigma)e^{-i\omega\sigma} \ .$$

$$t_0 \sum_{\tilde{n}=-\infty}^{\infty} \int ds\, g^2(s - \tilde{n}t_0)g(\tilde{n}t_0)g(\tilde{n}t_0)$$

Here the index $\tilde{n} = N - n$, the sum is over a finite number of terms, we have used again $g^\delta(t) = \delta^{-1/2}g(t\delta^{-1})$ and we have changed variables in the τ integration by letting $s = (\tau - t)(\epsilon\gamma)^{-1}$. Now only the term $m = 0$ survives in the sum because of the support properties of g and the result (17) follows from properties (5c) and (5d).

As we noted above the conditions (12), (18) and (19) on the local covariance have not been proven yet for reflected signals. Moreover, contrary to (14), the variance of the estimate (16) does not go to zero as ϵ goes to zero when $\delta = \gamma\epsilon$.

REFERENCES

[1] R. BURRIDGE, G. PAPANICOLAOU AND B. WHITE, *Statistics for pulse reflection from a randomly layered medium*, Siam J. Appl. Math. **47**, 146-168, 1987.

[2] R. BURRIDGE, G. PAPANICOLAOU, P. SHENG AND B. WHITE, *Probing a Random Medium with a Pulse*, Siam J. Appl. Math. **49**, 582-607, 1989.

[3] M. ASCH, G. PAPANICOLAOU, M. POSTEL, P. SHENG AND B. WHITE, *Frequency content of randomly scattered signals. Part I*, Wave Motion **12**, 429-450, 1990.

[4] G. PAPANICOLAOU, M. POSTEL, P. SHENG AND B. WHITE, *Frequency content of randomly scattered signals. Part II: Inversion*, Wave Motion **12**, 527-549, 1990.

[5] W. KOHLER, G. PAPANICOLAOU AND B. WHITE, *Reflection of waves generated by a point source over a randomly layered medium*, Wave Motion **13**, 53-87, 1991.

[6] P. SHENG, Z.-Q. ZHANG, B. WHITE AND G. PAPANICOLAOU, *Multiple scattering noise in one dimension: universality through localization length scales*, Phys. Rev. Letters **57**, 1000-1003, 1986.

[7] B. WHITE, P. SHENG, Z.-Q. ZHANG AND G. PAPANICOLAOU, *Wave localization characteristics in the time domain*, Phys. Rev. Letters **59**, 1918-1921, 1987.

[8] P. SHENG, B. WHITE, Z.Q. ZHANG AND G.PAPANICOLAOU, *Wave localization and multiple scattering in randomly-layered media*, in "Scattering and Localization of Classical Waves in Random Media", Ping Sheng editor, World Scientific , 563-619, 1990.

[9] B. WHITE, P. SHENG, M. POSTEL AND G. PAPANICOLAOU, *Probing through cloudiness: Theory of statistical inversion for multiply scattered data*, Phys. Rev. Letters **63**, 2228-2231, 1989.

[10] R. BURRIDGE, G. PAPANICOLAOU AND B. WHITE, *One dimensional wave propagation in a highly discontinuous medium*, Wave Motion, **10**, 19-44, 1988.

[11] W. KOHLER, G. PAPANICOLAOU, M. POSTEL AND B. WHITE, *Reflection of Pulsed Electromagnetic Waves from a Randomly Stratified Half Space*, to appear in the Journal of the Optical Society of America, 1991.

[12] M. ASCH, W. KOHLER, G. PAPANICOLAOU, M. POSTEL AND B. WHITE, *Frequence content of randomly scattered signals*, SIAM Review, Dec. 1991, to appear.

[13] G. PAPANICOLAOU AND S. WEINRYB, *A functional limit theorem for waves reflected by a random medium*, Appl. Math. and Optimization, submitted, 1990.

[14] K. AKI AND P.G. RICHARDS, *Quantitative Seismology*, W.H. Freeman, San Francisco, 1980.

[15] P.G. RICHARDS AND W. MENKE, *The apparent attenuation of a scattering medium* Bull. of Seismol. Soc. of Am. **73**, 1005-1021, 1983.

[16] W. MENKE, *A formula for the apparent attenuation of acoustic waves in randomly layered media*, Geophys. J. Res. Astr. Soc. **75**, 541-544, 1983.

[17] R. COURANT AND D. HILBERT, *Methods of Mathematical Physics, Volume I*, Interscience, New York, 1953.

[18] M.B. PRIESTLEY *Non-linear and non-stationary time series analysis*, Academic Press 1988.

[19] I. DAUBECHIES, *Orthonormal bases of compactly supported wavelets*, Comm. Pure Appl. Math. **41**, 909-996, 1988.

[20] Y. MEYER, *Ondelettes*, Herman, Paris 1989.

AUTOREGRESSIVE ESTIMATION OF THE PREDICTION MEAN SQUARED ERROR AND AN R² MEASURE: AN APPLICATION

R.J. BHANSALI*

Abstract. For predicting the future values of a stationary process, $\{x_t\}(t = 0, \pm 1, \pm 2, \dots)$, on the basis of its past, two key parameters are the h-step mean squared error of prediction, $V(h)$ $(h \geq 1)$, and $Z(h) = \{R(0) - V(h)\}/R(0)$, the corresponding measure, in an R^2 sense, of predictability of the process from its past alone, where $R(0)$ denotes the variance of x_t. The estimation of $V(h)$ and $Z(h)$ from a realization of T consecutive observations of $\{x_t\}$ is considered, without requiring that the process follows a finite parameter model. Three different autoregressive estimators are considered and their large sample properties discussed. An illustration of the results obtained with real data is given by applying the estimation procedures to the well-known Beveridge What Price Index, the Wolf Sunspot Series, as well as to a Referigerator Sales Series.

1. Introduction. Let $\{x_t\}$ $(t = 0, \pm 1, \pm 2, \dots)$ be a stationary process with mean 0, covariance function $R(u) = E(x_t x_{t+u})$ $(t, u = 0, \pm 1, \pm 2, \dots)$, correlation function $r(u) = R(u)/R(0)$ and the spectral density function

$$f(\mu) = (2\pi)^{-1} \sum_{s=-\infty}^{\infty} R(s) \exp(-is\mu).$$

Assume that the $R(u)$ are absolutely summable and $f(\mu)$ strictly positive, that is,

$$\sum_{u=-\infty}^{\infty} |R(u)| < \infty, \qquad f(\mu) > 0 \qquad (-\pi \leq \mu \leq \pi).$$

Then it is well known (Brillinger, 1975) that $\{x_t\}$ has the one-sided moving average representation

$$(1.1) \qquad x_t = \sum_{j=0}^{\infty} b(j)\varepsilon_{t-j}, \qquad b(0) = 1,$$

and the one-sided autoregressive representation

$$\sum_{j=0}^{\infty} a(j)x_{t-j} = \varepsilon_t, \qquad a(0) = 1,$$

where $\{\varepsilon_t\}$ is a sequence of uncorrelated random variables with mean 0 and variance σ^2 and the coefficients $b(j)$ and $a(j)$ are also absolutely summable, that is,

$$\sum_{j=0}^{\infty} |b(j)| < \infty, \sum_{j=0}^{\infty} |a(j)| < \infty.$$

*Department of Statistics and Computational Mathematics, The University of Liverpool, Victoria Building Brownlow Hill, P.O. Box 147 Liverpool L69 3BX, England

Suppose now that the infinite past, $\{x_t,\ t \leq 0\}$ say, is known and the future values, $\{x_h,\ h \geq 1\}$, are unknown. As is well-known (Whittle, 1963), the linear least-squares predictor of x_h is given by

$$(1.2) \qquad \overline{x}_0(h) = \sum_{j=h}^{\infty} b(j)\varepsilon_{h-j}$$

and the corresponding h-step mean squared error of prediction by

$$(1.3) \qquad V(h) = E[\{x_h - \overline{x}_0(h)\}^2] = \sigma^2 \sum_{j=0}^{h-1} b^2(j) \qquad (h \geq 1).$$

The ratio $W(h) = V(h)/R(0)$ measures the proportion of variance of x_h that is unexplained by $\overline{x}_0(h)$. Thus the quantity

$$(1.4) \qquad Z(h) = \{R(0) - V(h)\}/R(0) = 1 - W(h)$$

measures the proportion of variance of x_h that can be explained by $\overline{x}_0(h)$. By analogy with the classical regression theory, the $Z(h)$ provide, in an R^2 sense, a measure of predictability of the future values of the process from a knowledge of the infinite past.

We have, for all $h \geq 1$,

$$(1.5) \qquad 0 \leq Z(h) < 1,$$

and that $\{Z(h)\}^{\frac{1}{2}}$ is the correlation coefficient between x_h and $\overline{x}_0(h)$ (Jewell and Bloomfield, 1983).

Suppose that only $x_1, \ldots, x_T$ have been observed. We consider the question of estimating $V(h)$ and $Z(h)$ without assuming a finite parameter model for $\{x_t\}$. Three different autoregressive estimators of $V(h)$ are considered:

The basic idea behind the first two is to construct estimators of $V(h)$ by using its functional form (1.3) and replacing the unknown $b(j)$ and σ^2 by their respective estimates. The latter in turn are obtained from an initial autoregression fitted to the data by adopting two different methods considered by Bhansali (1989) for this purpose. As discussed there, the first method is implicit in the work of Hannan and Rissanen (1982), who deal with a different problem however. The second on the other hand provides 'plug in' estimators of the $b(j)$ from that of the estimated autoregressive coefficients.

The third approach by contrast is more direct and its motivating idea is to estimate $V(h)$ (and the prediction constants) by fitting a different autoregression for each h, see Shibata (1980), Gersch and Kitagawa (1983) and Findley (1983).

In all three approaches, the corresponding estimators of $Z(h)$ are constructed from (1.4) on replacing $V(h)$ and $R(0)$ by their estimates.

A discussion of the large sample properties of the estimators is given in section 4. Examples illustrating their application to the well-known Beveridge Wheat Price Index, the Sunspot Series and the Referigerator Sales Series of Miller and Wichern (1977) are given in Section 5.

2. Some uses of V(h) and Z(h) and a review of the previous work.

There are several reasons for estimating $V(h)$ and $Z(h)$:

First, for obtaining a prediction interval for the unknown x_h, $h \geq 1$, without assuming a finite parameter model for $\{x_t\}$. Shibata (1977), Bhansali (1978) and Lewis and Reinsel (1985) consider autoregressive estimation of the linear least-squares predictor, $\overline{x}_0(h)$, and derive an asymptotic expression for the increased mean squared error of prediction due to the estimation of the prediction constants. This expression however involves the unknown $V(h)$. Thus, as discussed by Stine (1987), under a Gaussian assumption, an approximate prediction interval for x_h may be obtained by substituting $V(h)$ by one of the estimates considered here. It should be noted, however, that Stine takes a 'parametric' approach since he assumes that x_t is a finite autoregression of known order. The argument just outlined may be used therefore to provide a 'non-parametric' alternative to his procedure.

Second, for judging the extent to which it is meaningful to forecast the future of $\{x_t\}$ from its past alone. Several authors have recommended that an appropriate R^2 measure like $Z(h)$ should be reported for this purpose; see Nelson (1976) and Good (1988). Pierce (1979) also recommends the computation of a measure analogous to $Z(h)$, but for the case where there are exogeneous variables. Moreover, the procedure suggested there first requires an estimation of $Z(h)$ for determining the amount of variability in the dependent variable that may be attributed to its past alone.

A related concept is that of the Memory-Type of a time series. A time series may be said to have a 'long memory' if it can be predicted far or indefinitely into the future; 'no memory' if the future cannot be predicted at all from the past; and 'short memory' if it is only partially predictable into the future. Parzen (1981) suggests the use of an estimate of $Z(h)$ in which $V(h)$ is estimated from the 'plug in' method discussed earlier as a diagnostic aid for determining the memory type of an observed time series.

Of course, like all measures of correlation between two variables, an interpretation of the $Z(h)$ is likely to be subjective, and it also depends on the nature of the observed time series. Interval estimates of $Z(h)$ may be obtained however by appealing to the asymptotic distribution of the estimates of $Z(h)$ derived by the author and discussed further in section 4. Furthermore, if as discussed in section 1, the $Z(h)$ are estimated by fitting a different autoregression for each h, then this procedure also answers the question of how far it is meaningful to forecast on the basis of the past alone, since if for a given h the past is unhelpful for forecasting then the corresponding autoregressive order selected should be zero.

Thirdly, for model comparison and for providing a yardstick against which the predictive power of various special models may be gauged. Hannan and Nicholls (1977) advocate the use of a smoothed periodogram estimator of $V(1) = \sigma^2$ for this purpose. It is well-recognised, however, that two or more models may have very similar values of $V(1)$ and yet differ very much in their physical adequacy and probably also in their values of $V(h)$ for $h > 1$, see Whittle (1963, p. 38) for an illustration of this point. Thus a more informative procedure is to consider $V(h)$ for $h \geq 1$.

Previous work on the question of estimating $V(h)$ may be grouped under the following headings:

1) The parametric estimation of $V(h)$ by fitting either a pure autoregressive or an autoregressive moving average, ARMA, model of known order, see Fuller and Hasza (1981), Ansley and Newbold (1981) and Stine (1987).

2) The bootstrap estimation of $V(h)$ for a finite autoregressive process of known order, see Findley (1986), Stine (1987) and Thombs and Schucany (1990).

3) Bayes estimation of $V(h)$, see Geweke (1988), although his procedure actually provides an estimator of the mean squared error of prediction with estimated coefficients.

4) Direct estimation of $V(h)$ by fitting a different autoregressive moving average model of known order for each h, see Kabaila (1981) and Stoica and Soderstrom (1984).

5) The 'non-parametric' estimation of $V(1) = \sigma^2$ from the raw or slightly smoothed periodogram, see Davis and Jones (1968) and Hannan and Nicholls (1977), amongst others. Shibata (1977) examines the autoregressive model fitting approach for estimating $V(1)$, while Bhansali (1974) discusses the estimation of $V(h)$, $h \geq 1$, by factorizing a 'window' estimate of $f(\mu)$.

By contrast, the sampling properties of estimates of $Z(h)$ have previously received little attention; see, however, Battaglia and Bhansali (1987) who investigate an index analogous to $Z(1)$ but for linear interpolation of x_t from $\{x_{t-j},\ j \neq 0\}$.

An advantage of the autoregressive method for estimating $V(h)$ and $Z(h)$ is that the estimates may be constructed without assuming a finite parameter model for the process and the order of autoregression to be fitted may be selected by minimizing the Akaike information criterion or one of its variants. A parametric method does not seem to be appropriate in this context because as discussed earlier a motivation for estimating $V(h)$ and $Z(h)$ is in part to provide a yardstick against which the predictive powers of various models may be gauged. For this purpose, one wishes to impose few prior restrictions on the process and this corresponds to using a 'nonparametric' approach.

The estimation of $V(h)$ and $Z(h)$ from the raw or smoothed periodogram may also be considered. It is clear that, under appropriate regularity conditions, asymptotic results analogous to those discussed in section 4 for the autoregressive estimates would also hold for the 'window' estimates so constructed. To save space, however, we only consider the autoregressive estimates here. The question of a comparison between the 'autoregressive' and 'window' approaches has received much attention in the literature under various headings; a clear answer concerning the superiority of one approach over the other has not emerged however.

3. The estimates. Given observations $X_1, \ldots, X_T$, the kth order Yule–Walker estimates, $\hat{a}_k(j)$ $(j = 1, \ldots, k)$, of the autoregressive coefficients are the solutions of the equations

$$\sum_{j=0}^{k} \hat{a}_k(j) R^{(T)}(u-j) = 0 \qquad (u = 1, \ldots, k),$$

where $\hat{a}_k(0) = 1$ and

(3.1)
$$R^{(T)}(u) = T^{-1} \sum_{t=1}^{T-|u|} X_t X_{t+|u|}$$

is a 'positive definite' estimator of $R(u)$.

Now, the $a(u)$, $R(u)$, σ^2 and $b(u)$ are related as

(3.2)
$$\sigma^2 = c(0),$$

(3.3)
$$b(u) = c(u)/c(0),$$

where

(3.4)
$$c(u) = \sum_{j=0}^{\infty} a(j) R(u+j) \qquad (u = 0, 1, \ldots).$$

Moreover, the $a(u)$ and $b(u)$ are also related more directly as

(3.5)
$$b(j) = -\sum_{u=1}^{j} a(u) b(j-u) \qquad (j = 1, 2, \ldots).$$

As discussed by Bhansali (1989), corresponding to the relations (3.2), (3.3) and (3.5), two different estimators of $V(h)$ based on the $\hat{a}_k(j)$ may be suggested. An estimator based on (3.2) and (3.3) is given by the following:

(3.6)
$$\widehat{V}_{1k}(h) = \hat{\sigma}^2(k) \sum_{j=0}^{h-1} \{\hat{b}_{1k}(j)\}^2 \qquad (h = 1, \ldots, H).$$

Here, with

$$\hat{c}_k(u) = \sum_{j=0}^{k} \hat{a}_k(j) R^{(T)}(u+j),$$

$\hat{\sigma}^2(k) = \hat{c}_k(0), \hat{b}_{1k}(j) = \hat{c}_k(j)/\hat{c}_k(0)$ and $H \geq 1$ is a pre-fixed integer. Note that the $\hat{c}_k(j)$ and $\hat{b}_{1k}(j)$ also occur in the work of Hannan and Rissanen (1982).

The corresponding 'plug in' estimator of $V(h)$ based on the relations (3.2) and (3.5) is given by

(3.7)
$$\widehat{V}_{2k}(h) = \hat{\sigma}^2(k) \sum_{j=0}^{h-1} \{\hat{b}_{2k}(j)\}^2 \qquad (h = 1, \ldots, H),$$

where $\hat{b}_{2k}(0) = 1$ and for $j \geq 1$, the $\hat{b}_{2k}(j)$ are obtained by solving (3.5), but the $\hat{a}_k(u)$ replacing the $a(u)$.

The corresponding estimators of $Z(h)$ are given by, with $i = 1, 2$ and $h = 1, \ldots, H$,

$$(3.8) \qquad \widehat{Z}_{ik}(h) = \{R^{(T)}(0) - \widehat{V}_{ik}(h)\}/R^{(T)}(0).$$

As discussed in section 1, a third estimator of $V(h)$ is obtained by fitting a different autoregression for each h, that is, by a least-squares regression of x_{t+h} on $x_t, x_{t-1}, \ldots, x_{t-k+1}$, where k is the order of the autoregression to be fitted. Let K_T be a preassigned upper bound for the autoregressive order. A direct kth order estimator of $V(h)$ is obtained by minimizing the following sum of squares (Shibata, 1980)

$$N^{-1} \sum_{t=K_T}^{T-h} \{x_{t+h} + \sum_{j=1}^{k} \hat{\alpha}_{hk}(j)x_{t+1-j}\}^2 \qquad (h = 1, \ldots, H)$$

with respect to the $\hat{\alpha}_{hk}(j)$, where $N = T - h - K_T + 1$.

Put $\widehat{\mathbf{D}}_h(k) = [D^{(T)}(u, v)]$ $(u, \ v = 1, \ldots, k)$, $\hat{\mathbf{d}}_h(k) = [d_h^{(T)}(1), \ldots, d_h^{(T)}(k)]'$, $\hat{\boldsymbol{\alpha}}_h(k) = -\widehat{\mathbf{D}}_h(k)^{-1}\hat{\mathbf{d}}_h(k) = [\hat{\alpha}_{hk}(1), \ldots, \hat{\alpha}_{hk}(k)]'$, where

$$D^{(T)}(u, v) = N^{-1} \sum_{t=K_T}^{T-h} X_{t+1-u}X_{t+1-v},$$

$$d_h^{(T)}(u) = N^{-1} \sum_{t=K_T}^{T-h} X_{t+h}X_{t+1-u}.$$

Then, with

$$\widehat{R}_h(0) = N^{-1} \sum_{t=K_T}^{T-h} X_{t+h}^2,$$

$$(3.9) \qquad \widehat{V}_{3k}(h) = \widehat{R}_h(0) + \hat{\mathbf{d}}_h(k)'\hat{\boldsymbol{\alpha}}_h(k) \qquad (h = 1, \ldots, H)$$

provides a kth order estimator of $V(h)$ obtained by fitting a different autoregression for each h. It may be noted that the $\hat{\alpha}_{hk}(j)$ provide the kth order estimators of the h-step prediction constants.

The corresponding estimator of $Z(h)$ is given by

$$(3.10) \qquad \widehat{Z}_{3k}(h) = \{\widehat{R}_h(0) - \widehat{V}_{3k}(h)\}/\widehat{R}_h(0) \qquad (h = 1, \ldots, H).$$

An appropriate value of k to be used here may be selected by minimizing the following criterion suggested by Shibata (1980):

$$(3.11) \qquad S_h(k) = \widehat{V}_{3k}(h)(N + 2k) \qquad (k = 0, 1, \ldots, K_T).$$

4. Sampling properties of the estimates. The author has investigated the asymptotic distribution of the $\widehat{V}_{ik}(h)$ by requiring that the order k of the fitted autoregression approaches ∞ simultaneously but sufficiently slowly with T. The results show that these three estimators are asymptotically equivalent in the sense that under appropriate regularity conditions as $T \to \infty$ they have the same asymptotic normal distribution. This finding, especially that pertaining to the $\widehat{V}_{3k}(h)$, is somewhat surprising because Kabaila (1981) has shown that for an ARMA process of known order, estimates of the prediction constants obtained by minimising the h-step mean squared error of prediction, $h > 1$, are asymptotically inefficient when compared with the corresponding maximum likelihood estimators, see also Stoica and Soderstrom (1984).

Let

$$y_t = \sum_{j=0}^{h-1} b(j)\, \varepsilon_{t-j}$$

be a moving average process of order $h-1$. We have $\text{var}(y_t) = V(h)$ and the problem of estimating $V(h)$ coincides with that of estimating $\text{var}(y_t)$. Now, if $\{y_t\}$ could have been observed, then a 'nonparametric' estimator of $\text{var}(y_t)$ would be given by $R_y^{(T)}(0)$, where $R_y^{(T)}(u)$ is defined by (3.1) but with y_t replacing x_t. However, $\{y_t\}$ has not be observed. Nevertheless, our asymptotic results show that, as $T \to \infty$, the $\widehat{V}_{ik}(h)$ and $R_y^{(T)}(0)$ are equivalent to terms of order $T^{-\frac{1}{2}}$ in probability. Thus, for a fixed h and each $i = 1,2,3$, the asymptotic distribution of $\sqrt{T}\{\widehat{V}_{ik}(h) - V(h)\}$ is the same as that of $\sqrt{T}\{R_y^{(T)}(0) - R_y(0)\}$, and the expression for the asymptotic variance of the former coincides with that for the limiting variance of the latter as given by Anderson (1971, p. 467), for example.

Thus the asymptotic results show that as $T \to \infty$ the effect of estimating the autoregressive coefficients on the sampling properties of the $\widehat{V}_{ik}(h)$ is negligible. This last observation helps to explain why $\widehat{V}_{3k}(h)$ is asymptotically equivalent to the $\widehat{V}_{1k}(h)$ and $\widehat{V}_{2k}(h)$. By contrast, we may expect the last two estimators to be asymptotically equivalent since as shown by Bhansali (1989), the corresponding moving average estimators, $\hat{b}_{1k}(j)$ and $\hat{b}_{2k}(j)$, are also asymptotically equivalent.

The asymptotic distribution of the $\widehat{Z}_{ik}(h)$ has also been investigated by the author. A main feature of the results is that the order of consistency of the estimates depends on whether or not $\{x_t\}$ is a moving average process of order $h - 1$. In the latter case, $Z(h) > 0$ and the order of consistency of the $\widehat{Z}_{ik}(h)$, like that of the $\widehat{V}_{ik}(h)$, is $T^{-\frac{1}{2}}$. By contrast, $Z(s) = 0$, $s \geq h$, in the former case and the order of consistency is $k^{\frac{1}{2}}/T$. We note that this result accords with a remark of Pierce (1979), although made in a slightly different context, see also Wishart (1931) and Battaglia and Bhansali (1987).

Thirdly the question of bias in estimating the $V(h)$ and $Z(h)$ has been investigated. The results show that if $o_p(k/T)$ terms are ignored, the the $\widehat{V}_{ik}(h)$ are biased downwards and

(4.1) $$\widetilde{V}_{ik}(h) = \{T/T - k)\}\widehat{V}_{ik}(h) \qquad (i = 1,2,3)$$

provide the corresponding 'bias corrected' estimators of $V(h)$.

For $h = 1$, the bias correction term (4.1) coincides with that of Akaike (1970), and our results show that the Akaike bias correction term also applies for 'non-parametric' estimation of $V(h)$, $h > 1$. Indeed, for $i = 1, 2$, the bias in estimating $V(h)$ is that due to the estimation of σ^2 by $\hat{\sigma}^2(k)$ and asymptotically the bias in estimating the factor $\Sigma\, b^2(j)$ is negligible; see also Ansley and Newbold (1981) on this point.

By contrast, the $\hat{Z}_{ik}(h)$ are biased upwards and they overestimate the value of $Z(h)$; the magnitude of bias depends inversely on that of $Z(h)$, the smaller the value of $Z(h)$, the greater the bias. If $Z(h) = 0$, the bias is k/T. If $o_p(k/T)$ terms are ignored.

$$(4.2) \qquad \tilde{Z}_{ik}(h) = \hat{Z}_{ik}(h) - (k/T)\{1 - \hat{Z}_{ik}(h)\}$$

provide the corresponding 'bias-corrected' estimators of $Z(h)$. A convenient method for computing the $\tilde{Z}_{ik}(h)$ is to substitute $\tilde{V}_{ik}(h)$ for $\hat{V}_{ik}(h)$, $i = 1, 2, 3$, in (3.8) and (3.10) respectively.

5. Applications. We illustrate the procedures discussed in section 3 by applying them to the well-known Beveridge Trend Free Wheat Price Index, 1500–1869, and the Wolf Annual Sunspot Series, 1700–1983, as well as the Refrigerator Sales Series of Miller and Wichern (1977, p. 378). The actual observations for the first series are given by Anderson (1971, pp. 623–626) and and the second by Wei (1990, pp. 446–447), and we in fact consider the series obtained by subtracting out the arithmetic means of 99.36 and 477.96 respectively. For the third data set, however, we consider the series of first differences, after subtracting out the arithmetic mean of 5.92.

For each series, the estimators $\hat{V}_{ik}(h)$ and $\hat{Z}_{ik}(h)$, $i = 1, 2, 3$, were obtained together with the bias-corrected versions $\tilde{V}_{ik}(h)$ and $\tilde{Z}_{ik}(h)$ respectively of these estimators.

For computing the $\hat{V}_{1k}(h)$ and $\hat{V}_{2k}(h)$, the Yule–Walker estimates of the autoregressive coefficients were used and the order k of the fitted autoregression was selected by the FPE_2 criterion of Bhansali and Downham (1977). By contrast, the $\hat{V}_{3k}(h)$ were computed by fitting a different autoregression for each h by least squares, in accordance with the procedure described in section 3. The order, $\tilde{k}_h$, say, of the fitted autoregression for each h was determined by the criterion (3.11) of Shibata (1980). The results are summarised below for each series separately.

Example 1: The Refrigerator Sales Series

The estimated values of the $\tilde{Z}_{ik}(h)$, $i = 1, 2, 3, h = 1, 2, \ldots, 10$ are shown in Table 1. To save space, the corresponding values of the $\hat{V}_{ik}(h)$, $\tilde{V}_{ik}(h)$ and $\hat{Z}_{ik}(h)$ are not shown, but these may be determined readily from Table 1 by noting that $R^{(T)}(0) = 16298.74$ and $\hat{R}_1(0) = 18071.91$. For $i = 1, 2$ the estimates were computed from an initial second order autoregressive model with the estimated coefficients, $\hat{a}_2(1) = 0.6547$ and $\hat{a}_2(2) = 0.3543$, which was the model selected by the FPE_2 criterion. The corresponding values of the different autoregressive orders selected by the Shibata criterion for each h are shown in Table 4. It may be gleaned that for $h = 1$ this criterion also selects an autoregression of order 2. However, for all $h > 1$, a model of order 0 is selected, except for $h = 5$, when a first order model is selected.

The $\widetilde{Z}_{ik}(h)$ are negative, close to 0 or actually equal to 0 for all $h > 1$ and $i = 1, 2, 3$, suggesting that the past is not helpful for prediction more than one step ahead, or equivalently that the forecasts of the series behave like that of a first order moving average process. A formal test of this null hypothesis was carried out by using an asymptotic result developed by the author.

The statistic

$$u = \{\widehat{Z}_{ik}(h+1) - (k/T)\}/[(2k/T^2) \sum_{j=-h}^{h} \{r^{(T)}(j)\}^2]^{\frac{1}{2}}$$

with $h = 1, k = 2, T = 51, r^{(T)}(1) = 0.48, i = 1$ and $\widehat{Z}_{1k}(2) = 0.00608$ was computed. Under the null hypothesis, u is asymptotically distributed as Normal with mean 0 and variance 1. The computed value of $|u|$ was 0.697, implying that the null hypothesis may not be rejected even at the 48% level of significance.

It should be noted that for this particular data set, a number of different model identification procedures lead to selecting a first order moving average model. Miller and Wichern (1977) use the Box–Jenkins approach of an initial analysis of the sample correlation and partial correlation functions and decide to fit a first order moving average model. Secondly, on using a procedure based on the sample inverse correlation function for discrimination between a pure autoregressive and a pure moving average model of unknown orders, Bhansali (1987) is also led to selecting a first order moving average model. Thirdly, an application of a consistent recursive order selection procedure suggested by Bhansali (1990) for determining the order of an ARMA model and that suggested by Hannan and Kavalieris (1986) lead to the same model. It should be noted, however, that the results of Table 1 and more generally the estimation procedure suggested in section 3, also provide estimates of the multi-step prediction variances of the data and the corresponding R^2 measures.

Example 2: The Beveridge Trend-Free Wheat Price Index

The computed values of the $\widetilde{Z}_{ik}(h)$, $i = 1, 2, 3; h = 1, 2, \ldots, 20$ are shown in Table 2. For $i = 1, 2$, the estimates were obtained from an initial 8th order autoregressive model with coefficients $\hat{a}_8(1) = -0.71, \hat{a}_8(2) = 0.34, \hat{a}_8(3) = -0.04, \hat{a}_8(4) = 0.03, \hat{a}_8(5) = -0.00023, \hat{a}_8(6) = 0.072, \hat{a}_8(7) = 0.025, \hat{a}_8(8) = 0.1344$, fitted to the data, which was the model selected by the FPE_2 criterion. The corresponding autoregressive orders selected by the Shibata criterion for each h are shown in Table 4.

The numerical values of the $\widetilde{Z}_{1k}(h)$ and $\widetilde{Z}_{2k}(h)$ decrease steadily as h increases and for $h \geq 10$ they take negative values. The values of $\widetilde{Z}_{3k}(h)$ for $h < 10$ are similar to those of $\widetilde{Z}_{1k}(h)$ and $\widetilde{Z}_{2k}(h)$ but for $h \geq 10$ the former either equal zero or take a value close to zero. The results suggest that the series has a short memory in the sense that it may not be predicted indefinitely into the future on the basis of the past alone. Indeed, for $h \geq 10$, the knowledge of the past appears to be practically unhelpful for prediction.

It may be noted that for some of the values of $h \geq 10$, the Shibata criterion selects an autoregression of order greater than 0. An implication of this result is that even for $h \geq 10$, the past may be used for prediction. This finding is not surprising since the results of Table 2 do not indicate that from the point of view of forecasting, the series behaves like a finite order moving average process. Rather the suggestion is that for $h \geq 10$ the estimated R^2 values are close to zero and the reduction in variance due to using the past for prediction at these lead times is very small.

Whittle (1952) and Akaike (1972) use 300 observations of this series to suggest a subset autoregressive model with lags 1,2 and 8. Also Bhansali (1987) analyses the full series of 371 observations and finds that the same subset model is selected by the FPE_3 criterion. It is unlikely however that from the point of view of h-step prediction with $h \geq 10$, this subset model, or any other linear model, would be more effective than the use of a constant mean value.

Example 3: The Wolf Sunspot Series

The values of $\widetilde{Z}_{ik}(h)$ for $i = 1, 2, 3$ and $h = 1, 2, \ldots, 20$ are shown in Table 3. The estimates for $i = 1, 2$ were computed from an initial 9th order autoregressive model with coefficients $\hat{a}_9(1) = -1.03$, $\hat{a}_9(2) = 0.303$, $\hat{a}_9(3) = 0.12$, $\hat{a}_9(4) = -0.1$, $\hat{a}_9(5) = 0.11$, $\hat{a}_9(6) = -0.01$, $\hat{a}_9(7) = -0.06$, $\hat{a}_9(8) = 0.07$, $\hat{a}_9(9) = 0.25$ selected by the FPE_2 criterion. The corresponding autoregressive orders selected by the Shibata criterion for each h and used for computing the $\widetilde{Z}_{3k}(h)$ are shown in Table 4.

The results suggest that the future values of the Sunspot series are highly predictable from the past, even up to 20 steps ahead. The numerical values of the three estimates however differ considerably as h increases and the values of the $\widetilde{Z}_{3k}(h)$ are generally higher than that of the $\widetilde{Z}_{1k}(h)$.

It appears unlikely that this series is generated by a linear mechanism and a non-linear model may be expected to yield better forecasts, see Priestley (1981). The procedure of fitting a different autoregression for each h then provides the linear least-squares forecasts within the class of linear autoregressive models, whereas the use of a single autoregressive model for multistep forecasting does not do so. The differences in the numerical values of the estimates may be indicative of the differing predictive powers of these two linear methods of generating the forecasts.

Table 1

VALUES OF THE $\widetilde{Z}_{ik}(h)$, $i = 1, 2, 3$, FOR THE REFRIGERATOR SALES DATA

h	$\widetilde{Z}_{1k}(h)$	$\widetilde{Z}_{2k}(h)$	$\widetilde{Z}_{3k}(h)$
1	0.3026	0.3026	0.3459
2	-0.0345	0.0036	0.0000
3	-0.0398	-0.0003	0.0000
4	-0.0469	-0.0237	0.0000
5	-0.0481	-0.0386	0.0458
6	-0.0705	-0.0393	0.0000
7	-0.1193	-0.0400	0.0000
8	-0.1254	-0.0407	0.0000
9	-0.1267	-0.0408	0.0000
10	-0.1292	-0.0408	0.0000

Table 2

VALUES OF THE $\widetilde{Z}_{ik}(h)$, $i = 1, 2, 3$, FOR THE
BEVERIDGE WHEAT PRICE INDEX

h	$\widetilde{Z}_{1k}(h)$	$\widetilde{Z}_{2k}(h)$	$\widetilde{Z}_{3k}(h)$
1	0.3998	0.3998	0.3998
2	0.0979	0.0975	0.0967
3	0.0828	0.0818	0.0837
4	0.0771	0.0773	0.0842
5	0.0648	0.0686	0.0777
6	0.0590	0.0656	0.0757
7	0.0517	0.0608	0.0741
8	0.0422	0.0525	0.0683
9	0.0165	0.0246	0.0390
10	-0.0071	-0.0012	0.0000
11	-0.0126	-0.0057	0.0000
12	-0.0126	-0.0060	0.0000
13	-0.0126	-0.0089	0.0115
14	-0.0134	-0.0114	0.0141
15	-0.0135	-0.0133	0.0164
16	-0.0138	-0.0151	0.0182
17	-0.0204	-0.0174	0.0120
18	-0.0279	-0.0191	0.0000
19	-0.0281	-0.0194	0.0064
20	-0.0325	-0.0195	0.0000

Table 3

VALUES OF THE $\widetilde{Z}_{ik}(h)$, $i = 1, 2, 3$, FOR THE SUNSPOT SERIES

h	$\widetilde{Z}_{1k}(h)$	$\widetilde{Z}_{2k}(h)$	$\widetilde{Z}_{3k}(h)$
1	0.793	0.793	0.797
2	0.574	0.571	0.601
3	0.459	0.450	0.502
4	0.436	0.423	0.488
5	0.434	0.421	0.486
6	0.432	0.419	0.485
7	0.423	0.410	0.478
8	0.417	0.402	0.473
9	0.416	0.397	0.471
10	0.402	0.394	0.461
11	0.347	0.361	0.415
12	0.240	0.292	0.327
13	0.144	0.231	0.201
14	0.112	0.200	0.194
15	0.109	0.194	0.193
16	0.107	0.194	0.192
17	0.101	0.187	0.192
18	0.095	0.175	0.190
19	0.090	0.169	0.188
20	0.090	0.169	0.187

Table 4

THE AUTOREGRESSIVE ORDER SELECTED BY THE SHIBATA
CRITERION FOR THE DATA IN EXAMPLES 1, 2 AND 3

h	EXAMPLE 1	ORDER SELECTED EXAMPLE 2	EXAMPLE 3
1	2	8	9
2	0	8	17
3	0	7	17
4	0	6	17
5	1	5	16
6	0	4	15
7	0	3	14
8	0	1	13
9	0	1	12
10	0	0	11
11	-	0	2
12	-	0	2
13	-	4	2
14	-	3	16
15	-	2	15
16	-	1	14
17	-	2	13
18	-	2	13
19	-	1	11
20	-	0	10

REFERENCES

AKAIKE, H, (1970), *Statistical predictor identification*, Ann. Inst. Statist. Math., **22**, 203–217.

AKAIKE, H. (1972), *Use of an information theoretic quantity for statistical model identification*, In Proceedings of the Fifth Hawaii Int. conf. on System Sciences 249–250, Western Periodicals Co..

ANDERSON T.W. (1971), *The Statistical Analysis of Time Series*, New York: Wiley.

ANSLEY, C.F. and NEWBOLD, P. (1981), *On the bias in estimates of forecast mean square error*, J. Amer. Statist. Assoc, **76**, 596–578.

BATTAGLIA, F. and BHANSALI, R.J. (1987), *Estimation of the interpolation error variance and an index of linear determinism*, Biometrika, **74**, 771–779.

BHANSALI, R.J. (1974), *Asymptotic properties of the Wiener–Kolmogorov Predictor - I*, J. Royal Statist. Soc., **B 36**, 61–73.

BHANSALI, R.J. (1978), *Linear prediction by autoregressive model fitting in the time domain*, Ann. Statist., **6**, 224–231.

BHANSALI, R.J. (1987), *The discrimination between autoregressive and moving average models from the estimated inverse correlations*, Metron **45**, 115–135.

BHANSALI, R.J. (1989), *Estimation of the moving average representation of a stationary process by autoregressive model fitting*, J. Time Series Anal., **10**, 215–232.

BHANSALI, R.J. (1990), *Consistent recursive estimation of the order of an autoregressive moving average process*, Int. Statist. Review (to appear).

BHANSALI, R.J. and DOWNHAM, D.Y. (1977), *Some properties of the order of an autoregressive model selected by a generalization of Akaike's FPE criterion*, Biometrika, **64**, 547–551.

BRILLINGER, D.R. (1975), *Time Series: Data Analysis and Theory*, New York: Holt, Rinehart and Winston.

DAVIS, H.T. and JONES, R.H. (1968), *Estimation of the innovation variance of a stationary time series*, J. Amer. Statist. Assoc., **63**, 141–149.

FINDLEY, D.F. (1983), *On using a different time series forecasting model for each forecast lead*, Research Report, Bureau of the Census, Washington, D.C.

FINDLEY, D.F. (1986), *On bootstrap estimates of forecast mean square errors for autoregressive processes*, In Computer Science and Statistics: The Interface (ed. D.M. Allen), pp. 11–17, Amsterdam: Elsevier.

FULLER, W.A. and HASZA, D.P. (1981), *Properties of predictors for autoregressive time series*, J. Amer. Statist. Assoc., 76, 155–161.

GERSCH, W. and KITTAGAWA, G. (1983), *The prediction of time series with trends and seasonalities*, J. Bus. and Eco. Statist. **1**, 253–264.

GEWEKE, J. (1988), *Antithetic acceleration of Monte Carlo integration in Bayesian inference*, J. Econometrics, **38**, 73–89.

GOOD, I.J. (1988), *The interface between Statistics and philosophy of science*, Statist. Sci., **3**, 386–412.

HANNAN, E.J. and KAVALIERIS, L. (1984), *A method for autoregressive moving average estimation*, Biometrika, **69**, 81–94.

HANNAN, E.J. and NICHOLLS, D.F. (1977), *The estimation of the prediction error variance*, J. Amer. Statist. Assoc., **72**, 834–840.

HANNAN, E.J. and RISSANEN, J. (1982), *Recursive estimation of mixed autoregressive-moving average order*, Biometrika, **69**, 81–94. Correction (1983), **70**, 303.

JEWELL, N.P. and BLOOMFIELD, P. (1983), *Canonical Correlations of past and future for time series: definitions and theory*, Ann. Statist., **11**, 837–847.

KABAILA, P.V. (1981), *Estimation based on one step ahead prediction versus estimation based on multi-step ahead prediction,*, **6**, 43–55.

LEWIS, R.A. and REINSEL, G.C. (1985), *Prediction of multivariate time series by autoregressive model fitting*, J. Mult. Anal., **16**, 393–411.

MILLER, R.B. and WICHERN, D.W. (1977), *Intermediate Business Statistics*, New York: Holt.

NELSON, C.R. (1976), *The interpretation of R^2 in autoregressive-moving average time series models*, Amer. Statist., **30**, 175–180.

PARZEN, E. (1981), *Time series model identification and prediction variance horizon*, In Applied Time Series Analysis II (ed. D. Findley), 415–447, New York: Academic Press.

PIERCE, D.A. (1979), *R^2 measures for time series*, J. Amer. Statist. Assoc., **74**, 901–910.

PRIESTLEY, M.B. (1981), *Spectral Analysis and Time Series*, Volume 2. New York: Academic Press.

SHIBATA, R. (1977), *Convergence fo least squares estimates of autoregressive parameters*, Aust. J. Statist., **19**, 226–235.

SHIBATA, R. (1980), *Asymptotically efficient selection of the order of the model for estimating parameters of a linar process*, Ann. Statist., **8**, 147–164.

STINE, R.A. (1987), *Estimating properties of autoregressive forecasts*, J. Amer. Statist. Assoc., **82**, 1072–1078.

STOICA, P. AND SODERSTROM, T. (1984), *Uniqueness of estimated k-step prediction models of ARMA processes*, Systems & Control Letters, **4**, 325–331.

THOMBS, L.A. AND SCHUCANY, W.R. (1989), *Bootstrap prediction intervals for autoregression*, J. Amer. Statist. Assoc., (To appear).

WEI, W.W.S. (1990), *Time Series Analysis*, Redwood City: Addison Wesley.

WHITTLE, P. (1952), *Tests of fit for time series*, Biometrika, **39**, 309–318.

WHITTLE, P. (1963), *Prediction and Regulation by Linear Least Squares Methods*, London: English University Press.

WISHART, J. (1931), *The mean and second moment of the multiple correlation coefficient in samples from a normal population*, Biometrika, **22**, 353.

ON BACKCASTING IN LINEAR TIME SERIES MODELS*

F. JAY BREIDT†, RICHARD A. DAVIS† AND WILLIAM DUNSMUIR‡

Abstract. This paper examines the role of backcasting in linear time series models. Backcasting had its genesis in Gaussian estimation as a means of getting a better approximation to the quadratic form in the likelihood. Though this approximation has been largely superseded by exact likelihood calculation, backcasting is sometimes a useful alternative, as it is for certain long-memory processes. We review these results and consider the performance and usefulness of backcasting in non-Gaussian likelihood approximation. Finally, we discuss the use of backcasting in a new method of bootstrapping autoregressions, and investigate the dependence structure of the residuals from backcasting.

1. Introduction. As a starting point for our discussions consider a discrete time series $\{X_t\}$ on which we have available observations $X_1, \dots, X_n$. We will, for the present, assume that $\{X_t\}$ follows an autoregressive-moving average (ARMA) model of the form

$$(1.1) \qquad X_t - \phi_1 X_{t-1} - \cdots - \phi_p X_{t-p} = Z_t + \theta_1 Z_{t-1} + \cdots + \theta_q Z_{t-q},$$

where $\phi(z) = 1 - \phi_1 z - \cdots - \phi_p z^p$ and $\theta(z) = 1 + \theta_1 z + \cdots + \theta_p z^p$ have no roots inside or on the unit circle, in which case (1.1) is causal and invertible. The Z_t's in (1.1) are assumed to be independent and identically distributed (iid) with density $f(z)$.

When f is the Gaussian density with mean zero and variance σ^2, the likelihood for $X_1, \dots, X_n$ is readily constructed. Even in cases where f is not Gaussian, the Gaussian likelihood is often used. Methods for constructing this criterion and maximizing it over the ϕ's, θ's, and σ^2 are well understood and are documented in [4].

Box and Jenkins [2] describe a method which ignores the determinant term in the Gaussian likelihood but which approximates the quadratic form appearing in the exponent by a sum of squares of forward forecast errors; these forecast errors are initiated by preperiod errors estimated by a method called "backcasting." This method is described in [2], and is in use in some popular statistical analysis packages including MINITAB. Briefly, it consists of running the model (1.1) backward in time and using it to forecast backward sufficiently many preperiod X_t's (and hence Z_t's).

In the Gaussian case the need to backcast is now overcome by using exact likelihood computation [4], although to our knowledge no complete finite sample comparison has been done of the "unconditional estimates" (backcasting) and the Gaussian estimates. A few facts are known from theory however:

---

*This research was supported in part by the Institute for Mathematics and its Applications with funds provided by the National Science Foundation.

†Department of Statistics, Colorado State University, Fort Collins, CO 80523 USA. Research supported by NSF Grant No. 8802559.

‡School of Information and Computing Sciences, Bond University, Gold Coast, Queensland 4229 Australia.

(1) If the determinant term in the Gaussian likelihood is omitted then:

   (a) For models with an autoregressive component there is no automatic bounding of the roots of $\phi(z)$ away from the unit circle [5].

   (b) For moving averages the sum of squares has no minimum unless $\theta(z)$ is restricted to the invertibility region.

(2) Inclusion of the determinant term in the moving average case ensures that the likelihood has a local optimum on the boundary of the parameter space [1].

(3) If the data are non-Gaussian then the model in (1.1) is *not* time-reversible [3] and the linear backcasts do not typically yield the optimal backcasts.

The first two of these facts are reviewed further in Section 2, while the third is discussed in Section 4.

Backcasting has also been used in the following ways:

- To initialize calculation of the likelihood for Laplacian (two-sided exponential) innovations or for least absolute deviation (L^1) estimation (see Section 3).

- To obtain bootstrap forecast intervals (see Section 4).

2. Computing the Gaussian likelihood. Often the parameters in a time series model are estimated by maximizing the Gaussian likelihood even when the underlying model is not assumed to be Gaussian. The reasons for using the Gaussian likelihood are twofold. First, the actual likelihood may not be computationally tractable or even completely specifiable. The latter occurs, for example, in an ARMA process when the underlying noise is assumed to be only iid with an unspecified distribution. Second, the increased performance of using the actual likelihood over the Gaussian likelihood is often only marginal. Because of this, maximum Gaussian likelihood estimates are also useful in more refined estimation procedures, such as adaptive methods, where good initial parameter estimates are required.

In time series analysis, we are often faced with the task of maximizing the Gaussian likelihood numerically— it is rare when the maximum Gaussian likelihood estimates can be found in closed form— which requires repeated evaluation of the likelihood. In this section we review a procedure for calculating the exact likelihood using the innovations algorithm and then discuss an approximation to the likelihood using backcasting.

The Gaussian likelihood based on the observed data $X_1, \dots, X_n$ is given by

$$(2.1) \qquad L(\Gamma_n) = (2\pi)^{-n/2} |\Gamma_n|^{-1/2} \exp\{-\frac{1}{2} \mathbf{X}_n' \Gamma_n^{-1} \mathbf{X}_n\}$$

where $\mathbf{X}_n = (X_1, \dots, X_n)'$ and $\Gamma_n = E(\mathbf{X}_n \mathbf{X}_n')$. (Here we have assumed that the mean is 0.) Define the one-step predictors and their mean-square errors by

$$\hat{X}_{t+1} = \begin{cases} 0, & \text{if } t = 0, \\ P(X_{t+1}|X_t, \dots, X_1), & \text{if } t > 0, \end{cases}$$

and

$$v_t = E(X_{t+1} - \hat{X}_{t+1})^2$$

where $P(\cdot|X_t,\ldots,X_1)$ denotes the orthogonal projection onto $\overline{\mathrm{sp}}\{X_1,\ldots,X_t\} = \overline{\mathrm{sp}}\{X_1, X_2 - \hat{X}_2,\ldots,X_t - \hat{X}_t\}$. The predictors $\hat{X}_{t+1}$ can be expressed as

$$\hat{X}_{t+1} = \theta_{t1}(X_t - \hat{X}_t) + \cdots + \theta_{tt}(X_1 - \hat{X}_1) \tag{2.2}$$

so that

$$\mathbf{X}_n = (\mathbf{X}_n - \hat{\mathbf{X}}_n) + \hat{\mathbf{X}}_n$$
$$= C(\mathbf{X}_n - \hat{\mathbf{X}}_n)$$

where

$$C = \begin{bmatrix} 1 & 0 & 0 & \cdots & 0 \\ \theta_{11} & 1 & 0 & \cdots & 0 \\ \theta_{21} & \theta_{22} & 1 & \cdots & 0 \\ \vdots & \vdots & \vdots & \ddots & \vdots \\ \theta_{n-1,n-1} & \theta_{n-1,n-2} & \theta_{n-1,n-3} & \cdots & 1 \end{bmatrix}.$$

The innovations algorithm (see p. 165 of [4]) can be used to compute the coefficients θ_{tj} and mean-square errors v_t recursively. Since the one-step prediction errors are uncorrelated, we have

$$\Gamma_n = E(\mathbf{X}_n \mathbf{X}_n') = CVC'$$

where $V = \mathrm{diag}(v_0,\ldots,v_{n-1})$. It follows that

$$|\Gamma_n| = v_0 v_1 \cdots v_{n-1} \tag{2.3}$$
$$\mathbf{X}_n' \Gamma_n^{-1} \mathbf{X}_n = (\mathbf{X}_n - \hat{\mathbf{X}}_n)' C'(CVC')^{-1} C(\mathbf{X}_n - \hat{\mathbf{X}}_n)$$
$$= (\mathbf{X}_n - \hat{\mathbf{X}}_n)' V^{-1}(\mathbf{X}_n - \hat{\mathbf{X}}_n)$$
$$= \sum_{t=1}^{n} \frac{(X_t - \hat{X}_t)^2}{v_{t-1}}, \tag{2.4}$$

so that the likelihood can be calculated as

$$L(\Gamma_n) = (2\pi)^{-n/2}(v_0 \cdots v_n)^{-1/2} \exp\left\{ -\frac{1}{2} \sum_{t=1}^{n} \frac{(X_t - \hat{X}_t)^2}{v_{t-1}} \right\}. \tag{2.5}$$

For a general time series this expression for computing the likelihood can be time-consuming. (The time required is roughly the same as that for calculating the likelihood straight from the definition (2.1).) However, for models with a simple structure, such as ARMA models, substantial savings in computation can be achieved using (2.4).

Consider the MA(q) process

$$X_t = Z_t + \theta_1 Z_{t-1} + \cdots + \theta_q Z_{t-q}$$

where $\{Z_t\} \sim \mathrm{IID}(0,\sigma^2)$. For this model, equation (2.2) simplifies to

$$\hat{X}_{t+1} = \theta_{t1}(X_t - \hat{X}_t) + \cdots + \theta_{tq}(X_{t+1-q} - \hat{X}_{t+1-q})$$

with

$$v_t = \sigma^2(1 + \theta_1^2 + \cdots + \theta_q^2) - \sum_{j=1}^{q} \theta_{tj}^2 v_{t-j}$$

where $\theta_{tj} := 0$ for $j > t$. Setting $r_t = v_t/\sigma^2$, which is independent of σ^2, the maximum Gaussian likelihood estimates of σ^2 and $\boldsymbol{\theta} = (\theta_1, \dots, \theta_q)$ satisfy

$$\hat{\sigma}^2 = \frac{1}{n} S(\hat{\boldsymbol{\theta}}),$$

where

$$S(\hat{\boldsymbol{\theta}}) = \sum_{t=1}^{n} \frac{(X_t - \hat{X}_t)^2}{r_{t-1}},$$

and $\hat{\boldsymbol{\theta}}$ minimizes the reduced likelihood,

$$\ell(\boldsymbol{\theta}) = \ln(n^{-1} S(\boldsymbol{\theta})) + n^{-1} \sum_{t=1}^{n} \ln r_{t-1}.$$

Provided the moving average is invertible, $1 + \theta_1 z + \cdots + \theta_q z^q \neq 0$ for $|z| \leq 1$, then

$$\theta_{tj} \to \theta_j \quad j = 1, \dots, q,$$
$$r_t \downarrow 1$$

as $t \to \infty$, and the rate of both convergences is geometric (see [4]). (In programming the likelihood, this result enables one to set $\theta_{tj} = \theta_j$ and $r_t = 1$ for even moderate values of t.) For an ARMA(p,q) process, the calculation of $\hat{X}_{t+1}$, r_t and the Gaussian likelihood is no more involved than it is for the MA(q) process—the number of operations is almost identical (see Section 5.3 of [4]).

For invertible ARMA processes and, more generally, invertible linear processes, the mean-square error of prediction

$$r_t = \frac{E(X_{t+1} - \hat{X}_{t+1})^2}{\sigma^2} \downarrow 1 \quad \text{as } t \uparrow \infty$$

so that the determinant term in the reduced likelihood

$$n^{-1} \sum_{t=1}^{n} \ln r_{t-1} \to 0.$$

This suggests approximating the reduced likelihood by ignoring the determinant term and hence estimating the parameters by minimizing the sum of squares

$$S(\boldsymbol{\beta}) = \sum_{t=1}^{n} \frac{(X_t - \hat{X}_t)^2}{r_{t-1}} = \mathbf{X}_n' G_n^{-1}(\boldsymbol{\beta}) \mathbf{X}_n$$

where $G_n(\boldsymbol{\beta}) = \sigma^{-2} \Gamma_n(\boldsymbol{\beta})$ and $\boldsymbol{\beta}$ is the parameter vector of interest. The resulting estimate, $\hat{\boldsymbol{\beta}}$, is called the *least squares estimate*. In many examples, the unconditional sum of squares, $\mathbf{X}_n' G_n^{-1}(\boldsymbol{\beta}) \mathbf{X}_n$, can be computed simply by backcasting—a method developed by Box and Jenkins. Suppose $\{X_t\}$ is the linear process

$$(2.6) \qquad X_t = \sum_{j=0}^{\infty} \psi_j Z_{t-j}$$

where $\{Z_t\}$ is iid and $\psi_j = \psi_j(\boldsymbol{\beta})$. Define

$$[Z_t] = P(Z_t | X_n, \dots, X_1).$$

Then if $\boldsymbol{\beta}$ is the true parameter value,

(2.7)
$$\sum_{t=-\infty}^{n} [Z_t]^2 = \mathbf{X}_n' G_n^{-1}(\boldsymbol{\beta}) \mathbf{X}_n.$$

To see this, let $\mathbf{Z}_{nQ} = (Z_n, \dots, Z_{-Q})'$ and note that

$$[\mathbf{Z}_{nQ}] := \begin{bmatrix} [Z_n] \\ \vdots \\ [Z_{-Q}] \end{bmatrix} = E(\mathbf{Z}_{nQ}\mathbf{X}_n')\Gamma_n^{-1}\mathbf{X}_n$$

whence

(2.8)
$$\sum_{t=-Q}^{n} [Z_t]^2 = \mathbf{X}_n' \Gamma_n^{-1} E(\mathbf{X}_n \mathbf{Z}_{nQ}') E(\mathbf{Z}_{nQ}\mathbf{X}_n')\Gamma_n^{-1}\mathbf{X}_n.$$

But by (2.6), we have

$$E(\mathbf{Z}_{nQ}\mathbf{X}_n') = \sigma^2 \begin{bmatrix} 0 & 0 & \cdots & 0 & 1 \\ 0 & 0 & \cdots & 1 & \psi_1 \\ 0 & 0 & \cdots & \psi_1 & \psi_2 \\ \vdots & \vdots & \ddots & \vdots & \vdots \\ 1 & \psi_1 & \cdots & \psi_{n-2} & \psi_{n-1} \\ \psi_1 & \psi_2 & \cdots & \psi_{n-1} & \psi_n \\ \vdots & \vdots & \vdots & \vdots & \vdots \\ \psi_{-Q+1} & \psi_{-Q+2} & \cdots & \psi_{-Q+n-1} & \psi_{-Q+n} \end{bmatrix}$$

so that the dot product of the ith and j^{th} columns of $E(\mathbf{Z}_{nQ}\mathbf{X}_n')$ converges to (as $Q \to \infty$)

$$\sigma^4 \sum_{k=0}^{\infty} \psi_k \psi_{k+|i-j|} = \sigma^2 E(X_i X_j).$$

This combined with (2.8) establishes (2.7). A different derivation of (2.7) for ARMA models is given in [2].

The calculation of $\sum_{t=-\infty}^{n} [Z_t]^2$ for a linear process via backcasting can be implemented as follows. First approximate $\{X_t\}$ by an MA(Q) process by truncating the sum in (2.6) at some large value Q. Using the backward representation of $\{X_t\}$ write

$$X_t = \sum_{j=0}^{Q} \psi_j W_{t+j}$$

where $\{W_t\}$ is a white noise sequence and set

(2.9) $$[W_t] = \begin{cases} 0, & \text{for } t > n \text{ or } t \le 0, \\ X_t - \sum_{j=1}^{Q} \psi_j[W_{t+j}], & \text{for } t = n, n-1, \ldots, 1. \end{cases}$$

The backcasts of $\{X_t\}$ are then approximated by

$$[X_t] = \sum_{j=0}^{Q} \psi_j[W_{t+j}], \quad \text{for } t = 0, -1, \ldots, -Q+1,$$

and finally the backcasts of $\{Z_t\}$ are found recursively from the relations

$$[Z_t] = \begin{cases} 0, & t < 1-Q, \\ [X_t] - \sum_{j=1}^{Q} \psi_j[Z_{t-j}], & t = -Q+1, \ldots, n. \end{cases}$$

To obtain a closer approximation to $S(\boldsymbol{\beta})$, this procedure can be iterated by forecasting the values of $X_{n+1}, \ldots, X_{n+Q}$ using the just computed $[Z_t]$'s, which leads to improved predictors of $[W_t], t = n+1, \ldots, n+Q$ in (2.9).

Alternatively, if $\{X_t\}$ admits an AR(∞) representation, then backcasting may be carried out for an AR(Q) approximation to $\{X_t\}$. This is the method adopted by Hosking [9] for fractionally integrated ARMA processes.

The backcasting method for approximating the Gaussian likelihood has two main drawbacks. First, it ignores the determinant term in the likelihood. While in many cases, the determinant term is negligible relative to the sum of squares, the effect of this factor is impossible to discern before a model is actually fitted to the data. For example, if $\{X_t\}$ is an ARMA(p, q) process, then the determinant of the covariance matrix becomes infinite as the zeros of the autoregressive polynomial approach the unit circle (see Lemma 2 of [5] for a general proof in the vector ARMA case). In effect, inclusion of the determinant term provides a built-in constraint on the zeros of the autoregressive polynomial. Also if $q > 0$, then the unconditional sum of squares $S(\boldsymbol{\beta})$ has no minimum unless one imposes the additional constraint that the moving average polynomial has no zeros inside the unit circle (see Problem 8.13 in [4]).

The second drawback of the backcasting procedure is determining a suitable value for Q. If a Newton-Raphson type procedure is implemented to minimize $S(\boldsymbol{\beta})$, then at each time the parameter estimates are updated, a new value of Q must be specified. Of course, this is not a problem if the process is a pure moving average, but for autoregressive processes, Q can be large especially when the zeros of the autoregressive polynomial are near the unit circle.

The innovations algorithm approach to likelihood calculation is exact and hence does not suffer from these shortcomings. However, unless the process can be reduced in some manner to one with a simple and easily computable autocovariance function, the innovations algorithm method may not be a viable option even for time series of moderate length. When this is the case, likelihood approximation via backcasting is often a feasible alternative.

3. Use of backcasting for non-Gaussian likelihood approximation. In the case where the errors $\{Z_t\}$ in the model (1.1) have a non-Gaussian density, the likelihood of a sample record $X_1, \ldots, X_n$ is generally difficult to calculate in closed form even for simple models, such as the lag one moving average $(p = 0, q = 1)$. Evaluation of the likelihood requires finite (pure MA case) or infinite (mixed ARMA case) convolutions of the density f.

For this reason, in the non-Gaussian case there is a much stronger incentive for seeking approximations to the initial distributions of the preperiod values, $(X_{-p+1}, \ldots, X_0, Z_{-q+1}, \ldots, Z_0)'$.

To fix ideas, we will focus on the multiplicative moving average model

$$(3.1) \qquad X_t = Z_t - \theta Z_{t-1} - \Theta Z_{t-12} + \theta\Theta Z_{t-13}.$$

We will also, in deriving estimates, assume that the Z_t's have the Laplacian density

$$f(z) = \frac{1}{2\gamma}\exp\left(-\frac{|z|}{\gamma}\right).$$

Then, the exact likelihood for $X_1, \ldots, X_n$ is

$$(3.2) \qquad L(\boldsymbol{\theta}, \gamma) = \int_{R^{13}} \prod_{t=1}^{n} f(e_t(\boldsymbol{\theta}; \mathbf{z}); \gamma) \prod_{j=-12}^{0} f(z_j; \gamma)\, dz_j,$$

where $\boldsymbol{\theta} = (\theta, \Theta)'$, $\mathbf{z} = (z_{-12}, \ldots, z_0)'$, and the $e_t(\boldsymbol{\theta}; \mathbf{z})$ are obtained from

$$e_t(\boldsymbol{\theta}; \mathbf{z}) = X_t + \theta e_{t-1}(\boldsymbol{\theta}; \mathbf{z}) + \Theta e_{t-12}(\boldsymbol{\theta}; \mathbf{z}) - \theta\Theta e_{t-13}(\boldsymbol{\theta}; \mathbf{z})$$

using initial values $\mathbf{z}$. Computation of (3.2) involves a 13-fold convolution for which there is no closed form and which proves to be computationally bothersome.

Two obvious approximations to (3.2) will be compared. Both approximate the densities of $Z_{-12}, \ldots, Z_0$ with point mass distributions:

Conditional estimates. Use $\mathbf{z} = (0, \ldots, 0)'$ (*i.e.*, the unconditional mean of the $e_j, j \le 0$). Minimize the resulting approximation to (3.2) and denote the estimates obtained by $\hat{\theta}_{\text{zero}}, \hat{\Theta}_{\text{zero}}$.

Backcast estimates. Use the *linear* prediction of $\mathbf{Z}$ given $X_1, \ldots, X_n, \boldsymbol{\theta}$ and use the model (3.1) in reverse time. Denote the estimates by $\hat{\theta}_{\text{back}}, \hat{\Theta}_{\text{back}}$. Two iterations of backcasting were found to be sufficient to assure convergence.

Of course, as can be deduced from Proposition 4.1 below, these backcasts are not the best estimates (in the L^1 or L^2 sense) of $\mathbf{Z}$ given $X_1, \ldots, X_n$. However, they are easily computable, unlike the conditional expectation of $\mathbf{Z}$.

Early experience indicated that these two methods gave substantially different estimates on real data. For example, in estimating the above model for approximately 100 data consisting of monthly rice sales differenced at lag twelve, we obtained $(\hat{\theta}_{\text{zero}}, \hat{\Theta}_{\text{zero}}) = (-0.07, 0.77)$, compared with $(\hat{\theta}_{\text{back}}, \hat{\Theta}_{\text{back}}) = (0.00, 0.63)$.

Our cause for concern was the difference in the estimates of the seasonal lag parameter Θ. For the same data a comparison of least squares using backcasting with least squares using conditional estimation gave a difference between the $\hat{\Theta}$'s of the same magnitude as the standard error of estimation.

Simulation studies, reported in detail in [7], confirmed our earlier experience. A selection of those results is presented here. For a record length of $n = 100$ and three error distributions (Gaussian, Laplacian, and Student's t with 2.1 d.f.), 400 replications from model (3.1) were generated. Table 1 gives the means, standard deviations (SD), and root mean-square errors (RMSE) for the two estimators defined above. Two values of the true parameters are also considered, namely $\theta = \Theta = 0.4$ and 0.8.

TABLE 1. *Means, standard deviations and root mean-square errors for*
400 replications of records of length $n = 100$ from the seasonal moving average model.

True	Normal			Laplace			$t(2.1$ d.f.$)$		
Parameter	Mean	SD	RMSE	Mean	SD	RMSE	Mean	SD	RMSE
$\theta=\Theta=0.8$									
$\hat{\theta}_{\text{back}}$	.66	.16	.22	.67	.13	.18	.70	.14	.17
$\hat{\theta}_{\text{zero}}$	.77	.10	.10	.77	.07	.08	.78	.07	.07
$\hat{\Theta}_{\text{back}}$	.76	.10	.10	.74	.09	.10	.73	.10	.12
$\hat{\Theta}_{\text{zero}}$	.65	.11	.18	.67	.09	.16	.68	.11	.16
$\theta=\Theta=0.4$									
$\hat{\theta}_{\text{back}}$	.34	.13	.14	.38	.09	.10	.38	.08	.08
$\hat{\theta}_{\text{zero}}$	.40	.12	.12	.41	.09	.09	.40	.07	.07
$\hat{\Theta}_{\text{back}}$	.58	.20	.27	.49	.14	.17	.46	.12	.14
$\hat{\Theta}_{\text{zero}}$	.39	.13	.13	.39	.09	.09	.39	.08	.08

The following statements hold uniformly across the three error distributions used:

(1) For both parameter values the best estimate of θ is $\hat{\theta}_{\text{zero}}$. For $\theta = 0.8$ this is particularly true and is due to the better bias *and* SD properties of $\hat{\theta}_{\text{zero}}$.

(2) For estimating $\Theta = 0.8$, $\hat{\Theta}_{\text{back}}$ clearly outperforms $\hat{\Theta}_{\text{zero}}$ (due to bias) and the reverse is true when $\Theta = 0.4$.

(3) For both parameter values $\hat{\theta}_{\text{back}}$ is closer to zero than $\hat{\theta}_{\text{zero}}$ while $\hat{\Theta}_{\text{zero}}$ is closer to zero than $\hat{\Theta}_{\text{back}}$.

As a consequence of the above observations we conclude that backcasting cannot be recommended as a uniformly better alternative than using zero initial conditions as an approximation to the likelihood in the case of Laplacian errors or as a way of initializing L^1 estimation for general error distributions.

An open research question is how best to approximate *simply* the distribution of the preperiod values in (3.2).

4. Backcasting in bootstrapping AR's for conditional predictive inference. Consider the problem of predicting X_{n+1} based on observations $X_1, \ldots, X_n$ from an AR(p) process

$$\phi(B)X_t = X_t - \phi_1 X_{t-1} - \cdots - \phi_p X_{t-p} = Z_t,$$

where $\phi(z) = 1 - \phi_1 z - \cdots - \phi_p z^p \neq 0$ for $|z| \leq 1$, B is the backward shift operator ($B^k X_t = X_{t-k}, k = 0, \pm 1, \ldots$), and $\{Z_t\}$ is iid with zero mean and common distribution function F. The best mean-square predictor of X_{n+1} is $\hat{X}_{n+1}(\boldsymbol{\phi}) = \phi_1 X_n + \cdots + \phi_p X_{n+1-p}$, so the structure of the AR(p) process suggests that inference for X_{n+1} conditioned on $\mathbf{X}_p := (X_{n+1-p}, \ldots, X_n)'$ is of most interest, as several authors have recently noted ([8], [10], [11]). Substituting some estimate $\hat{\boldsymbol{\phi}} = (\hat{\phi}_1, \ldots, \hat{\phi}_p)'$ for $\boldsymbol{\phi}$, we see that the prediction error

$$Y = X_{n+1} - \hat{X}_{n+1}(\hat{\boldsymbol{\phi}}) = Z_{n+1} + (\phi_1 - \hat{\phi}_1)X_n + \cdots + (\phi_p - \hat{\phi}_p)X_{n+1-p}$$

depends on both the noise and the parameter estimates.

If F and $\boldsymbol{\phi}$ are known, the distribution of $Y|\mathbf{X}_p$ can be obtained from simulation provided we can generate realizations of the AR(p) which pass through $\mathbf{X}_p$. One approach to generating such realizations is to initialize the AR(p) with the last p observations $\mathbf{X}_p$ and run the process backward in time using the backward representation:

$$(4.1) \qquad\qquad \phi(B^{-1})X_t = W_t,$$

where $\{W_t\}$ is the sequence defined by

$$(4.2) \qquad\qquad W_t = \frac{\phi(B^{-1})}{\phi(B)} Z_t$$

(see, for example, [8]). Since the spectral density of $\{W_t\}$ is

$$f_W(\lambda) = \frac{|\phi(e^{i\lambda})|^2}{|\phi(e^{-i\lambda})|^2} \frac{\sigma^2}{2\pi} = \frac{\sigma^2}{2\pi},$$

where σ^2 is the variance of Z_0, we see that $\{W_t\}$ is uncorrelated and has the same variance as the original white noise $\{Z_t\}$. We will refer to $\{W_t\}$ as the *backward noise*.

Though second-order properties do not distinguish between $\{Z_t\}$ and $\{W_t\}$, these sequences are probabilistically quite different in the case that Z_0 is non-Gaussian. In particular, the marginal distribution of W_0 can be completely unlike that of Z_0. Suppose, for example, that $\{Z_t\}$ is iid with $\mathrm{P}(Z_t = 0) = \mathrm{P}(Z_t = 1) = 1/2$, and that $\{X_t\}$ is an AR(1) process with $\phi_1 = 0.5$. Then

$$X_t = 0.5 X_{t-1} + Z_t = \sum_{j=0}^{\infty} (0.5)^j Z_{t-j},$$

which is distributed uniformly on $(0, 2)$, and in fact

$$W_t = X_t - 0.5X_{t+1}$$
$$= X_t - 0.5(0.5X_t + Z_{t+1})$$
$$= 0.75X_t - 0.5Z_{t+1}$$

(a linear combination of a uniform random variable and an independent Bernoulli), which has an absolutely continuous distribution! Typically, the marginal of W_0 will be "smoother" and "more nearly Gaussian" (see [6] and [11]) than that of Z_0, since W_0 is a convolution of the Z_t's. This can be seen again in Figure 4.1, where the distribution of Z_0 is a mixture of 90% Gaussian (-1,1) and 10% Gaussian (9,1), while the distribution of W_0 (from simulation) is quite symmetric.

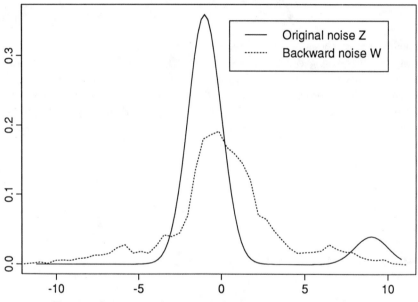

Fig. 4.1. Symmetry of marginal of W_0 compared to that of Z_0

The joint distributions of the W_t's and the Z_t's are also different. Findley [8] noted that in the AR(1) case, $\{W_t\}$, though uncorrelated, is not iid F and so cannot be simulated by iid draws from F. In [3], this result is proved in generality for ARMA processes: $\{W_t\}$ cannot be iid unless Z_0 is Gaussian. In fact, in [3] it is shown that the sequence $\{W_t\}$ does not even satisfy the weaker condition of *time-reversibility*, where a stationary process $\{W_t\}$ is defined to be time-reversible if for every integer n and for all integers $t_1, \ldots, t_n$, the vectors $(W_{t_1}, \ldots, W_{t_n})'$ and $(W_{-t_1}, \ldots, W_{-t_n})'$ have the same joint probability distribution [12]. This is evident in Figure 4.2, which shows a realization of backward noise for the AR(2) process

$$X_t = 0.5X_{t-1} + 0.3X_{t-2} + Z_t,$$

where $\{Z_t\}$ is iid with the contaminated Gaussian distribution described above. When hiking the sample path $(W_1, \ldots, W_{100})'$, we climb steep cliffs and descend jagged hills, but when hiking the time-reversed sample path $(W_{100}, \ldots, W_1)'$, we climb jagged hills and jump off cliffs!

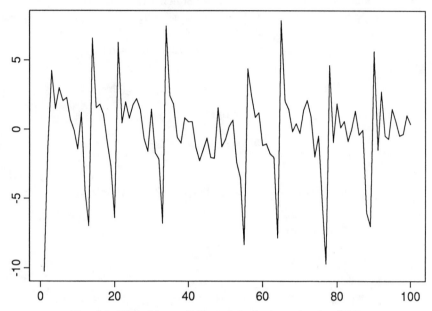

Fig. 4.2. Time-irreversibility of the backward noise $\{W_t\}$

Somewhere between iid and uncorrelated lies the martingale difference structure. Since Z_t is measurable with respect to the past σ-field $\mathcal{F}_t = \sigma(X_t, X_{t-1}, \ldots)$ and independent of $\mathcal{F}_{t-1}$, we have $\mathrm{E}[Z_t | \mathcal{F}_{t-1}] = 0$ almost surely (a.s.), and so $\{Z_t\}$ is a martingale difference with respect to the filtration $\{\mathcal{F}_t\}$. When does $\{W_t\}$ have this structure? Note that W_t is measurable with respect to the future σ-field $\mathcal{G}_t = \sigma(X_t, X_{t+1}, \ldots)$ and is uncorrelated with X_{t+h}, $h > 0$, since

$$
\begin{aligned}
\mathrm{E}[W_t X_{t+h}] &= \mathrm{E}[(X_t - \phi_1 X_{t+1} - \cdots - \phi_p X_{t+p}) X_{t+h}] \\
&= \gamma(h) - \phi_1 \gamma(h-1) - \cdots - \phi_p \gamma(h-p) \\
&= \mathrm{E}[(X_t - \phi_1 X_{t-1} - \cdots - \phi_p X_{t-p}) X_{t-h}] \\
&= \mathrm{E}[Z_t X_{t-h}] = 0,
\end{aligned}
$$

where $\gamma(h) = \mathrm{E}[X_t X_{t+h}]$. If Z_0 is Gaussian, then W_t is independent of $\mathcal{G}_{t+1}$ and $\mathrm{E}[W_t | \mathcal{G}_{t+1}] = 0$ a.s., so that $\{W_t\}$ is a backward martingale difference with respect to $\{\mathcal{G}_t\}$. (In fact, $\{W_t\}$ is an iid sequence.) The fact that the Gaussian distribution is unusual in this respect is the content of the following proposition.

PROPOSITION 4.1. *Let $\{X_t\}$ be a causal AR(p) process $\phi(B)X_t = Z_t$, where $\{Z_t\}$ is iid with zero mean and finite non-zero cumulant κ_r of order $r > 2$. Then*

$\{W_t\}$ *defined in (4.2) cannot be a backward martingale difference sequence with respect to* $\{\mathcal{G}_t\}$.

Proof. First note that $X_t = \sum_{j=0}^{\infty} \psi_j Z_{t-j}$, where the sequence $\{\psi_j\}$ is absolutely summable and has an infinite number of non-zero terms. Now suppose that $\{W_t\}$ is a backward martingale difference sequence with respect to $\{\mathcal{G}_t\}$. Then for $h \geq 0$,

$$\mathrm{E}[W_{t-h}|\mathcal{G}_{t+1}] = \mathrm{E}[\,\mathrm{E}[W_{t-h}|\mathcal{G}_{t-h+1}]\,|\mathcal{G}_{t+1}] = 0 \text{ a.s.},$$

so that

$$\mathrm{E}[\phi(B)W_t|\mathcal{G}_{t+1}] = \mathrm{E}[\phi(B^{-1})Z_t|\mathcal{G}_{t+1}] = 0 \text{ a.s.}$$

Since $\sigma(X_{t+h}) \subset \mathcal{G}_{t+1}$ for $h > 0$, we also have

$$\mathrm{E}[\phi(B^{-1})Z_t|X_{t+h}] = 0 \text{ a.s.}$$

and so

$$\mathrm{E}[\phi(B^{-1})Z_t X_{t+h}^k] = \mathrm{E}[X_{t+h}^k \mathrm{E}[\phi(B^{-1})Z_t|X_{t+h}]] = 0 \text{ a.s.}$$

for $k = 0, 1, \dots, r-1$. It follows that the rth-order cumulant $cum\,(\phi(B^{-1})Z_t, X_{t+h}, \dots, X_{t+h}) = 0$, and so with $\phi_0 = -1$, we have

$$
\begin{aligned}
0 &= cum\,(\phi(B^{-1})Z_t, X_{t+h}, \dots, X_{t+h}) \\
&= cum\left(\sum_{j=0}^{p}(-\phi_j)Z_{t+j}, \sum_{j_1=0}^{\infty}\psi_{j_1}Z_{t+h-j_1}, \dots, \sum_{j_{r-1}=0}^{\infty}\psi_{j_{r-1}}Z_{t+h-j_{r-1}} \right) \\
&= \sum_{j=0}^{p}\sum_{j_1=0}^{\infty}\cdots\sum_{j_{r-1}=0}^{\infty}(-\phi_j)\psi_{j_1}\cdots\psi_{j_{r-1}}cum\,(Z_{t+j}, Z_{t+h-j_1}, \dots, Z_{t+h-j_{r-1}}) \\
&= \sum_{j=0}^{p}(-\phi_j)\psi_{h-j}^{r-1}\kappa_r.
\end{aligned}
$$

Since $\kappa_r \neq 0$, it follows that

$$\psi_h^{r-1} = \sum_{j=1}^{p}\phi_j\psi_{h-j}^{r-1}$$

for $h \geq 1$. But starting with $\psi_0 = 1$ we have the standard recursion

$$\psi_h = \sum_{j=1}^{p}\phi_j\psi_{h-j}$$

for $h \geq 1$ (see Section 3.3 of [4]). Clearly $\psi_1^{r-1} = \phi_1 = \psi_1$. Assume $\psi_k^{r-1} = \psi_k$ for $k \leq h$. Then $\psi_{h+1}^{r-1} = \sum_{j=1}^{p}\phi_j\psi_{h+1-j}^{r-1} = \sum_{j=1}^{p}\phi_j\psi_{h+1-j} = \psi_{h+1}$, so by induction $\psi_h^{r-1} = \psi_h$ for $h = 1, 2, \dots$ Thus $\psi_h = -1, 0,$ or 1, and since $\sum_j |\psi_j| < \infty$, $\psi_j = 0$ for all large j— a contradiction. Therefore $\{W_t\}$ is not a backward martingale difference sequence with respect to $\{\mathcal{G}_t\}$. $\square$

In particular, this means that the best linear predictor $\phi_1 X_{t+1} + \cdots + \phi_p X_{t+p}$ of X_t given the future is not the best mean-square predictor; that is, the best linear backcast is not the optimal backcast. For example, in the Bernoulli-driven AR(1) given above, the best linear predictor of X_t in terms of its future is $0.5 X_{t+1}$, which has mean-square error $\text{Var}(W_t) = 0.25$, while the best mean-square predictor is the nonlinear function $(2X_{t+1}) \, mod \, 2$, which has mean-square error zero.

Though the distribution and the dependence structure of $\{W_t\}$ are complicated, the sequence can be simulated if we rewrite (4.2) as

$$\phi(B)W_t = \phi(B^{-1})Z_t$$

and note that $\{W_t\}$ is an ARMA(p,p) process driven by the iid sequence $\{Z_t\}$. The sequence $\{W_t\}$ is then easily generated by standard ARMA simulation techniques. Using the backward representation, the backward noise, and the initial values $\mathbf{X}_p$, we can generate the required sample paths passing through $\mathbf{X}_p$. For each such series, we estimate $\boldsymbol{\phi}$ by $\hat{\boldsymbol{\phi}}$ and sample from F to obtain a realization of Y conditioned on $\mathbf{X}_p$. This simulation procedure gives us a standard against which to compare prediction procedures and gives us a pattern to follow for bootstrapping.

Of course in a statistical setting we know neither $\boldsymbol{\phi}$ nor F. In this case $\boldsymbol{\phi}$ becomes a nuisance parameter and bootstrapping becomes appealing: estimate $\boldsymbol{\phi}$ by $\hat{\boldsymbol{\phi}}$ and F by $\hat{F}$, where $\hat{F}$ is the empirical distribution function of the (suitably centered and rescaled) residuals $\{\hat{\phi}(B)X_t\}, t = p+1, \ldots, n$. With these substitutions, use the above simulation procedure to bootstrap conditional sample paths $\{X_t^*\}$ which pass through $\mathbf{X}_p$. Specifically, take the following steps:

STEP 0. *Compute $\hat{\boldsymbol{\phi}}$ for the observed data $\{X_t\}_{t=1}^n$.*

STEP 1. *Compute residuals $\{\hat{\phi}(B)X_t\}_{t=p+1}^n$ and center and rescale to obtain $\{\hat{Z}_t\}_{t=p+1}^n$. Let $\hat{F}$ denote the empirical distribution function of the $\hat{Z}_t$'s.*

Remark. Center by subtracting off the sample mean of the residuals $\{\hat{\phi}(B)X_t\}_{t=p+1}^n$ and rescale by the factor $\sqrt{(n+p)/(n+2p)}$ as recommended in [10]. The mean of the distribution $\hat{F}$ is zero and the variance will be denoted by $\hat{\sigma}^2$.

STEP 2. *Generate a bootstrap replicate $\{W_t^*\}$ of the backward noise $\{W_t\}$ via*

$$\hat{\phi}(B)W_t^* = \hat{\phi}(B^{-1})\hat{Z}_t^*,$$

a causal ARMA(p,p), where $\{\hat{Z}_t^*\}$ is iid $\hat{F}$.

Remark. Since the spectral density of $\{W_t^*\}$ is

$$f_{W^*}(\lambda) = \frac{|\hat{\phi}(e^{i\lambda})|^2}{|\hat{\phi}(e^{-i\lambda})|^2} \frac{\hat{\sigma}^2}{2\pi} = \frac{\hat{\sigma}^2}{2\pi},$$

where $\hat{\sigma}^2$ is the variance of $\hat{Z}_0^*$, we see that $\{W_t^*\}$ is uncorrelated and has the same variance as the resampled noise $\{\hat{Z}_t^*\}$.

STEP 3. *Generate a bootstrap replicate* $\{X_t^*\}$ *of* $\{X_t\}$ *passing through* $\mathbf{X}_p$ *via*

$$X_t^* = X_t, \quad t = n, n-1, \ldots, n-p+1;$$
$$X_t^* = \hat{\phi}_1 X_{t+1}^* + \cdots + \hat{\phi}_p X_{t+p}^* + W_t^*, \quad t = n-p, n-p-1, \ldots, 1,$$

using the process $\{W_t^*\}$ *generated in Step 2 (retaining order).*

STEP 4. *Compute the estimate* $\hat{\boldsymbol{\phi}}^*$ *from the bootstrap replicate. Compute a bootstrapped conditional prediction error*

$$Y^* = \hat{Z}_{n+1}^* + (\hat{\phi}_1 - \hat{\phi}_1^*)X_n + \cdots + (\hat{\phi}_p - \hat{\phi}_p^*)X_{n+1-p},$$

where $\hat{Z}_{n+1}^*$ *is a random draw from* $\hat{F}$.

STEP 5. *If* R *bootstrap replicates (and conditional prediction errors) have been obtained, then stop; otherwise, repeat Steps 2–4.*

Figure 4.3 shows 75 observations of the AR(2) process $X_t - 0.75X_{t-1} + 0.5X_{t-2} = Z_t$, where $\{Z_t\}$ is iid contaminated Gaussian, and a bootstrapped replicate, which is conditioned to pass through the last two observations.

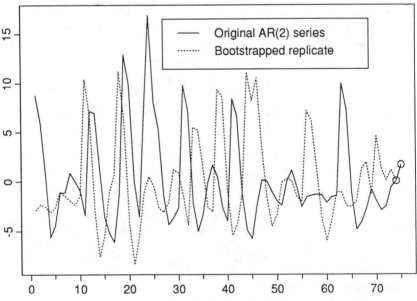

Fig. 4.3. AR(2) sample path and a conditionally bootstrapped replicate

Interestingly, previous attempts to bootstrap conditional AR(p) sample paths [11] have used resampled *backward* residuals $\{\hat{W}_t\} = \{\hat{\phi}(B^{-1})X_t\}$ in equation (4.1), though as Findley [8] has noted, this implies an iid structure for the random variables $\{W_t\}$, which are known to be merely uncorrelated. In [11], the authors do explore the possibility of expressing backward residuals in terms of forward residuals and resampling the latter, but the equation (2.7) which they derive has an explosive autoregressive part (a zero inside the unit circle), which may explain the excessive variability they found.

Resampling backward residuals as in [11] also tends to symmetrize the marginal distribution for the bootstrapped series, which we will denote by $\{Y_t^*\}$. The symmetry of $\{Y_t^*\}$ can be seen in Figures 4.4 and 4.5. These figures also show that the marginal distribution for the series $\{X_t^*\}$ bootstrapped using the method above matches the marginal distribution for the observed series $\{X_t\}$ closely. To make clear the difference between $\{X_t^*\}$ and $\{Y_t^*\}$ as $n \to \infty$, these plots are based on 10,000 observations of the AR(2) process $X_t - 0.75X_{t-1} + 0.5X_{t-2} = Z_t$, where $\{Z_t\}$ is iid contaminated Gaussian; the difference is noticeable even in small samples ($n =75$–100).

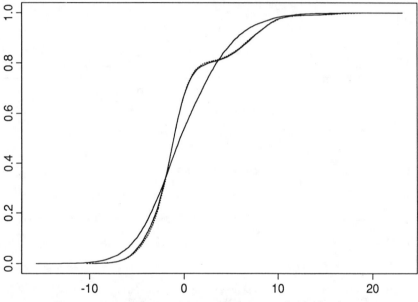

Fig. 4.4. Empirical distribution functions for $\{X_t\}$, $\{X_t^*\}$, and $\{Y_t^*\}$

In [8], Findley concludes that "generally satisfactory methods are lacking for obtaining bootstrap sample paths through the final observations" $\mathbf{X}_p$. The new method described above answers Findley's objections.

In further work we plan to explore the usefulness of this bootstrapping method for prediction purposes. Some preliminary analytic and simulation results indicate that nearly all one-step prediction error variability is due to noise and not parameter estimation (meaning bootstrapping is unnecessary). We hope to resolve this issue and then consider bootstrapping prediction error distributions for h-step predictors, where parameter variability should have more impact.

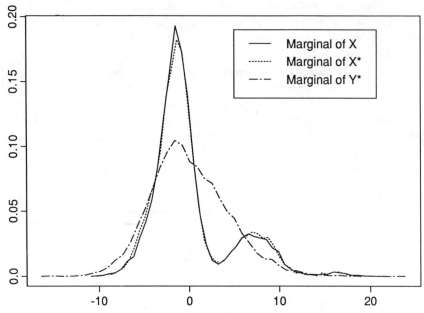

Fig. 4.5. Estimated densities for $\{X_t\}$, $\{X_t^*\}$, and $\{Y_t^*\}$

REFERENCES

[1] T. W. ANDERSON AND A. TAKEMURA, *Why do noninvertible estimated moving averages occur?*, Journal of Time Series Analysis, 7 (1986), pp. 235–254.

[2] G. E. P. BOX AND G. M. JENKINS, *Time Series Analysis: Forecasting and Control*, revised edition, Holden-Day, San Francisco, 1976.

[3] F. J. BREIDT AND R. A. DAVIS, *Time-reversibility, identifiability, and independence of innovations for stationary time series*, (submitted), 1990.

[4] P. J. BROCKWELL AND R. A. DAVIS, *Time Series: Theory and Methods*, Springer-Verlag, New York, 1987.

[5] M. DEISTLER, W. DUNSMUIR, AND E. J. HANNAN, *Vector linear time series models: corrections and extensions*, Adv. Appl. Prob., 10 (1978), pp. 360–372.

[6] D. DONOHO, *On minimum entropy deconvolution*, Applied Time Series Analysis **2**, ed. by D. F. Findley, Academic Press, New York, 1981, pp. 565–608.

[7] W. DUNSMUIR, *A simulation study of L^1 estimation of a seasonal moving average time series model*, Working Paper 1990-3-34/R, School of Information and Computing Sciences, Bond University, 1990.

[8] D. F. FINDLEY, *On bootstrap estimates of forecast mean square errors for autoregressive processes*, Computer Science and Statistics: The Interface, ed. by D. M. Allen, North-Holland, Amsterdam, 1986, pp. 11–17.

[9] J. R. M. HOSKING, *Modeling persistence in hydrological time series using fractional differencing*, Water Resources Research, 20 (1984), pp. 1898–1908.

[10] R. A. STINE, *Estimating properties of autoregressive forecasts*, J. Amer. Statist. Assoc., 82 (1987), pp. 1072–1078.

[11] L. A. THOMBS AND W. R. SCHUCANY, *Bootstrap prediction intervals for autoregression*, J. Amer. Statist. Assoc., 85 (1990), pp. 486–492.

[12] G. WEISS, *Time-reversibility of linear stochastic processes*, J. Appl. Prob., 12 (1975), pp. 831–836.

FOURIER AND LIKELIHOOD ANALYSIS IN NMR SPECTROSCOPY

DAVID R. BRILLINGER* AND REINHOLD KAISER†

Abstract. Nuclear magnetic resonance (nmr) is a quantum mechanical phenomenon that may be employed to study the structure of a variety of molecules, crystals and polymers. The time series data collected are traditionally Fourier transformed and the Fourier amplitudes examined for peaks. Higher-order transforms are sometimes employed. If the substance and relevant interactions are known one can set down a set of differential equations describing the temporal evolution of the state matrix that describes the system. These differential equations are bilinear in the input and the system state. The time series recorded in an experiment is, up to noise, a linear function of the entries of the state matrix. In the research to be presented, the Fourier techniques of analysis are compared with a maximum likelihood analysis based on the state matrix. Results are presented for an experiment involving 2,3-dibromothiophene.

Key words. bilinear system, Bloch equations, Fourier analysis, maximum likelihood estimation, m-sequence, residual analysis, signal-generated noise, system identification, transfer function

AMS(MOS) subject classifications. 62M15, 62P99

1. Introduction. The concerns of this paper are to provide an example of the maximum likelihood analysis of data collected in nuclear magnetic resonance (nmr) spectroscopy and to give some comparative discussion of maximum likelihood and Fourier based techniques. The nmr case is based upon revered theory allowing conceptual modelling and a state space formulation. The layout of the paper is: first some pertinent background concerning nmr is set down, next comes some formal development following the basic theoretical layout, then a discussion of the problem as one of system identification is presented. Sections 5 and 6 describe a particular laboratory experiment carried out and present an analysis of its results. The paper concludes with a discussion comparing and contrasting the various approaches and mentions some possible future work.

2. Nuclear magnetic resonance spectroscopy. Nuclear magnetic resonance spectroscopy is a quantum mechanical phenomenon that may be employed to study the structure of a variety of molecules, crystals and polymers. In the procedure, a sample of the material whose structure is to be investigated is placed in a strong magnetic field, 1.41 Tesla for our measurements. This field is constant in time and uniform in space throughout the volume occupied by the sample. It exerts a mechanical torque on those nuclei in the sample that carry a magnetic dipole moment, tending to align these nuclei in the direction of the field. However, the magnetic dipole of a nucleus is associated with an intrinsic spin angular momentum, and the torque consequently causes a gyroscopic precession of the nuclear

*Department of Statistics, University of California, Berkeley, CA 94720. Research supported by NSF Grant DMS–8900613

†Department of Physics, University of New Brunswick, Fredericton, CANADA E3B 5A3. Research supported by NSERC Grant A1565

spin axis about the direction of the magnetic field. The precession frequency is proportional to the magnetic field strength at the site of a nucleus and thus provides information about the nuclear environment. The nuclear precessional motion may be stimulated by applying to the sample a weak oscillating magnetic field directed at right angles to the strong constant magnetic field. A resonance effect occurs when the oscillation frequency of this weak field coincides with a nuclear precession frequency. The nuclear motion is sensed by monitoring the voltage that is induced by the moving nuclear magnetic dipoles in a coil that is wound at right angles to the strong magnetic field. An nmr spectrum for a particular sample is a graph showing the amplitude of this induced voltage as function of the oscillation frequency of the weak stimulating magnetic field.

FIGURE 1

2,3-dibromothiophene

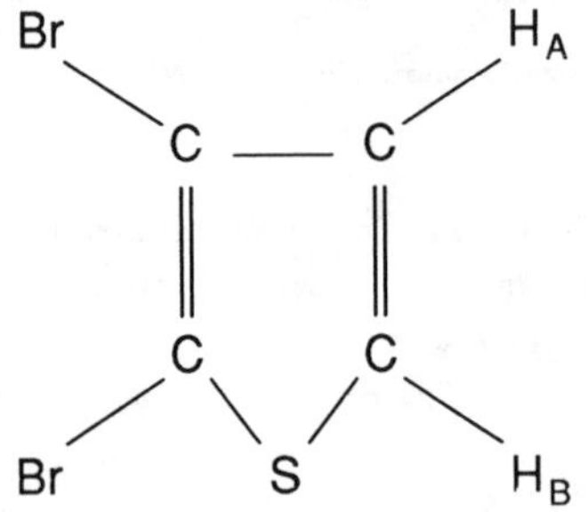

The data used for our work are derived from a sample of 2,3-dibromothiophene. This substance is liquid at room temperature, its molecules have the chemical structure shown in Figure 1. The naturally abundant isotopes of carbon (^{12}C) and sulfur (^{32}S) have zero magnetic dipole moment and are thus not observable by nmr methods. The stable bromine isotopes (^{79}Br) and (^{81}Br) both carry magnetic dipole moments, but they also carry a sizeable electric quadrupole moment which couples them to fluctuating electric fields in the sample and this also makes them inactive for our work. Our data thus arise from the magnetic resonances of the nuclei (protons) of the two hydrogen atoms labelled H_A and H_B in Figure 1. The nuclei are surrounded by molecular electrons that hold the atoms in the structure of Figure 1. These electrons tend to shield the nuclei from the strong applied field. The proximity of the sulfur atom causes the electronic shielding of nucleus H_A to differ slightly from the shielding available to H_B, thus leading to slightly different precession frequencies for H_A and H_B. In more detail, the precession frequency of hydrogen nuclei in our 1.41 Tesla field is 6.00E7 Hz, and the difference of resonance frequencies of H_A and H_B was found as 32.57 Hz, so the nmr "chemical shift" between H_A and H_B amounts to $32.57/6.00E7 = 0.543$ parts per million (ppm).

The nmr spectrum of 2,3-dibromothiophene is shown in Figure 3b below. The two doublets of resonance peaks are here separated by the 32.57 Hz chemical shift and each doublet arises from one hydrogen nucleus H_A or H_B, respectively. The doublet splitting of the resonances is caused by a coupling of H_A with H_B via the magnetic field set up at H_B by the magnetic moment of H_A, and vice versa. (More precisely, it is the part of this coupling that is not averaged to zero by the thermal tumbling of the molecule in the liquid.) Quantum mechanics yields for each hydrogen nucleus in a strong magnetic field, two stationary states with nuclear spin axis either parallel or antiparallel to the direction of the magnetic field. So, one H_B resonance peak arises from molecules having the H_A spin axis parallel to the magnetic field, the other from molecules having the H_A spin axis antiparallel to the field. The strength of this intramolecular nuclear coupling is measured by the doublet splitting as J=5.76 Hz for the case of Figure 3.

Other significant parameters are various relaxation times classified as either longitudinal T_1 or transverse T_2. The longitudinal T_1 is the time constant with which the nuclear magnetization eventually aligns itself with the direction of the strong magnetic field after the sample is placed into the magnet. The transverse T_2 are the time constants governing the "frictional damping" of the oscillatory components of the state vector that describe the nuclear precession. These relaxation times depend on thermal random motions in the sample. They are of several seconds duration for our sample.

A quantitative description of the nuclear magnetic spin dynamics in a macroscopic sample requires a quantum mechanical formulation in terms of the spin density operator $\boldsymbol{\rho}$, (see Slichter (1990) and Ernst et al. (1987)). The motion of this operator is described by the von Neumann equation

$$(2.1) \qquad \frac{d\boldsymbol{\rho}}{dt} = \mathbf{R}(\boldsymbol{\rho} - \boldsymbol{\rho}^T) + (\mathbf{H}\boldsymbol{\rho} - \boldsymbol{\rho}\mathbf{H})/i\hbar$$

Here, $\mathbf{R}$ is a superoperator describing relaxation of the density operator towards its thermal equilibrium value $\boldsymbol{\rho}^T$, the symbol $\hbar$ designates Planck's constant/2π, and $i = \sqrt{-1}$. (Superoperators are discussed in Ernst et al. (1987)). The symbol $\mathbf{H}$ designates the Hamilton operator for the energy of the nuclear magnetic dipoles in the magnetic fields that are applied to the sample. It can be separated into two parts,

$$(2.2) \qquad \mathbf{H} = \mathbf{H}_0 + \mathbf{H}_1(t)$$

such that $\mathbf{H}_0$ describes the interactions of the nuclei with the strong magnetic field and the intramolecular coupling, and $\mathbf{H}_1$ describes the interaction with the oscillating magnetic field that is used to stimulate the precession motion. The first term, $\mathbf{H}_0$, is constant in time and strong compared to the second term, $\mathbf{H}_1(t)$, which is proportional to the stimulating field.

The operators in (2.1) are linear and it is convenient to represent them by matrices operating in the space spanned by the eigenvectors of $\mathbf{H}_0$. For our sample which contains only two nmr active hydrogen nuclei, this space is of dimension 4.

Furthermore, it is numerically helpful to freeze the fast precession by transforming to a physical x, y, z-space that rotates about the z direction of the strong magnetic field at 6E7 revolution/sec. Equations (2.1) then take the form shown in Figure 2 for the 16 entries ρ_{jk} representing the density operator $\boldsymbol{\rho}$ in matrix form. Only ten equations are written out in Figure 2, the remaining six are complex conjugates of the last six since $\boldsymbol{\rho}$ is hermitian. The relaxation operator $\mathbf{R}$ has been represented by relaxation times T.

FIGURE 2

$$\frac{d\rho_{11}}{dt} = \frac{i}{2}\{(\rho_{12} - \rho_{21}) + (\rho_{13} - \rho_{31})(c - s)\}\gamma x - (\rho_{11} - \rho_{11}^{T})/T_1$$

$$\frac{d\rho_{22}}{dt} = \frac{i}{2}(\rho_{21} - \rho_{12} + \rho_{24} - \rho_{42})(c + s)\gamma x - (\rho_{22} - \rho_{22}^{T})/T_1$$

$$\frac{d\rho_{33}}{dt} = \frac{i}{2}(\rho_{31} - \rho_{13} + \rho_{34} - \rho_{43})(c - s)\gamma x - (\rho_{33} - \rho_{33}^{T})/T_1$$

$$\frac{d\rho_{44}}{dt} = \frac{i}{2}\{(\rho_{42} - \rho_{24})(c + s) + (\rho_{43} - \rho_{34})(c - s)\}\gamma x - (\rho_{44} - \rho_{44}^{T})/T_1$$

$$\frac{d\rho_{12}}{dt} = \frac{i}{2}\rho_{12}(J - \sqrt{\ } + \omega_A + \omega_B) + \frac{i}{2}\{(\rho_{11} - \rho_{22} + \rho_{14})(c + s) -$$
$$\rho_{32}(c - s)\}\gamma x - \rho_{12}/T_{2B}$$

$$\frac{d\rho_{13}}{dt} = \frac{i}{2}\rho_{13}(J + \sqrt{\ } + \omega_A + \omega_B) + \frac{i}{2}\{(\rho_{11} - \rho_{33} + \rho_{14})(c - s) -$$
$$\rho_{23}(c + s)\}\gamma x - \rho_{13}/T_{2A}$$

$$\frac{d\rho_{14}}{dt} = i\rho_{14}(\omega_A + \omega_B) + \frac{i}{2}\{(\rho_{12} - \rho_{24})(c + s) + (\rho_{13} - \rho_{34})(c - s)\}\gamma x - \rho_{14}/T_{2D}$$

$$\frac{d\rho_{23}}{dt} = i\rho_{23}\sqrt{\ } + \frac{i}{2}\{(\rho_{21} + \rho_{24})(c - s) - (\rho_{13} + \rho_{43})(c + s)\}\gamma x - \rho_{23}/T_{2Z}$$

$$\frac{d\rho_{24}}{dt} = \frac{i}{2}\rho_{24}(-J + \sqrt{\ } + \omega_A + \omega_B) + \frac{i}{2}\{(\rho_{22} - \rho_{44} - \rho_{14})(c + s) + \rho_{23}(c - s)\}\gamma x -$$
$$\rho_{24}/T_{2A}$$

$$\frac{d\rho_{34}}{dt} = \frac{i}{2}\rho_{34}(-J - \sqrt{\ } + \omega_A + \omega_B) + \frac{i}{2}\{(\rho_{33} - \rho_{44} - \rho_{14})(c - s) + \rho_{32}(c + s)\}\gamma x -$$
$$\rho_{34}/T_{2B}$$

The four diagonal elements ρ_{jj} may be interpreted as probabilities normalized by $Tr(\boldsymbol{\rho}) = 1$, to have the spin system in each of the four eigenstates of the Hamiltonian $\mathbf{H}_0$. Without stimulus and in thermal equilibrium, these are the only nonzero elements with values

$$(2.3) \qquad \rho_{11}^{T} = (1 + \varepsilon)/4 \qquad \rho_{22}^{T} = \rho_{33}^{T} = 1/4 \qquad \rho_{44}^{T} = (1 - \varepsilon)/4$$

with $\varepsilon \ll 1$ depending on sample temperature and magnetic field strength. The off-diagonal elements describe the precessional motion and may be interpreted as coherences between eigenstates of $\mathbf{H}_0$. The stimulus $X(t)$ links them to the diagonal elements since $\mathbf{H}_1$ does not commute with $\mathbf{H}_0$. They may be classified by the quantum jump involved in the coherence as follows. The set $\rho_{12}, \rho_{13}, \rho_{24}, \rho_{34}$ is responsible for the output signal induced by the precessing nuclear magnetic moments in the receiver coil. Depending on whether this coil is wound around the sample in the x or y direction, the output is

(2.4)
$$Y_x = (c+s)\operatorname{Re}(\rho_{12} + \rho_{24}) + (c-s)\operatorname{Re}(\rho_{13} + \rho_{34})$$
$$Y_y = (c+s)\operatorname{Im}(\rho_{12} + \rho_{24}) + (c-s)\operatorname{Im}(\rho_{13} + \rho_{34})$$

Here and in Figure 2 the meaning of symbols is:

J is the coupling strength of H_A with H_B

ω_A, ω_B are the precession frequencies in the rotating coordinate frame of H_A and H_B, respectively

$$\delta = |\omega_A - \omega_B| = \text{chemical shift}$$

$$\theta = \arctan(J/\delta) \qquad c = \cos(\theta/2) \qquad s = \sin(\theta/2) \qquad \sqrt{\ } = \sqrt{J^2 + \delta^2}$$

γ is a scale factor for the input amplitude

The equations of Figure 2 shows that, in the absence of a stimulus X, the four elements $\rho_{12}, \rho_{13}, \rho_{24}, \rho_{34}$ precess in the rotating coordinate frame like $\exp\{-1/T_2 + i\omega\}$ with frequencies

(2.5)
$$\omega = (\pm J \pm \sqrt{J^2 + (\omega_a - \omega_B)^2})/2 + (\omega_A + \omega_B)/2$$

These are the four nmr resonance frequencies. The superposition of the four damped oscillations that are observed in some simulated output $Y(t)$, $t, = 0, \ldots, T-1$, in response to an impulse stimulus, is shown in Figure 3a below. The corresponding Fourier spectrum

(2.6)
$$\left| \sum_{t=0}^{T-1} Y(t) e^{-i\lambda t} \right|$$

is shown in Figure 3b. These four $\boldsymbol{\rho}$ elements are referred to as single quantum coherences because they are associated with emission or absorption by a molecule of a single quantum $\hbar$ of angular momentum. Of the remaining off-diagonal elements, ρ_{23} is a zero quantum coherence and ρ_{14} is a double quantum coherence. They are not directly observable in the output Y.

FIGURE 3

Pulse Response

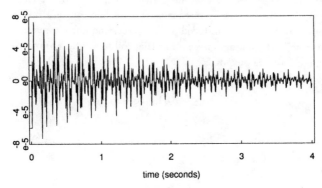

time (seconds)

Fourier Amplitude

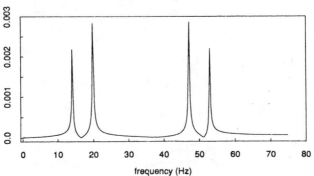

frequency (Hz)

The 16 equations of motion for the ρ_{jk} may be written in vector form as

$$(2.7) \qquad \frac{dS(t)}{dt} = a + AS(t) + BS(t)X(t)$$

with S a 16-vector containing the ρ_{jk}, a a 16-vector holding the four ρ_{jj}^T/T_1 thermal equilibrium terms and zero otherwise, A a 16×16 matrix collecting the diagonal terms of Figure 2, and B a 16×16 matrix holding the off-diagonal terms. B is symmetric with entries purely imaginary, it does not commute with A. The output equations (2.4) may be written

$$(2.8) \qquad Y(t) = \mathrm{Re}(c^\tau S(t))$$

with suitable 16-vectors c.

3. A problem of system identification. The transition equation (2.7) and the measurement equation (2.8) together describe a system carrying input signals,

X, over into corresponding output signals, Y. The concern is to determine the unknown parameters of the system. A variety of techniques have been proposed. In the previous section it was seen how Fourier transforming the response to a pulse could lead to estimates of the "frequency" parameters, and indeed this technique has long been employed in nmr spectroscopy, see eg. Becker and Farrar (1972). Information can also be gained via a succession of two-pulse experiments and those experiments have the possibility of displaying the presence of cross-coupling of frequencies for "large" input, X. The practice of nmr spectroscopy has moved on to the employment of multipulse sequences, eg. Kay. *et al.* (1990) where sequences of 10 pulses are employed.

In the case of step-function input, one can set down explicit representations for the solutions of (2.6), see for example Brillinger (1985, 1990). Suppose, now, that

$$(3.1) \qquad\qquad X(t) = 0 \qquad \text{for } s \leq t < u$$

$$X(t) = x \qquad \text{for } u \leq t < v$$

$$X(t) = 0 \qquad \text{for } v \leq t$$

This is a pulse of height x and width $v - u$. Writing $\mathbf{D} = \mathbf{A} + \mathbf{B}x$, the solution to (2.6) is given by

$$\mathbf{S}(t) = e^{\mathbf{A}(t-s)}\mathbf{S}(s) + \mathbf{A}^{-1}[e^{\mathbf{A}(t-s)} - \mathbf{I}]\mathbf{a} \qquad \text{for } s \leq t < u$$

$$\mathbf{S}(t) = e^{\mathbf{D}(t-u)}\mathbf{S}(u) + \mathbf{D}^{-1}[e^{\mathbf{D}(t-u)} - \mathbf{I}]\mathbf{a} \qquad \text{for } u \leq t < v$$

$$(3.2) \qquad \mathbf{S}(t) = e^{\mathbf{A}(t-v)}\mathbf{S}(v) + \mathbf{A}^{-1}[e^{\mathbf{A}(t-v)} - \mathbf{I}]\mathbf{a} \qquad \text{for } v \leq t$$

referring respectively, to the periods before, during and after the pulse.

Figure 3a gives an example of the response, Y, to a pulse (3.1) applied to the system in thermal equilibrium, employing parameter values appropriate to a 2,3-DBT sample. (See Appendix A.1.) 600 points have been plotted. The bottom graph gives the modulus of the Fourier transform of this output as computed by (2.6). Four substantial peaks are seen to be present in the latter. This could have been anticipated from the form of the matrix $\mathbf{A}$ and expression (3.2). The frequencies are those of (2.5).

FIGURE 4

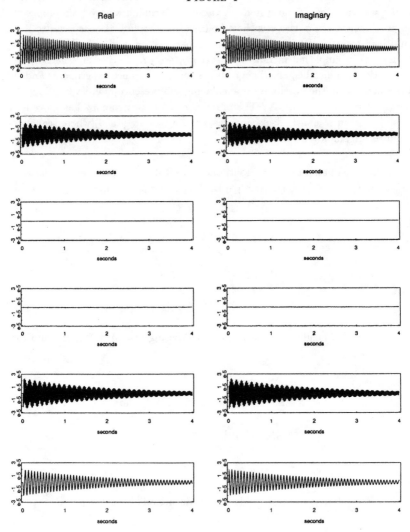

Figure 4 gives the evolution of the nondiagonal entries of the state matrix following the pulse input. The time series remaining 0 correspond to ρ_{23} and ρ_{14}. The series fluctuates with the frequencies of the elements of $\mathbf{A}$. An interesting phenomenon is present in the plots of the second and fifth rows. There seem to be beats, suggesting the presence of two frequencies. To better understand this phenomenon, Figure 5 provides a graph of 50 points from the function

$$e^{-\alpha t}\sin(\beta t)$$

with parameter values corresponding to those of the numerical experiment. The phenomenon is now seen to arise from the particular relationship of the sampling rate and the base frequency. Typically but 1 or 2 points are being plotted between the zero crossings and hence the deceptive appearance.

FIGURE 5
Frequency = 52.94 Hz, sampling rate 150 Hz

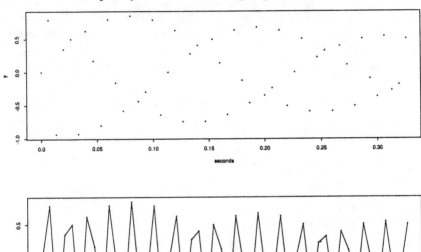

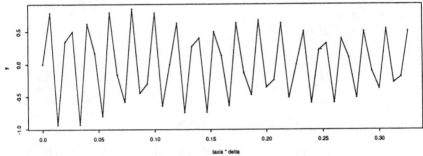

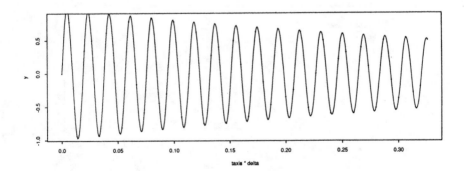

Next suppose, a pulse further to the pulse (3.1), is input to the system commencing at time u_1. Suppose it is has amplitude x_1 and width $\nu_1 - u_1$, then the equations of subsequent evolution are

$$\mathbf{S}(t) = \exp\{\mathbf{D}_1(t - u_1)\}\mathbf{S}(u_1) + \mathbf{D}_1^{-1}[\exp\{\mathbf{D}_1(t - u_1)\} - \mathbf{I}]\mathbf{a} \quad \text{for } u_1 < t \le \nu_1$$

$$(3.3) \quad \mathbf{S}(t) = \exp\{\mathbf{A}(t - \nu_1)\}\mathbf{S}(\nu_1) + \mathbf{A}^{-1}[\exp\{\mathbf{A}(t - \nu_1) - \mathbf{I}]\mathbf{a} \quad \text{for } \nu_1 < t$$

Here $\mathbf{D}_1 = \mathbf{A} + \mathbf{B}x_1$.

If $|\gamma x|$ is small, then the system is approximately linear and the output, Y, will show simply the frequencies of $\rho_{12}, \rho_{24}, \rho_{13}, \rho_{34}$ following Figure 2. If $|\gamma x|$ is large, nonlinear phenomena may show themselves. To illustrate this, let $\Delta(t)$ denote a pulse starting at time 0 having width σ. Consider a suite of two-pulse experiments with input $X(t) = \Delta(t) + \Delta(t - s)$, for a succession of values s. That, for example, the frequency of ρ_{12} interacts with that of ρ_{14}, may be seen from the fifth equation of Figure 2.

FIGURE 6

Two-pulse Fourier amplitude - 1 degree flip angle

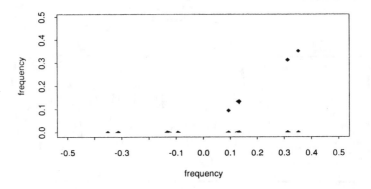

Two-pulse Fourier amplitude - 90 degree flip angle

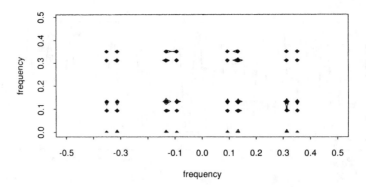

Simulations of this technique were carried out. Figure 6 presents the absolute value of the two dimensional Fourier transform of the outputs, Fourier transforming with respect to s, the interval between pulses and u, time since the second pulse finished. In the case of small input, (1 degree flip angle, see Appendix A.1 for definition), there are only off-axis peaks apparent along the diagonal $\lambda = \mu$. For large input, (90 degree flip angle), a host of off-diagonal peaks appear. The peaks along the horizontal axis occur in the manner of expression (4.9) below.

4. System identification by cross-correlation. Nmr spectroscopy has on occasion employed stochastic or pseudo-stochastic input, see eg. Ernst (1970), Kaiser (1970), Blümich (1985). One cross-correlates the input and output and then Fourier transforms the result. This may be motivated as follows: successive substitutions into the equation (2.7), assuming $\mathbf{B}$ or X small, leads to

$$(4.1) \quad S(t) = -\mathbf{A}^{-1}\mathbf{a} + \int^{t} e^{\mathbf{A}(t-s)}\mathbf{C}X(s)ds + \iint^{t\,s} e^{\mathbf{A}(t-s)}\mathbf{B}e^{\mathbf{A}(s-r)}\mathbf{C}X(r)X(s)drds + \cdots$$

with $\mathbf{C} = -\mathbf{A}^{-1}\mathbf{a}$, see eg, Blümich and Ziessow (1983), Banks (1988), Brillinger (1990). The linear, quadratic and third-order *asymmetric* (or triangular) transfer functions here are

$$(4.2) \qquad\qquad (i\lambda\mathbf{I} - \mathbf{A})^{-1}\mathbf{C}$$

$$(4.3) \qquad\qquad (i(\lambda + \mu)\mathbf{I} - \mathbf{A})^{-1}\mathbf{B}(i\lambda\mathbf{I} - \mathbf{A})^{-1}\mathbf{C}$$

$$(4.4) \qquad (i(\lambda + \mu + \nu)\mathbf{I} - \mathbf{A})^{-1}\mathbf{B}(i(\lambda + \mu)\mathbf{I} - \mathbf{A})^{-1}\mathbf{B}(i\lambda\mathbf{I} - \mathbf{A})^{-1}\mathbf{C}$$

with similar expressions for the higher-order cases. It is to be noted that peaks will occur in the absolute values of the linear transfer function at the resonance frequencies indicated earlier. For the quadratic and higher-order terms, a matching of frequencies connected by $\mathbf{B}$ is needed. Workers in nmr spectroscopy speak of "coupling" in this type of circumstance.

In the kernel approach to nonlinear systems analysis, it is usual to employ *symmetric* transfer functions in expansions such as (4.1), writing for example

$$(4.5) \qquad\qquad S(t) \approx$$

$$\int \mathbf{a}_1(u)X(t-u)du + \iint \mathbf{a}_2(u,\nu)X(t-u)X(t-\nu)dud\nu + \iiint \mathbf{a}_3(u,\nu,w)$$
$$X(t-u)X(t-\nu)X(t-w)dud\nu dw$$

with $\mathbf{a}_2$ and $\mathbf{a}_3$ symmetric in their arguments. In the case that stationary Gaussian input has been employed and the system is quadratic, (i.e. $\mathbf{a}_3 = 0$ in (4.5)), the linear transfer function is given by

$$(4.6) \qquad\qquad A_1(\lambda) = \mathbf{f}_{SX}(\lambda)/f_{XX}(\lambda)$$

white the quadratic one is given by

$$(4.7) \qquad\qquad A_2(-\lambda, -\mu) = \mathbf{f}_{XXS}(\lambda,\mu)/2f_{XX}(\lambda)f_{XX}(\mu)$$

with f_{XX} the input power spectrum, with f_{SX} the cross-spectrum of the input and output and with f_{XXS} a cross-bispectrum of the input and output, see eg. Tick (1961). These equations suggest how to estimate $\mathbf{A}_1$ and $\mathbf{A}_2$. Extensions exist to the higher-order terms, see Wiener (1958), Brillinger (1970), Marmarelis and Marmarelis (1978), Blümich (1985). In the case of pseudorandom input, expressions (4.6) and (4.7) hold approximately, see eg. Marmarelis and Marmarelis (1978).

The following hybrid technique shows how elementary cross-correlation techniques may be used to display the presence of cross-coupling. Suppose N denotes a white noise sequence with variance σ^2. Consider the suite of experiments in which the input is taken to be $N(t) + N(t - s)$ for a succession of values s. (Such an experiment was discussed in Blümich (1981).) For each individual experiment estimate the cross-spectrum, $f_{SN}(\lambda, s)$. Next Fourier transform this with respect to s to obtain a function of two frequencies. Off-diagonal peaks will be indicative of the presence of cross-coupling. To be specific, suppose that $\mathbf{S}(t)$ is given by (4.5) and that N is white noise with third cumulant $\kappa_3 = 0$ and fourth κ_4. By elementary computations (see Appendix B) one can show that for given lag s, $f_{SN}(\lambda, s)$ is given by

$$(4.8) \qquad \frac{\sigma^2}{2\pi}(1 + e^{-i\lambda s})\mathbf{A}_1(\lambda) + \left(\frac{\sigma^2}{2\pi}\right)^2 (1 + e^{-i\lambda s})3 \int \mathbf{A}_3(\lambda, \nu, -\nu)|1 + e^{i\nu s}|^2 d\nu$$

plus a term in κ_4. Here $\mathbf{A}_3$ is assumed symmetric in its arguments. (It will be obtained by permuting the arguments of (4.4) and averaging.) Taking the Fourier transform of (4.8) with respect to s and denoting the corresponding argument by μ, leads to

$$(4.9) \qquad \sigma^2 A_1(\lambda)[\delta(\mu) + \delta(\mu + \lambda)] + \frac{\sigma^4}{\pi}3 \int \mathbf{A}_3(\lambda, \nu, -\nu)d\nu[\delta(\mu) + \delta(\mu + \lambda)]$$

$$+ \frac{\sigma^4}{\pi}3[\mathbf{A}_3(\lambda, \mu, -\mu) + \mathbf{A}_3(\lambda, \mu + \lambda, -\mu - \lambda)]$$

plus a term in κ_4 and with δ the Dirac delta function. The delta functions are seen to lead to ridges about the lines $\mu = 0$ and $\mu = -\lambda$. Focusing on the last term of (4.9) and following expression (4.4) the terms

$$(4.10) \qquad (i\lambda \mathbf{I} - \mathbf{A})^{-1}\mathbf{B}(i(\lambda + \mu)\mathbf{I} - \mathbf{A})^{-1}\mathbf{B}(i\lambda \mathbf{I} - \mathbf{A})^{-1}\mathbf{C}$$

and their permuted variants will appear. The matrix $\mathbf{A}$ is diagonal hence peaks will appear at appropriate locations (λ, μ).

A numerical simulation experiment was carried out to examine the above technique. The noise process N consisted of pulses of amplitudes ± 1 the values being independent and equiprobable. The flip angle was 10 degrees and otherwise the parameters were as in Appendix A.1. Figure 7 presents the results of the analysis. Figure 7a provides the absolute value of $\mathbf{c}^\tau(4.10)$ supplemented by the 5 other permuted terms. A variety of peaks and cross-peaks are seen to appear. Figure 7b is the result of estimating (4.9). Peaks and cross-peaks occur as well as indications of the presence of the delta functions. The procedure does appear practical. In the estimation the s values ran from 1 to 128 and the cross-spectrum was estimated at 128 frequencies by averaging cross-periodograms of 100 successive stretches of data.

FIGURE 7

Transfer function amplitude - 10 degree flip angle

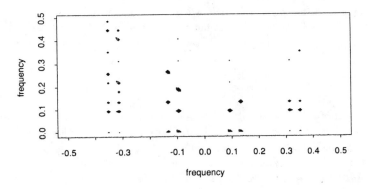

Hybrid analysis Fourier amplitude - 10 degree flip angle

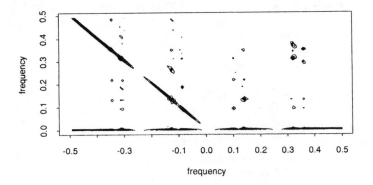

There are contrasting circumstances for emplying pulse input and "noise" input. The noise input has the advantage that input power required is low (but applied over a longer time). It has the further advantage that the response need be measured only once and thereafter can be subjected to a number of modelling analyses. The pulse input requires a high power and in consequence can be hard to produce and even damaging. However the approach is flexible, with specific pulse sequences able to be tailored to specific purposes.

Next consideration turns to data collected in a laboratory experiment.

5. Experimental details. The equations shown in Figure 2 describe the dynamics of the nuclear spin system in a x, y, z coordinate system that rotates at 6E7 rev/sec about the z direction of the strong magnetic field. The nmr spectrometer is fixed in the laboratory, and the rotation is simulated electronically by providing interaction with the sample via a 60 MHz radio frequency (rf) carrier sinusoid. This carrier is generated by multiplying with 6 the frequency of the output sinusoid of a 10 MHz quartz crystal oscillator that serves as master clock for the spectrometer. The input stimulus X modulates this carrier by means of a balanced modulator, and the nuclear output voltage Y is demodulated in a phase sensitive detector which is referenced to the carrier sinusoid. The phase angle of this reference depends on cable lengths, time delays in electronic components, etc, and the measured output is thus a projection in some direction ϕ in the x, y plane of the rotating coordinate frame,

$$(5.1) \qquad\qquad Y(t) = \cos \phi Y_x(t) + \sin \phi Y_y(t) + \ noise$$

This output of the phase sensitive detector is low-pass filtered by passing it through a 4-pole Bessel filter set to a 150 Hz corner frequency in order to reduce high frequency noise that would be folded into the Nyquist bandwidth by the sampling described below. The level of the input stimulus exceeds the nuclear response signal by some 100 dB, and to avoid direct crosstalk from the input to the output, the modulated carrier is gated to produce short rf pulses, and the detector output is sampled between input pulses. This time sharing system is governed by a digital pulse generator that is driven from the 10 MHz spectrometer clock. The data studied were generated by deriving the input stimulus from a 12-stage binary shift register with feedback such as to produce the m-sequence

$$(5.2) \qquad\qquad x_j = x_{j-1} x_{j-4} x_{j-6} x_{j-12}$$

starting from $x_j = -1$ for $j = 1, 2, \ldots 12$. The balanced modulator converts the $+1, -1$ levels of the shift register to 0 degree, 180 degree phase shifts of the 60 MHz carrier. The shift register was driven from the pulse generator at a rate to produce 150 bits/sec. The advantage of employing an m-sequence is that input data need not be recorded. The m-sequence has period 4095 bits corresponding to 27.3 seconds, and received data were recorded for one such period after the nuclear spin system had run through at least one prior period to reach a steady state. The pulse generator was programmed to use the dwell time of each bit, 1/150 sec $=6667 \, \mu$ sec, for the time sharing of input and output as follows.

Time 0   : open rf gate for 30 μ sec to apply input pulse to nuclear sample;

   77 μ sec: shift m-sequence to next bit;

3567 μ sec: sample output of antialiasing filter;

6667 μ sec: loop back to time 0.

The gated and modulated rf carrier was amplified such that a 30 μ sec pulse produced a 3.6 degree flip angle, (see Appendix A.1 for the definition). It was later learned that the 150 Hz antialiasing filter causes a time delay of 1/300 sec so that the nuclear output was actually sampled 204 μ sec after the end of the input pulse. However, crosstalk was still suppressed.

The 150 Hz sampling rate of the output corresponds to a Nyquist frequency window from 0 to 75 Hz, and the precession frequencies of the hydrogen nuclei must be within this window in the rotating coordinate frame, i.e. within $(6E7 \pm 75)$Hz in the fixed laboratory frame. Since this frequency is proportional to the intensity of the strong magnetic field, the 1.41 Tesla electromagnet must be controlled to considerably better than 1 part per million. To this end, our sample held a homogeneous mixture of 0.24 ml 2,3-dibromothiophene with 0.18 ml dimethylsulfoxide-d6. The latter compound has six nmr-active deuterium nuclei per molecule, and these yield a single nmr resonance near 9.21 MHz in a 1.41 Tesla field. A second spectrometer was set up to operate with the same probe using a rf carrier derived from the 10 MHz system clock by means of a frequency synthesizer such as to be adjustable near 9.21 MHz. The output of this second spectrometer is used in a feedback loop to hold the strong magnetic field on the peak of the deuterium resonance of the sample.

6. RESULTS

6.1. Fourier analysis. The top graph of Figure 8 displays the stretch of output corresponding to the first 600 points of the m-sequence input. The bottom graph gives an estimate of $|c^\tau A_1(\lambda)|$ as computed in the manner of expression (4.6). The second-order spectra that appear were estimated by structuring the data into 15 segments of 512 successive points, the segments were overlapped by 256 points. Prior to the Fourier transform the values were tapered. The cross- and ordinary periodograms were computed for each segment, and these then averaged to obtain estimates of f_{YX} and f_{XX}, and thereby an estimate of $|c^\tau A_1|$. This graph displays four substantial peaks, in the manner of Figure 3. From the locations of these peaks the parameters ω_A, ω_B, J may be estimated by inspection.

6.2. Maximum likelihood analysis. In practice the electronics of the measurement process leads to a measurement equation more complicted than (2.8). A further parameter needs to be introduced. It is an unknown, but small, time delay, τ, relative to the input pulse timing at which the sampled values are recorded. The output is still given by (5.1) but with τ built into Y_x and Y_y. For given initial values, $S(0)$, for the state vector and given parameter values, the signal $S(t)$ may be evaluated recursively following expressions (3.2) and (3.3). If the noise in (5.1) is Gaussian white, then one has a nonlinear regression problem and is led to estimate the unknowns by minimizing

$$(6.1) \qquad \sum_{j=1}^{n} |Y(j) - \beta[\cos \phi Y_x(j) + \sin \phi Y_y(j)]|^2$$

with j indexing the times of measurements and with β a further parameter introduced to handle the unknown scaling of the measurement process. Appendix A.2 presents some further computational details.

FIGURE 8

Noise Response

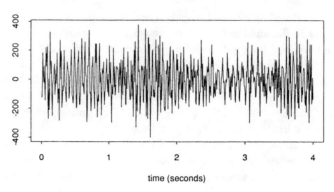

time (seconds)

Modulus Transfer Function

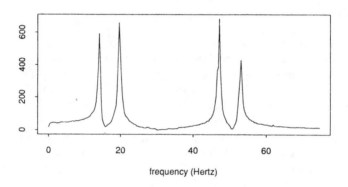

frequency (Hertz)

6.3 Results of the analyses. In the results to be presented, the thermal equilibrium values were taken to be (2.3). Here ε is a small positive quantity, that is effectively a scaling variable. Maximum likelihood fitting was carried out for the full set of 4095 data points and separately for four successive stretches of 1000 points. The unknown parameters estimated were: J, ω_A, ω_B, T_1, T_2, ϕ, τ, β and the unknown $\rho_{jk}(0)$, in total 24 unknowns. Figure 9 and Figures 10, 11, 12, 13 provide the results. The first panel, in each case is a scatter plot of the fitted versus the corresponding observed values. The second and third panels are respectively plots of the logarithms of the absolute values of the Fourier transforms of the residual and observed series. The final panel is a scatter plot of the two log $|FT|$'s versus each other.

The complete data set analysis, displayed in Figure 9, resulted in a correlation of .81 between fitted and observed data values. The second through fourth panels in this Figure suggest the presence of signal-generated noise. Specifically note the parallel shapes in the second and third and the scatter parallel to the diagonal line in the fourth panel. It is interesting to note how the peak just above 60 Hz. in the third panel has emerged from the "noise" level.

FIGURE 9 - COMPLETE DATA SET

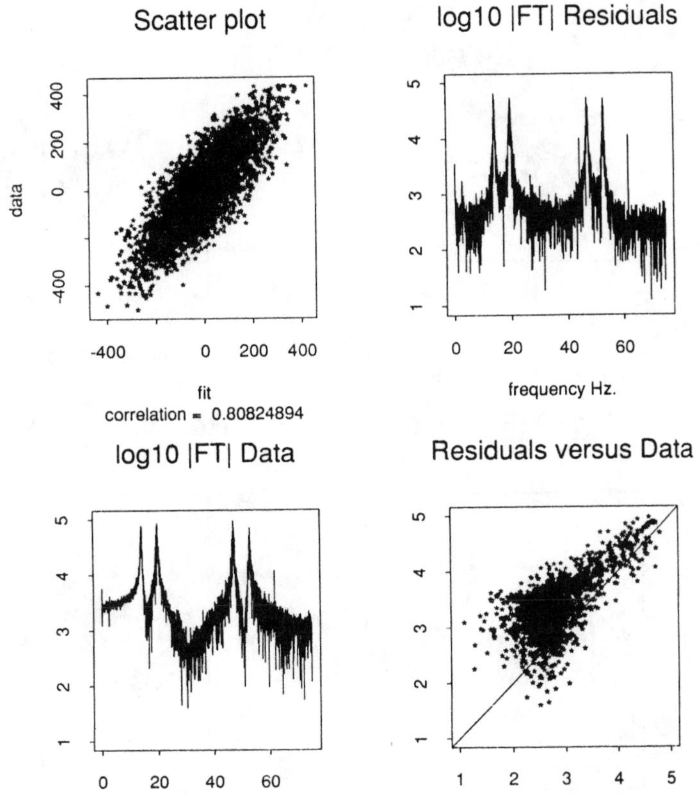

Scatter plot

log10 |FT| Residuals

log10 |FT| Data

Residuals versus Data

fit
correlation = 0.80824894

frequency Hz.

The results are similar, see Figures 10, 11, 12, 13, for the separate stretches of 1000. The correlation coefficients are considerably higher, .93, .90, .96, .96. For the fourth stretch it is notable how the "birdie" near 60 Hz has emerged from the noise. The following table presents the estimates for the principal parameters. The relaxation times were poorly determined.

	J	ω_A	ω_B	$\omega_A - \omega_B$
1–4095	5.728	49.718	17.154	32.564
1–1000	5.714	49.465	16.830	32.635
1001–2000	5.848	49.784	17.323	32.461
2001–3000	5.742	49.795	17.233	32.562
3001–4000	5.733	49.661	17.028	32.633

FIGURE 10 - FIRST 1000 POINTS

Scatter plot

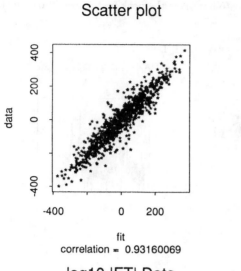

fit
correlation = 0.93160069

log10 |FT| Residuals

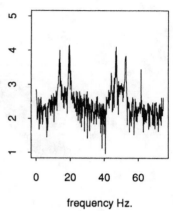

frequency Hz.

log10 |FT| Data

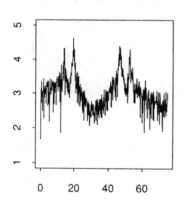

Residuals versus Data

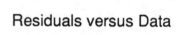

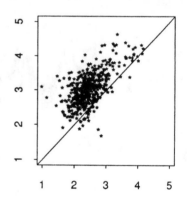

FIGURE 11 - SECOND 1000 POINTS

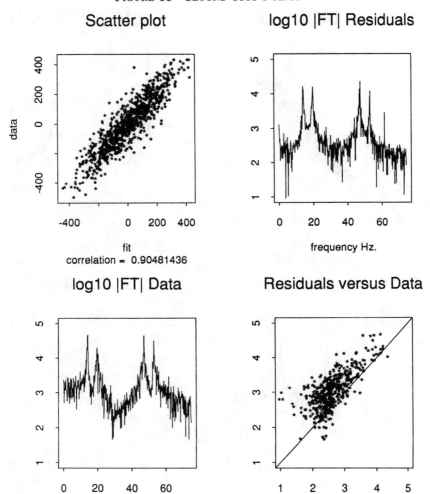

Scatter plot

log10 |FT| Residuals

fit
correlation = 0.90481436

frequency Hz.

log10 |FT| Data

Residuals versus Data

FIGURE 12 - THIRD 1000 POINTS

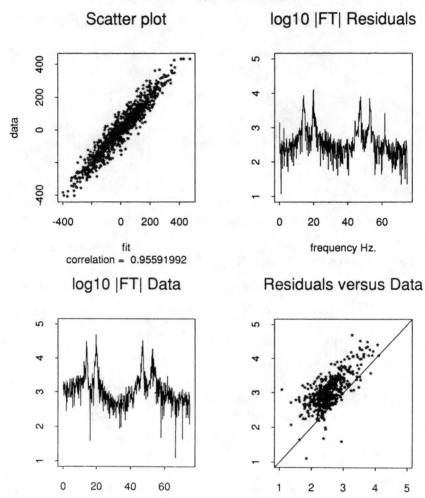

FIGURE 13 - FOURTH 1000 POINTS

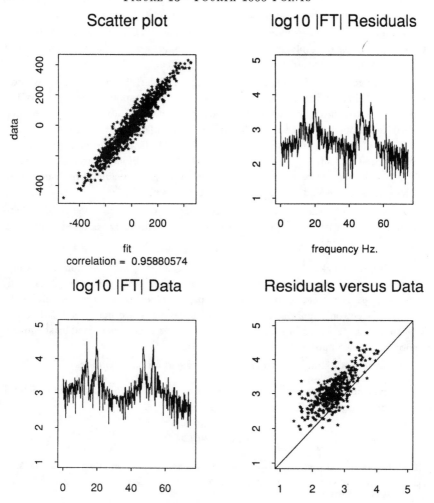

7. **Discussion.** Three techniques for estimating the parameters of nmr spectroscopy will now be discussed. The first technique is to compute spectra of some order following pulse or noise input. The second is to fit sums of exponential cosines following pulse input. The third is to build a conceptual model and employ maximum likelihood. This last was the special concern of this paper. Each technique has advantages and disadvantages.

Advantages of the Fourier-based techniques are that they are direct, robust and do not need full models. Disadvantages are that some parameters cannot be estimated and rephasing may be needed to reduce leakage between frequencies.

An advantage of the exponential-sinusoid model, as implemented in Miller and Greene (1989) for example, is that a full model is not needed. However pulse input is needed and some parameters cannot be estimated.

Advantages of the full modelling approach include: the parameters are interpretable, efficient fitting methods are available with corresponding estimates of uncertainty, there is flexibility in parametrization, the state variables may be estimated and tracked, coupling/phasing/nonlinearities are handled as a matter of course. An advantage of ± 1 input is that only 3 matrix exponentials need be computed. Disadvantages are that: a full model is needed and this will be difficult for large molecules, which parameters can be effectively estimated and when remains to be understood, initial values are needed for the optimization routines.

A hybrid approach in which a simple model is fit and then the residuals are examined by Fourier techniques for peaks would seem likely to be effective in a variety of circumstances.

8. Future work. The source of the signal-generated noise remains a mystery. There are several ways to approach the problem. One is to better model the anti-aliasing filter. A second is to sample the output values more often. A third is to seek other electronic noise sources. If these approaches are unsuccessful then the least squares criterion (6.1) would be replaced by one from generalized least squares incorporating the apparent noise spectrum form.

In future work standard errors will be provided for the estimates.

Acknowledgement. We would like to thank Reuben Hale for a helpful conversation concerning the "beating" in Figure 4.

REFERENCES

[1] S.P. BANKS, *Mathematical Theories of Nonlinear Systems*, Prentice Hall, London, 1988.

[2] E.D. BECKER AND T.C. FARRAR, *Fourier transform spectroscopy*, Science, 178 (1972), pp. 351–368.

[3] B. BLÜMICH, *Nichtlineare Rauschanalyse in der magnetischen Kernresonanzspektroskopie*, Dissertation, TU Berlin (1981).

[4] B. BLÜMICH, *Multidimensional spectroscopy II. Analysis of 3D lineshapes obtained from stochastic N.M.R. of two level systems*, Molecular Physics, 48 (1983), pp. 969–980.

[5] B. BLÜMICH, *Stochastic nmr spectroscopy*, Bull. Magnet. Resonance, 7 (1985), pp. 5–26.

[6] B. BLÜMICH AND D. ZIESSOW, *Nonlinear noise analysis in nuclear magnetic resonance spectroscopy. 1D, 2D, and 3D spectra*, J. Chem. Phys. 78 (1983), pp. 1059–1076.

[7] D.R. BRILLINGER, *The identification of polynomial systems by means of higher order spectra*, J. Sound Vib. 12 (1970), pp. 301–313.

[8] D.R. BRILLINGER, *Some statistical aspects of nmr spectroscopy*, Actas del Segundo Congresso Latinoamericano de Probabilidad y Estatistica Matematica, Caracas (1985), pp. 9–17.

[9] D.R. BRILLINGER, *A study of second- and third-order spectral procedures and maximum likelihood in the identification of a bilinear system*, IEEE Trans ASSP, 38 (1990), pp. 1238–1245.

[10] R.R. ERNST, *Magnetic resonance with stochastic excitation*, J. Magnetic Resonance, 3 (1970), pp. 10–27.

[11] R.R. ERNST, G. BODENHAUSEN AND A. WOKAUN, *Principles of Nuclear Magnetic Resonance in One and Two Dimensions*, Clarendon press, Oxford (1987).

[12] R. KAISER, *Coherent spectrometry with noise signals*, J. Magnetic Resonance, 3 (1970), pp. 28–43.

[13] L.E. KAY et al., *Four dimensional heteronuclear triple-resonance nmr spectroscopy of inter-leukin-1β in solution*, Science, 249 (1990), pp. 411–414.

[14] P. MARMARELIS AND V.Z. MARMARELIS, *Analysis of Physiological Systems: the White Noise Approach*, Plenum Press, New York (1978).

[15] M.I. MILLER AND A.S. GREENE, *Maximum-likelihood estimation for nuclear magnetic resonance spectroscopy*, J. Magnetic Resonance 83 (1989), pp. 525–548.

[16] C. MOLLER AND C.VANLOAN, *Nineteen dubious ways to compute the exponential of a matrix*, SIAM Review, 20 (1978), pp. 801–836.

[17] C.P. SLICHTER, *Principles of Magnetic Resonance*, Springer Verlag, Berlin 3rd edition (1990).

[18] L.J. TICK, *The estimation of transfer functions of quadratic systems*, Technometrics, 3 (1961), pp. 563–567.

[19] N. WIENER, *Nonlinear Problems in Random Theory*, MIT Press, Cambridge, 1958.

APPENDIX

A.1 Simulation Details.   For the simulations described in Sections 3 and 4, the following parameter values were employed: $\omega_A = 49.8 Hz, \omega_B = 17.2 Hz, J = 5.77 Hz, T_1 = 4.0$ seconds and all the other T's 2.0 seconds. The diagonal elements of $\rho(0)$ were $(1+\varepsilon)/4$, $1/4$, $1/4$, $(1-\varepsilon)/4$ respectively with $\varepsilon = 1.0$ and the off diagonals were all 0. , the spacing between samples was $1/150$ seconds, the pulse width was taken to be $30/6667$ of the sampling interval, with the pulses applied $3100/6667$ of the sampling interval after a measurement. These values were meant to mimic the laboratory experiment.

The amplitude of the applied stimulus is typically described in terms of a *flip angle*. In the case of a pulse of width σ time units and amplitude γx the flip angle is given by $\gamma x \sigma$.

A.2 Computational Details.   A few computational aspects of the study will be mentioned. The computations were via FORTRAN complex double precisioon on Sun and Sparc workstations. In order to generate the signal, matrix exponentials are needed. Computing such things is not elementary, see eg. Moler and Van Loan (1978). The procedure adopted in the present work was to first obtain the decomposition, $\mathbf{A} = \mathbf{U}\mathbf{\Lambda}\mathbf{U}^{-1}$, via a NAG routine and from that obtain the matrix exponential, $\exp\{\mathbf{A}\} = \mathbf{U}\exp\{\mathbf{\Lambda}\}\mathbf{U}^{-1}$. The Harwell routine VA09A was employed to minimize the sum of squares in the parameters appearing in nonlinear fashion. The initial state parameter, $\mathbf{S}(0)$, appears in a linear fashion, so in the minimization it was first "determined" for given values of the other parameters, then these parameters were determined in turn. Iteration was continued until apparent convergence. Several positive parameters were expressed as exponentials to stabilize the computations.

B. The Derivation of (4.8) and (4.9). It is convenient to approach the problem via the Cramér representation

$$N(t) = \int e^{it\lambda} dZ_N(\lambda)$$

$$\mathbf{S}(t) = \int e^{it\lambda} d\mathbf{Z_S}(\lambda)$$

where

$$cov\{dZ_N(\lambda), dZ_N(\mu)\} = \delta(\lambda - \mu) f_{NN}(\lambda) d\lambda d\mu$$
$$cov\{d\mathbf{Z_S}(\lambda), dZ_N(\mu)\} = \delta(\lambda - \mu) \mathbf{f_{SN}}(\lambda) d\lambda d\mu$$

Here $cov\{X, Y\} = E\{X\overline{Y}\}$ for zero mean complex variables and $f_{NN} = \sigma^2/2\pi$ as N is white noise.

Next note that from $X(t) = N(t) + N(t - s)$

$$dZ_X(\lambda) = [1 + e^{-i\lambda s}] dZ_N(\lambda)$$

Writing expression (4.5) in the frequency domain

$$d\mathbf{Z_S}(\lambda) = \mathbf{A}_1(\lambda) dZ_X(\lambda) + \int \mathbf{A}_2(\lambda - \beta, \beta) dZ_X(\lambda - \beta) dZ_X(\beta) +$$
$$\iint \mathbf{A}_3(\lambda - \beta - \gamma, \beta, \gamma) dZ_X(\lambda - \beta - \gamma) dZ_X(\beta) dZ_X(\gamma)$$

Supposing the third cumulant κ_3 of N to be zero, the second term here may be ignored in the computations to come. The last expression is then

$$\mathbf{A}_1(\lambda)[1 + e^{-\lambda s}] dZ_N(\lambda) + \iint \mathbf{A}_3(\lambda - \beta - \gamma, \beta, \gamma)[1 + e^{-is(\lambda - \beta - \gamma)}][1 +$$
$$e^{-is\beta}][1 + e^{-is\gamma}] dZ_N(\lambda - \beta - \gamma) dZ_N(\beta) dZ_N(\gamma)$$

And so

$$\mathbf{f_{SN}}(\lambda)\delta(0) = E\{d\mathbf{Z_S}(\lambda) dZ_N(-\lambda)\} = \mathbf{A}_1(\lambda)[1 + e^{-i\lambda s}]\delta(0)\frac{\sigma^2}{2\pi} +$$

$$\iint \mathbf{A}_3(\lambda - \beta - \gamma, \beta, \gamma)[1 + e^{-is(\lambda - \beta - \gamma)}][1 + e^{-is\beta}][1 + e^{-is\gamma}]\left[\{\delta(\beta + \gamma)^2 + \right.$$
$$\left. \delta(\lambda - \beta)^2 + \delta(\lambda - \gamma)^2\}\frac{\sigma^4}{(2\pi)^2} + \delta(0)\frac{\kappa_4}{(2\pi)^3}\right] d\beta d\gamma$$

From the assumed symmetry of $\mathbf{A}_3$, this gives (4.8).

Finally, to get expression (4.9) expand $|1 + e^{-i\nu s}|^2$, multiply by $e^{-i\mu s}$, and integrate using $\int e^{-i\mu s} ds = 2\pi \delta(\mu)$.

TRANSFER-FUNCTION MODELS WITH
NON-STATIONARY INPUT

P.J. BROCKWELL†, R.A. DAVIS† AND H. SALEHI‡

Abstract. We consider the problem of parameter estimation for the transfer function model of Box and Jenkins (1976), relating the zero-mean stationary input series $\{X_t, t = 1, \ldots, n\}$ to the zero-mean stationary output series $\{Y_t, t = 1, \ldots, n\}$. We suppose that the series $\{\boldsymbol{\eta}_t = (X_t, Y_t)'\}$ is obtained from the *observed* input-output series $\{\boldsymbol{\rho}_t = (R_t, S_t)'\}$ by differencing and mean correction, or more specifically that $\boldsymbol{\eta}_t = (1 - B)^d(1 - B^s)^\delta \boldsymbol{\rho}_t - \boldsymbol{\mu}$, where B denotes the backward shift operator. We use a state-space representation of $\boldsymbol{\eta}_t$ to generate a corresponding representation for $\boldsymbol{\rho}_t$ and use the latter to compute the Gaussian likelihood of the observations (conditional on the first $d + s\delta$ values of $\boldsymbol{\rho}_t$). In addition to facilitating the calculation of the exact Gaussian likelihood and best linear predictors based on finitely many observations, the state-space representation is particularly valuable in allowing us to deal exactly with missing values of either $\{R_t\}$ or $\{S_t\}$. The results are illustrated with reference to the Leading Indicator – Sales data of Box and Jenkins.

Key words. Transfer function model, State-Space Representation, Gaussian Likelihood.

1. INTRODUCTION

The term transfer-function modelling, as employed by Box and Jenkins (1976) (see also Priestley (1981) and Brockwell and Davis(1987)), refers to the selection of a model of the form,

$$(1.1) \qquad X_t = \phi_1^{-1}(B)\theta_1(B)Z_t,$$

$$(1.2) \qquad Y_t = \phi^{-1}(B)\theta(B)X_t + N_t,$$

$$(1.3) \qquad N_t = \phi_2^{-1}(B)\theta_2(B)W_t,$$

to represent a bivariate stationary series $\{(X_t, Y_t)', t = 0, \pm 1, \pm 2, \cdots\}$ of *inputs* X_t and *outputs* Y_t. The inference is based on observations of the two series at times $t = 1, \cdots, n$. In the model defined by (1.1) - (1.3), B denotes the backward shift operator, $\{Z_t\}$ and $\{W_t\}$ are uncorrelated zero-mean white-noise sequences, $E(Z_t^2) = \sigma_1^2$, $E(W_t^2) = \sigma_2^2$ and ϕ, ϕ_1, ϕ_2, θ, θ_1, θ_2 are polynomials of degrees p, p_1, p_2, q, q_1 and q_2 respectively. It is assumed that $\phi(z)$, $\phi_1(z)$ and $\phi_2(z)$ are non-zero for all $z \in \mathbb{C}$ such that $|z| \leq 1$ and (without loss of generality) that $\phi(0) = \phi_1(0) = \phi_2(0) = \theta_1(0) = \theta_2(0) = 1$. Notice however that no assumption is made about $\theta(0)$ and that if $\{Y_t\}$ lags behind $\{X_t\}$ by at least one time unit then $\theta(0)$ will be zero.

The usual approach to transfer-function modelling first fits an ARMA model of the form (1.1) to the input series $\{X_t\}$ using a criterion such as the AIC statistic for order selection. This gives estimates $\hat{\phi}_1$ and $\hat{\theta}_1$ for the polynomials in equation (1.1).

†Department of Statistics, Colorado State University, Fort Collins, CO, 80523 USA. Research supported by NSF grants DMS 8820559 and DMS 9006422.

‡Department of Statistics and Probability, Michigan State University, East Lansing, MI, 48824 USA.

Application of the operator $\theta_1^{-1}(B)\phi_1(B)$ to each side of (1.2) gives

$$(1.4) \qquad\qquad Y_t^* = \phi^{-1}(B)\theta(B)Z_t + N_t^*$$

where $\{Y_t^*\}$, $\{N_t^*\}$ are the transformed output and noise sequences respectively and $\{Z_t\}$ is uncorrelated with $\{N_t^*\}$.

It is clear from (1.4) that the coefficient t_j of B^j in the power series expansion of $\phi^{-1}(B)\theta(B)$ is

$$(1.5) \qquad\qquad t_j = \mathrm{cov}(Y_t^*, Z_{t-j})/\mathrm{var}(Z_t).$$

These coefficients can therefore be estimated by applying the *estimated* operator, $\hat{\theta}_1^{-1}(B)\hat{\phi}_1(B)$ to the input and output sequences to get estimates of $\{Z_t\}$ and $\{Y_t^*\}$ respectively and then to use these estimated sequences to compute estimates $\hat{t}_j$ from (1.5).

The dependence of the estimated coefficients $\hat{t}_j$ on j is then used to suggest possible polynomials ϕ and θ such that the coefficients of z^j in the power series expansion of $\phi^{-1}(z)\theta(z)$ are approximately $\hat{t}_j$, $j = 0, 1, \cdots$. This provides us with preliminary estimates of the polynomials appearing in (1.2).

More efficient estimation of the parameters appearing in (1.2) and (1.3) is then carried out using least squares as described for example in Brockwell and Davis (1987), Section 12.2. Prediction with the fitted model can be done very simply using large-sample approximations.

The procedure described above is satisfactory for long series with no missing values, however it makes use of a number of approximations which for short series can lead to substantial errors in both model-fitting and prediction. The state-space formulation of the model (1.1) - (1.3) developed in Brockwell, Davis and Salehi (1990) permits an exact treatment of the estimation problem using the Gaussian likelihood of $\{(X_t, Y_t), t = 1, \cdots, n\}$. This representation (one of many possible state-space representations) is given in Section 2, together with the associated Kalman prediction recursions and the innovation form of the exact Gaussian likelihood. In Section 3 we show how to derive a corresponding state-space representation in the case when the observations must be differenced and mean-corrected in order to generate the series $\{\eta_t = (X_t, Y_t)'\}$. This representation is of critical importance when the observed non-stationary series $\{\rho_t = (R_t, S_t)'\}$ has missing values, since the Kalman recursions can easily be adapted to generate exact linear one-step predictors (and hence the Gaussian likelihood) in this case.

2. THE STATE SPACE MODEL FOR THE DIFFERENCED SERIES η_t

Following Brockwell, Davis and Salehi (1990), we can rewrite the transfer function model (1.1) for $\eta_t = (X_t, Y_t)'$ in the state-space form,

$$(2.1) \qquad \begin{bmatrix} X_t \\ Y_t \end{bmatrix} = \begin{bmatrix} C_1 & 0 & 0 \\ D_0 C_1 & C_0 & C_2 \end{bmatrix} \begin{bmatrix} x_t \\ y_t \\ n_t \end{bmatrix} + \begin{bmatrix} D_1 & 0 \\ D_0 D_1 & D_2 \end{bmatrix} \begin{bmatrix} Z_t \\ W_t \end{bmatrix},$$

where $\{\boldsymbol{\xi}_t := (\boldsymbol{x}_t', \boldsymbol{y}_t', \boldsymbol{n}_t')'\}$ is the unique stationary solution of

$$(2.2) \qquad \begin{bmatrix} \boldsymbol{x}_{t+1} \\ \boldsymbol{y}_{t+1} \\ \boldsymbol{n}_{t+1} \end{bmatrix} = \begin{bmatrix} A_1 & 0 & 0 \\ B_0 C_1 & A_0 & 0 \\ 0 & 0 & A_2 \end{bmatrix} \begin{bmatrix} \boldsymbol{x}_t \\ \boldsymbol{y}_t \\ \boldsymbol{n}_t \end{bmatrix} + \begin{bmatrix} B_1 & 0 \\ B_0 D_1 & 0 \\ 0 & B_2 \end{bmatrix} \begin{bmatrix} Z_t \\ W_t \end{bmatrix}.$$

The matrices A_0, B_0, C_0 and D_0 in the representation (2.1) and (2.2) are defined as follows:

$$A_0 = \begin{bmatrix} 0 & 1 & 0 & \cdots & 0 \\ 0 & 0 & 1 & \cdots & 0 \\ \vdots & \vdots & \vdots & \ddots & \vdots \\ 0 & 0 & 0 & \cdots & 1 \\ -\phi_r & -\phi_{r-1} & -\phi_{r-2} & \cdots & -\phi_1 \end{bmatrix}, \quad B_0 = \begin{bmatrix} b_1 \\ b_2 \\ \vdots \\ b_{r-1} \\ b_r \end{bmatrix},$$

$$C_0 = [1 \quad 0 \quad \cdots \quad 0] \qquad \text{and} \qquad D_0 = b_0.$$

Here $r = max(p, q)$ and $\phi_j, b_j, j = 0, \cdots, r$, are the coefficients of $z^0, \cdots, z^r$ in the power series expansions of $\phi(z)$ and $\theta(z)/\phi(z)$ respectively.

The matrices $A_i, B_i, C_i, D_i, i = 1, 2$, are defined analogously in terms of the polynomials ϕ_i and θ_i appearing in (1.1) and (1.3). We shall write (2.1) and (2.2) in the more compact form,

$$(2.3) \qquad \boldsymbol{\eta}_t = C\boldsymbol{\xi}_t + D\boldsymbol{\epsilon}_t,$$

where $\{\boldsymbol{\xi}_t\}$ is the unique stationary solution of the "state equation"

$$(2.4) \qquad \boldsymbol{\xi}_{t+1} = A\boldsymbol{\xi}_t + B\boldsymbol{\epsilon}_t.$$

The covariance matrix Σ of $\boldsymbol{\epsilon}_t := (Z_t, W_t)'$ is given by

$$(2.5) \qquad \Sigma = \begin{bmatrix} \sigma_1^2 & 0 \\ 0 & \sigma_2^2 \end{bmatrix},$$

and from (2.4) it is easy to compute the covariance matrix Π of the state-vector $\boldsymbol{\xi}_t$ as

$$(2.6) \qquad \Pi = \sum_{j=0}^{\infty} A^j B \Sigma B' (A')^j.$$

We also find from (2.3) that the covariance matrix Γ of $\boldsymbol{\eta}_t := (X_t, Y_t)'$ is

$$(2.7) \qquad \Gamma = C\Pi C' + D\Sigma D'.$$

Example 1 In Example 12.2.1 of Brockwell and Davis (1987), the Leading Indicator - Sales Data of Box and Jenkins (1976) was differenced at lag 1 and mean-corrected. The resulting series $\{(X_t, Y_t)', t = 1, \cdots, 149\}$ was then fitted by a model of the form (1.1) - (1.3), namely,

$$(2.8) \qquad X_t = (1 - .474B)Z_t, \quad \{Z_t\} \sim WN(0, .0779),$$

$$(2.9) \qquad Y_t = 4.717B^3(1 - .724B)^{-1}X_t + N_t,$$

$$(2.10) \qquad N_t = (1 - .582B)W_t, \quad \{W_t\} \sim WN(0, .0486).$$

Constructing a state-space model as described above, we obtain a five-dimensional state vector $\boldsymbol{\xi}_t = [\boldsymbol{x}_t', \boldsymbol{y}_t', \boldsymbol{n}_t']'$ in which $\boldsymbol{x}_t$ and $\boldsymbol{n}_t$ are each one-dimensional. Specifically we find that

$$(2.11) \qquad \begin{bmatrix} X_t \\ Y_t \end{bmatrix} = \begin{bmatrix} 1 & 0 & 0 & 0 & 0 \\ 0 & 1 & 0 & 0 & 1 \end{bmatrix} \begin{bmatrix} \boldsymbol{x}_t \\ \boldsymbol{y}_t \\ \boldsymbol{n}_t \end{bmatrix} + \begin{bmatrix} 1 & 0 \\ 0 & 1 \end{bmatrix} \begin{bmatrix} Z_t \\ W_t \end{bmatrix}$$

where $\{\boldsymbol{\xi}_t = (\boldsymbol{x}_t', \boldsymbol{y}_t', \boldsymbol{n}_t')'\}$ is the unique stationary solution of

$$(2.12) \qquad \begin{bmatrix} \boldsymbol{x}_{t+1} \\ \boldsymbol{y}_{t+1} \\ \boldsymbol{n}_{t+1} \end{bmatrix} = \begin{bmatrix} 0 & 0 & 0 & 0 & 0 \\ 0 & 0 & 1 & 0 & 0 \\ 0 & 0 & 0 & 1 & 0 \\ 4.717 & 0 & 0 & .724 & 0 \\ 0 & 0 & 0 & 0 & 0 \end{bmatrix} \begin{bmatrix} \boldsymbol{x}_t \\ \boldsymbol{y}_t \\ \boldsymbol{n}_t \end{bmatrix} + \begin{bmatrix} -.474 & 0 \\ 0 & 0 \\ 0 & 0 \\ 4.717 & 0 \\ 0 & -.582 \end{bmatrix} \begin{bmatrix} Z_t \\ W_t \end{bmatrix},$$

and $\{Z_t\}$, $\{W_t\}$ are uncorrelated white-noise sequences as in (2.8) and (2.10) with variances $\sigma_1^2 = .0779$ and $\sigma_2^2 = .0486$ respectively.

Prediction using the state-space model. A primary purpose of transfer-function modelling is to compute the best linear mean-square predictors $P_t Y_{t+h}$, $h = 1, 2, \cdots$, where P_t denotes projection on span$\{X_s, Y_s, 1 \le s \le t\}$. In the notation of (2.3) and (2.4), it suffices to compute $P_t \boldsymbol{\eta}_{t+h}$, $h = 1, 2, \cdots$, where

$$P_t \boldsymbol{\eta}_{t+h} := \begin{bmatrix} P_t X_{t+h} \\ P_t Y_{t+h} \end{bmatrix}, \quad h = 1, 2, \cdots.$$

The Kalman recursions for the state-space model (2.3), (2.4) (see e.g. Aoki (1987)) enable us to write

$$(2.13) \qquad P_t \boldsymbol{\xi}_{t+1} = A P_{t-1} \boldsymbol{\xi}_t + K_t(\boldsymbol{\eta}_t - P_{t-1}\boldsymbol{\eta}_t),$$

$$(2.14) \qquad P_t \boldsymbol{\xi}_{t+h} = A^{h-1} P_t \boldsymbol{\xi}_{t+1}, \quad h = 2, 3, \cdots,$$

$$(2.15) \qquad P_t \boldsymbol{\eta}_{t+h} = C P_t \boldsymbol{\xi}_{t+h}, \quad h = 1, 2, \cdots,$$

where the Kalman gain, K_t, and the error covariance matrices, $\Delta_t := E[(\boldsymbol{\eta}_t - P_{t-1}\boldsymbol{\eta}_t)(\boldsymbol{\eta}_t - P_{t-1}\boldsymbol{\eta}_t)']$ and $\Omega_t := E[(\boldsymbol{\xi}_t - P_{t-1}\boldsymbol{\xi}_t)(\boldsymbol{\xi}_t - P_{t-1}\boldsymbol{\xi}_t)']$ are found recursively from the equations,

$$(2.16) \qquad \Omega_1 = \Pi,$$

$$(2.17) \qquad \Delta_t = C\Omega_t C' + D\Sigma D',$$

$$(2.18) \qquad K_t = (A\Omega_t C' + B\Sigma D')\Delta_t^{-1},$$

$$(2.19) \qquad \Omega_{t+1} = \Pi - A(\Pi - \Omega_t)A' - K_t \Delta_t K_t',$$

and the covariance matrices Σ and Π were defined in (2.5) and (2.6). The one-step predictors $\hat{\boldsymbol{\xi}}_{t+1} := P_t \boldsymbol{\xi}_{t+1}$ and $\hat{\boldsymbol{\eta}}_{t+1} := P_t \boldsymbol{\eta}_{t+1}$, $t = 0, 1, \cdots$, are found from the recursions,

$$(2.20) \qquad \hat{\boldsymbol{\xi}}_{t+1} = A\hat{\boldsymbol{\xi}}_t + K_t(\boldsymbol{\eta}_t - \hat{\boldsymbol{\eta}}_t),$$

$$(2.21) \qquad \hat{\boldsymbol{\eta}}_{t+1} = C\hat{\boldsymbol{\xi}}_{t+1},$$

with initial conditions $\hat{\boldsymbol{\xi}}_1 = \hat{\boldsymbol{\eta}}_1 = \mathbf{0}$. The h-step predictors, $P_t X_{t+h}$ and $P_t Y_{t+h}$, $h = 1, 2, \cdots$, are easily found from the one-step predictors using (2.14) and (2.15).

Maximum likelihood estimation. Consider the model (1.1)–(1.3) with parameters $\sigma, \nu, \theta_{11}, \cdots, \theta_{1q_1}, \phi_{11}, \cdots, \phi_{1p_1}, \xi_1, \cdots, \xi_r, \phi_1, \cdots, \phi_p, \theta_{21}, \cdots, \theta_{2q_2}, \phi_{21},$ $\cdots, \phi_{2p_2}$, where

$$\theta_i(z) = 1 + \theta_{i1}z + \cdots + \theta_{iq_i}z^{q_i}, \ i = 1, 2,$$
$$\phi_i(z) = 1 + \phi_{i1}z + \cdots + \phi_{ip_i}z^{p_i}, \ i = 1, 2,$$
$$\theta(z) = z^{b-1}(\xi_1 z + \cdots + \xi_r z^r) = z^{b-1}\xi(z), \ b \geq 0, \ r \geq 1,$$

and

$$\phi(z) = 1 + \phi_1 z + \cdots + \phi_p z^p.$$

To estimate the parameters of the model with specified orders r, p_1, p_2, q, q_1, q_2, and b (the *delay* parameter), we maximize the Gaussian likelihood of $\{\boldsymbol{\eta}_t, t = 1, \cdots, n\}$. The orders are be chosen so as to minimize the AIC statistic.

In order to calculate the exact Gaussian likelihood function based on the observations $\{\boldsymbol{\eta}_1, \cdots, \boldsymbol{\eta}_n\}$, we express it in terms of the innovations as in Schweppe (1965) to obtain

$$(2.22) \qquad L(\boldsymbol{\beta}) = (2\pi)^{-n} \Big(\prod_{j=1}^{n} \det \Delta_j \Big)^{-1/2} \exp\Big[-\frac{1}{2} \sum_{j=1}^{n} (\boldsymbol{\eta}_j - \hat{\boldsymbol{\eta}}_j)' \Delta_j^{-1} (\boldsymbol{\eta}_j - \hat{\boldsymbol{\eta}}_j) \Big],$$

where $\boldsymbol{\beta}$ denotes the vector of parameters consisting of all the coefficients and white-noise variances appearing in the model (1.1) - (1.3).

The Gaussian likelihood $L(\boldsymbol{\beta})$ is easily computed for any prescribed set of parameter values from (2.22) and the recursions (2.16)-(2.21). Maximization with respect to the parameters is carried out numerically using a non-linear optimization algorithm to give the maximum likelihood estimator $\hat{\boldsymbol{\beta}}$. (For large n a very close approximation to the maximum likelihood estimator $\hat{\boldsymbol{\beta}}$ can be found by maximizing the marginal likelihood of $X_1, \cdots, X_n$ with respect to $\boldsymbol{\beta}_1$, where $\boldsymbol{\beta}_1$ denotes the parameters in (1.1), and then, with this fixed value for $\boldsymbol{\beta}_1$, maximizing (3.1) with respect to the remaining parameters $\boldsymbol{\beta}_0$ and $\boldsymbol{\beta}_2$ appearing in (1.2) and (1.3) respectively. This procedure amounts to fitting a model to the input series and then maximizing the conditional likelihood of the output series given the input series. It is not completely equivalent to maximizing (2.22) since there is some dependence on $\boldsymbol{\beta}_1$ of the conditional likelihood of $Y_1, \cdots, Y_n$ given $X_1, \cdots, X_n$.)

The likelihood is maximized for a variety of values of $p, p_1, p_2, r, q_1, q_2, b$, and the model selected is the one which minimizes the AIC statistic,

$$(2.23) \qquad AIC = -2\ln L(\hat{\boldsymbol{\beta}}) + 2(1 + p + p_1 + p_2 + r + q_1 + q_2),$$

where $\hat{\boldsymbol{\beta}}$ is the maximum likelihood estimator of $\boldsymbol{\beta}$.

3. THE STATE–SPACE REPRESENTATION FOR THE NON-STATIONARY SERIES $\{\boldsymbol{\rho}_t\}$

We consider now the construction of a state-space model for the *original* observations $\{\boldsymbol{\rho}_t = (R_t, S_t)', \ t = -d - \delta s + 1, \ldots, n\}$, where

$$\boldsymbol{\eta}_t = (1 - B)^d (1 - B^s)^\delta \boldsymbol{\rho}_t - \boldsymbol{\mu}, \ t = 1, \ldots, n,$$

and $\boldsymbol{\eta}_t$ has the state-space representation defined by (2.3) and (2.4). Example 1 is of this type with $d = 1$, $\delta = 0$, $\boldsymbol{\mu} = (.0228, .420)'$ and $n = 149$.

The first step is to define a new set of observation vectors,

$$(3.1) \qquad \boldsymbol{\eta}_t^* = \boldsymbol{\rho}_t - \boldsymbol{\mu} t^{d+\delta} s^{-\delta}/(\delta + d)! \ .$$

Then

$$(3.2) \qquad (1 - B)^d (1 - B^s)^\delta \boldsymbol{\eta}_t^* = \boldsymbol{\eta}_t.$$

Observe next from (2.3) that we can write a new observation equation,

$$(3.3) \qquad \boldsymbol{\eta}_t^* = C\boldsymbol{\xi}_t + \sum_{j=1}^{d+s\delta} h_j \boldsymbol{\eta}_{t-j}^* + D\boldsymbol{\epsilon}_t = C^* \boldsymbol{\xi}^* + D^* \boldsymbol{\epsilon}_t,$$

where $D^* = D$, $C^* = [C \quad h_1 I \quad \cdots \quad h_{d+s\delta} I]$, I is the 2×2 identity matrix, $(1 - B)^d (1 - B^s)^\delta = 1 - \sum_{j=1}^{d+s\delta} h_j B^j$, and the new state vectors, $\boldsymbol{\xi}_t^* :=$ $(\boldsymbol{\xi}_t', \boldsymbol{\eta}_{t-1}^{*'}, \ldots, \boldsymbol{\eta}_{t-d-s\delta}^{*'})'$ satisfy

$$(3.4) \qquad \boldsymbol{\xi}_{t+1}^* = A^* \boldsymbol{\xi}_t^* + B^* \boldsymbol{\epsilon}_t, \ t = 1, 2, \ldots,$$

where

$$A^* = \begin{bmatrix} A & 0 & 0 & \cdots & 0 \\ C & h_1 I & h_2 I & \cdots & h_{d+s\delta} I \\ 0 & I & 0 & \cdots & 0 \\ \vdots & \vdots & \ddots & \ddots & \vdots \\ 0 & 0 & \cdots & I & 0 \end{bmatrix} \text{ and } B^* = \begin{bmatrix} B \\ D \\ 0 \\ \vdots \\ 0 \end{bmatrix}.$$

The initial condition for the state equation is

$$(3.5) \qquad \boldsymbol{\xi}_1^* = \begin{bmatrix} \boldsymbol{\xi}_1 \\ \boldsymbol{\eta}_0^* \\ \vdots \\ \boldsymbol{\eta}_{1-d-s\delta}^* \end{bmatrix} = \begin{bmatrix} \sum_{j=0}^{\infty} A^j B\boldsymbol{\epsilon}_{-j} \\ \boldsymbol{\eta}_0^* \\ \vdots \\ \boldsymbol{\eta}_{1-d-s\delta}^* \end{bmatrix},$$

and we assume that the orthogonality conditions,

$$\begin{bmatrix} \boldsymbol{\eta}_0^* \\ \vdots \\ \boldsymbol{\eta}_{1-d-s\delta}^* \end{bmatrix} \perp \begin{bmatrix} Z_t \\ W_t \end{bmatrix}, \ t = 0, \pm 1, \pm 2, \ldots,$$

are satisfied. (Orthogonality of the random column vectors $\boldsymbol{X}$ and $\boldsymbol{Y}$ means that $E(\boldsymbol{X}\boldsymbol{Y}') = 0$.)

To determine the best predictors $P_n^*(\boldsymbol{\eta}_{n+h}^*)$ of $\boldsymbol{\eta}_{n+h}^*$, and in particular the one-step predictors, $\hat{\boldsymbol{\xi}}_{n+1}^* = P_n^* \boldsymbol{\xi}_{n+1}^*$, based on linear combinations of the components of $\boldsymbol{\eta}_t^*$, $t = 1 - d - s\delta, \ldots, n$, we simply apply the Kalman recursions (2.13)–(2.21) with

$A, B, C, D, \boldsymbol{\xi}$ and $\boldsymbol{\eta}$ replaced by $A^*, B^*, C^*, D^*, \boldsymbol{\xi}^*$ and $\boldsymbol{\eta}^*$ respectively. The initial conditions for the recursions are found from (3.4). Thus

$$(3.6) \qquad \hat{\boldsymbol{\xi}}_1^* = \begin{bmatrix} \mathbf{0} \\ \boldsymbol{\eta}_0^* \\ \vdots \\ \boldsymbol{\eta}_{1-d-s\delta}^* \end{bmatrix}$$

and

$$(3.7) \qquad \Omega_1^* := E[(\boldsymbol{\xi}_1^* - \hat{\boldsymbol{\xi}}_1^*)(\boldsymbol{\xi}_1^* - P_0^* \boldsymbol{\xi}_1^*)'] = \begin{bmatrix} \Pi & 0 \\ 0 & 0 \end{bmatrix},$$

where Π is the covariance matrix (see (2.6)) of $\boldsymbol{\xi}_t$. Finally, from (3.1), the best predictor of $\boldsymbol{\rho}_{n+h}$ using linear combinations of the constant 1 and the components of $\boldsymbol{\rho}_t$, $t = 1 - d - s\delta, \ldots, n$ is $P_n^*(\boldsymbol{\eta}_{n+h}^*) + \boldsymbol{\mu}(n+h)^{d+\delta} s^{-\delta}/(\delta + d)!$.

Conditional likelihood calculation. The conditional Gaussian likelihood of $\{\boldsymbol{\eta}_t^*, \ t = 1, \ldots, n\}$ given $\{\boldsymbol{\eta}_t^*, \ t = 1-d-s\delta, \ldots, 0\}$ is found from (2.22) on replacing $\Delta_j, \boldsymbol{\eta}_j$ and $\hat{\boldsymbol{\eta}}_j$ by $\Delta_j^*, \boldsymbol{\eta}_j^*$ and $\hat{\boldsymbol{\eta}}_j^*$ respectively. In the case when there are no missing values, maximization of this conditional likelihood with respect to $\boldsymbol{\beta}$ is precisely equivalent to maximization of (2.22). When there are missing values in either or both of the series $\{R_t\}$ and $\{S_t\}$ however, it will not be possible to evaluate the right-hand side of (2.22). We consider next the problem of maximum likelihood estimation in this case. (The same problem arises in general when fitting ARIMA processes to data with missing values. See Ansley and Kohn (1985) and Bell and Hillmer (1990) for a different state-space approach to this problem.)

Missing values. Missing values in one or both of the series $\{R_t, \ t = 1, \ldots, n\}$, $\{S_t, \ t = 1, \ldots, n\}$ can be handled using the following device. Define the 2×2 matrices,

$$I = \begin{bmatrix} 1 & 0 \\ 0 & 1 \end{bmatrix}, \quad E_1 = \begin{bmatrix} 1 & 0 \\ 0 & 0 \end{bmatrix}, \quad E_2 = \begin{bmatrix} 0 & 0 \\ 0 & 1 \end{bmatrix}, \quad O = \begin{bmatrix} 0 & 0 \\ 0 & 0 \end{bmatrix},$$

and let

$$M_t := \begin{cases} I & \text{if } \boldsymbol{\rho}_t = (R_t, S_t)' \text{ is missing,} \\ E_1 & \text{if } R_t \text{ is missing and } S_t \text{ is not,} \\ E_2 & \text{if } S_t \text{ is missing and } R_t \text{ is not,} \\ O & \text{otherwise.} \end{cases}$$

With $\{\boldsymbol{\xi}_t^*, \ t = 1, 2, \ldots\}$ determined as before by (3.4) and (3.6), we now define a new sequence of observation vectors,

$$(3.8) \qquad \boldsymbol{\eta}_t^{(m)} = C_t \boldsymbol{\xi}_t^* + D_t \boldsymbol{\epsilon}_t + M_t \boldsymbol{\delta}_t,$$

where $C_t := (I - M_t)C^*$, $D_t := (I - M_t)D^*$ and $\{\boldsymbol{\delta}_t\}$ is an iid sequence of bivariate $N(\mathbf{0}, I)$ random vectors independent of $\{\boldsymbol{\epsilon}_t\}$. The sequence $\{\boldsymbol{\eta}_t^{(m)}\}$ is then precisely the sequence defined by the observation equation (3.3), except that unobserved values are replaced by independent standard normal random variables. To within multiplication by a parameter-independent factor, the likelihood of the sequence defined by (3.8) is therefore the same as the likelihood of the *observed* values of R_t and S_t. Since the realized numerical values of the vectors $\boldsymbol{\delta}_t$ enter the likelihood as multiplicative parameter-independent constants only, we can assume (for likelihood maximization) that they are all zero. Thus we replace all missing values by zeroes in order to generate a complete set of "observations".

The conditional Gaussian likelihood of the original data, given $\{\boldsymbol{\rho}_t, \ t = 1 - d - s\delta, \dots, 0\}$, can therefore be calculated (to within multiplication by a parameter-independent factor) by replacing all missing values with zeroes and using the observation equation (3.8) and state equation (3.4) to generate the one step predictors $\hat{\boldsymbol{\eta}}_t^{(m)}$, $t = 1, \dots, n$, and corresponding error covariance matrices $\Delta_t^{(m)}$, which are then substituted into the right-hand side of (2.22). The required Kalman recursions for computing $\hat{\boldsymbol{\eta}}_t^{(m)}$ and $\Delta_t^{(m)}$ from (3.4) and (3.8) are as follows.

(3.9) $$\Delta_t^{(m)} = C_t \Omega_t^* C_t' + D_t \Sigma D_t' + M_t M_t',$$

(3.10) $$K_t = (A^* \Omega_t^* C_t' + B^* \Sigma D_t') \Delta_t^{(m)-1},$$

(3.11) $$\Omega_{t+1}^* = \Omega_1^* - A^*(\Omega_1^* - \Omega_t^*)A^{*'} - K_t \Delta_t^{(m)} K_t'$$

where the covariance matrices Σ and Ω_1^* were defined in (2.5)and (3.7) respectively. The one-step predictors $\hat{\boldsymbol{\xi}}_{t+1}^*$ and $\hat{\boldsymbol{\eta}}_{t+1}^{(m)}$, $t = 0, 1, \cdots$, are found from the recursions,

(3.12) $$\hat{\boldsymbol{\xi}}_{t+1}^* = A^* \hat{\boldsymbol{\xi}}_t^* + K_t(\boldsymbol{\eta}_t^{(m)} - \hat{\boldsymbol{\eta}}_t^{(m)})$$

and

(3.13) $$\hat{\boldsymbol{\eta}}_{t+1}^{(m)} = C_{t+1} \hat{\boldsymbol{\xi}}_{t+1}^*$$

with initial conditions $\hat{\boldsymbol{\xi}}_1^* = \hat{\boldsymbol{\eta}}_1^{(m)} = 0$.

(As in the case when there are no missing values, a very good approximation to the maximum likelihood estimators can be obtained by first maximizing the likelihood of the input series with respect to $\boldsymbol{\beta}_1$ (dealing with missing values as described in Brockwell and Davis (1987), Section 12.3) and then maximizing the Gaussian likelihood of the bivariate series with respect to $\boldsymbol{\beta}_0$ and $\boldsymbol{\beta}_2$, keeping $\boldsymbol{\beta}_1$ fixed.)

In the case when there are missing values among the first $d + s\delta$ observations $\{\boldsymbol{\rho}_t = (R_t, S_t)', \ t = -d - \delta s + 1, \dots, 0\}$, the missing values can be treated as unknown parameters and included among the variables with respect to which the likelihood is maximized.

Example 1 (ctd)   Deleting the last three sales figures from the Sales - Leading Indicator Data of Box and Jenkins (1976), we can use the technique described above to maximize the Gaussian likelihood of the remaining observations, treating the last three sales figures as missing observations. Using the same values for

r, q, p_1, q_1, p_2, q_2 and b as before, we obtain the maximum likelihood model for the mean-corrected differenced observations,

$$X_t = (1 - .476B)Z_t, \quad \{Z_t\} \sim \text{WN}(0, .0768),$$
$$Y_t = 4.707B^3(1 - .725B)^{-1}X_t + N_t,$$
$$N_t = (1 - .624B)W_t, \quad \{W_t\} \sim \text{WN}(0, .0462).$$

This model can now be used to find the best linear least squares estimates (based on all of the available observations) of both the missing three sales observations and of future values. The following table shows these estimates as computed from the missing-value model. The last column of the table shows the *actual* values of the sales figures at times $t = 148, 149$ and 150, as well as the predicted sales values for a further three time units, obtained from the maximum likelihood model for the original data with no missing observations. The consistency between the two sets of predicted values demonstrates that little would be lost (as far as sales prediction is concerned) if the last three sales figures in this example were missing.

Table 1. Best linear estimates of sales

t	Missing values at $t = 148, 149, 150$	No missing values
148	261.62	261.80
149	262.00	262.20
150	262.67	262.70
151	262.83	262.89
152	264.18	264.23
153	263.43	263.47

REFERENCES

[1] C. F. ANSLEY AND R. KOHN, *On the estimation of ARIMA models with missing observations*, Time Series Analysis of Irregularly Observed Data, Springer Lecture Notes in Statistics, Vol 25, ed. by E. Parzen, Springer-Verlag, New York, 1985, pp. 9–37.

[2] M. AOKI, *State Space Modeling of Time Series*, Springer-Verlag, Berlin.

[3] W. BELL AND S. HILLMER, *Initializing the Kalman filter for non-stationary time series models*, U.S. Census Bureau.

[4] G.E.P. BOX AND G.M. JENKINS, *Time Series Analysis, Forecasting and Control*, revised edition, Holden Day, San Francisco, 1976.

[5] P.J. BROCKWELL AND R.A. DAVIS, *Time Series : Theory and Methods*, Springer-Verlag, New York, 1987.

[6] P.J. BROCKWELL, R.A. DAVIS AND H. SALEHI, *A state-space approach to transfer function modelling*, Inference from Stochastic Processes, I.V. Basawa and N.U. Prabhu (eds), Marcel Dekker, New York, 1990.

[7] M.B. PRIESTLEY, *Spectral Analysis and Time Series*, Vols 1 and 2, Academic Press, New York, 1981.

[8] F.C. SCHWEPP, *Evaluation of likelihood functions for Gaussian signals*, IEEE Transactions on Information Theory, IT-11 (1965), pp. 61–70.

RESAMPLING TECHNIQUES FOR STATIONARY TIME-SERIES: SOME RECENT DEVELOPMENTS*

E. CARLSTEIN†

Abstract. A survey is given of resampling techniques for stationary time-series, including algorithms based on the jackknife, the bootstrap, the typical-value principle, and the subseries method. The techniques are classified as "model-based" or "model-free", according to whether or not the user must know the underlying dependence mechanism in the time-series. Some of the techniques discussed are new, and have not yet appeared elsewhere in the literature.

Key words. subsampling, jackknife, bootstrap, typical-values, subseries, nonparametric, dependence, mixing

AMS(MOS) subject classifications. 62G05, 62M10

1. Introduction. Resampling techniques enable us to address the following class of statistical problems: A sample series of n random variables $(Z_1, Z_2, \ldots, Z_n) =: \vec{Z}_n^0$ is observed from the strictly stationary sequence $\{Z_i : -\infty < i < +\infty\}$, and a statistic $t_n(\vec{Z}_n^0) =: t_n^0$ is computed from the observed data. The objective is to describe the sampling distribution of the statistic t_n^0, using only the data $\vec{Z}_n^0$ at hand.

The scenario is nonparametric, i.e., the marginal distribution F of Z_i is unknown to the statistician; also, the statistic defined by the function $t_n(\cdot) : \mathbf{R}^n \mapsto \mathbf{R}^1$ may be quite complicated (e.g., an adaptively defined statistic, or a robustified measure of location, dispersion, or correlation in the sample series). Therefore, direct analytic description of t_n^0's sampling distribution may be impossible. These analytic difficulties will be further exacerbated by nontrivial dependence in $\{Z_i\}$. Resampling techniques substitute nonparametric sample-based numerical computations in place of intractable mathematics. Moreover, resampling algorithms are "omnibus," i.e., they are phrased in terms of a general statistic $t_n(\cdot)$ so that each new situation does not require the development of a new procedure. In keeping with the spirit of omnibus nonparametric procedures, it is also desirable for resampling techniques to require only minimal technical assumptions.

Depending on the particular application, there are various different features of t_n^0's sampling distribution that one may wish to describe via resampling. For example, the goal might be to obtain point estimates of t_n^0's moments (e.g., variance, skewness, or bias). Or, the focus might be on estimating the percentiles of t_n^0's sampling distribution. In some cases, the sampling distribution of t_n^0 can be used in constructing confidence intervals on an unknown target parameter θ. As a diagnostic tool, one may want to determine whether the statistic t_n^0 has an approximately normal sampling distribution, and, if not, how it departs from normality. For each of these objectives there are appropriate resampling algorithms.

*Research supported by NSF Grant #DMS-8902973, and by the Institute for Mathematics and its Applications with funds supported by the NSF.
†Department of Statistics, University of North Carolina, Chapel Hill, N.C. 27599

The fundamental strategy in resampling is to generate "replicates" of the statistic t from the available data $\vec{Z}_n^0$, and then use these replicates to model the true sampling distribution of t_n^0. The choice of a particular resampling algorithm for generating replicates depends upon the intended application (e.g., moment estimation, percentile estimation, confidence intervals, or diagnostics, as discussed above) and upon the structure in the original data $\vec{Z}_n^0$ (e.g., independence versus time-series versus regression).

Resampling algorithms for time-series are intuitively motivated by analogy to the established resampling algorithms for independent observations. Therefore, Section 2 reviews the jackknife, the bootstrap, and the typical-value principle – three specific resampling algorithms for generating replicates from independent observations. When the original observations are serially dependent, these resampling algorithms must be appropriately modified in order to yield valid replicates of the statistic t. The modified resampling algorithms can be "model-based" (i.e., they can exploit an assumed dependence mechanism in $\{Z_i\}$) or they can be "model-free" (i.e., no knowledge of the dependence mechanism in $\{Z_i\}$ is needed). Section 3 surveys the model-based resampling algorithms for time-series, including the Markovian bootstrap and bootstrapping of residuals; Section 4 surveys the model-free approaches, including the blockwise jackknife, the blockwise bootstrap, the linked blockwise bootstrap, and the subseries method.

The survey given in Sections 2, 3, and 4 is expository, relying mostly on intuitive explanations of the resampling algorithms (for precise technical conditions the reader is directed to the original references). For each resampling algorithm, a natural question to ask is "Does it work?" Specifically, for which statistics $t_n(\cdot)$ and marginal distributions F does the resampling algorithm provide replicates which adequately model the desired feature of t_n^0's sampling distribution? In many cases, the answer to this question will be inextricably tied to the issue of t_n^0's asymptotic normality.

2. Resampling algorithms for independent observations. This review of the jackknife, the bootstrap, and the typical-value principle is meant to provide an intuitive foundation – in the independent case – for the time-series resampling techniques which will be discussed in Sections 3 and 4. Therefore, the focus of this Section is on the seminal works in resampling for independent observations. An exhaustive review of these resampling algorithms is not attempted here; indeed, there have been more than 400 publications on the bootstrap alone since its introduction just over a decade ago.

Throughout this Section, assume that the random variables $\{Z_i : -\infty < i < +\infty\}$ are independent, and that $t_m(\cdot), m \geq 1$, is symmetric in its m arguments.

2.1 Jackknife. The jackknife algorithm generates replicates of the statistic t by deleting observations from the sample $\vec{Z}_n^0$, and then computing the statistic on the remaining data. Thus, the ith "jackknife replicate" of the statistic t is

$$t_n^{(i)} := t_{n-1}(Z_1, Z_2, \ldots, Z_{i-1}, Z_{i+1}, \ldots, Z_n)$$

for $i \in \{1, 2, \ldots, n\}$.

To estimate $\mathbf{V}\{t_n^0\}$, the variance of t_n^0's sampling distribution, Tukey [65] proposed

$$\hat{\mathbf{V}}_J\{t_n^0\} := \sum_{i=1}^{n} \frac{(t_n^{(i)} - \bar{t}_n)^2}{n} \cdot (n-1),$$

where $\bar{t}_n := \sum_{i=1}^{n} t_n^{(i)}/n$. This "jackknife estimate of variance" $\hat{\mathbf{V}}_J\{t_n^0\}$ uses the variability amongst the jackknife replicates to model the true sampling variability of t_n^0. Since the $t_n^{(i)}$s share so many observations, they do not exhibit as much variability as would n independent realizations of t_n^0; the "extra" factor of $(n-1)$ accounts for this effect by inflating the variance estimate.

A similar approach can be used to estimate the bias ($\mathbf{E}\{t_n^0\} - \theta$) of t_n^0. In fact, Quenouille [46, 47] originated jackknifing for this purpose. His bias estimate, $(\bar{t}_n - t_n^0) \cdot (n-1)$, uses the average of the jackknife replicates to model the true expectation of t_n^0.

In order for $\hat{\mathbf{V}}_J\{t_n^0\}$ to be an asymptotically unbiased and consistent estimator of $\mathbf{V}\{t_n^0\}$, it is necessary that the statistic t_n^0 have an asymptotically normal sampling distribution (van Zwet [66]). Asymptotic normality of the general statistic t_n^0 and its jackknife replicates has also been studied by Hartigan [30]. Note that the jackknife estimate of variance fails when t_n^0 is the sample median, even though t_n^0 is asymptotically normal (see Efron [20] for this example, as well as a thorough analysis of the jackknife). The jackknife method does allow deletion of more than one observation when computing the jackknife replicates; this is explored by Shao and Wu [52].

2.2 Typical-Values. Hartigan [29] introduced the typical-value principle for constructing nonparametric confidence intervals on an unknown parameter θ. A collection of random variables $\{V_1, V_2, \ldots, V_k\}$ are "typical-values for θ" if each of the $k+1$ intervals between the ordered random variables

$$-\infty \equiv V_{(0)} < V_{(1)} < V_{(2)} < \cdots < V_{(k)} < V_{(k+1)} \equiv +\infty$$

contains θ with equal probability. Given such a collection of typical-values, confidence intervals on θ may be constructed by taking the union of adjacent intervals; in particular, $\mathbf{P}\{V_{(1)} < \theta < V_{(k)}\} = 1 - 2/(k+1)$.

The connection between typical-values and resampling is this: Suppose that the statistic t estimates the unknown parameter θ. For any subset S of the data indices, i.e., $S \equiv \{i_1, i_2, \ldots, i_m\} \subseteq \{1, 2, \ldots, n\}$, compute the corresponding "subset replicate" of t:

$$t[S] := t_m(Z_{i_1}, Z_{i_2}, \ldots, Z_{i_m}).$$

In many situations, the collection of subset replicates $\{t[S_1], t[S_2], \ldots, t[S_k]\}$ are actually typical-values for θ. This is true, for example, whenever F is a continuous symmetric distribution with center of symmetry θ, the statistic t is the sample mean, and the S_js are the $k = 2^n - 1$ possible non-empty subsets of $\{1, 2, \ldots, n\}$. Notice that t_n^0 need not be asymptotically normal in this case, e.g., if F is Cauchy. Valid typical-values can be obtained from certain smaller collections of subset replicates, where the S_js may be chosen deterministically or randomly (see also Efron [20]). Furthermore, typical-values can be obtained from subset replicates of statistics t other than the sample mean, e.g., the sample median (Efron [20]), M-estimates (Maritz [42]), and general asymptotically normal statistics (Hartigan [30]).

2.3 Bootstrap. Efron's [19] bootstrap algorithm uses F_n, the observed empirical distribution of $\vec{Z}_n^0$, in place of the unknown distribution F. Given F_n, which assigns mass $1/n$ at each Z_i ($1 \leq i \leq n$), a "bootstrap sample" $(Z_1^*, Z_2^*, \ldots, Z_m^*) =: \vec{Z}_m^{0*}$ is generated by i.i.d. sampling from F_n. The corresponding "bootstrap replicate" of t is

$$t_m^{0*} := t_m(\vec{Z}_m^{0*}).$$

For fixed data $\vec{Z}_n^0$, it is possible to generate arbitrarily many bootstrap replicates of t, by repeatedly drawing bootstrap samples from F_n. The conditional distribution (i.e., given $\vec{Z}_n^0$) of these bootstrap replicates is used to model the true sampling distribution of t_n^0. For example, to estimate $\mathsf{V}\{t_n^0\}$, the "bootstrap estimate of variance" is

$$\hat{\mathsf{V}}_B\{t_n^0\} := \mathsf{E}\{(t_m^{0*} - \mathsf{E}\{t_m^{0*}|\vec{Z}_n^0\})^2 \mid \vec{Z}_n^0\}.$$

In many situations, the bootstrap replicates can actually be used to estimate the entire distribution function of t_n^0. To obtain a valid estimate, appropriate choices must be made for the bootstrap sample size $m \equiv m_n$ and for the standardizations $(a_n, b_n; a_m^*, b_m^*)$ of t_n^0 and t_m^{0*}. Specifically, the bootstrap estimate of $\mathsf{P}\{\tilde{t}_n^0 \leq x\}$ is

$$\mathsf{P}\{\tilde{t}_m^{0*} \leq x \mid \vec{Z}_n^0\},$$

where $\tilde{t}_n^0 := a_n(t_n^0 - b_n)$ and $\tilde{t}_m^{0*} := a_m^*(t_m^{0*} - b_m^*)$ are the standardized versions. This bootstrap estimate is said to be "strongly uniformly consistent" if

$$\sup_{x \in R} \left| \mathsf{P}\{\tilde{t}_m^{0*} \leq x|\vec{Z}_n^0\} - \mathsf{P}\{\tilde{t}_n^0 \leq x\} \right| \overset{a.s.}{\to} 0 \quad \text{as } n \to \infty.$$

When the statistic t is the sample mean, strong uniform consistency of the bootstrap has been established in several scenarios: Bickel and Freedman [6] deal with the asymptotically normal case, as does Singh [53] who actually shows that the bootstrap estimate can be second-order correct (i.e., strong uniform consistency holds with a rate factor of $\sqrt{n}$); Arcones and Giné [1] treat the asymptotically α-stable case ($0 < \alpha < 2$). The bootstrap is strongly uniformly consistent when t is a von Mises functional statistic, both in the asymptotically normal case (Bickel and Freedman [6]) and in the asymptotically non-normal case (Bretagnolle [10]). Bickel and Freedman [6] also show that the bootstrap is strongly uniformly consistent when t is the [asymptotically normal] sample median. Swanepoel [58] establishes strong uniform consistency for the bootstrap when t is the sample maximum [having an exponential limiting distribution]. Thus the bootstrap algorithm is valid in a wide variety of i.i.d. situations.

3. Model-based resampling algorithms for time-series. If the dependence mechanism in $\{Z_i\}$ is known, then that mechanism can be incorporated into the resampling algorithm and hence into the generated replicates of t. This Section discusses in detail two such model-based modifications of the bootstrap algorithm. [Note that model-based modifications of the bootstrap algorithm have also been studied in the case where $\{Z_i\}$ is a non-stationary $AR(1)$ time-series and t is a least-squares estimator of the AR parameter (see Basawa, Mallik, McCormick, and Taylor [4] and Basawa, Mallik, McCormick, Reeves, and Taylor [5]).]

3.1 Bootstrapping residuals. Here it is assumed that the dependence structure in $\{Z_i\}$ satisfies

$$Z_i = g(Z_{i-1}, Z_{i-2}, \ldots, Z_{i-k}; \vec{\beta}) + \varepsilon_i,$$

where $g(\cdot)$ is a known function, $\vec{\beta}$ is a vector of unknown parameters, and the additive unobservable errors $\{\varepsilon_i : -\infty < i < +\infty\}$ are i.i.d. with unknown distribution F_ε having mean zero.

The algorithm begins by calculating $\hat{\beta}$, a data-based estimate of $\vec{\beta}$. Using this $\hat{\beta}$, residuals

$$\hat{\varepsilon}_i := Z_i - g(Z_{i-1}, Z_{i-2}, \ldots, Z_{i-k}; \hat{\beta}), \quad i \in \{k+1, k+2, \ldots, n\},$$

are computed. The residuals in turn are used to construct an estimate of F_ε, namely $\hat{F}_\varepsilon$ which puts mass $1/(n-k)$ on each centered residual $\hat{\varepsilon}_i - \sum_{j=k+1}^{n} \hat{\varepsilon}_j/(n-k)$, $i \in \{k+1, k+2, \ldots, n\}$. The bootstrap sample $\vec{Z}_m^{0*}$ is now generated using the estimated dependence mechanism, that is

$$Z_i^* := g(Z_{i-1}^*, Z_{i-2}^*, \ldots, Z_{i-k}^*; \hat{\beta}) + \varepsilon_i^*, i \in \{1, 2, \ldots, m\},$$

where $\{\varepsilon_i^* : 1 \le i \le m\}$ are i.i.d. from $\hat{F}_\varepsilon$ and $Z_i^* \equiv Z_{i+k}$ for $i \in \{-(k-1), -(k-2), \ldots, 0\}$. The corresponding bootstrap replicate of t is simply

$$t_m^{0*} := t_m(\vec{Z}_m^{0*}).$$

The literature has concentrated on the special case where $\{Z_i\}$ is autoregressive, i.e., $g(z_1, z_2, \ldots, z_k; (\beta_0, \beta_1, \beta_2, \ldots, \beta_k)) = \beta_0 + \beta_1 z_1 + \beta_2 z_2 + \cdots + \beta_k z_k$, and where t is a least-squares estimator of the autoregressive parameter having a limiting normal distribution. In this case, the bootstrap is strongly uniformly consistent (Freedman [25]) and furthermore can be second-order correct (Bose [7]); see Efron and Tibshirani [21] and Freedman and Peters [26] for further discussion and applications. Swanepoel and van Wyk [59] apply the bootstrap of autoregressive residuals to constructing a confidence interval on the power spectrum.

The related problem of obtaining prediction intervals for autoregressive time-series has been addressed via bootstrap algorithms by Stine [56] and by Thombs and Schucany [62, 63]; see also Findley [22]. Extensions to the case of ARMA time-series are considered by Thombs [60,61].

3.2 Markovian bootstrap. Suppose now that $\{Z_i\}$ has first-order Markovian dependence structure with unknown transition density $f(z_1|z_0)$. The algorithm begins by computing $\hat{f}(\cdot|\cdot)$, a sample-based estimate of $f(\cdot|\cdot)$; specifically, this $\hat{f}(\cdot|\cdot)$ may be computed using kernel density estimation techniques. The bootstrap sample $\vec{Z}_m^{0*}$ is generated from the estimated dependence mechanism as follows: Select Z_1^* at random from F_n, and draw Z_i^* from the distribution with density $\hat{f}(\cdot|Z_{i-1}^*), i \in \{2, 3, \ldots, m\}$. The corresponding bootstrap replicate of t is again $t_m^{0*} := t_m(\vec{Z}_m^{0*})$.

Rajarshi [48] introduced this bootstrap algorithm and established strong uniform consistency when t is the [asymptotically normal] sample mean. An extension of this algorithm to Markovian random fields is studied by Lele [38]. The case of Markov chains is considered by Kulperger and Prakasa Rao [34], Athreya and Fuh [2], and Basawa, Green, McCormick, and Taylor [3].

3.3 Other algorithms. Bose [9] studies the bootstrap for moving average models; related work can be found in Bose [8]. A bootstrap method for state space models is investigated by Stoffer and Wall [57]. Franke, Härdle, and Kreiss [24] apply the bootstrap to M-estimates of ARMA parameters. Solow [54] proposes a bootstrap algorithm which first estimates the pairwise correlations in the data and then transforms to approximate independence.

Several approaches to jackknifing time-series data have been suggested, beginning with Brillinger [11]. He suggests that, in computing the ith jackknife replicate, the deleted observation Z_i should be treated as a "missing value" in the time-series and should be replaced via interpolation using, say, an ARMA model (Brillinger [12]); thus the complete time-series structure is retained by each replicate. When the statistic t is itself expressible as a function of ARMA residuals (e.g., $\sum_i(\hat{\varepsilon}_i)^2/n$), then Davis [18] considers applying the i.i.d. jackknife algorithm directly by deleting residuals. Gray, Watkins, and Adams [28] give an extension of the jackknife to piecewise continuous stochastic processes.

The present survey emphasizes the intuition of resampling algorithms in the time-domain. There are, however, several frequency-domain resampling algorithms: A bootstrap method designed for Gaussian time-series is studied by Stine [55], Ramos [50], and Hurvich and Zeger [32]; Hartigan [31] introduces a "perturbed periodogram" method; Hurvich, Simonoff, and Zeger [33] propose a bootstrap method for moving average data as well as a jackknife method; also see Franke and Härdle [23] and Thomson and Chave [64].

4. Model-free resampling algorithms for time-series. The drawback of the model-based techniques discussed in Section 3 is that they require the user to know the correct underlying dependence mechanism in $\{Z_i\}$ (this point is emphasized by Freedman [25] in the case of autoregression). The strength of the i.i.d. bootstrap, jackknife, and typical-value principles is that they are nonparametric (i.e., the marginal distribution of Z_i is unknown); therefore it is more appropriate to develop analogous resampling techniques for stationary data that are similarly free of assumptions regarding the dependence mechanism (i.e., the joint distribution of the Z_is is allowed to be unknown). In practice, the joint probability structure of $\{Z_i\}$ is more obscure than the marginal probability structure of Z_i, so it is un-realistic to assume that the former is known when the latter is unknown. These considerations motivate the model-free algorithms discussed in this Section.

In this model-free scenario, it is convenient to measure the strength of dependence in $\{Z_i\}$ by a "mixing coefficient"

$$\alpha(r) := \sup_{A \in \mathcal{F}\{...,Z_{-1},Z_0\}, B \in \mathcal{F}\{Z_r,Z_{r+1},...\}} |\mathbf{P}\{A \cap B\} - \mathbf{P}\{A\}\mathbf{P}\{B\}|,$$

as introduced by Rosenblatt [51]. In order for the model-free resampling algorithms to be valid, it is typically assumed that $\alpha(r) \to 0$ at some appropriate rate as $r \to \infty$. Intuitively, this says that observations which are separated by a long time-lag behave approximately as if they were independent.

In the absence of assumptions about the dependence mechanism in $\{Z_i\}$, it is natural to focus attention on the "blocks" of sample data

$$\vec{Z}_\ell^i := (Z_{i+1}, Z_{i+2}, \ldots, Z_{i+\ell}), \quad 0 \le i < i + \ell \le n.$$

These blocks automatically retain the correct dependence structure of $\{Z_i\}$. For asymptotic validity, it is usually required that $\ell \to \infty$ as $n \to \infty$, so that the blocks ultimately reflect the dependencies at all lags.

4.1 Blockwise jackknife. In its simplest form, the blockwise jackknife generates replicates of the statistic t by deleting blocks of ℓ observations from $\vec{Z}_n^0$, and then computing the statistic on the remaining data. Thus, the ith "blockwise jackknife replicate" of t is

$$t_n^{<i>} := t_{n-\ell}(\vec{Z}_n^0 \setminus \vec{Z}_\ell^i)$$

for $i \in \{0, 1, \ldots, n - \ell\}$. The resulting estimate of $\mathbf{V}\{t_n^0\}$ is

$$\hat{\mathbf{V}}_{BJ}\{t_n^0\} := \sum_{i=0}^{n-\ell} \frac{(t_n^{<i>} - \bar{t}_n^{<\cdot>})^2}{n - \ell + 1} \cdot c_{n,\ell},$$

where $\bar{t}_n^{<\cdot>} := \sum_{i=0}^{n-\ell} t_n^{<i>}/(n - \ell + 1)$ and $c_{n,\ell}$ is an appropriate standardizing constant. This "blockwise jackknife estimate of variance" was proposed by Künsch [35]; he showed that $\hat{\mathbf{V}}_{BJ}\{t_n^0\}$ is consistent when t belongs to a certain class of asymptotically normal functional statistics (including the sample mean). A generalization of the blockwise jackknife is investigated by Politis and Romano [43].

4.2 Blockwise bootstrap. This method extends the bootstrap to dependent data by resampling the blocks. The algorithm is essentially as follows: For fixed ℓ, construct the "empirical ℓ-dimensional marginal distribution" $F_{\ell,n}$, i.e., the distribution putting mass $1/(n-\ell+1)$ on each sample block $\vec{Z}_\ell^i$, $i \in \{0, 1, \ldots, n-\ell\}$. Now, assuming $k := n/\ell$ is an integer, generate k "bootstrap blocks" by i.i.d. random resampling of blocks from $F_{\ell,n}$. Denote these bootstrap blocks as

$$\vec{Z}_{\ell,j}^* \equiv (Z_{(j-1)\ell+1}^*, Z_{(j-1)\ell+2}^*, \ldots, Z_{j\ell}^*), j \in \{1, 2, \ldots, k\}.$$

The blockwise bootstrap sample $\vec{Z}_n^{0*}$ is then constructed by appending these blocks together, i.e.,

$$\vec{Z}_n^{0*} := (\vec{Z}_{\ell,1}^*, \vec{Z}_{\ell,2}^*, \ldots, \vec{Z}_{\ell,k}^*).$$

Thus $\vec{Z}_n^{0*}$ inherits the correct dependence structure – at least within blocks. The corresponding "blockwise bootstrap replicate" of t is $t_n^{0*} := t_n(\vec{Z}_n^{0*})$.

This algorithm was introduced by Künsch [35] (see also Liu and Singh [41]). He shows that, when t is an asymptotically normal sample mean, the blockwise bootstrap is strongly uniformly consistent; second-order correctness is studied by Götze and Künsch [27] and by Lahiri [37]. A generalization of the blockwise bootstrap is considered by Politis and Romano [43] and by Politis, Romano, and Lai [44].

4.3 Linked blockwise bootstrap. The blockwise bootstrap sample $\vec{Z}_n^{0*}$ (obtained above) is not a good replicate of the original data $\vec{Z}_n^0$ in the following sense: The dependence structure near block "endpoints" is incorrect. For example, the bootstrap observations $\{Z_{j\ell}^*, Z_{j\ell+1}^*\}$ are adjacent in time but are [conditionally] independent! Graphically, the bootstrap sample $\vec{Z}_n^{0*}$ will exhibit anomalous behavior at the block endpoints. Künsch and Carlstein [36] propose the following modification of the blockwise bootstrap in order to correct this problem.

The "linked blockwise bootstrap" still selects the first block $\vec{Z}_{\ell,1}^*$ at random from the empirical ℓ-dimensional marginal distribution $F_{\ell,n}$. Now look at the final observation Z_ℓ^* in this first block, and identify its p "nearest neighbors" among the set of original observations $\{Z_1, Z_2, \ldots, Z_{n-\ell}\}$. Randomly select one of these p nearest neighbors. The selected observation – say, Z_v – is the "link". The second bootstrap block is then taken to be $\vec{Z}_{\ell,2}^* = \vec{Z}_\ell^v$, the block of original data immediately following the link. The $(j+1)^{\text{th}}$ bootstrap block $\vec{Z}_{\ell,j+1}^*$ is similarly obtained by randomly linking to the final observation $Z_{j\ell}^*$ from the jth bootstrap block.

This linked blockwise bootstrap is still based on blocks; hence $\vec{Z}_n^{0*}$ still has exactly the correct dependence structure within blocks – without requiring any knowledge of the underlying dependence mechanism. The linked blockwise bootstrap improves on the blockwise bootstrap by guaranteeing a more natural transition from one bootstrap block to the next.

4.4 Subseries. The most simplistic way to generate replicates of t from the blocks is by calculating the statistic on the individual blocks themselves. Thus, for fixed ℓ, the ith "subseries replicate" of t is

$$t_\ell^i := t_\ell(\vec{Z}_\ell^i), \quad i \in \{0, 1, \ldots, n-\ell\}.$$

These subseries replicates can be used to construct typical-values, to estimate moments of t, and for diagnostics on t's sampling distribution.

When t estimates a parameter θ, and $k := n/\ell$ is an integer, then the random variables

$$V_i := (t_\ell^0 + t_\ell^{i\ell})/2, \quad i \in \{1, 2, \ldots, k-1\}$$

can behave asymptotically like typical-values for θ (Carlstein [16]). This approach is valid for a large class of asymptotically normal statistics t, including the sample mean and sample percentiles (see also Carlstein [13]).

The pth moment of t's sampling distribution can be estimated via the pth empirical moment of the subseries replicates:

$$\sum_{i=0}^{k-1}(t_\ell^{i\ell})^p/k.$$

This method is consistent in a broad range of situations, and does not generally require the statistic t to be asymptotically normal (Carlstein [14, 15]). For an extension of this technique to spatial processes, see Possolo [45].

The subseries replicates can also be used for diagnostics, e.g., to graphically assess non-normality or skewness in t's sampling distribution. This suggestion is theoretically justified by a strong uniform consistency result for the empirical distribution of the subseries replicates (Carlstein [17]).

4.5 Other algorithms. For statistics obtained from estimating-equations, a jackknife algorithm (Lele [39]) and a bootstrap algorithm (Lele [40]) have been developed. Venetoulias [67] introduces a technique for generating replicates in image data. Rajarshi [49] proposes a "direct" method for estimating the variance of t; although his method actually does not involve any resampling, it is in the same spirit as the model-free algorithms of this Section.

REFERENCES

[1] M.A. ARCONES AND E. GINÉ, *The bootstrap of the mean with arbitrary bootstrap sample size*, Ann. Inst. H. Poincaré, 25 (1989), pp. 457–481.

[2] K.B. ATHREYA AND C.D. FUH, *Bootstrapping Markov chains: countable case*, Tech. Rep. (1989); Dept. of Statist., Iowa State U., Ames.

[3] I.V. BASAWA, T.A. GREEN, W.P. McCORMICK AND R.L. TAYLOR, *Asymptotic bootstrap validity for finite Markov chains*, Tech. Rep. (1989); Dept. of Statist., U. of Georgia, Athens.

[4] I.V. BASAWA, A.K. MALLIK, W.P. McCORMICK AND R.L. TAYLOR, *Bootstrapping explosive autoregressive processes*, Ann. Statist., 17 (1989), pp. 1479–1486.

[5] I.V. BASAWA, A.K. MALLIK, W.P. McCORMICK, J.H. REEVES AND R.L. TAYLOR, *Bootstrapping unstable autoregressive processes*, Tech. Rep., 122 (1989); Dept. of Statist., U. of Georgia, Athens.

[6] P.J. BICKEL AND D.A. FREEDMAN, *Some asymptotic theory for the bootstrap*, Ann. Statist., 9 (1981), pp. 1196–1217.

[7] A. BOSE, *Edgeworth correction by bootstrap in autoregressions*, Ann. Statist., 16 (1988), pp. 1709–1722.

[8] ————, *Bootstrap in time series models*, in Proceedings of the Winter Simulation Conference, M. Abrams, P. Haigh, and J. Comfort (eds), IEEE, San Francisco, 1988, pp. 486–490.

[9] ————, *Bootstrap in moving average models*, Ann. Instit. Statist. Math. (1989) (to appear).

[10] J. BRETAGNOLLE, *Lois limites du Bootstrap de certaines fonctionnelles*, Ann. Inst. H. Poincaré, 19 (1983), pp. 281–296.

[11] D.R. BRILLINGER, Discussion of *A generalized least-squares approach to linear functional relationships*, J. Royal Statist. Soc. (Series B), 28 (1966), p. 294.

[12] ————————, Discussion of *Influence functionals for time series*, Ann. Statist., 14 (1986), pp. 819–822.

[13] E. CARLSTEIN, *Asymptotic normality for a general statistic from a stationary sequence*, Ann. Probab., 14 (1986), pp. 1371–1379.

[14] ———, *The use of subseries values for estimating the variance of a general statistic from a stationary sequence*, Ann. Statist., 14 (1986), pp. 1171–1179.

[15] ———, *Law of large numbers for the subseries values of a statistic from a stationary sequence*, Statist., 19 (1988), pp. 295–299.

[16] ———, *Asymptotic multivariate normality for the subseries values of a general statistic from a stationary sequence – with applications to nonparametric confidence intervals*, Probab. Math. Statist., 10 (1989), pp. 191–200.

[17] ———, *Using subseries values to describe the distribution of a general statistic from a stationary sequence*, Tech. Rep. in preparation (1990); Dept. of Statist., U. of N. Carolina, Chapel Hill.

[18] W. DAVIS, *Robust interval estimation of the innovation variance of an ARMA model*, Ann. Statist., 5 (1977), pp. 700–708.

[19] B. EFRON, *Bootstrap methods: another look at the jackknife*, Ann. Statist., 7 (1979), pp. 1–26.

[20] ———, *The Jackknife, the Bootstrap and Other Resampling Plans*, SIAM (1982), Philadelphia.

[21] B. EFRON AND R. TIBSHIRANI, *Bootstrap methods for standard errors, confidence intervals, and other measures of statistical accuracy*, Statistical Science, 1 (1986), pp. 54–77.

[22] D.F. FINDLEY, *On bootstrap estimates of forecast mean square errors for autoregressive processes*, in *Computer Science and Statistics: The Interface*, D.M. Allen (ed.), Elsevier (North-Holland), Amsterdam, 1986, pp. 11–17.

[23] J. FRANKE AND W. HÄRDLE, *On bootstrapping kernel spectral estimates*, Ann. Statist. (1990) (to appear).

[24] J. FRANKE, W. HÄRDLE AND J. KREISS, *Applications of the bootstrap in time series analysis*, Instit. Math. Statist. Bulletin, 19 (1990), pp. 427–428.

[25] D. FREEDMAN, *On bootstrapping two-stage least-squares estimates in stationary linear models*, Ann. Statist., 12 (1984), pp. 827–842.

[26] D. FREEDMAN AND S. PETERS, *Bootstrapping a regression equation: some empirical results*, J. Amer. Statist. Assoc., 79 (1984), pp. 97–106.

[27] F. GÖTZE AND H.R. KÜNSCH, *Blockwise bootstrap for dependent observation: higher order approximations for studentized statistics*, Instit. Math. Statist. Bulletin, 19 (1990), p. 443.

[28] H.L. GRAY, T.A. WATKINS AND J.E. ADAMS, *On the jackknife statistic, its extensions, and its relation to e_n-transformations*, Ann. Math. Statist., 43 (1972), pp. 1–30.

[29] J.A. HARTIGAN, *Using subsample values as typical values*, J. Amer. Statist. Assoc., 64 (1969), pp. 1303–1317.

[30] ———, *Necessary and sufficient conditions for asymptotic joint normality of a statistic and its subsample values*, Ann. Statist., 3 (1975), pp. 573–580.

[31] ———, *Perturbed periodogram estimates of variance*, Internat. Statist. Rev., 58 (1990), pp. 1–7.

[32] C.M. HURVICH AND S.L. ZEGER, *Frequency domain bootstrap methods for time series*, Tech. Rep. 87-115 (1987); Statist. & Oper. Res., Grad. School of Busi., New York U., New York.

[33] C.M. HURVICH, J.S. SIMONOFF, AND S.L. ZEGER, *Variance estimation for sample autocovariances: direct and resampling approaches*, Austral. J. Statist. (1990) (to appear).

[34] R. KULPERGER AND B.L.S. PRAKASA RAO, *Bootstrapping a finite state Markov chain*, Sankhya (Series A), 51 (1989), pp. 178–191.

[35] H.R. KÜNSCH, *The jackknife and the bootstrap for general stationary observations*, Ann. Statist., 17 (1989), pp. 1217–1241.

[36] H.R. KÜNSCH AND E. CARLSTEIN, *The linked blockwise bootstrap for serially dependent observations*, Tech. Rep. in preparation (1990); Dept. of Statist., U.of N. Carolina, Chapel Hill.

[37] S.N. LAHIRI, *Second order optimality of stationary bootstrap*, Tech. Rep. (1990); Dept. of Statist., Iowa State U., Ames.

[38] S. LELE, *Non-parametric bootstrap for spatial processes*, Tech. Rep. 671 (1988); Dept. of Biostatist., Johns Hopkins U., Baltimore.

[39] ———, *Jackknifing linear estimating equations: asymptotic theory and applications in stochastic processes*, J.Roy. Statist. Soc. (Series B) (1990) (to appear).

[40] ———, *Resampling using estimating equations*, Tech. Rep. 706 (1990); Dept. of Biostatist., Johns Hopkins U., Baltimore.

[41] R.Y. LIU AND K. SINGH, *Moving blocks jackknife and bootstrap capture weak dependence*, Tech. Rep. (1988); Dept. of Statist., Rutgers U., New Brunswick.

[42] J.S. MARITZ, *A note on exact robust confidence intervals for location*, Biometrika, 66 (1979), pp. 163–166.

[43] D.N. POLITIS AND J.P. ROMANO, *A general resampling scheme for triangular arrays of α-mixing random variables with application to the problem of spectral density estimation*, Tech. Rep. 338 (1989); Dept. of Statist., Stanford U., Stanford.

[44] D.N. POLITIS, J.P. ROMANO AND T. LAI, *Bootstrap confidence bands for spectra and cross-spectra*, Tech. Rep. 342 (1990); Dept. of Statist., Stanford U., Stanford.

[45] A. POSSOLO, *Subsampling a random field*, Tech. Rep., 78 (1986); Dept. of Statist., U. of Washington, Seattle.

[46] M. QUENOUILLE, *Approximate tests of correlation in time series*, J. Roy. Statist. Soc. (Series B), 11 (1949), pp. 18–84.

[47] —————, *Notes on bias in estimation*, Biometrika, 43 (1956), pp. 353–360.

[48] M.B. RAJARSHI, *Bootstrap in Markov sequences based on estimates of transition density*, Ann. Instit. Statist. Math. (1989) (to appear).

[49] —————, *On estimation of the variance of a statistic obtained from dependent observations*, Tech. Rep. (1990); Dept. of Statist., U. of Poona, Pune.

[50] E. RAMOS, *Resampling methods for time series*, Ph.D. Thesis (1988); Dept. of Statist., Harvard U., Cambridge.

[51] M. ROSENBLATT, *A central limit theorem and a strong mixing condition*, Proc. Nat. Acad. Sci., 42 (1956), pp. 43–47.

[52] J. SHAO AND C.F. J. WU, *A general theory for jackknife variance estimation*, Ann. Statist., 17 (1989), pp. 1176–1197.

[53] K. SINGH, *On the asymptotic accuracy of Efron's bootstrap*, Ann. Statist., 9 (1981), pp. 1187–1195.

[54] A.R. SOLOW, *Bootstrapping correlated data*, Math. Geology, 17 (1985), pp. 769–775.

[55] R.A. STINE, *A frequency domain bootstrap for time series*, Tech. Rep. (1985); Dept. of Statist., Wharton School, U. of Pennsylvania, Philadelphia.

[56] —————, *Estimating properties of autoregressive forecasts*, J. Amer. Statist. Assoc., 82 (1987), pp. 1072–1078.

[57] D.S. STOFFER AND K.D. WALL, *Bootstrapping state space models: Gaussian maximum likelihood estimation and the Kalman filter*, Tech. Rep. (1990); Dept. of Math. & Statist., U. of Pittsburgh, Pittsburgh.

[58] J.W.H. SWANEPOEL, *A note on proving that the (modified) bootstrap works*, Commun. Statist. Theory Meth., 15 (1986), pp. 3193–3203.

[59] J.W.H. SWANEPOEL AND J.W.J. VAN WYK, *The bootstrap applied to power spectral density function estimation*, Biometrika, 73 (1986), pp. 135–141.

[60] L.A. THOMBS, *Prediction for ARMA(p,q) series using the bootstrap*, Instit. Math.Statist. Bulletin, 16 (1987), p. 225.

[61] —————, *Model dependent resampling plans for time series*, Tech. Rep. (1987); Dept. of Statist., U. of S. Carolina, Columbia.

[62] L.A. THOMBS AND W.R. SCHUCANY, *Bootstrap prediction intervals for autoregression*, Proceedings of the Decision Sciences Institute (1987), pp. 179–181.

[63] —————, *Bootstrap prediction intervals for autoregression*, J. Amer. Statist. Assoc., 85 (1990), pp. 486–492.

[64] D. THOMSON AND A. CHAVE, *Jackknifed error estimates for spectra, coherences, and transfer functions*, Tech. Rep. (1988); AT&T Bell Laboratories, Murray Hill.

[65] J. TUKEY, *Bias and confidence in not quite large samples*, Ann. Math. Statist., 29 (1958), p. 614.

[66] W.VAN ZWET, *Second order asymptotics and the bootstrap*, Lect. Notes (1990); Dept. of Statist., U. of N. Carolina, Chapel Hill.

[67] A. VENETOULIAS, *Artificial replication in the analysis of image data*, Instit. Math. Statist. Bulletin, 19 (1990), p. 345.

STATE SPACE MODELING AND CONDITIONAL MODE ESTIMATION FOR CATEGORICAL TIME SERIES*

LUDWIG FAHRMEIR**

Abstract. This paper deals with the following state space models for time series of ordered or unordered categorical responses: Conditional upon time-varying parameters or states, observation models are dynamic versions of categorical response models for cross-sectional data. State transition models may be Gaussian or not.

As an alternative to conditional mean filtering and smoothing, which generally will require repeated multidimensional integrations, estimation is based on modes of the posterior distribution. Factorization of the conditional information matrix leads to Gauss-Newton iterations which are closely related to extended Kalman filtering for non-Gaussian data. Estimation of unknown hyperstructural parameters is discussed and the methods are illustrated by some examples. Relations to semiparametric approaches as well as extensions to general non-Gaussian state space models are indicated.

Key words. categorical time series, state space models, conditional mode filtering.

1. Introduction. State space models relate time series observations $\{y_t\}$ to unobserved states $\{\beta_t\}$ by an observation model for y_t given β_t and a transition model for $\{\beta_t\}$. Gaussian linear state space models are defined by a linear observation equation

$$(1.1) \qquad y_t = Z'_t\beta_t + \varepsilon_t, \quad t = 1, 2, \ldots,$$

and a linear transition equation

$$(1.2) \qquad \beta_t = T_t\beta_{t-1} + v_t, \quad t = 1, 2, \ldots,$$

together with the usual assumptions on the Gaussian noise processes (e.g. Sage and Melsa, 1971, ch. 7, Anderson and Moore, 1979, ch. 2). Given the observations $y_1, \ldots, y_t$, estimation of β_t (filtering) and of $\beta_0, \ldots, \beta_{t-1}$ (smoothing), together with corresponding error covariance matrices, is of primary interest. Optimal conditional mean estimates are provided recursively by the linear Kalman filter and smoother. Linear state space models have been successfully applied to linear structural time series models, where states are unobserved trend or seasonal components or, possibly time-varying, effects of explanatory variables x_t (e.g. Kitagawa and Gersch, 1984, Harvey, 1989, ch. 4). As a basic example, let us consider

$$(1.3) \qquad y_t = \mu_t + x'_t\alpha_t + \varepsilon_t, \quad t = 1, 2, \ldots,$$

*Presented at the Institute for Mathematics and its Applications Workshop "New Directions in Time Series Analysis", Minneapolis, Juli 1-29, 1990. Research supported in part by the Deutsche Forschungsgemeinschaft. The paper is dedicated to the memory of Heinz Leo Kaufmann (†), friend and coauthor for a long time.

**Institut für Statistik and Wissenschaftstheorie, Universität München Ludwigstr. 33 D-8000 München 22.

with a trend component μ_t and covariate effects α_t. Simple yet rather effective nonstationary trend models are the *first order random walk model*

$$(1.4) \qquad \mu_t = \mu_{t-1} + u_t,$$

and the *second order random walk model*

$$(1.5) \qquad \mu_t = 2\mu_{t-1} - \mu_{t-2} + u_t,$$

where $\{u_t\}$ is Gaussian white noise with var $u_t = \sigma_\mu^2$. Covariate effects α_t may be constant or time-varying. Assuming a first order random walk model $\alpha_t = \alpha_{t-1} + w_t$, $w_t \sim N(0, \Sigma_\alpha)$, the case of constant parameters $\alpha_0 = \alpha_1 = \cdots = \alpha_t$ is covered by setting $\Sigma_\alpha = 0$. Another common model is the local linear trend model

$$\mu_t = \mu_{t-1} + \lambda_t + u_t, \quad \lambda_t = \lambda_{t-1} + v_t, \quad v_t \sim N(0, \sigma_\lambda^2),$$

with the local slope component λ_t. Thinking of trends varying continuously over time, one may also wish to consider corresponding models. Continuous time versions of random walks are (integrated) Wiener processes or, more general, diffusion processes. Wahba (1978) showed that such models for $\{\beta_t\}$ are closely related to spline smoothing, and Wecker and Ansley (1983), Kohn and Ansley (1987) exploit this connection for non- and semiparametric spline smoothing by Kalman filtering and smoothing.

Structural time series models are written in state space form by appropriate specification of β_t, Z_t, T_t etc. in (1.1), (1.2): For example, model (1.3), (1.4) is obtained by setting $\beta_t = (\mu_t, \alpha_t), v_t = (u_t, w_t), T_t = I$ and

$$(1.6) \qquad Z_t = \begin{bmatrix} 1 \\ x_t \end{bmatrix}, \quad \mathrm{cov}\, v_t = \begin{bmatrix} \sigma_\mu^2 & 0 \\ 0 & \Sigma_\alpha \end{bmatrix}.$$

For the second order random walk model, one defines $\beta_t = (\mu_t, \mu_{t-1}, \alpha_t)$ and

$$(1.7) \qquad T_t = \begin{bmatrix} 2 & -1 & 0 \\ 1 & 0 & \\ 0 & & I \end{bmatrix}, Z_t = \begin{bmatrix} 1 \\ 0 \\ x_t \end{bmatrix}, \mathrm{cov}\, v_t = \begin{bmatrix} \sigma_\mu^2 & 0 & 0 \\ 0 & 0 & \\ 0 & & \Sigma_\alpha \end{bmatrix}.$$

In Section 2, we introduce state space models for uni- and multivariate time series of categorical responses, where categories may be unordered or ordered. The linear observation model (1.1) is replaced by dynamic extensions of categorical response models, such as logit and probit models, cumulative models, sequential models, and others. Since the observation models are conditional upon current states or parameters, covariates and, possibly, past responses, the models can be considered as "parameter-driven" as well as "observation-driven". We will retain the assumption of a linear Gaussian transition model (1.2) as long as this is compatible with the categorical observation model. This has the advantage that existing stochastic models for trend and seasonality, known from linear structural time series analyses, can be combined explicitly with categorical observation models. However, the approach is not restricted to this assumption.

Since we abandon the assumption of normality and linearity, exact conditional mean filters and smoothers are generally no longer available in closed form. Apart from some simple models which admit conjugate prior-posterior solutions (Harvey and Fernandes, 1989), conditional means and conditional densities $p(\beta_s|y_1,\ldots,y_t)$ would have to be computed by numerical integration. For scalar observations and states such an approach is presented by Kitagawa (1987). For multidimensional problems, the computational burden due to repeated multiple integrations can become infeasible. West, Harrison and Migon (1985), West and Harrison (1989) combine conjugate analysis with linear Bayes approximations. However, their method raises some problems, in particular in extending it to multicategorical data. Therefore, as an alternative to conditional mean and variance estimation, we propose conditional mode filtering and smoothing by maximizing conditional densities. Section 3 presents Gauss-Newton and Fisher scoring algorithms, which make efficient use of LDL'-factorizations (e.g. Golub and van Loan, 1989) of block-tridiagonal information matrices. For exponential family observation models and linear Gaussian transition models a detailed exposition has been given in Fahrmeir and Kaufmann (1991). In addition, it turns out that generalized extended Kalman filtering and smoothing (Fahrmeir 1990) can be understood as a simplified version of these algorithms, and it is proposed to use it as a starting solution for Fisher scoring or Gauss-Newton algorithms. Practical experience indicates that for many of the common models only very few iterations are necessary, or that they may even be omitted at all. In Section 4 we report on some applications to real data sets.

In the concluding remarks, it is pointed out how the conditional filtering and smoothing algorithms of this paper could be used for non- and semiparametric methods, and how the approach can be extended to general non-Gaussian state space models without major difficulties. A case of specific interest could be conditional mode filtering for robust state space models with non-Gaussian error distribution.

2. State space models for categorical time series. In this section we first consider some models for univariate and multivariate categorical time series in more detail, and mention a few other ones. The focus is on dynamic categorical observation models, since we prefer to exploit the simplicity of linear transition models as far as this is possible. A rather general model, subsuming all specific models, is formulated at the end of the section together with some basic assumptions.

Univariate categorical time series. Let us first consider the simplest yet basic type of an individual categorical time series $\{y_t, t \geq 1\}$, where y_t is a single observation at time t on a univariate categorical response variable y with $m \geq 2$ categories. Setting $q = m - 1$, observations can be described by the vector $y_t = (y_{t1},\ldots,y_{tq})'$ where

$$(2.1) \qquad y_{tj} = \begin{cases} 1, & \text{category } j \text{ has been observed} \\ 0, & \text{otherwise}, \end{cases} \qquad j = 1,\ldots,q.$$

In parallel a sequence $\{x_t, t \geq 1\}$ of covariate vectors may be available, and $\{\beta_t, t \geq 0\}$ is a sequence of unobservable p-dimensional state or parameter vectors. Let

$$y_t^* = (y_1',\ldots,y_t')', \quad x_t^* = (x_1',\ldots,x_t')', \quad \beta_t^* = (\beta_0',\beta_1',\ldots,\beta_t')'$$

denote histories of responses, covariates and states up to time t.

State space observation models specify the conditional distribution of y_t, given the current state β_t, covariates x_t, and – possibly – past responses. Individual categorical responses y_t as above are completely characterized by the vector $\pi_t = (\pi_{t1}, \ldots, \pi_{tq})'$ of conditional probabilities

$$\pi_{tj} = Pr(y_{tj} = 1 | \beta_t, x_t^*, y_{t-1}^*), \quad j = 1, \ldots, q.$$

We allow for the following general specification of *observation models* for individual univariate categorical time series:

$$\pi_{tj} = h_{tj}(\beta_t, x_t^*, y_{t-1}^*), \quad j = 1, \ldots, q,$$

or in vector notation

(2.2) $$\pi_t = h_t(\beta_t, x_t^*, y_{t-1}^*), \quad t = 1, 2, \ldots,$$

where $h_t = (h_{t1}, \ldots, h_{tq})'$ is a measurable function of the conditioning variables, subject to $h_{tj} \in (0,1), j = 1, \ldots, q$ and $h_{t1} + \cdots + h_{tq} < 1$. Further more we will require that h_t is twice continuously differentiable with respect to β_t, with the possible exception of a finite or countable subset of $\mathbf{R}^p$.

This observation model is supplemented by a *Markovian transition model* for $\{\beta_t\}$, i.e. by specifying a transition density $p(\beta_t | \beta_{t-1})$ or, more general, a conditional transition density $p(\beta_t | \beta_{t-1}, y_{t-1}^*)$. For simplicity, we will assume linear transition models whenever they are compatible with the observation model. However, this may not always be the case, compare e.g. the cumulative logit model below.

A number of interesting models fit into this general framework. Perhaps most important are *dynamic regression models for categorical time series*. This family is obtained by specializing the observation model (2.2) to

(2.3) $$\pi_t = h(Z_t'\beta_t), \quad t = 1, 2, \ldots,$$

where $h : \mathbf{R}^r \to (0,1)^q$ is the *response function*, and Z_t is a $(p \times r)$-*design matrix* depending on lagged values y_{t-1}^* and on covariates. This design matrix is predetermined, since we assume that it is known before observation y_t is made. Dynamic modeling of $\{Z_t\}$ can be performed in the same way as for models with constant parameters, compare e.g. Fahrmeir and Kaufmann (1987), Kaufmann (1987), Zeger and Qaqish (1988, for binary time series). Let us consider some specific examples.

Logit models for unordered categories

The basic model is defined by

(2.4) $$\pi_{tj} = \frac{\exp(z_t'\beta_{tj})}{1 + \sum\limits_{k=1}^{q} \exp(z_t'\beta_{tk})},$$

where $\beta_{tj}, j = 1, \ldots, q$ are alternative-specific parameter vectors, and the design vector z_t is a function of covariates and a fixed number of past responses. Setting

$z'_t = (1, x'_t, y_{-1}, \ldots,)$ or $z'_t = (1, 0, x'_t, y_{t-1}, \ldots)$, similarly as in (1.6), (1.7), one obtains (autoregressive) logistic observation models. Moreover, interaction terms may be included. It should be noted that not only interactions between past responses alone can be of interest, but that interactions among covariates and past responses may be important as well. As an example consider logistic modeling for transition probabilities of nonstationary first order Markov chains as it has been used in the constant parameter setting e.g. by Garber (1989): Admitting time-varying parameters, this model is defined by specifying a logistic model for transitions from state i to state j, i.e.

$$(2.5) \qquad Pr(y_{tj} = 1 | y_{t-1,i} = 1, x_t) = \exp(x'_t \alpha_{tij}) / \left[1 + \sum_{k=1}^{q} \exp(x'_t \alpha_{tik}) \right],$$

where $\alpha_{tij}, i, j = 1, \ldots, q$ are transition-specific parameters. Setting $z'_t = (x_t \otimes y_{t-1})'$, where, $\otimes$ is the Kronecker symbol, and $\beta'_{tj} = (\alpha'_{t1j}, \ldots, \alpha'_{tqj})'$, the transition model (2.5) can be rewritten in the form (2.4).

The dynamic logit model (2.4) can be brought into the general form (2.3) by defining

$$\beta_t = \begin{bmatrix} \beta_{t1} \\ \vdots \\ \beta_{tq} \end{bmatrix}, Z'_t = \begin{bmatrix} z'_t & & 0 \\ & \ddots & \\ 0 & & z'_t \end{bmatrix},$$

and the response function in accordance with (2.4). The model can easily be extended to discrete choice situations, where alternatives are influenced by alternative-specific covariates $w_{tj}, j = 1, \ldots, q$, as follows:

$$\pi_{tj} = \exp(z'_t \beta_{tj} + w'_{tj} \gamma_t) / \left[1 + \sum_{k} \exp(z'_t \beta_{tk} + w'_{tk} \gamma_t) \right].$$

Enlarging β_t above by γ_t, and defining

$$Z'_t = \begin{bmatrix} z'_t & & 0 & w'_{t_1} \\ & \ddots & & \\ 0 & & z'_t & w'_{tq} \end{bmatrix},$$

such dynamic discrete choice models can again be stated in the form (2.3).

Cumulative models for ordered categories

These models can be derived from a threshold mechanism for an underlying linear dynamic model. The resulting response probabilities are

$$(2.6) \qquad \pi_{tj} = F(\mu_{tj} - z'_t \alpha_t) - F(\mu_{t,j-1} - z'_t \alpha_t), \quad j = 1, \ldots, q,$$

with ordered threshold parameters $-\infty = \mu_{t0} < \mu_{t1} < \cdots < \mu_{tq} < \infty$ and a global parameter vector α_t, a design vector z_t defined similarly as above, and F a known distribution function, e.g. the logistic or standard Gaussian distribution function. For identifiability reasons z_t must not contain the constant 1. Defining

$$\beta_t = \begin{bmatrix} \mu_{t1} \\ \vdots \\ \mu_{tq} \\ \alpha_t \end{bmatrix}, Z_t' = \begin{bmatrix} 1 & & z_t' \\ & \ddots & \vdots \\ & 1 & z_t \end{bmatrix},$$

and the response function in accordance with (2.6), the model is again in the form (2.3). Threshold parameters may be constant or time-varying. For $q \geq 2$, the order restriction $\mu_{t1} < \cdots < \mu_{tq}$ has to be kept in mind. If thresholds vary according to one of the stochastic trend models in the introduction, this ordering can be destroyed with positive probability. This is not a problem in practice as long as thresholds are clearly separated and variances of the errors are small. However, the problem can be overcome by an appropriate reparameterization: Introducing the parameter vector $\tilde{\mu}_t = (\tilde{\mu}_{t1}, \ldots, \tilde{\mu}_{tq})'$, thresholds may be reparameterized by

$$\mu_{t1} = \tilde{\mu}_{t1}, \quad \mu_{tr} = \mu_{t1} + \sum_{s=2}^{r} \exp(\tilde{\mu}_{ts}), \quad r = 2, \ldots, q,$$

or equivalently by

$$\tilde{\mu}_{t1} = \mu_{t1}, \quad \tilde{\mu}_{tr} = \ln(\mu_{tr} - \mu_{t,r-1}).$$

Then $\tilde{\mu}_t$ may vary free in $\mathbf{R}^q$, and the order restriction for μ_t is fulfilled automatically. Another possibility would be to model the differences $\mu_{tr} - \mu_{t,r-1}$ directly by some nonnegative transition density.

Sequential models

Models of this type are well-known in probabilistic choice situations, where the decision for a certain alternative consists of a finite number of hierarchically ordered steps, see e.g. Maddala (1983, ch. 3). For simplicity, let us consider two-step models: Suppose that the set of categories $\{1, \ldots, m\}$ is partitioned into a smaller number of classes, say $C_1, \ldots, C_r$, $r < m$, where classes may e.g. contain "similar" categories. For example, in Markov chain terminology, $C_1, \ldots, C_r$ may be recurrent or transient classes (compare Dersimonian and Baker, 1988, in the context of discrete time competing risks models). In two-step models the first decision for one of the classes C_i, $i = 1, \ldots, r$ follows a conventional categorical response model (e.g. an unordered or ordered logit model). Conditional upon choice of C_i, the second step consists of a categorical response model for the categories within C_i. The resulting two-step response mechanism is also a categorical response model of the form (2.3), but we omit further details.

All the models discussed above share the common property of (2.3) that parameters β_t appear in form of a finite number of linear predictors $\eta = (\eta_1, \ldots, \eta_q) = Z_t'\beta_t$. The general model (2.2) does not make such a restriction. Therefore it covers also models where predictors are nonlinear functions of parameters, or models with additional parameters as e.g. nested logit models or generalized extreme-value models with "similarity" parameters.

Up to now we have assumed that y_t is a single categorical observation on the response variable y. In many applications y_t consists of a number n_t of aggregated repeated observations on y with the same value x_t for the covariates. In this case we describe the data by the vector $y_t = (y_{t1}, \ldots, y_{tq})'$ of relative frequencies of observations falling into the respective categories together with the known repetition number n_t.

The simplest observation models for time series of repeated categorical responses are *multinomial models*: If repeated single observations at time t are (conditionally) independent and have the same (conditional) response probabilities π_t, then the distribution of the vector y_t of relative frequencies is (scaled) multinomial with parameters $\pi_{t1}, \ldots, \pi_{tq}, n_t$, and it suffices again to specify π_t as in (2.2) or (2.3).

However, the multinomial assumption will not always be appropriate, as e.g. in the presence of overdispersion. The simplest model is obtained by introducing an additional multiplicative overdispersion parameter ϕ to the common covariance matrix of the multinomial distribution. Alternatively more sophisticated approaches are available: In particular, compound multinomial distributions (Johnson and Kotz, 1969) or an extension of univariate double exponential families (Efron, 1988) to multicategorical observations should be of potential use for modeling the conditional observation distribution $p(y_t|\beta_t, y_{t-1}^*, x_t^*)$. Then it will *not* suffice to specify response probabilities π_t by (2.2), and additional modeling, e.g. of the variance function, will be necessary.

Multivariate categorical time series. Let now $y_t = (y_{t1}, \ldots, y_{ti}, \ldots, y_{tn})$ denote a vector of multicategorical responses observed at time t. Depending on the situation, different models can be useful.

If the components $y_{ti}, i = 1, \ldots, n$ are observations on the same response variable y for a cross-section of units $i, i = 1, \ldots, n$, as e.g. in panel or discrete time survival data, it can be reasonable to work with the assumption of *conditional independence* of individual responses y_{ti} within y_s, given β_t, y_{t-1}^* and x_t^*, i.e.

$$(2.7) \qquad p(y_t|\beta_t, y_{t-1}^*, x_t^*) = \prod_i p(y_{it}|\beta_t, y_{t-1}^*, x_t^*), \quad t = 1, 2, \ldots$$

This conditional independence assumption corresponds to independence assumptions on observations in purely cross-sectional situations. For longitudinal data, (2.7) allows for interaction via parameters, covariates, and the common "history" y_{t-1}^*.

If the components of y_t are conditionally correlated, as e.g. in a cluster of categorical observations, *conditional* categorical response models may be considered. For binary data and constant parameters, conditional logit models have been proposed e.g. by Connolly and Liang (1988) and Liang and Zeger (1989). In these models, the conditional distribution of y_{it}, given the remaining current responses $y_{-ti} = \{y_{tj}, j \neq i\}$, as well as past responses and covariates, is specified. Unfortunately, the joint conditional distribution $p(y_t|\beta_t, y_{t-1}^*, x_t^*)$ becomes rather complex due to a complicated norming constant involving β_t. Therefore, a pseudolikelihood estimation procedure is proposed, where the genuine likelihood is replaced by the product of the conditional densities of the y_{ti}'s.

Both approaches above are symmetric with respect to responses y_{ti} in y_t. In other situations, where components are hierarchically ordered in a natural way, asymmetric modelling may be preferable. As an example consider bivariate responses (y_{t1}, y_{t2}), where a decision for $y_{t1} = i, y_{t2} = j$ may be the result of two steps: first a decision for $y_{t1} = i$ is made, followed by a decision for one of the categories of y_{t2}, conditional upon $y_{t1} = i$. This corresponds to sequential models considered in (2.1), and asymmetric models for multivariate categorical responses can be constructed by the same principles.

A general state space model. In any case, all models have the same general structure: The *observation model* is specified by conditional densities

$$(2.8) \qquad p(y_t|\beta_t, y_{t-1}^*, x_t^*), \quad t = 1, 2, \ldots,$$

and the *transition model* is defined by transition densities

$$(2.9) \qquad p(\beta_t|\beta_{t-1}y_{t-1}^*, x_t^*), \quad t = 1, 2, \ldots,$$

and an initial density $p(\beta_0)$. More common but less general, transition densities are assumed to be unconditional with respect to y_{t-1}^*, x_t^*, i.e. they are defined by $p(\beta_t|\beta_{t-1})$. As already mentioned earlier, we consider linear transition equations as the most important specification of (2.9). More general than in (1.2), the errors v_t may have a (possibly) non-Gaussian density $f(v)$, with $E(v) = 0$. The transition density is then given by

$$(2.10) \qquad p(\beta_t|\beta_{t-1}) = f(\beta_t - T_t\beta_{t-1}), \quad t = 1, 2, \ldots$$

To specify the bivariate process $\{y_t, \beta_t\}$ completely in terms of genuine joint densities, additional basic assumptions, typical for state space models in general, are required.

A1. Conditional on β_t and (y_{t-1}^*, x_t^*), current observations y_t are independent of β_{t-1}^*, i.e.

$$p(y_t|\beta_t^*, y_{t-1}^*, x_t^*) = p(y_t|\beta_t, y_{t-1}^*, x_t^*), \quad t = 1, 2, \ldots$$

This conditional independence assumption is basic in state space models. In the more conventional conditionally Gaussian setting (e.g. Sage and Melsa, 1971) it is implied by the error structure of the parameter and observation model. It states that, given y_{t-1}^*, x_t^*, the current parameter β_t contains the same information on y_t as the complete sequence β_t^*.

A2. Conditional on y_{t-1}^*, x_{t-1}^*, covariates x_t are independent of β_{t-1}^*, i.e.

$$p(x_t|\beta_{t-1}^*, y_{t-1}^*, x_{t-1}^*) = p(x_t|y_{t-1}^*, x_{t-1}^*), \quad t = 1, 2, \ldots$$

Loosely speaking, A2. means that the covariate process contains no information on the parameter process. It can be omitted for deterministic covariates.

In (conditionally) Gaussian filtering the noise processes of the observation and the transition equations (1.1), (1.2) are usually assumed to be independent. In our more general setting, this assumption has to be replaced by a further conditional independence assumption:

A3. $$p(\beta_t|\beta_{t-1}^*, x_t^*, y_{t-1}^*) = p(\beta_t|\beta_{t-1}, x_t^*, y_{t-1}^*), \quad t = 1, 2, \ldots$$

i.e. the state process is conditionally Markovian.

To describe the conditional mode filtering and smoothing algorithms of the next section in condensed form, we introduce the following notation. It is tacitly assumed that the usual regularity and smoothness conditions appropriate in the context of maximum likelihood estimation hold. The conditional loglikelihood contribution of y_t (or only its kernel) is

(2.11) $$\ell_t(\beta_t) = \log p(y_t|\beta_t, y_{t-1}^*, x_t^*),$$

(2.12) $$r_t(\beta_t) = \partial \ell_t(\beta_t)/\partial \beta_t$$

is the score function, and

(2.13) $$R_t(\beta_t) = -\partial^2 \ell_t(\beta_t)/\partial \beta_t \partial \beta_t' \text{ or } E(-\partial^2 \ell_t(\beta_t)/\partial \beta_t \partial \beta_t'|\beta_t, y_{t-1}^*, x_t^*)$$

is a generic symbol for the random information or the conditional information. The latter is often easier to evaluate and has the advantage of being positive semidefinite under regularity assumptions. For univariate multinomial models with response probabilities defined by (2.2), one obtains

(2.14) $$r_t(\beta_t) = H_t(\beta_t)\Sigma_t^{-1}(\beta_t)(y_t - \pi_t(\beta_t)),$$

with $\pi_t(\beta_t) := h_t(\beta_t, y_{t-1}^*, x_t^*), H_t(\beta_t) = \partial h_t(\beta_t)/\partial \beta_t$ and $\Sigma_t(\beta_t)$ the covariance matrix of the multinomial distribution, inserting $\pi_t = \pi_t(\beta_t)$. The conditional information is given by

(2.15) $$R_t(\beta_t) = H_t(\beta_t)\Sigma_t^{-1}(\beta_t)H_t'(\beta_t).$$

For conditionally independent multivariate models it follows from (2.7) that the contribution of y_t is simply the sum of individual contributions $y_{ti}, i = 1, \ldots, n$:

$$\ell_t(\beta_t) = \sum_i \ell_{ti}(\beta_t), \quad r_t(\beta_t) = \sum_i r_{ti}(\beta_t), \quad R_t(\beta_t) = \sum_i R_{ti}(\beta_t).$$

For product-multinomial models the individual score functions and information matrices have the forms (2.14), (2.15).

Similar notation will be convenient for the transition part of the model: Suppressing y_{t-1}^*, x_t^* notationally, loglikelihoods (or only kernels) are denoted by

(2.16) $$a_t(\beta_t, \beta_{t-1}) = \log p(\beta_t|\beta_{t=1}), a_0(\beta_0) = \log p(\beta_0).$$

Furthermore we define

(2.17)
$$
\begin{aligned}
c_t(\beta_t, \beta_{t-1}) &= -\partial \log p(\beta_t|\beta_{t-1})/\partial\beta_t \\
e_t(\beta_{t+1}, \beta_t) &= -\partial \log p(\beta_{t+1}|\beta_t)/\partial\beta_t \\
C_t(\beta_t, \beta_{t-1}) &= -\partial^2 \log p(\beta_t|\beta_{t-1})/\partial\beta_t\partial\beta_t' \\
E_t(\beta_{t+1}, \beta_t) &= -\partial^2 \log p(\beta_{t+1}|\beta_t)/\partial\beta_t\partial\beta_t'.
\end{aligned}
$$

For linear transition models (2.10) the loglikelihoods are given by

$$
a_t(\beta_t, \beta_{t-1}) = \log f(\beta_t - T_t\beta_{t-1}).
$$

Defining

(2.18)
$$
g(u) = -\partial \log f(u)/\partial u, \; G(u) = -\partial^2 \log f(u)/\partial^2 u,
$$

the derivatives are obtained immediately from (2.17), e.g.

$$
c_t(\beta_t, \beta_{t-1}) = g(\beta_t - T_t\beta_{t-1}), \; C_t(\beta_t, \beta_{t-1}) = G(\beta_t - T_t\beta_{t-1}), \; \text{etc.}
$$

For linear Gaussian transition models with positive definite covariance matrices $Q_t, t \geq 0$, the loglikelihood kernels are

(2.19)
$$
\begin{aligned}
a_t(\beta_t, \beta_{t-1}) &= -\frac{1}{2}(\beta_t - T_t\beta_{t-1})'Q_t^{-1}(\beta_t - T_t\beta_{t-1}) \\
a_0(\beta_0) &= -\frac{1}{2}(\beta_0 - \xi_0)'Q_0^{-1}(\beta_0 - \xi_0),
\end{aligned}
$$

and the derivatives are

(2.20)
$$
\begin{aligned}
c_t &= Q_t^{-1}(\beta_t - T_t\beta_{t-1}), t \geq 1, c_0 = Q_o^{-1}(\beta_0 - \xi_0) \\
e_t &= -T_{t+1}'Q_{t+1}^{-1}(\beta_{t+1} - T_t\beta_t) \\
C_t &= Q_t^{-1}, E_t = T_t'Q_t^{-1}T_t.
\end{aligned}
$$

3. Conditional mode filtering and smoothing. In addition to states $\{\beta_t\}$, the models of Section 2 generally contain "hyperparameters" θ, i.e. unknown non-random quantities, such as initial values, transition and covariance matrices of the transition model.

We first consider estimation of β_t^* for known or given hyperstructural parameters, since it forms an essential part of the joint estimation procedure for β_t^* and θ. Given the data y_t^*, x_t^*, estimates of β_t^* will be based on the conditional density $f(\beta_t^*|y_t^*, x_t^*)$, suppressing θ notationally. Since this density is not available in closed form Bayesian analysis would require numerical integration. With current techniques (e.g. Kitagawa, 1987, for scalar states and observations), this approach becomes computationally crucial for larger data sets, even for moderate dimensions p of β_s. In analogy to the estimation of random-effects models for categorical data (e.g. Stiratelli, Laird and Ware, 1984), we therefore propose to estimate parameters by maximization of the conditional density. Repeated application of Bayes' theorem yields

$$p(\beta_t^* | y_t^*, x_t^*) =$$

$$\prod_{s=1}^{t} p(y_s | \beta_s^*, y_{s-1}^*, x_s^*) \prod_{s=1}^{t} p(\beta_s | \beta_{s-1}^*, y_{s-1}^*, x_s^*) \prod_{s=1}^{t} p(x_s | \beta_{s-1}^*, y_{s-1}^*, x_s^*) / p(y_t^*, x_t^*).$$

Using A1, A2, A3 we obtain

$$p(\beta_t^* | y_t^*, x_t^*) \propto \prod_{s=1}^{t} p(y_s | \beta_s, y_{s-1}^*, x_s^*) \prod_{s=1}^{t} p(\beta_s | \beta_{s-1}).$$

Maximization of the conditional density is thus equivalent to maximizing the log-posterior

(3.1)
$$L_t^*(\beta_t^*) = \sum_{s=1}^{t} l_s(\beta_s) + a_0(\beta_0) + \sum_{s=1}^{t} a_s(\beta_s, \beta_{s-1}),$$

where the contributions l_s, a_s are defined by (2.11), (2.16). The criterion L_t^* can be interpreted as a penalized loglikelihood, with the sum of the log-priors a_s acting as a roughness penalty.

Score function and information matrix. For estimation of $\beta_t^* = (\beta_o', \ldots, \beta_t')'$, the score function $u_t^*(\beta_t^*) = \partial L_t^*(\beta_t^*)/\partial \beta_t^*$ and the random information $U_t^*(\beta_t^*) = -\partial^2 L_t^*(\beta_t^*)/\partial \beta_t^* \partial \beta_t^{*'}$, or its conditional expectation, are of interest. Due the recursive nature of state space models, u_t^* and U_t^* have a special structure. The score function can be partitioned as

$$u_t^* = (u_o', \ldots, u_t')'.$$

The subvectors are given by

(3.2)
$$u_o = -c_o - e_1$$
$$u_s = r_s - c_s - e_{s+1}, \quad s = 1, \ldots, t-1$$
$$u_t = r_t - c_t,$$

with r_s, c_s, e_s defined by the derivatives in (2.12), (2.17). The subvectors u_s, $s = 1, \ldots, t-1$, depend only on β_{s-1}, β_s and β_{s+1}, u_o only on β_o, β_1, and u_t only on β_{t-1}, β_t. The information matrix therefore is block-tridiagonal and symmetric:

(3.3)
$$U_t^* = \begin{bmatrix} U_{00} & U_{01} & & & \\ & & \ddots & & 0 \\ U_{01}' & U_{11} & & \ddots & \\ & \ddots & & \ddots & \\ & & & & U_{t-1,t} \\ 0 & & \ddots & & \\ & & U_{t-1,t}' & & U_{tt} \end{bmatrix}$$

with blocks given by

$$U_{oo} = C_o + E_1, \quad U_{tt} = R_t + C_t$$
$$U_s = R_s + C_s + E_{s+1}, \quad s = 1, \ldots, t-1,$$
$$U_{s-1,s} = -\partial^2 \log p(\beta_s|\beta_{s-1})/\partial\beta_{s-1}\partial\beta'_s \quad s = 1, \ldots t.$$

Assuming the existence of all derivatives as well as positive definiteness of C_s, E_s, the information matrix is positive definite. Then U_t^* can be uniquely factorized into

(3.4)
$$\begin{bmatrix} I & & & \\ -B'_1 & I & & 0 \\ & -B'_2 & \ddots & \\ 0 & & \ddots & \\ & & -B'_t & I \end{bmatrix} \begin{bmatrix} D_0 & & & 0 \\ & D_1 & & \\ & & \ddots & \\ 0 & & & D_t \end{bmatrix} \begin{bmatrix} I & -B_1 & & 0 \\ & I & -B_2 & \\ & & \ddots & \\ 0 & & & \ddots & -B_t \\ & & & & I \end{bmatrix}$$

with $D_o, \ldots, D_t$ positive definite, compare Golub and van Loan (1989, ch. 4). Multiplying out and comparing with corresponding blocks in U_t^*, it is easily seen that $D_o = U_{oo}$ and

(3.5)
$$-B_s = D_{s-1}^{-1}U_{s-1,s}$$
$$D_s = U_{ss} + B'_sU_{s-1,s}.$$

In Fahrmeir and Kaufmann (1991), B_s and D_s are computed in a seemingly different way by a covariance matrix recursion. Due to the uniqueness of the factorization (3.4), this is equivalent to the recursion (3.5). Moreover, their Proposition 1 (iii) remains valid, so that the diagonal blocks $A_{ss}, s = 1, \ldots, t$ of the inverse $(U_t^*)^{-1}$ can be computed by backward recursions:

(3.6)
$$A_{tt} = D_t^{-1}, A_{s-1,s-1} = D_{s-1}^{-1} + B_sA_{ss}B'_s, \quad s = t, \ldots, 1.$$

Gauss-Newton and Fisher scoring iterations. A maximizer of $L_t(\beta_t^*)$ can be found by Gauss-Newton or Fisher scoring iterations. Let $\hat{\beta}_t^*$ denote the current iterate. Then the next iterate is $\hat{\beta}_t^* + \delta_t^*$, where $\delta_t^* = (\delta'_o, \ldots, \delta'_t)'$ solves

(3.7)
$$[U_t^*(\hat{\beta}_t^*)]\delta_t^* = u_t^*(\hat{\beta}_t^*).$$

If U_t^* is the random information, then (3.7) is a Gauss-Newton, otherwise a Fisher scoring iteration. Due to the decomposition (3.4), (3.7) can be solved by forward-backward recursions which avoid explicit inversion of U_t^* and results in an algorithm of complexity $0(t)$. First one solves

$$\begin{bmatrix} I & & & \\ -B'_1 & I & & \\ & -B'_2 & \ddots & \\ 0 & & \ddots & \\ & & & -B'_t \end{bmatrix} \begin{bmatrix} \epsilon_o \\ \epsilon_1 \\ \vdots \\ \epsilon_t \end{bmatrix} = \begin{bmatrix} u_0 \\ u_1 \\ \vdots \\ u_t \end{bmatrix}$$

for the auxiliary vector $\epsilon_t^* = (\epsilon_o', \ldots, \epsilon_t')'$ by forward recursions, and then

$$\begin{bmatrix} D_o & -D_oB_1 & & 0 \\ & D_1 & \ddots & \\ & & \ddots & -D_{t-1}B_t \\ 0 & & & -D_t \\ 0 \end{bmatrix} \begin{bmatrix} \delta_o \\ \delta_1 \\ \vdots \\ \delta_t \end{bmatrix} = \begin{bmatrix} \epsilon_o \\ \epsilon_1 \\ \vdots \\ \epsilon_t \end{bmatrix}$$

by backward recursions. This results in the following *Gauss-Newton (Fisher scoring) step*:

1. Initialization: $\epsilon_o = u_o, D_o = U_{oo}$,

2. Forward recursions: For $s = 1, \ldots, t$

$$B_s = -D_{s-1}^{-1}U_{s-1,s}$$
$$D_s = U_{ss} + B_s'U_{s-1,s}$$
$$\epsilon_s = u_s + B_s'\epsilon_{s-1}.$$

3. Filter correction:

$$\delta_t = D_t^{-1}\epsilon_t.$$

4. Smoother corrections: For $s = t, \ldots, 1$

$$\delta_{s-1} = D_{s-1}^{-1}\epsilon_{s-1} + B_s\delta_s.$$

After convergence, a numerical approximation $\hat{\beta}_{0|t}, \ldots, \hat{\beta}_{s|t}, \ldots, \hat{\beta}_{t|t}$ to the conditional mode is obtained. Additionally, corresponding curvatures $V_{0|s}, \ldots, V_{s|t}, \ldots, V_{t|t}$, replacing conditional error covariance matrices, are available by setting

$$V_{s|t} := A_{ss}, \quad s = 0, \ldots, t,$$

and computing A_{ss} by the backward recursions (3.6).

Extended Kalman filtering and smoothing. Fahrmeir (1990) proposes a generalization of the well-known extended Kalman filter for (conditionally) Gaussian observation models to exponential family observation models, retaining a linear Gaussian transition model. It can be derived as a recursive gradient algorithm for maximizing $L_t^*(\beta_t^*)$, following the line of arguments of Sage and Melsa (1971, pp. 447) in their derivation of the common extended Kalman filter for Gaussian observations. In addition, Fahrmeir and Kaufmann (1991) show that it can be considered as a simplifying approximation of the Fisher scoring iterations above. Arguments carry over to state space models considered in this paper, including non-exponential family observations. For linear transition models (2.10) the final result is the following *generalized extended Kalman filter and smoother*:

1. Initialization: For $t = 1, 2, \ldots$

$$\hat{\beta}_{o|o} = \alpha_o, V_{o|o} = Q_o.$$

2. Filter prediction steps:

$$\hat{\beta}_{t|t-1} = T_t\hat{\beta}_{t-1|t-1}$$
$$V_{t|t-1} = T_tV_{t-1|t-1}T'_t + Q_t,$$

where Q_o, Q_t are covariance matrices or curvatures corresponding to $p(\beta_o)$, $f(u)$.

3. Filter correction steps:

$$V_{t|t} = (V^{-1}_{t|t-1} + R_t)^{-1}$$
$$\hat{\beta}_{t|t} = \hat{\beta}_{t|t-1} + V_{t|t}r_t$$

4. Smoother steps: For $s = t, \ldots, 1$

$$\hat{\beta}_{s-1|t} - \hat{\beta}_{s-1|s-1} = B_s(\hat{\beta}_{s|t} - \hat{\beta}_{s|s-1})$$
$$V_{s-1|t} - V_{s-1|s-1} = B_s(V_{s|t} - V_{s|s-1})B'_s$$

with

$$B_s = V_{s-1|s-1}T'_sV^{-1}_{s|s-1}.$$

The score function r_t and the information matrices R_t, defined by (2.12), (2.13), are evaluated at $\hat{\beta}_{t|t-1}$, as well as Q_o and Q_t.

In our experience this algorithm is a good compromise between computational simplicity and numerical approximation quality for common categorical response models, which posses rather smooth response functions h. We recommend the following procedure: First apply the extended Kalman filter and smoother. Then use it as the initial solution for the Gauss-Newton (Fisher scoring) iterations. In many situations, the algorithm will stop after only one or two iterations, compare the examples in Section 4.

In our derivation of conditional mode estimators we have assumed positive definiteness of second derivatives C_t, E_t in (2.17). For linear Gaussian transition models this is equivalent to positive definite covariance matrices $Q_t, t \geq 0$. This restriction would rule out e.g. semiparametric models with a trend component following a second order random walk and with constant covariate effects. For both algorithms, the restriction to positive definite covariance matrices made above can be dropped in the following way: Without loss of generality, suppose that covariance matrices are block structured as

$$Q_t = \begin{bmatrix} \tilde{Q}_t & 0 \\ 0 & 0 \end{bmatrix}, \tilde{Q}_t \text{ positive definite }.$$

Define

(3.8) $$Q_t(\delta) = \begin{bmatrix} \tilde{Q}_t & 0 \\ 0 & \delta I \end{bmatrix}, \quad \delta > 0$$

and apply the algorithms, using $Q(\delta)$ instead of Q. Carrying out the limit operation $\delta \to 0$, algorithms for Q_t positive semidefinite can be obtained. For the extended Kalman filter, the algorithm above remains unchanged, but the restriction to positive definite Q_t can be dropped. Alternatively, Gauss-Newton iterations may be carried out numerically with sufficiently small δ in (3.8), compare Section 4.

Estimation of hyperparameters. We restrict discussion to the simpler situation of linear Gaussian transition models with unknown initial values ξ_o, Q_o and unknown error covariance matrix $Q_t = Q$ (independent of t). Following analogous proposals made in the related situation of empirical Bayes estimation in random effect models for categorical data (Stiratelli, Laird and Ware, 1984), we suggest to use a variant of the EM algorithm for joint estimation of β_t^*, ξ_o, Q_o and Q, thereby substituting conditional means and covariance matrices by conditional modes and curvatures obtained from the filtering and smoothing algorithm above. For the time series situation the resulting estimation algorithm is used also by Fahrmeir (1990) and has been implemented and studied in detail by Goss (1990). The joint estimation procedure can be summarized as follows:

1. Choose appropriate starting values $\xi_o^{(o)}, Q_o^{(o)}, Q^{(o)}$. For $k = 0, 1, 2, \ldots$

2. Smoothing: Compute $\hat{\beta}_{s|t}^{(k)}, V_{s|t}^{(k)}, s = 1, \ldots, t$ by extended Kalman filtering and smoothing, with unknown parameters ξ_o, Q_o, Q replaced by $\xi_o^{(k)}, Q_o^{(k)}, Q^{(k)}$.

3. EM-step: Compute $\xi_o^{(k+1)}, Q_o^{(k+1)}, Q^{(k+1)}$ by

$$\xi_o^{(k+1)} = \hat{\beta}_{o|t}^{(k)}, \quad Q_o^{(k+1)} = V_{o|t}^{(k)},$$

$$Q^{(k+1)} = \frac{1}{t} \sum_{s=1}^{t} \left[\left(\hat{\beta}_{s|t}^{(k)} - T_s \hat{\beta}_{s-1|t}^{(k)} \right) \left(\hat{\beta}_{s|t}^{(k)} - T_s \hat{\beta}_{s-1|t}^{(k)} \right)' + V_{s|t}^{(k)} \right.$$

$$\left. - T_s B_s^{(k)} V_{s|t}^{(k)} - V_{s|t}^{(k)'} B_s^{(k)'} T_s' + T_s V_{s-1|t}^{(k)} T_s' \right]$$

with $B_s^{(k)}$ defined as in the smoothing steps.

Other approaches to estimation of hyperparameters, e.g. cross-validation, may be considered as well. This is deferred to subsequent work. In the empirical examples of the next section hyperparameters have been estimated by the above algorithm.

4. Some applications. For illustration and for comparison, we first apply conditional mode filtering and smoothing to binomial time series already analyzed in the literature by different methods.

Rainfall Data (from Kitagawa, 1987). Figure 1 displays the number of occurences of rainfall in the Tokyo area for each calendar day during the years 1983–1984. To obtain a smoothed estimate of the probability π_t of occurence of rainfall on calendar day $t, t = 1, \ldots, 366$, Kitagawa chose the following simple binomial logit model:

$$\pi_t = h(\beta_t) = \exp(\beta_t)/(1 + \exp(\beta_t)), \quad n_t = 2$$
$$\beta_{t+1} = \beta_t + v_t, \quad v_t \sim N(0, \sigma^2).$$

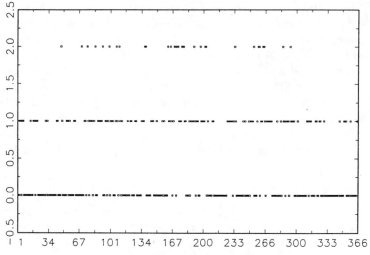

Figure 1: Rainfall data, Tokyo, 1983–84.

Figure 2 shows corresponding estimates $\hat{\pi}_t = h(\hat{\beta}_{t|366})$, resulting from conditional mode estimates $\hat{\beta}_{t|366}$ of the generalized extended Kalman smoother (dotted line) and the Gauss-Newton smoother, which stopped after two further iterations. Comparison with Kitagawa's conditional mean estimates reveals only minor departures for the initial time period, while estimates are almost identical later. This leads to the conjecture that posterior distributions become approximately normal for longer observation periods.

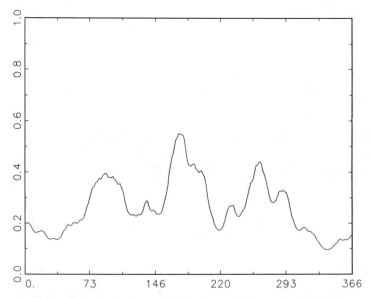

Figure 2: Smoothed probabilities $\hat{\pi}_t$,.. .... "extended" Kalman smoother, —— Gauss-Newton smoother.

103

Advertising data. West et al. (1985) analyzed a time series of numbers y_t of those people, out of a sample of $n = 66$, who gave a positive response to the weekly advertising of a chocolate bar in a TV commercial. As a measure of influence, an "adstock coefficient" serves as a covariate x_t. Figure 3 contains the data and fitted values, obtained from conditional mode smoothing with the binomial logit model

$$\pi_t = \exp(\mu_t + x_t\alpha_t)/(1 + \exp(\mu_t + x_t\alpha_t)), \quad n_t = 66,$$
$$\beta_t = (\mu_t, \alpha_t)' = \beta_{t-1} + v_t, \quad v_t \sim N(0, \text{diag}(\sigma_\mu^2, \sigma_\alpha^2)).$$

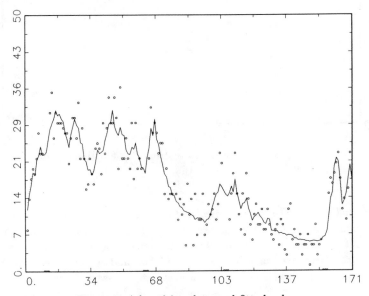

Figure 3: Advertising data and fitted values.

This model is closely related to the model used by West et al. The conditional mode smoother of μ_t in Figure 4 shows a slight decreasing trend, whereas the positive advertising effect α_t is almost constant in time. Therefore it seems reasonable to reanalyze the data with the observation model

$$\pi_t = \exp(\mu_t + x_t\alpha)/(1 + \exp(\mu_t + x_t\alpha)),$$

where the advertising effect is constant in time. The Gauss-Newton iterations are carried out by introducing an artificial random walk model for α with extremely small $\sigma_\alpha^2 = 10^{-14}$, compare Section 3. Figure 5 shows the estimate of α and μ_t. Compared to Figure 4, variations of α_t are now transferred to μ_t, leading to enlarged variations of the trend.

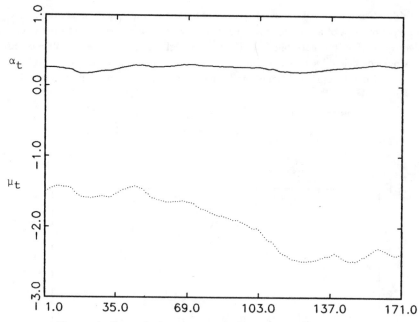

Figure 4: Smoothed estimates of trend μ_t and time-varying adstock effect α_t.

Longitudinal analysis of categorical business test data. In this example, we analyze multivariate categorical time series of monthly microdata, collected by the Ifo-Institut for economic research. The questionnaire of the "Ifo business test" describes the successive change of realizations, plans and expectations on variables like production, orders in hand, state of business, etc. Answers are categorical, most of them trichotomous with categories like "increase", "decrease", and "no change". The data consist in the categorical time series of 60 building trade firms, collected in the industrial branch "Steine and Erden", for the periods January 1980 to March 1988.

The response variable is formed by the production plans P_t. Its conditional distribution is assumed to depend on the covariates "orders in hand" A_t and "expected development of the state of business for the next 6 months" G_t, as well as on the production plans P_{t-1} of the previous month. (P, A and G are German mnemonics). No interaction effects are included. This choice is motivated by previous results of König, Nerlove, Oudiz (1982) and Nerlove (1983, p. 1273), using loglinear probability models. Each trichotomous ($m = 3$) variable is described by two ($q = 2$) dummy variables, with "decrease" as the reference category. Thus (1,0), (0,1) and (0,0) stand for the responses "increase", "no change" and "decrease". The cumulative logistic model (2.6) together with the conditional independence assumption (2.7) was chosen as a simple but plausible observation model. The vector β_t of structural time-varying parameters has dimension $p = 8$: it contains

105

2 threshold parameters μ_{t1}, μ_{t2} and 6 parameters corresponding to the dummies $G_t^1, G_t^2, A_t^2 A_t^2, P_{t-1}^1, P_{t-1}^2$, describing G_t, A_t, P_{t-1}. Parameter transitions are modelled by a random walk model with ξ_o, Q_o and $Q = Q_1 = \cdots = Q_t$ unknown, and Q restricted to be diagonal. Thus the EM steps in the joint estimation procedure can be performed as described at the end of the last section.

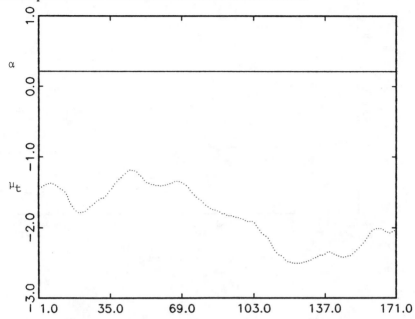

Figure 5: Smoothed estimates of trend μ_t and constant adstock effect α.

The following table contains the estimated variances for the parameters

thresholds		G_t^1	G_t^2	P_{t-1}^1	P_{t-1}^2	A_t^1	A_t^2
0.48	0.18	0.25	0.001	0.005	0.001	0.005	0.003

Smoothing estimates of the parameters corresponding to G_t^1, G_t^2 etc are displayed in Figure 6. It shows that an expected increase of production in the previous month and a positive state of business have a significant positive overall effect on current production plans, whereas the effect of increasing orders in hand is relatively small. Compared to the remaining effects, the parameter corresponding to the increase category G_t^1 of the covariate "state of business" has a remarkable temporal variation. The separate plot in Figure 7 exhibits a clear decline to a minimum at the beginning, and a distinct increase for the next 12 – 14 months. The second half of the increase period coincides with the first months of the new German government in autumn 1982, ending up with the elections to the German parliament in 1983. The growing positive effect of a positive state of business to the "increase" category of production plans indicates positive reactions of firms to the change of government.

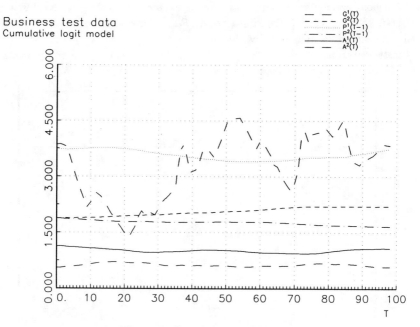

Figure 6: Covariate parameters.

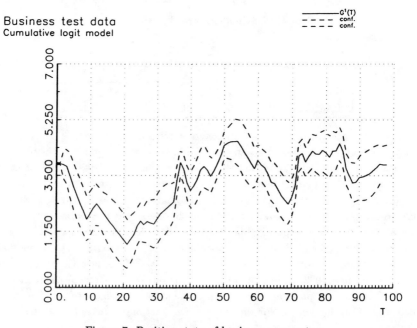

Figure 7: Positive state of business parameter.

Both threshold parameters (Figure 8) have a distinct seasonal pattern corresponding to successive years. Threshold parameter μ_{t1} has peaks about December/January, and lows in summer. An explanation may be the following: Firms in this industry branch manufacture initial products for the building industry. The variable "expected state of business" G_t serves as a substitute for the variable "expected demand", which is not contained in the questionnaire. To be able to satisfy increasing demand with the beginning of the building season in late winter, production plans are increased several weeks earlier with comparably high probability. Similarly, decreasing values of μ_{t1} in spring and low values in summer reflect the tendency not to increase the level of production any further. The seasonal variation of the second threshold parameter can be explained similarly.

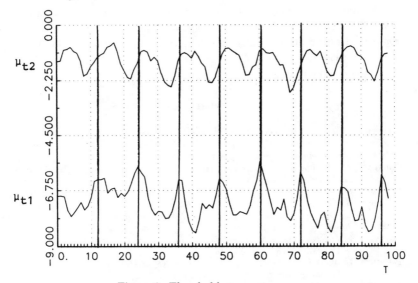

Figure 8: Threshold parameters.

5. Concluding remarks. There rest a number of remaining loose ends or interesting topics which may be considered for future research. Let me mention the following ones:

(i) There exist close connections to penalized likelihood estimation of semiparametric generalized linear models using spline functions, compare e.g. Green and Yandell (1985) and Green (1987). It is well known for linear Gaussian models how Kalman filtering and smoothing can be exploited to obtain efficient spline smoothing algorithms, compare e.g. Wecker and Ansley (1983), Kohn and Ansley (1987). It seems that the algorithms of this paper might have similar potential in the non-Gaussian case.

(ii) Going through Section 3, it is seen that essential use is made only of the general state space model at the end of Section 2, together with regularity assumptions which are typical for maximum likelihood estimation. This means that the conditional mode filtering and smoothing algorithms remain valid for general non-Gaussian state space models. A particular example are robustified linear state space models: For robustified observation equations with robust loglikelihood $l_t(\beta_t)$, the well-known ACM-filters and smoothers (compare Martin and Raferty, 1987) are identical to the extended Kalman filters and smoothers of this paper. This sheds new light on ACM filters: They can be understood as approximate conditional *mode* filters. An application of the Gauss-Newton algorithm to robustified state space models leads to conditional mode filters and smoothers which seem to be new and may be more accurate.

(iii) Experience with simulated and real data indicates that under appropriate assumptions and conditional densities $f(\beta_s|y_t^*, x_t^*)$ should be asymptotically normal. To my knowledge a rigorous proof of this conjecture is still missing.

(iv) Conditional *mode* estimators could be helpful starting solutions to obtain conditional mean estimators by numerical integration procedures, e.g. Gauss-Hermite Gauss quadrature (Schnatter, 1990), or Monte Carlo integration (e.g. Zeger and Karim, 1990, for random effects models, West, 1990, Carlin et al., 1990). In view of (iii) this might improve estimation for shorter time series.

REFERENCES

ANDERSON, B.D.D. AND MOORE, J.B., *Optimal filtering*, Prentice Hall, Englewood Cliffs (1979).

CARLIN, B. POLSON, N. AND STOFFER, D., *A Monte Carlo approach to nonnormal and nonlinear state space modelling*, Pittsburgh, preprint (1990).

CONNOLLY, M. AND LIANG, K., *Conditional logistic regression models for correlated binary data*, Biometrika 75 (1988), pp. 501–506.

DERSIMONIAN, R. AND BAKER, S., *Two-process models for discrete-time serial categorical response*, Statistics in Medicine 7 (1988), pp. 965–974.

EFRON, B.E., *Double exponential families and their use in generalized regression*, J. Amer. Statist. Ass. 81 (1988), pp. 709–721.

FAHRMEIR, L. AND KAUFMANN, H., *Regression models for nonstationary categorical time series*, J. Time Ser. Anal. 8 (1987), pp. 147–160.

FAHRMEIR, L., *Posterior mode estimation by extended Kalman filtering for multivariate dynamic generalized linear models*, to appear in JASA (1990).

FAHRMEIR, L. AND KAUFMANN, H., *On Kalman filtering, posterior mode estimation and Fisher scoring in dynamic exponential family regression*, Metrika (1991), pp. 37–60.

GARBER, A.M., *A discrete-time model of the aquisition of antibiotic-resistant infections in hospitalized patients*, Biometric 45 (1989), pp. 797–816.

GOLUB, H.G. AND VAN LOAN C.F., *Matrix Computation (2nd edition)*, North Oxford Academic, Oxford (1989).

GOSS, M., *Schätzung und Identifikation von Struktur-und Hyperstrukturparametern in dynamischen generalisierten linearen Modellen*, Dissertation, Universität Regensburg (1990).

GREEN, P., *Penalized likelihood for generalized semi-parametric regression models*, International Statistical Review 55 (1987), pp. 245–249.

GREEN, P. AND YANDELL, B.S., *Semi-parametric generalized linear models*, In: Gilchrist, R., Francis, B. and Whittaker, J. (ed.).; *Generalized linear models*, Lecture Notes in Statistics, Springer (1985).

HARVEY, A.C., *Forecasting, structural time series models and the Kalman filter*, Cambridge University Press, Cambridge (1989).

HARVEY, A.C. AND FERNANDES, C., *Time series models for count or qualitative observations*, Journal of Business and Economic Statistics 7 (1988), pp. 407–422.

JOHNSON, N. AND KOTZ S., *Discrete distributions*, Houghton Miffin, Boston (1969).

KAUFMANN, H., *Regression models for nonstationary categorical time series: Asymptotic estimation theory*, The Annals of Statistics 15 (1987), pp. 79–98.

KITAGAWA, G. AND GERSCH, W., *Smoothness priors – state space modelling of time series with trend and seasonality*, Journal of the American Statistical Association Vol. 79 (1984), pp. 378–389.

KITAGAWA, G., *Non-Gaussian state-space modelling of nonstationary time series (with comments)*, Journal of the American Statistical Association 82 (1987), pp. 1032–1063.

KÖNIG, H., NERLOVE, M., OUDIZ, G., *Analyse mikroökonomischer Konjunkturtest-Daten mit loglinearen Wahrscheinlichkeitsmodellen: Eine Einführung*, IFO-Studien 28 (1982), pp. 155–191.

KOHN, R. AND ANSLEY, C., *A new algorithm for spline smoothing based on smoothing a stochastic process*, SIAM Journal Sci. Stat. Comput. 8 (1987), pp. 33–48.

LIANG, K. AND ZEGER, S.L., *A class of logistic regression models for multivariate binary time series*, Journal of the American Statistical Association 84 (1989), pp. 447–451.

MADDALA, G., *Limited-dependent and qualitative variables in econometrics*, Cambridge University Press, Cambridge (1983).

MARTIN, R.D. AND RAFTERY, A.E., *Robustness, computation, and non-euclidian models (comment)*, Journal of the American Statistical Association 82 (1987), pp. 1044–1050.

NERLOVE M., *Expectations, plans and realizations in theory and practice*, Econometrica 51 (1983), pp. 1251–1279.

SAGE, A.P. AND MELSA, J.L., *Estimation theory with applications to communications and control*, McGraw Hill, New York (1971).

SCHNATTER, S., *Approximate inference with a dynamic generalized linear trend model*, Techn. Universität Wien, preprint (1990).

STIRATELLI, R., LAIRD, N. AND WARE, J.H., *Random-effect models for serial observations with binary response*, Biometrics 40 (1984), pp. 961–971.

WAHBA, G., *Improper prior, spline smoothing and the problem of guarding against model errors in regression*, Journal of the Royal Statistical Society, B,44, No. 3 (1978), pp. 364–372.

WECKER, W.E. AND ANSLEY, C.F., *The signal extraction approach to nonlinear regression and spline smoothing*, Journal of the American Statistical Association Vol. 78 (1983), pp. 81–89.

WEST, M., *Bayesian computations: Sequential analyses and dynamic models*, Duke University, preprint (1990).

WEST, M. AND HARRISON, J., *Bayesian forecasting and dynamic models*, Springer, New York (1989).

WEST, M., HARRISON, R.J., MIGON, H.S., *Dynamic generalized linear models and Bayesian forecasting*, Journal of the American Statistical Association 80 (1985), pp. 73–97.

ZEGER, S.L. AND QAQISH, B., *Markov regression models for time series: A quasi-likelihood approach*, Biometrics 44 (1988), pp. 1019–1031.

ZEGER, S.L. AND KARIM, R., *Generalized linear models with random effects: A Gibb's sampling approach*, Journal of the American Statistical Association 86 (1991), pp. 79–86.

RANK TESTS FOR TIME SERIES ANALYSIS
A SURVEY

MARC HALLIN* AND MADAN L. PURI†

Abstract. Rank-based testing procedures have proven quite efficient in classical linear models with independent observations; they long ago have entered daily practice in such fields as biostatistics or experimental planning. Still, despite the fact that many "historical" rank statistics (runs, signs of differences, turning point, Wald and Wolfowitz's rank autocorrelation coefficient, ...) actually were devised for time series situations, and despite the recognized need for nonparametric, robust or non Gaussian methods in the area, most time series analysts completely ignore rank–based procedures. Our objective with the present survey is to show that rank–based techniques very successfully can handle most of the testing problems occurring in time series context, such as testing for white noise, testing ARMA (p,q) dependence against ARMA $(p+d, q+d)$ dependence, or testing linear hypotheses about the coefficients of an ARMA model—always, under unspecified innovation density. The resulting tests of course are distribution–free. But they also are as powerful as (often, strictly more powerful than) their classical, correlogram–based counterparts. In addition, they are considerably more robust: whereas classical parametric methods can yield extremely misleading diagnostic information when the data have outliers, atypical startup behavior or heavy–tailed distributions, rank–based tests exhibit much better resistance to aberrations of this type. All these properties should make them extremely attractive, e.g. in the identification and diagnostic checking process, where the conclusions drawn from rank–based techniques are likely to be much more reliable than those resulting from an inspection of traditional correlograms.

We start (Section 1) with a general introduction, where we show how invariance arguments naturally lead to (signed or unsigned) rank–based methods. Section 2 provides a bibliographical survey of rank–based methods in serial dependence problems. The so–called linear serial rank statistics and their asymptotic distributions are introduced in Section 3. Section 4 deals with the basic theoretical result from which the optimal testing procedures described in the subsequent sections follow: local asymptotic normality of ARMA likelihood families and the local sufficiency of ranks. Locally most powerful and locally maximin rank–based tests are investigated in Section 5. Their asymptotic performance is discussed; explicit ARE (asymptotic relative efficiency) values (with respect to Gaussian parametric methods) are provided. Finally, Section 6 deals with the important problem of aligned rank tests: asymptotic invariant aligned rank tests, asymptotically most stringent within the class of all asymptotically similar tests at given probability level are derived for linear hypotheses about the coefficients of an ARMA model (with unspecified innovation density). The ARE of these tests with respect to the corresponding Gaussian Lagrange multiplier method is still the same as in Section 5, and can be as high as 2, e.g. when using median–test (or Laplace) scores under double–exponential innovation density.

AMS(MOS) subject classifications. Primary 62G10, 62M10

Key words and phrases. Rank tests, signed rank tests, invariance, aligned rank tests, permutation tests, nonhomogeneous white noise, linear trend, serial and nonserial linear rank statistics, local asymptotic normality, local asymptotic sufficiency, rank autocorrelations, locally optimal rank tests, asymptotic relative efficiencies, asymptotic invariance, asymptotically similar tests.

1. INVARIANCE, THE HYPOTHESIS OF WHITE NOISE AND RANKS

1.1 Testing for white noise: invariance arguments. Consider the null hypothesis H under which a series $\mathbf{X} = (X_1, \ldots, X_n)$ of observations is white

---

*Département de Mathématique and Institut de Statistique, C.P. 210, Université Libre de Bruxelles, 1050 Bruxelles, Belgium

†Department of Mathematics, Indiana University, Swain Hall East, Bloomington, Indiana 47405 U.S.A. Research supported by the Office of Naval Research Contract N00014–91–J–1020

noise. More precisely, let H denote the hypothesis under which observations $X_1, \ldots,$ $X_t, \ldots, X_n$ are independent and identically distributed, with unspecified density (all densities here are with respect to the Lebesgue measure on $\mathbf{R}$). As an alternative, take the extremely vast and general collection K of all n-dimensional distributions under which the observed X_t's are not white noise anymore—either because of distributional heterogeneity, serial dependence, or both.

The whole theory of rank tests originates in this almost trivial observation that, letting $g : \mathbf{R} \to \mathbf{R}$ denote an arbitrary, continuous, one–to–one (hence monotonic: without any loss of generality, it can be assumed order–preserving) transformation, $\mathbf{X} = (X_1, \ldots, X_n)$ is white noise if and only if $g(\mathbf{X}) = (g(X_1), \ldots, g(X_n))$ is. More rigorously, it is easy to see that the problem of testing H against K is invariant with respect to the group $\mathcal{G}$ of continuous, order–preserving transformations $g : \mathbf{x} \mapsto g(\mathbf{x})$.

The classical attitude in such a situation consists in restricting oneself to the consideration of invariant tests, i.e. tests which are measurable with respect to some maximal invariant statistic. And a maximal invariant statistic for H consists of the vector of ranks $\mathbf{R} = (R_1, \ldots, R_n)$, where R_t denotes the rank of X_t among $X_1, \ldots, X_n$. This is how ranks come into the picture: invariance is the theoretical cornerstone of rank–based inference. Distribution–freeness (hence similarity and unbiasedness), increased power, robustness, $\ldots$, however attractive they are, are just by–products, and generally can be reached through other, more specific techniques.

As a consequence, there is no point in using ranks if, under the null hypothesis to be tested, the quantities from which ranks are computed do not constitute a white noise series (or, at least, an exchangeable series).

The above invariance argument is well accepted in such statistical areas as life testing or experimental planning. It probably sounds highly unfamiliar, theoretical and abstract to most time series analysts. Time series analysis indeed is very deeply marked with hidden Gaussian assumptions — more deeply, perhaps, than any other statistical area: the $\mathcal{L}^2$ approach, pervasive use of autocovariances and autocorrelations, Gaussian likelihood or least squares estimates, Gaussian Lagrange multipliers, $\ldots$ all arise from (explicit or implicit) Gaussian assumptions. Though most of such Gaussian methods generally remain *asymptotically* valid under fairly general assumptions, it is usually agreed that the data preferably should be at least *approximately* Gaussian. Whenever they really do not look Gaussian , various preliminary transformations are generally performed before the analysis is started: logarithms, Box–Cox transformations, etc... Now, if the objective is "to normalize" (under H) a series $X_1, \ldots, X_n$, the exact transformation to be applied is $\Phi^{-1} \circ F$, where F denotes the cumulated distribution function of the X_t's and Φ stands for the standard normal distribution function. Of course, the trouble is that F in practice is unknown. However, if the empirical distribution function $\widehat{F}$ is substituted for the unknown one, $\Phi^{-1} \circ \widehat{F}$ computed at X_t is nothing else than $\Phi^{-1}(R_t/(n+1))$, where R_t is the rank of X_t among $X_1, \ldots, X_n$; $\widehat{F}(X_t)$ here is taken as $R_t/(n+1)$ in order to avoid trivial complications in the definition of $\Phi^{-1} \circ \widehat{F}$. The transformed series then is measurable with respect to the vector of ranks, and any subsequent inference procedure (of the van der Waerden type, in rank–order theory terminology) will be rank–based, hence invariant. The idea of invariance thus is not that remote from time series practice where the parameters e.g. of a Box–Cox transformation are also estimated from the data.

1.2 Testing for white noise: unbiasedness arguments. Still, one might be reluctant in enforcing the invariance principle underlying the use of rank–based tests. A more commonly accepted hypothesis testing principle then is the principle of unbiasedness. It is well known that (under fairly general continuity assumptions) a necessary condition for unbiasedness is similarity. Whenever a sufficient, boundedly complete statistic exists for the submodel consisting of the common border $\overline{H} \cap \overline{K}$ between the null hypothesis and the alternative (in any topology for which expectation is continuous with respect to the distributions in H and K), α-similarity in turn is equivalent to Neyman's α–structure property, with respect to the latter statistic. And test statistics possessing this latter property are obtained by conditioning upon the sufficient statistic at hand, then considering conditional tests.

Clearly, in the present problem of testing H against K, $\overline{H} \cap \overline{K} = H$, and a sufficient, complete statistic is the order statistic $\mathbf{X}_{(.)} = (X_{(1)} \le X_{(2)} \le \cdots \le X_{(n)})$. Conditioning upon $\mathbf{X}_{(.)}$ yields the class of *permutation tests*.

Whereas invariance arguments lead to rank–based tests, unbiasedness arguments thus lead to permutation tests. Since rank tests are a particular case of permutation tests, unbiasedness conditions are uniformly weaker than invariance ones.

As we shall see, the class of rank–based tests and hence, a fortiori, that of permutation tests are locally asymptotically essentially complete for problems dealing e.g. with ARMA models, in the sense that they both contain locally asymptotically optimal tests for all usual testing problems in the area.

1.3 Hypotheses reducing to white noise. Aligned ranks. The hypothesis of white noise of course plays a very fundamental role, in its own right, in a variety of classical problems, such as testing against location shifts in two–or k-sample situations, testing against trend or serial dependence, etc. This role is even more fundamental in view of the host of problems which after adequate transformation reduce to that of testing for white noise.

In most statistical models, an observed series $\mathbf{X} = (X_1, \ldots, X_n)$ indeed is such that, denoting by $\boldsymbol{\theta}$ some parameter, the transformed series $\mathbf{Z}(\boldsymbol{\theta}) = (Z_1, \ldots, Z_m) = \mathcal{F}_{\boldsymbol{\theta}}(\mathbf{X})$—call it a residual series—is white noise. In a "parametric" setting, the distribution of this residual series is assumed to be known (in most cases, it is assumed to be Gaussian); very often however, a "nonparametric" approach, where the residual density remains unspecified, would be much more realistic. The problem of testing $H_0 : \boldsymbol{\theta} = \boldsymbol{\theta}_0$ then reduces to that of testing for white noise (with unspecified density) in terms of the residual series $\mathbf{Z}(\boldsymbol{\theta}_0)$.

Accordingly, if invariance arguments are to be considered, rank tests should be used, where the ranks R_t are computed from the residual series $\mathbf{Z}(\boldsymbol{\theta}_0)$. Similarly, permutation tests, satisfying weaker unbiasedness requirements, should be based on the $m!$ permutations of $\mathbf{Z}(\boldsymbol{\theta}_0)$.

In the specific area of time series analysis, ARMA models, or linear models with ARMA error terms, or bilinear models, fall within the above class of statistical models whenever the innovation density remains unspecified. Consider for example the case of an autoregressive process with linear trend.

Example 1.1. AR(1) model with linear trend. The observations X_t, $t = 0, 1, \ldots$, n, satisfy

$$X_t = \alpha + \beta c_t + e_t, \qquad t = 0, 1, \ldots, n$$

where α and β are regression parameters, $c_0, c_1, \ldots, c_n$ are known regression constants, and e_t is some solution of

$$e_t - \rho e_{t-1} = \varepsilon_t, \qquad t = 1, \ldots, n,$$

with $\{\varepsilon_t\}$ an iid process with unspecified density, mean zero and variance one, and $|\rho| < 1$. Clearly, the null hypothesis

$$H_0 : (\alpha, \beta, \rho) = (\alpha_0, \beta_0, \rho_0)$$

is equivalent to

$$H : \mathbf{Z}(\alpha_0, \beta_0, \rho_0) \quad \text{is white noise (unspecified density)}$$

where

$$\mathbf{Z}(\alpha_0, \beta_0, \rho_0) = (Z_1, \ldots, Z_n),$$

with

$$Z_t = X_t - \rho_0 X_{t-1} - \alpha_0(1 - \rho_0) - \beta_0(c_t - \rho_0 e_{t-1}), \qquad t = 1, \ldots, n.$$

Still, hypotheses to be tested in practice seldom are of the simple form $H_0 : \boldsymbol{\theta} = \boldsymbol{\theta}_0$, under which the parameter $\boldsymbol{\theta}$ has a completely specified value $\boldsymbol{\theta}_0$. In most problems, null hypotheses take the more general form $H_0 : \boldsymbol{\theta} \in \Theta_0$, where Θ_0 is some subset of the parametric space Θ. Common examples in time series analysis are: testing an ARMA (p, q) model (with unspecified innovation density and unspecified autoregressive and moving average coefficients) against alternatives of ARMA $(p + \pi, q + \pi)$ dependence (π a given positive integer), testing for the absence of trend in an ARMA (p, q) process with unspecified covariance structure and unspecified innovation density, testing for ARMA (p, q) dependence (specified orders p and q, unspecified coefficients, unspecified innovation density) against bilinear dependence, etc.

Though a transformation $\mathcal{F}_{\boldsymbol{\theta}}$ still exists which transforms the observed series $\mathbf{X}$ into a residual white noise $\mathbf{Z}(\boldsymbol{\theta})$, the value of $\boldsymbol{\theta}$ remains unspecified also under H_0.

A tempting, intuitively natural idea then consists in substituting some "good" estimate $\widehat{\boldsymbol{\theta}}$ for the unknown paramether $\boldsymbol{\theta}$, and an "estimated" residual series $\mathbf{Z}(\widehat{\boldsymbol{\theta}})$

for the "true" one $\mathbf{Z}(\boldsymbol{\theta})$. By a "good" estimate $\widehat{\boldsymbol{\theta}}$, it is meant that $\widehat{\boldsymbol{\theta}}$ has values in Θ_0 and enjoys, under H_0, such standard properties as being root n-consistent. Estimated residuals can be expected to be close to the exact ones. Treating them as the latter yields *aligned ranks* $\widehat{R}_t$ and *aligned rank tests*, or *aligned permutation tests*.

The big trouble with aligned ranks is that they generally do not enjoy (even "approximately") the fundamental invariance properties of exact ranks: the estimation of $\boldsymbol{\theta}$ indeed most of time destroys the exchangeability structure of $\mathbf{Z}(\boldsymbol{\theta})$ which constitutes the justification of rank–based methods, so that from a decision–theoretical point of view, there is no point anymore in using (aligned) ranks. By the way, the distribution–freeness property of rank–based statistics under the null hypothesis, which was a consequence of invariance, also disappears, hence the similarity and unbiasedness features of (genuinely distribution-free) rank tests. As a consequence, the computation of exact critical values for aligned rank tests is generally impossible. Even worse: the asymptotic results available for rank statistics (based on exact ranks R_t) generally cannot be used for the corresponding aligned ones (based on the aligned ranks $\widehat{R}_t$) since in general exact and aligned rank statistics are not even asymptotically equivalent.

From a theoretical point of view as from the point of view of practice, aligned rank methods (as well as aligned permutation tests) thus are settling quite a number of nontrivial problems. Section 6 below is devoted to providing some methodology in solving these problems in time series context.

1.4 Testing for symmetric white noise. The density of the observations X_t (of the residuals Z_t in section 1.3) under the null hypothesis H so far has been assumed to be completely unspecified (except perhaps for moment conditions such as, e.g. the existence of a finite variance or finite Fisher information). In many practical problems, this density further can be assumed to be symmetric (with respect to some known median, which can be set equal to zero without any loss of generality). Denote by H' this hypothesis of symmetric white noise.

The vector of ranks $\mathbf{R}$ then loses its maximal invariant status for the benefit of the couple $(\mathbf{s}, \mathbf{R}_+)$, where $\mathbf{s} = (s_1, \ldots, s_n)$ is the vector of signs $s_t = \mathrm{sign}\,(X_t)$ and $\mathbf{R}_+ = (R_{+,1}, \ldots, R_{+,n})$ is the vector of the ranks $R_{+,t}$ of the absolute values $|X_t|$ among $|X_1| \ldots |X_n|$ (invariance here is with respect to the group of continuous, even, order-preserving transformations).

Statistics which are measurable with respect to this maximal invariant are known as *signed–rank statistics*.

The unbiasedness approach to the same problem leads to conditioning upon the order statistic of absolute values $|\mathbf{X}|_{(\cdot)} = (|X|_{(1)} \leq \cdots \leq |X|_{(n)})$, which is sufficient complete here, yielding a class of sign-assignment – absolute value permutation tests, based on the $2^n n!$ possible combination of a permutation of $|\mathbf{X}|_{(\cdot)}$ and a vector of n signs.

1.5 Testing for nonhomogeneous white noise. A further case frequently occurs in applications, in which the observations, or adequate residual values, under the null hypotheses to be tested, are still independent and symmetrically distributed, though their distributions possibly might not be identical anymore—denote this by H''. This hypothesis H'' of nonhomogeneous symmetric white noise allows for heteroskedasticity, contamination, etc. Symmetric discrete distributions are also allowed.

A maximal invariant (with respect to the group of componentwise continuous, even and order–preserving transformations of $\mathbf{X}$) is the vector $\mathbf{s}$ of signs, leading to sign tests.

The unbiasedness approach—conditioning upon the vector of absolute values $(|X_1|, \ldots, |X_n|)$—yields the broader class of tests which are *conditionally* measurable with respect to the vector signs.

2. NONPARAMETRIC TESTS FOR SERIAL DEPENDENCE: A BIBLIOGRAPHICAL SURVEY

The nonparametric, rank–based approach to time series analysis problems actually has a pretty long history, since rank tests against serial dependence and runs tests, which are a particular case of rank tests, can be traced back to the very beginnings of rank–based inference. Wald and Wolfowitz (1943) already suggest to substitute the ranks, or some function thereof, for the observations in the problem of testing randomness against serial dependence. Tests based on runs, runs up and down, signs of first differences or turning points already had been considered, for the same problem, in Fisher (1926), Kermack and McKendrick (1937), Mood (1940) and Wallis and Moore (1941); they subsequently have been developed in Moore and Wallis (1943), Levene and Wolfowitz (1944), Wolfowitz (1944), David (1947), Goodman (1958), Edgington (1961) and Granger (1963).

Jogdeo (1968) derives asymptotical results for a very general class of rank statistics, which includes the rank autocorrelation coefficients to be used repeatedly in subsequent sections. The conditions he puts on the score functions however are too restrictive for most purposes (excluding, e.g. the so–called van der Waerden autocorrelations).

In none of these early papers is any particular alternative considered, nor any optimality question addressed. The first attempts to investigate the power of serial rank procedures against specific alternatives are due to Beran (1972), who introduces rank analogues of integrated periodogram spectral processes, and, in a more applied context, to Knoke (1977), who conducts a Monte Carlo study of the asymptotic relative efficiencies of several tests based on serial rank statistics (namely, the Wald–Wolfowitz rank autocorrelation coefficient, the turning point statistic and a Kolmogorov–Smirnov one) with respect to the classical first–order sample autocorrelation coefficient, for first–order autoregressive alternatives. This, in some sense, was a first step towards the introduction of rank methods in time series practice—but still the statistics studied there are not new, are not specifically devised against any particular alternative, and cannot handle the more general problem of testing, e.g. an ARMA model against other ARMA models (as described in section 1.3).

Locally most powerful rank tests for randomness against Gaussian autoregressive or moving average alternatives are derived in Gupta and Govindarajulu (1980). The test statistics depend on ranks via the expected values of products of order statistics, and are asymptotically equivalent to our van der Waerden autocorrelation coefficients. Aiyar (1981) also investigates first–order statistics of the Wald–Wolfowitz and van der Waerden types, and derives their asymptotic relative efficiencies with respect to the corresponding Gaussian procedure, still under first–order autoregressive alternatives.

In a similar situation, Dufour (1981) suggests (traditional nonserial) signed–rank tests for H' based on the signs and ranks of products of the form $X_t X_{t-k}$. This way of reducing a serial problem to a nonserial one however may result in a loss of relevant information and the ranks used there are not those following from invariance arguments (except in the case of nonhomogeneous white noise—see Section 1.5). In a more applied context, Bartels (1982) introduces a rank–based version of von Neumann's ratio statistic which is asymptotically equivalent to the Wald–Wolfowitz autocorrelation coefficient of order one, and investigates the power of the resulting test through Monte Carlo techniques.

Runs also have been used for testing the absence of cross–correlation between two series: see Goodman and Grunfeld (1961) and Yang and Schreckengost (1981). In the more complex domain of dynamic econometric models, Campbell and Dufour (1991) present a promising rank-based approach to the Mankiw–Shapiro rational expectation model. Switzer (1984) proposes a Wald–Wolfowitz version of the spatial variogram—a rather isolated attempt to introduce ranks in the area of spatial processes.

In a somewhat different, nonserial context, rank tests against trend alternatives have been intensively investigated (see Mann (1945) or Savage (1957) for early papers on the subject, Aiyar, Guillier and Albers (1979) for a more recent one), as well as the asymptotic behavior of nonserial rank statistics, including multivariate ones, under various mixing conditions (Serfling (1968), Albers (1978), Tran (1988), Harel (1988) and Harel and Puri (1989a,b; 1990a; 1991; 1992)).

For more details, we refer to the review papers by Dufour, Lepage and Zeidan (1982) and Bhattacharyya (1984); see also Govindarajulu (1983).

Up to this point, and in spite of a long history, the subject of rank–based inference for time series problems thus remained largely unexplored. Quite a number of partial results were scattered around, but no coherent and structured theory was available. The few existing central limit theorems for serial rank statistics were too restrictive for most purposes; except for a few exceptions, the only problem that had been considered was that of testing for white noise; optimality problems remained essentially untouched... the rank–order section of the time series analyst's tool kit was pretty limited and nearly empty.

A systematic and coherent treatment of time series problems, based on a LeCam–Hájek approach, since then has been undertaken by the authors in a series of papers, starting with Hallin, Ingenbleek and Puri (1985, 1987), where the problem of testing for white noise against ARMA alternatives is considered in its full generality (yielding locally maximin, rank–based portmanteau tests). The long term objective is to obtain a logically consistent methodology for classical time series problems (identification, diagnostic checking,...), relying on adequate, rank based substitutes for time series analysis familiar tools such as correlograms, partial correlograms, Lagrange multipliers, etc.

Locally asymptotically maximin rank–based tests for testing an ARMA model (with unspecified innovation density) against other ARMA models are derived in Hallin and Puri (1988). The particular case of first–order autoregressive models is treated in detail in Dufour and Hallin (1987). The small sample performance of rank–based tests is investigated in Hallin and Mélard (1989), and appears to be surprisingly good, even for pretty short series. Signed rank techniques, for ARMA models with symmetric, otherwise unspecified innovation densities, are considered in Hallin and Puri (1991a) and Hallin, Laforet and Mélard (1989). Optimal runs tests, allowing for nonhomogeneous innovation processes, are treated in Dufour and Hallin (1990a), the problem of testing multivariate white noise against alternatives of (multivariate) ARMA dependence in Hallin, Ingenbleek and Puri (1989) and Hallin and Puri (1991b), and that of testing white noise against first–order, superdiagonal bilinear dependence in Benghabrit and Hallin (1992). Finally, the very general case of ARMA models with a regression trend (also known as the dynamic regression model) is studied in Hallin and Puri (1991c), where locally optimal, aligned, rank and signed–rank, tests are derived for a variety of problems, including that of testing an ARMA (p, q) model with unspecified trend component, unspecified ARMA coefficients and unspecified innovation density, against ARMA $(p + \pi, q + \pi)$ alternatives.

A review of the main results in the above papers is the subject of the present survey.

Related results on the asymptotic distribution of serial rank statistics under dependence can be found in Harel and Puri (1990b,c), Tran (1990), Nieuwenhuis and Ruymgaart (1990). Permutation tests against serial dependence are considered in David and Fix (1966), Ghosh (1954), Dufour and Roy (1985, 1986), Dufour and Hallin (1990b) and Hallin, Mélard and Milhaud (1992). A somewhat hybrid test, with mixed parametric and nonparametric (rank-based) features has been considered by Kreiss (1990a) for testing AR models.

Rank–based techniques have proven very efficient in classical linear model problems (regression, trend, analysis of variance,...), and have entered daily practice in biostatistics and experimental planning. When the required technology is available, there is no reason for time series analysis to escape this rule—even more so in view of the recognized need for robust and non–Gaussian methods in the area.

3. LINEAR RANK STATISTICS

3.1 Serial and nonserial linear rank statistics. Denote by $\mathbf{Z}^{(n)} = Z_1^{(n)}, \ldots,$ $Z_t^{(n)}, \ldots, Z_n^{(n)})$ a series of length n. Depending on the problem, $\mathbf{Z}^{(n)}$ either may be the original observation, or some residual series. Let $R_t^{(n)}$ be the rank of $Z_t^{(n)}$ among $Z_1^{(n)}, \ldots, Z_n^{(n)}$. If $\mathbf{Z}^{(n)}$ is white noise, then, with probability one, $\mathbf{R}^{(n)} = (R_1^{(n)}, \ldots, R_n^{(n)})$ is uniformly distributed over the $n!$ permutations of $(1, 2, \ldots, n)$, whatever the underlying density of the $Z_t^{(n)}$'s.

It is well known (Hájek and Šidák 1967; Puri and Sen 1971 and 1985) that locally asymptotically best tests for linear models (two or k samples, regression, analysis of variance,...) can be based upon the class of (*simple nonserial*) *linear rank statistics*, of the form

$$(3.1) \qquad S^{(n)} = n^{-1} \sum_{t=1}^{n} c_t a^{(n)}(R_t^{(n)}) \, ,$$

where $a^{(n)}$ denotes some *score function*, and the c_t's are known (*regression*) constants. In a time series context, they also can be used in testing against trend (Mann 1945; Savage 1957). Such nonserial statistics however cannot capture serial dependence features: actually, it can be shown that their asymptotic distribution is the same under white noise as under local alternatives of serial dependence. Hallin et al. (1985) therefore suggest the consideration of *serial linear rank statistics*, of the form

$$(3.2) \qquad S^{(n)} = (n-k)^{-1} \sum_{t=k+1}^{n} a^{(n)}(R_t^{(n)}, R_{t-1}^{(n)}, \ldots, R_{t-k}^{(n)}) \, ,$$

where the score function $a^{(n)}$ depends on the values of $(k+1)$ successive ranks: (3.2) is a serial linear rank statistic *of order* k, and can be expected to be sensitive to serial dependencies of orders 1 through k. Nonserial statistics can be considered as a particular case ($k = 0$) of the serial ones (or linear combinations thereof).

Classical examples are the *runs statistic* (where runs are taken with respect to the median), or order one, with scores

$$a^{(n)}(i_1, i_2) = I[(2i_1 - n - 1)(2i_2 - n - 1) < 0] \, ,$$

the *turning point* statistic, or order three, with scores

$$a^{(n)}(i_1, i_2, i_3) = I[i_1 > i_2 < i_3 \quad \text{or} \quad i_1 < i_2 > i_3]$$

or Wald and Wolfowitz' rank autocorrelation coefficient of order k (a serial version of Spearman's statistic, here, unlike in Wald and Wolfowitz (1943)'s paper, in a noncircular version) with scores

$$(3.3) \qquad a^{(n)}(i_1, \ldots, i_{k+1}) = i_1 i_{k+1} \, .$$

The mean $m^{(n)}$ of (3.2) under the assumption that $\mathbf{Z}^{(n)}$ is white noise is easy to compute:

$$(3.4) \qquad m^{(n)} = [n(n-1) \ldots (n-k)]^{-1} \sum \sum_{i_1 \neq \cdots \neq i_{k+1}} a^{(n)}(i_1, \ldots, i_{k+1}) \, ,$$

where $\sum \cdots \sum\limits_{i_1 \neq \cdots \neq i_{k+1}}$ stands for a summation running over all $(k+1)$–tuples of *dis-tinct* integers between 1 and n. The variance of (3.2) also is easy to obtain from combinatorial arguments—though a general closed form is rather tedious.

Most serial statistics of practical interest can be decomposed into a linear combination of simpler statistics of the form

$$(3.5) \qquad S^{(n)} = (n-k)^{-1} \sum_{t=k+1}^{n} a^{(n)}(R_t^{(n)}) b^{(n)}(R_{t-k}^{(n)}) .$$

The runs statistic for example, is of the form $S_1^{(n)} + S_2^{(n)}$, where $S_1^{(n)}$ and $S_2^{(n)}$ rely on the scores

$$a_1^{(n)}(i_1, i_2) = I[2i_1 < n+1].I[2i_2 > n+1]$$

and

$$a_2^{(n)}(i_1, i_2) = I[2i_2 > n+1].I[2i_1 < n+1] .$$

A statistic of the form (3.5) will be called a *simple* serial linear rank statistic of order k. The Wald–Wolfowitz autocorrelation coefficient, as well as the f–rank autocorrelation coefficients to be introduced later, are *simple* rank statistics.

Letting

$$S_{\ell m}^{(n)} = \sum_{i=1}^{n} [a^{(n)}(i)]^{\ell} [b^{(n)}(i)]^{m} ,$$

we have, for the mean $m^{(n)}$ of the simple statistic $S^{(n)}$ and the variance $(\sigma^{(n)})^2$ of $(n-k)^{\frac{1}{2}} S^{(n)}$ under white noise,

$$(3.6) \qquad m^{(n)} = [n(n-1)]^{-1} [S_{10}^{(n)} S_{01}^{(n)} - S_{11}^{(n)}]$$

and

$$(3.7) \qquad (\sigma^{(n)})^2 = [n(n-1)]^{-1} [S_{20}^{(n)} S_{02}^{(n)} - S_{22}^{(n)}]$$

$$+ 2 \frac{(n-2k)^+}{n-k} [n(n-1)(n-2)]^{-1} [S_{10}^{(n)} S_{01}^{(n)} S_{11}^{(n)} - S_{21}^{(n)} S_{01}^{(n)} - S_{12}^{(n)} S_{10}^{(n)} - (S_{11}^{(n)})^2 + 2S_{22}^{(n)}]$$

$$+ \frac{(n-k)(n-k-1) - 2(n-2k)^+}{n-k} [n(n-1)(n-2)(n-3)]^{-1}$$

$$\times [(S_{10}^{(n)} S_{01}^{(n)})^2 + 2(S_{11}^{(n)})^2 + S_{20}^{(n)} S_{02}^{(n)} - 6S_{22}^{(n)}$$

$$- 4S_{11}^{(n)} S_{10}^{(n)} S_{01}^{(n)} - S_{20}^{(n)} (S_{01}^{(n)})^2 - S_{02}^{(n)} (S_{10}^{(n)})^2 + 4S_{21}^{(n)} S_{01}^{(n)} + 4S_{12}^{(n)} S_{10}^{(n)}]$$

$$- (n-k)(m^{(n)})^2, \qquad \text{where } (\cdot)^+ = \max(., 0).$$

Accordingly, $(n-k)^{\frac{1}{2}} (S^{(n)})/\sigma^{(n)}$ is exactly standardized (still, under the hypothesis that $\mathbf{Z}^{(n)}$ is white noise).

3.2 Serial and nonserial linear signed–rank statistics. As mentioned in Section 1.4, whenever symmetry assumptions can be made, the vector of ranks loses its maximal invariance properties for the benefit of $(\mathbf{s}^{(n)}, \mathbf{R}_+^{(n)})$, where $\mathbf{s}^{(n)} = (s_1, \ldots, s_n)$ is the vector of signs $s_t = \text{sign}(Z_t^{(n)}) = Z_t^{(n)}/|Z_t^{(n)}|$ (with the convention $0/0 = 1$), and $\mathbf{R}_+^{(n)}$ denotes the vector of the ranks $R_{+,t}^{(n)}$ of absolute values $|Z_t^{(n)}|$ among $|Z_1^{(n)}|, \ldots, |Z_n^{(n)}|$. If $\mathbf{Z}^{(n)}$ is absolutely continuous symmetric white noise, then $\mathbf{s}^{(n)}$ and $\mathbf{R}_+^{(n)}$ are independently distributed, $\mathbf{s}^{(n)}$ uniformly over the 2^n elements of $\{-1, 1\}^n$, and $\mathbf{R}_+^{(n)}$ uniformly over the $n!$ permutations of $\{1, \ldots, n\}$.

Here again, locally best tests for linear models with independent and identically distributed symmetric error (unspecified density) terms can be based on (nonserial) linear, signed–rank statistics, of the form

$$(3.8) \qquad S_+^{(n)} = n^{-1} \sum_{t=1}^{n} c_t a_+^{(n)}(s_t R_{+,t}^{(n)}) \, ,$$

where the c_t's are known constants, and $a_+^{(n)}$ denotes a score function, defined over $\{\pm 1, \pm 2, \ldots, \pm n\}$. Time series problems require the more general class of serial linear signed–rank statistics (Hallin and Puri 1991a), of the form

$$(3.9) \qquad S_+^{(n)} = (n - k)^{-1} \sum_{t=k+1}^{n} a_+^{(n)}(s_t R_{+,t}^{(n)}, s_{t-1} R_{+,t-1}^{(n)}, \ldots, s_{t-k} R_{+,t-k}^{(n)}).$$

Simple examples are Goodman (1958)'s simplified runs test (with $a_+^{(n)}(i_1, i_2) = I[i_1 i_2 < 0]$), which coincides (up to additive and multiplicative constants) with Dufour (1981)'s runs test, or the signed version of the Spearman–Wald–Wolfowitz serial correlation tests, with $a_+^{(n)}(i_1, \ldots, i_{k+1}) = i_1 i_{k+1}$.

A linear signed–rank statistic will be called *simple* if it can be written as

$$(3.10) \qquad S_+^{(n)} = (n - k)^{-1} \sum_{t=k+1}^{n} s_t s_{t-k} a_+^{(n)}(R_{+,t}^{(n)}) b_+^{(n)}(R_{+,t-k}^{(n)}) \, .$$

The signed Spearman–Wald–Wolfowitz autocorrelations, as well as the signed f-rank autocorrelations to be described later, belong to this family of simple serial statistics. The mean of (3.10) under the hypothesis of symmetric white noise obviously is zero, whereas, due to the fact that all cross–products have expectation zero, the variance of $(n - k)^{\frac{1}{2}} S_+^{(n)}$ takes the very simple form

$$(3.11) \qquad (\sigma_+^{(n)})^2 = [n(n - 1)]^{-1} \sum_{i_1 \neq i_2} (a_+^{(n)}(i_1) b_+^{(n)}(i_2))^2$$

$$= [n(n - 1)]^{-1} \left\{ \sum_i [a_+^{(n)}(i)]^2 \sum_i [b_+^{(n)}(i)]^2 - \sum_i [a_+^{(n)}(i) b_+^{(n)}(i)]^2 \right\} \, .$$

Here again, $(n - k)^{\frac{1}{2}} S_+^{(n)}/\sigma_+^{(n)}$ is exactly standardized (provided that $\mathbf{Z}^{(n)}$ is symmetric white noise).

3.3 Asymptotic normality. Assuming that the series $\mathbf{Z}^{(n)}$ from which the ranks are taken is white noise, consider the serial statistic $S^{(n)}$ in (3.2). A real–valued function J, defined over the $(k+1)$-dimensional unit square $(0,1)^{k+1}$, will be called a *score-generating function* for $S^{(n)}$ if

$$(3.12) \qquad \int_{(0,1)^{k+1}} J^{2+\delta}(u_1,\ldots,u_{k+1})du_1 \ldots du_{k+1} < \infty$$

for some $\delta > 0$ and

$$(3.13) \qquad \lim_{n\to\infty} E\left[(a^{(n)}(R_{U,1}^{(n)},\ldots,R_{U,k+1}^{(n)}) - J(U_1,\ldots,U_{k+1}))^2\right] = 0 \;,$$

where $R_{U,1}^{(n)} \ldots, R_{U,k+1}^{(n)}$ denote the ranks of $U_1,\ldots U_{k+1}$ in a random sample $U_1,\ldots, U_n$ of independent and identically distributed, uniform (over [0,1]) random variables.

Define

$$(3.14) \quad J^*(u_1,\ldots,u_{k+1}) = J(u_1,\ldots,u_{k+1}) - \sum_{i=1}^{k+1} E[J(U_1,\ldots,U_{i-1},u_1,U_i,\ldots,U_k)]$$
$$+ k\; E[J(U_1,\ldots,U_{k+1})] \;.$$

Then, relying on a central limit theorem (e.g., Yoshihara, 1976) for U–statistics under absolutely regular processes, the following asymptotic normality result has been proved (cf. Hallin et al., 1985).

PROPOSITION 3.1. *Let J denote a score–generating function for the serial rank statistic $S^{(n)}$ in (3.2). Then $(n-k)^{\frac{1}{2}}(S^{(n)} - m^{(n)})$ is asymptotically normal (under white noise), with mean zero and variance V^2, where*

$$(3.15) \qquad V^2 = \int_{[0,1]^{k+1}} [J^*(u_1,\ldots,u_{k+1})]^2 du_1 \ldots du_{k+1}$$
$$+ 2\sum_{i=1}^{k} \int_{[0,1]^{k+i+1}} J^*(u_1,\ldots,u_{k+1})J^*(u_{1+i},\ldots,u_{k+1+i})du_1 \ldots du_{k+1+i} \;.$$

This result allows for explicit normal approximations, under the null hypotheses of white noise, for P–values and critical points of tests based on linear serial rank statistics.

Example 3.1. The Spearman autocorrelation coefficient of order one is often defined (cf. e.g. Kendall and Stuart, 1968) as

$$r_S^{(n)} = \frac{(n-1)^{-1}\Sigma_{t=2}^{n}R_t^{(n)}R_{t-1}^{(n)} - (n+1)^2/4}{(n^2-1)/12} \;,$$

though an exactly standardized version (see Hallin and Mélard 1988) is more convenient for practical purposes. Letting $S^{(n)} = (n-1)^{-1} \sum_{t=2}^{n} R_t^{(n)} R_{t-1}^{(n)}/(n+1)^2$, with $m^{(n)} = (3n+2)/12(n+1)$, it is easy to check that

$$(3.16) \qquad r_S^{(n)} - 12(S^{(n)} - m^{(n)}) = 24(n-1)^{-1}(S^{(n)} - m^{(n)}) - (n-1)^{-1} .$$

$S^{(n)}$ is a linear serial rank statistic of order one. A score–generating function for $S^{(n)}$ is $J(u,v) = uv$, yielding

$$J^*(u,v) = uv - u + \frac{1}{4} .$$

It follows from Proposition 3.1 that $(n-1)^{\frac{1}{2}}(S^{(n)} - m^{(n)})$ is asymptotically normal, with mean zero and variance

$$V^2 = \int_{[0,1]^2} \left(uv - u + \frac{1}{4} \right)^2 \, du\,dv = 1/144 .$$

This, with (3.16), entails that $r_S^{(n)} - 12(S^{(n)} - m^{(n)})$ is $o_P(n^{-\frac{1}{2}})$, as $n \to \infty$, and confirms the classical result that $(n-1)^{\frac{1}{2}} r_S^{(n)}$ (as well as $12(n-1)^{\frac{1}{2}}(S^{(n)} - m^{(n)})$) is asymptotically standard normal.

As in the nonserial case, (3.13) can be shown to hold for J satisfying (3.12) and scores defined by

$$a^{(n)}(i_1, \ldots, i_{k+1}) = E\left[J(U_1, \ldots, U_{k+1}) \big| R_1^{(n)} = i_1, \ldots, R_{k+1}^{(n)} = i_{k+1} \right]$$

(*exact scores*) or, for J monotone with respect to each argument, satisfying (3.12), and scores

$$a^{(n)}(i_1, \ldots, i_{k+1}) = J\left(\frac{i_1}{n+1}, \ldots, \frac{i_{k+1}}{n+1} \right)$$

(*approximate scores*).

The case of signed–rank statistics is roughly similar. A *score–generating function* J^+ for $S_+^{(n)} = (n-k)^{-1} \Sigma a_+^{(n)}(s_t R_t^{(n)}, \ldots, s_{t-k} R_{t-k}^{(n)})$ is a real–valued function, defined over the $(k+1)$–dimensional open square $(-1,1)^{k+1}$, such that, denoting by $V_1, \ldots, V_n$ an n–tuple of independent and identically distributed rectangular $[-1,1]$ variables,

$$(3.17) \qquad E[|J_+(V_1, V_2, \ldots, V_{k+1})|^{2+\delta}] < \infty$$

for some $\delta > 0$, and

$$(3.18) \quad \lim_{n \to \infty} E\{[a_+^{(n)}(\mathrm{sgn}(V_1)R_{+,1}^{(n)}, \ldots, \mathrm{sgn}(V_{k+1})R_{+,k+1}^{(n)}) - J_+(V_1, \ldots, V_{k+1})]^2\} = 0 ,$$

as $n \to \infty$. $R_{+,i}^{(n)}$ here denotes the rank of $|V_i|$ among $|V_1|, \ldots, |V_n|$; the notation sgn(V_i) is used in an obvious fashion. Associated with J_+, define J_+^* as

$$(3.19) \quad J_+^*(v_1, \ldots, v_{k+1}) = J_+(v_1, \ldots, v_{k+1})$$

$$- 2^{-(k+1)} \sum_{s \in \{-1,1\}} \sum_{\ell=0}^{k} \int_{[-1,1]^k} J_+(w_1, \ldots, w_\ell, sv_1, w_{\ell+1}, \ldots, w_k) d\mathbf{w}$$

$$+ 2^{-(k+1)} k \int_{[-1,1]^k} J_+(w_1, w_2, \ldots, w_{k+1}) d\mathbf{w} ,$$

where $\int_{[-1,1]^k} (\ldots) d\mathbf{v}$ stands for $\int_{-1}^{1} \ldots \int_{-1}^{1} (\ldots) \text{sgn}(v_1) \ldots \text{sgn}(v_k) dv_1 \ldots dv_k$.

PROPOSITION 3.2. *Let J_+ denote a score–generating function for the signed–rank statistic $S_+^{(n)}$ in (3.9). Then $(n-k)^{\frac{1}{2}} S_+^{(n)}$ is asymptotically normal (under symmetric white noise), with mean zero and variance V_+^2, where (same notation as in (3.18))*

$$(3.20) \quad V_+^2 = E[(J_+^*(V_1, \ldots, V_{k+1}))^2] + 2 \sum_{i=1}^{k} E[J_+^*(V_1, \ldots, V_{k+1}) J_+^*(V_{1+i}, \ldots, V_{k+1+i})].$$

Example 3.2. Parallel to the unsigned statistic considered in Example 3.1, a signed version of the Spearman autocorrelation coefficient of order one can be defined as

$$r_{+,S}^{(n)} = \frac{(n-1)^{-1} \sum_{t=2}^{n} s_t s_{t-1} R_{+,t}^{(n)} R_{+,t-1}^{(n)}}{(n+1)(2n+1)/6}$$

(though an exactly standardized version might be preferable for practical purposes—see Hallin et al. 1990). Letting

$$S_+^{(n)} = (n-1)^{-1} \sum_{t=2}^{n} s_t s_{t-1} R_{+,t}^{(n)} R_{+,t-1}^{(n)} / (n+1)^2,$$

we have $r_{+,S}^{(n)} = 6 S_+^{(n)} (n+1)^2 / (n+1)(2n+1)$. A score–generating function for $S_+^{(n)}$ is $J_+(u,v) = uv$. Since $E[V_1 V_2 | |V_\ell| = |v|] = 0, \ell = 1, 2, v \in [0,1]$ and $E[V_1 V_2] = 0$, $J_+^*(u,v) = J_+(u,v)$ (under symmetric white noise), and

$$V_+^2 = \frac{1}{4} \int_{[-1,1]^2} (uv)^2 du \, dv = 1/9 .$$

It follows that $(n-1)^{\frac{1}{2}} S^{(n)}$ is asymptotically normal, with mean zero and variance $1/9$, and that

$$r_{+,S}^{(n)} - 3 S_+^{(n)} = S_+^{(n)} \left[6 \frac{(n+1)^2}{(n+1)(2n+1)} - 3 \right] = o_P(n^{-\frac{1}{2}}) .$$

Consequently, $(n-1)^{\frac{1}{2}}r^{(n)}_{+,S}$ is asymptotically standard normal (under symmetric white noise). This statistic, to the best of our knowledge, never had been considered in the literature. Its asymptotic performance is investigated in Hallin et al. (1990).

As in the unsigned case, (3.18) holds, for $(2+\delta)$-integrable score-generating functions J_+ and *exact scores*

$$a_+^{(n)}(i_1,\ldots,i_{k+1}) = E\left[J_+(V_1,\ldots,V_{k+1})\big|\,\text{sgn}(V_1)R^{(n)}_{+,1} = i_1,\ldots,\text{sgn}(V_{k+1})R^{(n)}_{+,k+1}\right],$$

or, provided in addition that J_+ is monotone with respect to each argument, with *approximate scores*

$$a_+^{(n)}(i_1,\ldots,i_{k+1}) = J_+\left(\frac{i_1}{n+1},\ldots,\frac{i_{k+1}}{n+1}\right).$$

4. LOCAL ASYMPTOTIC NORMALITY OF ARMA PROCESSES AND THE LOCAL SUFFICIENCY OF RANKS

4.1 Local asymptotic normality results for ARMA models. The asymptotic results of Section 3 are valid under white noise assumptions, and thus allow for constructing invariant, distribution–free tests based on signed or unsigned serial linear rank statistics, for a variety of time series problems. If, however, power or optimality issues are to be addressed, more information is needed on the structure of the specific problem at hand. Since uniformly most powerful tests cannot be expected to exist—such strong optimality results in general are not available, in time series analysis, even in a more restricted parametric, Gaussian context—weaker optimality properties have to be considered. These weaker properties, of a local and asymptotic nature, rely on the local structure of the families of likelihood functions involved.

This local structure has been studied by several authors, who under various technical assumptions have established the locally asymptotically normal (LAN) structure (LeCam, 1960) of families of ARMA likelihood functions. Davies (1973) and Dzhaparidze (1986) investigate the LAN property for Gaussian ARMA processes. Akritas and Johnson (1982) deal with AR processes only. Hallin et al. (1985) and Hallin and Puri (1987) established a LAN result for general ARMA processes, using classical, Cramér-type, technical assumptions. Swensen (1985) considers the case of AR processes with a regression trend, and Garel (1989) that of MA ones, still with trend. Kreiss (1987) deals with ARMA processes without trend, and Kreiss (1990b) with AR(∞) process. The results of Swensen and Kreiss rely on martingale central limit theorems, and those by Akritas and Johnson on quadratic mean differentiability conditions which are less restrictive than the Cramér–type assumptions made by Hallin et al. (1985), Hallin and Puri (1987) and Garel (1989).

We shall not attempt here to describe in detail the technical conditions under which the LAN property holds. The assumptions we are giving here are not the weakest; but they are quite simple, and they are satisfied by most densities used in practice (not all of them: e.g. Cauchy).

4.2 Notation and main assumptions. All densities below—denoted by f, g, with distribution functions F, G—are of the form

$$g(x) = g_\sigma(x) = \sigma^{-1} g_1(x/\sigma) > 0 , \quad x \in \mathbf{R}$$

with

$$\int x \, g_1(x) dx = 0 \quad \text{and} \quad \int x^2 g_1(x) dx = 1 .$$

The variance σ^2 will always remain unspecified; since however it plays no role in the sequel, it will be dropped in the notation. Still for simplicity, we shall assume that g is absolutely continuous, so that the derivative $\dot{g}(x) = dg(x)/dx$ exists for almost all x. Defining $\varphi_g(x) = -\dot{g}(x)/g(x)$, we also assume that the Fisher information

$$\int \varphi_g^2(x) g(x) dx = \sigma^2 I(g) = I(g_1) = \int \varphi_{g_1}^2(x) g_1(x) dx$$

is finite. Since $g(x)$ is always strictly positive, G is strictly increasing, and the inverse G^{-1} is well defined.

Consider the stochastic difference equation (ARMA (p_1, q_1) model)

$$(4.1) \qquad X_t - \sum_{i=1}^{p_1} A_i X_{t-i} = \varepsilon_t + \sum_{i=1}^{q_1} B_i \varepsilon_{t-i} ,$$

in short $A(L)X_t = B(L)\varepsilon_t$, where L stands for the lag operator and $A(z) = 1 - \Sigma A_i z^i$, $B(z) = 1 + \Sigma B_i z^i$. In all ARMA models below, it is always assumed that $A(z)$ and $B(z)$, $z \in \mathbf{C}$ have no common roots, and satisfy the usual invertibility and causality conditions. We denote by $H_g^{(n)}(A, B)$ the hypothesis under which an observed series $\mathbf{X}^{(n)} = (X_1^{(n)}, \ldots, X_t^{(n)}, \ldots, X_n^{(n)})$ is a finite realization of some solution of (4.1), where $\{\varepsilon_t\}$ is a white noise process (i.e., an iid process) with density g. The notation $H^{(n)}(A, B) = \cup_g H_g^{(n)}(A, B)$ is used whenever g remains completely unspecified (except for the few general technical conditions above); the notation $H_+^{(n)}(A, B) = \cup_g \{H_g^{(n)}(A, B) | g$ symmetric with respect to zero$\}$ is used whenever g is symmetric but otherwise unspecified (still, apart from the few general technical conditions).

4.3 Signed and unsigned f–rank autocorrelations. Just as parametric autocorrelation coefficients play an essential role in the classical parametric analysis of time series models, a special case of serial rank statistics will play an essential role in our nonparametric approach. These nonparametric counterparts of usual autocorrelation coefficients have been introduced in Hallin et al. (1987) as *f–rank autocorrelations* and in Hallin and Puri (1991a) as *signed f–rank autocorrelations*.

Denote by $f, \varphi = -\dot{f}/f$ and F a probability density function, the corresponding score function and distribution function, respectively. The f–rank and signed f–rank autocorrelations (see below for a definition) being invariant with respect to scale transformations, f can be chosen as f_1, so that $\int x^2 f(x)dx = 1$. The (unsigned) f–rank autocorrelation of lag i is then defined as

$$(4.2) \quad r_{i;f}^{(n)} = (n-i)^{-1} \left[\sum_{t=i+1}^{n} \varphi \left(F^{-1} \left(\frac{R_t^{(n)}}{n+1} \right) \right) F^{-1} \left(\frac{R_{t-i}^{(n)}}{n+1} \right) - m^{(n)} \right] / s^{(n)} ,$$

where $m^{(n)}$ and $s^{(n)}$ are given by (3.6) and (3.7) with

$$S_{\ell m}^{(n)} = \sum_{i=1}^{n} \varphi \left(F^{-1} \left(\frac{i}{n+1} \right) \right)^{\ell} \left(F^{-1} \left(\frac{i}{n+1} \right) \right)^{m} .$$

Particular cases are

(a) the *van der Waerden autocorrelation coefficients*, associated with Gaussian densities,

$$r_{i;vdW}^{(n)} = \left[(n-i)^{-1} \sum_{t=i+1}^{n} \Phi^{-1} \left(\frac{R_t^{(n)}}{n+1} \right) \Phi^{-1} \left(\frac{R_{t-i}^{(n)}}{n+1} \right) - m_{vdW}^{(n)} \right] / s_{vdW}^{(n)}$$

(as usual $\Phi(x) = (2\pi)^{-\frac{1}{2}} \int_{-\infty}^{x} e^{-y^2/2} dy$ denotes the standard normal distribution function; note that, due to the symmetry of the Gaussian distribution, quite a number of terms in (3.6) and (3.7) cancel out),

(b) the *Wilcoxon autocorrelation coefficients*, associated with logistic densities,

$$r_{i;W}^{(n)} = \left[(n-i)^{-1} \sum_{t=i+1}^{n} \left(\frac{R_t^{(n)}}{n+1} - \frac{1}{2} \right) \log \frac{R_{t-i}^{(n)}}{n+1-R_{t-i}^{(n)}} - m_W^{(n)} \right] / s_W^{(n)} ,$$

(c) the *Laplace autocorrelation coefficients*, associated with double exponential densities,

$$r_{i;L}^{(n)} = \left[(n-i)^{-1} \sum_{t=i+1}^{n} \operatorname{sgn} \left(\frac{R_t^{(n)}}{n+1} - \frac{1}{2} \right) \left[\log(2 \frac{R_{t-i}^{(n)}}{n+1}) I \left[R_t^{(n)} \geq \frac{n+1}{2} \right] \right.\right.$$
$$\left.\left. - \log \left(2 - 2 \frac{R_{t-i}^{(n)}}{n+1} \right) I \left[R_t^{(n)} < \frac{n+1}{2} \right] \right] - m_L^{(n)} \right] / s_L^{(n)} .$$

Note that only the van der Waerden statistic can be considered, *stricto sensu*, as an autocorrelation coefficient. Actually, up to normalizing constants, $r_{i;vdW}^{(n)}$ is the classical autocorrelation coefficient resulting from substituting the transformed observations ($\Phi^{-1}(R_t^{(n)}/n+1)$) for the original one ($X_t^{(n)}$ or $Z_t^{(n)}$). The symmetric structure of this measure of serial dependence, where the past and future play exchangeable roles, is a consequence of the fact that (except for a few, very particular MA cases) Gaussian ARMA processes are the only time–reversible ones (Weiss, 1975; Hallin et al., 1988). The past and future, in non Gaussian processes, do not play symmetric roles; accordingly, they do not play symmetric roles in the definition of f–rank autocorrelations either.

The importance of f–rank autocorrelations will follow from the asymptotic decomposition of log likelihood ratios (section 4.4 below).

Under symmetric densities, a signed version of f–rank autocorrelations, involving signed ranks, can be used with the same notation as above. Letting $F_+ = 2F - 1$, define the signed f–rank autocorrelation of lag i as

$$(4.3)\ r_{i;f}^{(n)+} = (n-i)^{-1} \sum_{t=i+1}^{n} \mathrm{sgn}(Z_t^{(n)} Z_{t-i}^{(n)}) \varphi\left(F_+^{-1}\left(\frac{R_{+,t}^{(n)}}{n+1}\right)\right) F_+^{-1}\left(\frac{R_{+,t-i}^{(n)}}{n+1}\right) / s_{+,f}^{(n)}$$

where $\{Z_t^{(n)}\}$ is the series from which the ranks are computed, and

$$(s_{+,f}^{(n)})^2 = [n(n-1)]^{-1}\left\{ \sum_{i=1}^{n} \varphi\left(F_+^{-1}\left(\frac{i}{n+1}\right)\right)^2 \sum_{i=1}^{n}\left(F_+^{-1}\left(\frac{i}{n+1}\right)\right)^2 \right.$$
$$\left. - \sum_{i=1}^{n}\left[\varphi\left(F_+^{-1}\left(\frac{i}{n+1}\right)\right) F_+^{-1}\left(\frac{i}{n+1}\right)\right]^2 \right\}.$$

Here again, provided the $Z_t^{(n)}$'s are symmetric white noise, $(n-i)^{\frac{1}{2}} r_{i;f}^{(n)+}$ is exactly standardized. Particular cases are

(d) the *signed van der Waerden autocorrelation coefficients*

$$r_{i;vdW}^{(n)+} = (n-i)^{-1} \sum_{t=i+1}^{n} \mathrm{sgn}(Z_t^{(n)} Z_{t-i}^{(n)}) \Phi^{-1}\left(\frac{n+1+R_{+,t}^{(n)}}{2(n+1)}\right) \Phi^{-1}\left(\frac{n+1+R_{+,t-i}^{(n)}}{2(n+1)}\right)$$
$$\times [n(n-1)]^{\frac{1}{2}} \left\{ \left[\sum_{i=1}^{n}(\Phi^{-1}(i/(n+1)))^2\right]^2 - \sum_{i=1}^{n}(\Phi^{-1}(i/(n+1)))^4 \right\}^{-\frac{1}{2}},$$

(e) the *signed Wilcoxon autocorrelation coefficients*

$$r_{i;W}^{(n)+} = (n-i)^{-1} \sum_{t=i+1}^{n} \mathrm{sgn}(Z_t^{(n)} Z_{t-i}^{(n)}) R_{+,t}^{(n)} \log\left(\frac{n+1+R_{+,t-i}^{(n)}}{n+1-R_{+,t-i}^{(n)}}\right) / s_{+,W}^{(n)},$$

(f) the *signed Laplace autocorrelation coefficients*

$$r_{i;L}^{(n)+} = -(n-i)^{-1} \sum_{t=i+1}^{n} \mathrm{sgn}(Z_t^{(n)} Z_{t-i}^{(n)}) \log\left(1 - \frac{R_{+,t-i}^{(n)}}{n+1}\right) \Big/ \left\{ n^{-1} \sum_{i=1}^{n}\left[\log\left(\frac{i}{n+1}\right)\right]^2 \right\}^{\frac{1}{2}}.$$

A signed van der Waerden autocorrelation is thus the usual autocorrelation computed from the series of "signed standard normal quantiles" $\mathrm{sgn}(Z_t^{(n)})\Phi^{-1}$ $\left(\frac{1}{2} + R_t^{(n)}/2(n+1)\right)$ associated with the residual series $\mathbf{Z}^{(n)}$. Wilcoxon and Laplace autocorrelations constitute weighted versions of the traditional Wilcoxon signed rank and runs test statistics respectively.

The following asymptotic results can be established for all f–rank autocorrelations. If $H^{(n)}(A, B)$ (resp. $H_+^{(n)}(A, B)$) is the hypothesis of interest, the ranks (resp. signed ranks) to be used are those of the residuals $Z_t^{(n)} = [A(L)/B(L)]X_t^{(n)}$ (the "starting values" used in the inversion of $B(L)$ can be chosen arbitrarily.)

PROPOSITION 4.1. (Hallin et al., 1987; Hallin and Puri, 1991).

(i) Under $H^{(n)}(A, B)$, $(n - i)^{\frac{1}{2}} r_{i;f}^{(n)}$ and $(n - j)^{\frac{1}{2}} r_{j;f}^{(n)}$, $i \neq j$, are asymptotically jointly normal, with mean zero and unit covariance matrix.

(ii) Under $H_+^{(n)}(A, B)$, $(r_{i;f}^{(n)+} - r_{i;f}^{(n)})$ is $o_P(n^{-\frac{1}{2}})$; accordingly, $(n - i)^{\frac{1}{2}} r_{i;f}^{(n)+}$ and $(n - j)^{\frac{1}{2}} r_{j;f}^{(n)+}$, $i \neq j$ also are asymptotically jointly normal, with mean zero and unit covariance matrix.

4.4 Local asymptotic normality. Let $p_2 \geq p_1$, $q_2 \geq q_1$, $\mathbf{A} = (A_1 \ldots A_{p_1}$ $0 \ldots 0)' \in \mathbf{R}^{p_2}$, $\mathbf{B} = (B_1 \ldots B_{q_1} 0 \ldots 0)' \in \mathbf{R}^{q_2}$, $\boldsymbol{\theta} = (\mathbf{A}', \mathbf{B}')' \in \mathbf{R}^{p_2 + q_2}$. Denote by $\boldsymbol{\Theta}$ the (open) subset of $\mathbf{R}^{p_2 + q_2}$ for which (4.1) constitutes a valid, causal and invertible, ARMA model of orders p_1 and q_1 (i.e. $A_{p_1} \neq 0 \neq B_{q_1}$, no common roots, causality and invertibility). Similarly, let $\boldsymbol{\gamma} \in \mathbf{R}^{p_2}$, $\boldsymbol{\delta} = \mathbf{R}^{q_2}$, $\boldsymbol{\tau} = (\boldsymbol{\gamma}, \boldsymbol{\delta}')' \in \mathbf{R}^{p_2 + q_2}$. $H_g^{(n)}(\boldsymbol{\theta} + n^{-\frac{1}{2}} \boldsymbol{\tau})$ accordingly constitutes a sequence of hypotheses, approaching, in some sense to be made clearer in the sequel, to $H_g^{(n)}(\boldsymbol{\theta})$. Under this sequence $\mathbf{X}^{(n)}$ is generated by some ARMA (p, q) model, $p_1 \leq p \leq p_2$, $q_1 \leq q \leq q_2$, with coefficients of the form $\mathbf{A} + n^{-\frac{1}{2}} \boldsymbol{\gamma}, \mathbf{B} + n^{-\frac{1}{2}} \boldsymbol{\delta}$ and innovation density g.

Denote by $L_{\boldsymbol{\theta};g}^{(n)}$ the likelihood function of $\mathbf{X}^{(n)}$ under $H_g^{(n)}(\boldsymbol{\theta})$, hence by $L_{\boldsymbol{\theta}+n^{-\frac{1}{2}} \boldsymbol{\tau};g}^{(n)}$ the likelihood of $\mathbf{X}^{(n)}$ under $H_g^{(n)}(\boldsymbol{\theta} + n^{-\frac{1}{2}} \boldsymbol{\tau})$. Define the random variable

$$\Lambda_{\boldsymbol{\theta};\boldsymbol{\tau};g}^{(n)}(\mathbf{X}^{(n)}) = \log \left[L_{\boldsymbol{\theta}+n^{-\frac{1}{2}} \boldsymbol{\tau};g}^{(n)}(\mathbf{X}^{(n)}) / L_{\boldsymbol{\theta};g}^{(n)}(\mathbf{X}^{(n)}) \right]$$

(whenever $L_{\boldsymbol{\theta}+n^{-\frac{1}{2}} \boldsymbol{\tau};g}^{(n)}$ and $L_{\boldsymbol{\theta};g}^{(n)}$ are zero, Λ can be left arbitrary).

Denote by g_u and h_u the Green's functions associated with the difference operators $A(L)$ and $B(L)$ respectively (i.e. characterized by

$$\sum_{u=0}^{\infty} g_u L^u = [A(L)]^{-1} \quad \text{and} \quad \sum_{u=0}^{\infty} h_u L^u = [B(L)]^{-1}) .$$

Let

(4.4a)
$$a_i = \sum_{j=1}^{\min(p_2, i+p_1-1)} \gamma_j g_{i-j}$$

and

(4.4b)
$$b_i = \sum_{j=1}^{\min(q_2, i+q_1-1)} \delta_j h_{i-j},$$

so that

$$a(L) = \sum_{i=1}^{\infty} a_i L^i = \left[\sum_{i=1}^{p_2} \gamma_i L^i \right] / A(L)$$

and

$$b(L) = \sum_{i=1}^{\infty} b_i L^i = \left(\sum_{i=1}^{q_2} \delta_i L^i \right) / B(L).$$

It follows from the causality and invertibility properties of (4.1) that the sequences $\mathbf{a} = (a_i)$ and $\mathbf{b} = (b_i)$ are absolutely summable: denote by $\|\mathbf{a} + \mathbf{b}\|$ their ℓ^2 norm $\left[\sum_{i=1}^{\infty} (a_i + b_i)^2 \right]^{\frac{1}{2}}$. Also, letting $\pi = \max(p_2 - p_1, q_2 - q_1)$, introduce the $(\pi + p_1 + q_1) \times (p_2 + q_2)$ matrix

$$(4.5)\ \mathbf{M}(\boldsymbol{\theta}) = \begin{pmatrix} 1 & 0 & \cdots & 0 & 1 & 0 & \cdots & 0 \\ g_1 & 1 & & & h_1 & 1 & & \\ \vdots & & \ddots & \vdots & \vdots & & \ddots & \vdots \\ g_{p_2-1} & & \cdots & 1 & h_{q_2-1} & & \cdots & 1 \\ g_{p_2} & & \cdots & g_1 & h_{q_2} & & \cdots & h_1 \\ \vdots & & & \vdots & \vdots & & & \vdots \\ g_{\pi+p_1+q_1-1} & & \cdots & g_{\pi+p_1+q_1-p_2} & h_{\pi+p_1+q_1-1} & & \cdots & h_{\pi+p_1+q_1-q_2} \end{pmatrix}.$$

Finally, denote by $\{\psi_t^{(1)} \ldots \psi_t^{(p_1+q_1)}\}$ an arbitrary fundamental system of solutions of the homogeneous equation (or order $p_1 + q_1$) $A(L)B(L)\psi_t = 0$, $t \in \mathbf{Z}$; by $\mathbf{C}_\psi$ the Casorati matrix

$$(4.6) \qquad \mathbf{C}_\psi = \begin{pmatrix} \psi_{\pi+1}^{(1)} & \cdots & \psi_{\pi+1}^{(p_1+q_1)} \\ \psi_{\pi+2}^{(1)} & \cdots & \psi_{\pi+2}^{(p_1+q_1)} \\ \vdots & & \vdots \\ \psi_{\pi+p_1+q_1}^{(1)} & \cdots & \psi_{\pi+p_1+q_1}^{(p_1+q_1)} \end{pmatrix}$$

and by $n^{\frac{1}{2}} \mathbf{T}_{\psi;g}^{(n)}$ the vector of rank statistics

$$(4.7) \qquad n^{\frac{1}{2}} \mathbf{T}_{\psi;g}^{(n)} = \begin{pmatrix} (n-1)^{\frac{1}{2}} r_{1;g}^{(n)} \\ \vdots \\ (n-\pi)^{\frac{1}{2}} r_{\pi;g}^{(n)} \\ \sum_{i=\pi+1}^{n-1} (n-i)^{\frac{1}{2}} \psi_i^{(1)} \ r_{i;g}^{(n)} \\ \vdots \\ \sum_{i=\pi+1}^{n-1} (n-i)^{\frac{1}{2}} \psi_i^{(p_1+q_1)} \ r_{i;g}^{(n)} \end{pmatrix},$$

where the ranks are those of the residuals $Z_t^{(n)} = [A(L)/B(L)]X_t^{(n)}$. The covariance matrix of $n^{\frac{1}{2}}\mathbf{T}_{\psi;g}^{(n)}$ under $H^{(n)}(A,B)$ is

(4.8)
$$\mathbf{W}_\psi^{(n)} = \left(\begin{array}{c|c} \mathbf{I} & \mathbf{0} \\ \hline \mathbf{0} & \mathbf{w}_\psi^{(n)} \end{array} \right) ,$$

with
$$w_{\psi;k\ell}^{(n)} = \sum_{i=\pi+1}^{n-1} \psi_i^{(k)}\psi_i^{(\ell)} , \quad k,\ell = 1,\ldots,p_1+q_1 ,$$

which converges, as $n \to \infty$, to

(4.9)
$$\mathbf{W}_\psi = \left(\begin{array}{c|c} \mathbf{I} & \mathbf{0} \\ \hline \mathbf{0} & \mathbf{w}_\psi \end{array} \right) ,$$

with $w_{\psi,k\ell} = \sum_{i=\pi+1}^{\infty} \psi_i^{(k)}\psi_i^{(\ell)}$ (convergence again follows from the causality and invertibility of $A(L)$ and $B(L)$).

We may now state the main LAN result.

PROPOSITION 4.2. *For each $\boldsymbol{\theta} \in \Theta$ (and under mild technical assumptions on g)*

(i)
$$\Lambda_{\boldsymbol{\theta};\boldsymbol{\tau}^{(n)};g}^{(n)}(\mathbf{X}^{(n)}) = n^{\frac{1}{2}}(\boldsymbol{\tau}^{(n)})'\mathbf{M}'(\boldsymbol{\theta}) \left(\begin{array}{c|c} \mathbf{I}_{\pi\times\pi} & \mathbf{0} \\ \hline \mathbf{0} & \mathbf{C}_\psi'^{-1} \end{array} \right) \mathbf{T}_{\psi;g}^{(n)}(I(g_1))^{\frac{1}{2}}$$

$$-\frac{1}{2} (\boldsymbol{\tau}^{(n)})'\mathbf{M}'(\boldsymbol{\theta}) \left(\begin{array}{c|c} \mathbf{I}_{\pi\times\pi} & \mathbf{0} \\ \hline \mathbf{0} & \mathbf{C}_\psi'^{-1}\mathbf{w}_\psi\mathbf{C}_\psi^{-1} \end{array} \right) \mathbf{M}(\boldsymbol{\theta})\boldsymbol{\tau}^{(n)} \; I(g_1) + o_P(1) ,$$

under $H_g^{(n)}(\boldsymbol{\theta})$, as $n \to \infty$, for $\boldsymbol{\tau}^{(n)}$ such that $\sup_n (\boldsymbol{\tau}^{(n)})'\boldsymbol{\tau}^{(n)} < \infty$.

(ii) *for all $\boldsymbol{\tau} \in \mathbb{R}^{p_2+q_2}$, $n^{\frac{1}{2}}\boldsymbol{\tau}'\mathbf{M}'(\boldsymbol{\theta}) \left(\begin{array}{c|c} \mathbf{I}_{\pi\times\pi} & \mathbf{0} \\ \hline \mathbf{0} & \mathbf{C}_\psi'^{-1} \end{array} \right) \mathbf{T}_{\psi;g}^{(n)}$ is asymptoti-cally normal, under $H^{(n)}(\boldsymbol{\theta})$, as $n \to \infty$, with mean zero and variance*

$$\boldsymbol{\tau}'\mathbf{M}'(\boldsymbol{\theta}) \left(\begin{array}{c|c} \mathbf{I}_{\pi\times\pi} & \mathbf{0} \\ \hline \mathbf{0} & \mathbf{C}_\psi'^{-1}\mathbf{w}_\psi\mathbf{C}_\psi^{-1} \end{array} \right) \mathbf{M}(\boldsymbol{\theta})\boldsymbol{\tau} \; I(g_1) = \|\mathbf{a}+\mathbf{b}\|^2 I(g_1).$$

(iii) *Denote by $r_{i;g}^{(n)}(\boldsymbol{\tau})$ the f-rank autocorrelations computed from the residuals*

$$Z_t^{(n)}(\boldsymbol{\tau}) = \left[\sum_{i=1}^{p_2}(A_i + n^{-\frac{1}{2}}\gamma_i)L^i / \sum_{i=1}^{q_2}(B_i + n^{-\frac{1}{2}}\delta_i)L^i \right] X_t^{(n)} .$$

Then, under $H_f^{(n)}(\boldsymbol{\theta})$, as $n \to \infty$,

(4.10) $\quad n^{\frac{1}{2}}(r_{i;g}^{(n)}(\boldsymbol{\tau}) - r_{i,g}^{(n)}) =$

$$-(a_i + b_i)\left\{\int_0^1 \varphi_f(F^{-1}(u))\varphi_g(G^{-1}(u))du \int_0^1 F^{-1}(u)G^{-1}(u)du\right\}I(g_1)^{-\frac{1}{2}} + o_P(1)$$

$$= -(a_i + b_i)I(g|f) + o_P(1)$$

with $I(g|f) = \left\{\int_0^1 \varphi_f(F^{-1}(u))\varphi_g(G^{-1}(u))du \int F^{-1}(u)G^{-1}(u)du\right\}I(g_1)^{-\frac{1}{2}}$.

It follows from (i) and (ii) that $\Lambda_{\boldsymbol{\theta};r;g}^{(n)}(\mathbf{X}^{(n)})$ is asymptotically normal, under $H_g^{(n)}(\boldsymbol{\theta})$, with mean $-\frac{1}{2}\|\mathbf{a} + \mathbf{b}\|^2 I(g_1)$ and variance $\|\mathbf{a} + \mathbf{b}\|^2 I(g_1)$.

The family of likelihoods $\{L_{\boldsymbol{\theta};g}, \boldsymbol{\theta} \in \Theta\}$ is thus locally asymptotically normal (more precisely, the family $\{L_{\boldsymbol{\theta};g}, \boldsymbol{\theta} \in \mathbb{R}^{p_2+q_2}\}$ is *restricted locally asymptotically normal* for $\boldsymbol{\theta} \in \Theta$) in the sense of LeCam (1960)'s conditions (DN1 to DN6), and $\mathbf{T}_{\psi;f}^{(n)}$—hence the corresponding f–rank autocorrelations, or the ranks themselves are locally sufficient. Note that the dimension of the locally sufficient statistic $\mathbf{T}_{\psi;f}^{(n)}$ is $\pi + p_1 + q_1 = \max(p_1 + q_2, p_2 + q_1)$, whereas the dimension of the parameter space is $p_2 + q_2$. This corresponds to the well-known fact that the information matrix of an ARMA model is singular.

In view of Proposition 4.1, Proposition 4.2 remains valid, under symmetric densities, if signed ranks and signed rank autocorrelations are substituted for the unsigned ones. This latter fact is of particular interest.

The type of ranks (signed or unsigned) to be adopted—if rank–based techniques are to be considered—indeed depends, as explained in Section 1, on the invariance features of the testing problem at hand. Now, the effectiveness of the choice between signed unsigned ranks, in the classical symmetric i.i.d. case, is obscured, in practice, by the fact that (unsigned) ranks are totally insensitive to a variety of alternatives. Testing the slope of a regression line e.g. can be achieved (in a strictly unbiased manner) by means of either unsigned or signed rank techniques, whereas testing the intercept using unsigned ranks is impossible. A highly perverse consequence of this fact is that the type of alternative at hand in most cases apparently dictates which type of ranks (unsigned ones for the slope, signed ones for the intercept) should be adopted, and thus which invariance argument is relevant—see e.g. Puri and Sen (1985, Sections 5.2 and 5.3) for a typical example of this questionable attitude. Serial dependence problems and, more particularly, that of testing an ARMA model with unspecified innovation density, provide an interesting instance where an effective choice between signed and unsigned ranks cannot be eluded, and where power considerations do not supersede the much more fundamental invariance principles underlying this choice.

The following consequences of Proposition 4.2 will be particularly useful in the sequel.

PROPOSITION 4.3. *(i) Let J denote a score–generating function for the serial rank statistic $S^{(n)}$ in (3.2). Then, under $H_g^{(n)}(\boldsymbol{\theta}+n^{-\frac{1}{2}}\boldsymbol{\tau})$, $(n-k)^{\frac{1}{2}}(S^{(n)}-m^{(n)})$ is asymptotically normal, as $n \to \infty$, with mean $-\frac{1}{2}\sum_{i=1}^{k}(a_i+b_i)C_i$, where*

$$C_i = \sum_{j=0}^{k-i} \int_{[0,1]^{k+1}} J^*(u_1,\ldots,u_{k+1})\varphi_g(G^{-1}(u_{1+j}))G^{-1}(u_{1+j+i})du_1 \ldots du_{k+1} ,$$

and variance V^2 given in (3.15).

(ii)   *Let J_+ denote a score–generating function for the serial signed–rank statistic $S_+^{(n)}$ in (3.9). Then, under $H_g^{(n)}(\boldsymbol{\theta} + n^{-\frac{1}{2}}\boldsymbol{\tau})$ (g, a symmetric density function), $(n-k)^{\frac{1}{2}}S_+^{(n)}$ is asymptotically normal, as $n \to \infty$, with mean $\sum_{i=1}^{k}(a_i+b_i)C_i^+$, where*

$$C_i^+ = \sum_{j=0}^{k-i} \int_{[0,1]^{k+1}} J_+(2u_1-1,\ldots,2u_{k+1}-1)\varphi_g(G^{-1}(u_{1+j}))G^{-1}(u_{1+j+i})du_1 \ldots du_{k+1} ,$$

and variance V_+^2 given in (3.18) (due to the fact that

$$\int_0^1 [J_+(2u-1,\ldots,2u_{k+1}-1)-J_+^*(2u_1-1,\ldots,2u_{k+1}-1)\varphi_g(G^{-1}(u_\ell))G^{-1}(u_{\ell'})du_1 \ldots du_{k+1}=0$$

for all $1 \le \ell \ne \ell' \le k+1$, whether J_+ or J_+^* is used does not affect the value of C_i^+).

(iii)   *Both $(n-i)^{\frac{1}{2}}r_{i;f}^{(n)}$ and $(n-i)^{\frac{1}{2}}r_{i,f}^{(n)+}$ (which are asymptotically standard normal under $H^{(n)}(\boldsymbol{\theta})$ and $H_+^{(n)}(\boldsymbol{\theta})$, respectively), are asymptotically normal under $H_g^{(n)}(\boldsymbol{\theta}+n^{-\frac{1}{2}}\boldsymbol{\tau})$, as $n \to \infty$, with mean $(a_i+b_i)I(f|g)$, where*

$$I(f|g) = \left\{ \int_0^1 \varphi_f(F^{-1}(u))\varphi_g(G^{-1}(u))du \int_0^1 F^{-1}(u)G^{-1}(u)du \right\} [I(f_1)]^{-\frac{1}{2}}$$

and variance one (for $r_{i;f}^{(n)+}$, of course, f and g are to be symmetric).

PROPOSITION 4.4. *(i) $n^{\frac{1}{2}}\mathbf{T}_{\psi;f}^{(n)}$ is asymptotically normal, as $n \to \infty$, with mean $\mathbf{0}$ under $H^{(n)}(\boldsymbol{\theta})$, mean*

(4.11)
$$\begin{pmatrix} a_1 + b_1 \\ \vdots \\ a_\pi + b_\pi \\ \sum_{i=\pi+1}^{\infty}(a_i+b_i)\psi_i^{(1)} \\ \vdots \\ \sum_{i=\pi+1}^{\infty}(a_i+b_i)\psi_i^{(p_1+q_1)} \end{pmatrix} I(f|g) = \boldsymbol{\mu}_\psi(\boldsymbol{\tau})I(f|g)$$

under $H_g^{(n)}(\boldsymbol{\theta} + n^{-\frac{1}{2}}\boldsymbol{\tau})$ and full–rank covariance matrix $\mathbf{W}_\psi$ (see (4.9)) under both.

(ii) The quadratic rank statistic $n(\mathbf{T}_{\psi;f}^{(n)})' \begin{pmatrix} \mathbf{I} & \mathbf{0} \\ \mathbf{0} & \mathbf{w}_\psi^{-1} \end{pmatrix} \mathbf{T}_{\psi;f}^{(n)}$ does not depend on the particular fundamental system $\{\psi_t^{(1)}, \dots, \psi_t^{(p_1+q_1)}\}$ adopted; it is asymptotically chi–square, with $\pi + p_1 + q_1 = \max(p_1 + q_2, p_2 + q_1)$ degrees of freedom under $H^{(n)}(\boldsymbol{\theta})$, and asymptotically noncentral chi–square, still with $\max(p_1 + q_2, p_2 + q_1)$ degrees of freedom, but with noncentrality parameter $\frac{1}{2}[\|\mathbf{a} + \mathbf{b}\| I(f|g)]^2$ under $H_g^{(n)}(\boldsymbol{\theta} + n^{-\frac{1}{2}}\boldsymbol{\tau})$.

The consequences of Proposition 4.2 are of primary importance for all inference problems in the area: hypothesis testing, estimation, model selection and identification, among others. In the hypothesis testing context, which we develop in some detail in Sections 5 and 6, Proposition 4.2, in a very intuitive interpretation, implies that, asymptotically and locally, testing problems about $\boldsymbol{\theta}$ can be treated as testing problems about the mean $\boldsymbol{\mu}_\psi$ of the central sequence of statistics $\mathbf{T}_{\psi;f}^{(n)}$. A complex problem about ARMA parameters $\boldsymbol{\theta}$ thus turns out to be asymptotically equivalent to a hopefully much simpler one about the mean of a multinormal statistic with locally constant covariance structure.

5. LOCALLY ASYMPTOTICALLY OPTIMAL RANK TESTS

5.1 Locally asymptotically most powerful tests. The theoretical results of Section 4.4 allow for the construction of locally asymptotically optimal rank tests for a variety of problems. The problems treated in the present section are those dealing with ARMA models which, under the null hypothesis and except for the innovation density, are completely specified. Depending upon the alternative (which may have a one–dimensional or multidimensional structure, locally (one–sided) most powerful or locally maximin rank–based tests can be derived.

Denote by $H^{(n)}$ and $K^{(n)}$ two sequences of hypotheses, i.e. two sequences of non-overlapping subsets of some parameter space. A sequence $\phi_*^{(n)}$ of tests is called *asymptotically most powerful for* $H^{(n)}$ against $K^{(n)}$ at probability level α if

$$(5.1) \qquad \limsup_{n \to \infty} \left[E_{\boldsymbol{\theta}^{(n)}}(\phi_*^{(n)}) - \alpha \right] \leq 0, \quad \boldsymbol{\theta}^{(n)} \in H^{(n)}$$

and, for any sequence $\phi^{(n)}$ satisfying (5.1),

$$(5.2) \qquad \liminf_{n \to \infty} \left[E_{\boldsymbol{\theta}^{(n)}}(\phi_*^{(n)} - \phi^{(n)}) \right] \geq 0, \quad \boldsymbol{\theta}^{(n)} \in K^{(n)}.$$

Whenever $K^{(n)}$ can be considered as a *local alternative* with respect to $H^{(n)}$ (e.g. $K^{(n)}$ contiguous to $H^{(n)}$), $\phi_*^{(n)}$ will also be termed *locally asymptotically most powerful*.

PROPOSITION 5.1. *The (sequence of) rank–based test(s)*

$$(5.3) \qquad \phi^{(n)} = 1 \quad \text{iff} \quad \sum_{i=1}^{n-1} (n-i)^{\frac{1}{2}} (a_i + b_i) r_{i;f}^{(n)} > k_{1-\alpha} \left[\sum_{i=1}^{n-1} (a_i + b_i)^2 \right]^{\frac{1}{2}},$$

where $k_{1-\alpha} = \Phi^{-1}(1-\alpha)$ denotes the $(1-\alpha)$–standard normal quantile,

(i) is locally asymptotically most powerful (at probability level α) for testing $H^{(n)}(\boldsymbol{\theta})$ against $H_f^{(n)}(\boldsymbol{\theta} + n^{-\frac{1}{2}} k\boldsymbol{\tau})$, $k > 0$ arbitrary.

(ii) has asymptotic power

$$(5.4) \qquad 1 - \Phi(k_{1-\alpha} - \|\mathbf{a} + \mathbf{b}\| \, I(f|g))$$

against $H_g^{(n)}(\boldsymbol{\theta} + n^{-\frac{1}{2}}\boldsymbol{\tau})$.

Note that (as expected) the asymptotic power of the rank–based test (5.3) against $H_f^{(n)}(\boldsymbol{\theta} + n^{-\frac{1}{2}}\boldsymbol{\tau})$ —namely, $1 - \Phi(k_{1-\alpha} - \|\mathbf{a} + \mathbf{b}\| [I(f_1)]^{\frac{1}{2}})$ —equals that of the Neyman test $\phi_{NP}^{(n)}$ (for $H_f^{(n)}(\boldsymbol{\theta})$ against $H_f^{(n)}(\boldsymbol{\theta} + n^{-\frac{1}{2}}\boldsymbol{\tau})$). The latter indeed consists in rejecting $H_f^{(n)}(\boldsymbol{\theta})$ whenever $\Lambda_{\boldsymbol{\theta};\boldsymbol{\tau};f}^{(n)}$ is "too large", i.e., in view of Proposition 4.2, asymptotically reduces to

$$(5.5) \qquad \phi_{NP}^{(n)} = 1 \text{ iff } \Lambda_{\boldsymbol{\theta};\boldsymbol{\tau};f}^{(n)} + \frac{1}{2}\|\mathbf{a} + \mathbf{b}\|^2 I(f_1) > k_{1-\alpha}\|\mathbf{a} + \mathbf{b}\|[I(f_1)]^{\frac{1}{2}} \ .$$

The asymptotic distribution of $\Lambda_{\boldsymbol{\theta};\boldsymbol{\tau};f}^{(n)}$ under $H_f^{(n)}(\boldsymbol{\theta} + n^{-\frac{1}{2}}\boldsymbol{\tau})$ can be shown to be normal, with mean $\frac{1}{2}\|\mathbf{a} + \mathbf{b}\|^2 I(f_1)$ and the same variance as under $H_f^{(n)}(\boldsymbol{\theta})$. The asymptotic power of (5.5) therefore is $1 - \Phi(k_{1-\alpha} - \|\mathbf{a} + \mathbf{b}\|[I(f_1)]^{\frac{1}{2}})$. The norm $\|\mathbf{a}+\mathbf{b}\|$ accordingly can be interpreted as a "natural" distance between the sequences $H^{(n)}(\boldsymbol{\theta})$ and $H_f^{(n)}(\boldsymbol{\theta} + n^{\frac{1}{2}}\boldsymbol{\tau})$, equivalent to the L^1–distance $2 \sup_{\alpha}\{1 - \alpha - \Phi(k_{1-\alpha} - \|\mathbf{a} + \mathbf{b}\|[I(f_1)]^{-\frac{1}{2}})\}$.

Here again, in case the innovation densities are restricted to symmetric ones, signed ranks and signed autocorrelation coefficients can be substituted for the unsigned ones: Proposition 5.1 can be formulated without modification with $r_{i;f}^{(n)+}$ instead of $r_{i;f}^{(n)}$ (and, of course, symmetric densities f and g). If a parametric, Gaussian approach, is adopted (all densities then, under the null hypothesis as well as under the alternative, are assumed to be Gaussian), Proposition 5.1 takes the form

PROPOSITION 5.2. Denote by $r_i^{(n)}$ the classical autocorrelation coefficient of order i. The parametric test

$$\phi^{(n)} = 1 \text{ iff } \sum_{i=1}^{n-1}(n - i)^{\frac{1}{2}}(a_i + b_i) \, r_i^{(n)} > k_{1-\alpha}\left[\sum_{i=1}^{n-1}(a_i + b_i)^2\right]^{\frac{1}{2}}$$

(i) is locally most powerful (at probability level α) for testing $H^{(n)}(\boldsymbol{\theta})$ (or $H_+^{(n)}(\boldsymbol{\theta})$) against $H_N^{(n)}(\boldsymbol{\theta} + n^{-\frac{1}{2}} k\boldsymbol{\tau})$ (where N stands for a normal density with mean zero and arbitrary variance), $k > 0$ arbitrary.

(ii) has asymptotic power

$$(5.6) \qquad 1 - \Phi(k_{1-\alpha} - \|\mathbf{a} + \mathbf{b}\|)$$

against $H_g^{(n)}(\boldsymbol{\theta} + n^{-\frac{1}{2}}\boldsymbol{\tau})$, whatever g may be.

This confirms the analogy between f–rank autocorrelations (under innovation density f) and usual autocorrelation coefficients (under Gaussian innovation densities). The eventual superiority of f–rank autocorrelations comes from the $\mathbf{I}(\mathbf{f}|\mathbf{g})$ factor appearing in (5.4), but not in (5.6).

Example 5.1. Consider the problem of testing the null hypotheses $H^{(n)}(0)$ that $\mathbf{X}^{(n)}$ is white noise (with unspecified density) against an alternative of possible ARMA (p,q) dependence, $\max(p,q) = 1$, i.e. $X_t - AX_{t-1} = \varepsilon_t + B\varepsilon_{t-1}$, $A + B > 0$. Here $Z_t^{(n)} = X_t^{(n)}$ and the ranks are those of the observed series $\mathbf{X}^{(n)}$ itself. The rank–based test rejecting $H^{(n)}(0)$ whenever $(n-1)^{\frac{1}{2}} r_{1;f}^{(n)} > k_{1-\alpha}$ is locally most powerful (against $X_t - n^{-\frac{1}{2}}\alpha X_{t-1} = \varepsilon_t + n^{-\frac{1}{2}}\beta\varepsilon_{t-1}$, $\alpha + \beta > 0$ arbitrary) if ε_t, under the alternative, has density f (with arbitrary variance). If Gaussian (logistic, double exponential) alternatives are to be privileged or are to be feared most, then a van der Waerden (Wilcoxon, Laplace) autocorrelation $r_{1;vdW}^{(n)}$ $(r_{1;W}^{(n)}, r_{1;L}^{(n)})$ should be adopted. If the density under $H^{(n)}(0)$ can be assumed symmetric, then signed ranks and signed autocorrelations can be substituted for the unsigned ones.

Example 5.2. Consider the problem of testing $\rho = \rho_0$ against $\rho > \rho_0$ in the $AR(1)$ model $X_t - \rho X_t = \varepsilon_t$ with unspecified but symmetric innovation density. A signed–rank, genuinely distribution–free test can be performed, which is locally most powerful against Gaussian alternatives, among all tests at (asymptotic) probability level α, by rejecting $\rho = \rho_0$ whenever

$$(1 - \rho_0^2)^{\frac{1}{2}} \sum_{i=1}^{n-1} (n-i)^{\frac{1}{2}} \rho_0^{i-1} r_{i;f}^{(n)+} > k_{1\alpha} .$$

The signs and ranks here are those of the residuals $Z_t^{(n)} = X_t^{(n)} - \rho_0 X_{t-1}^{(n)}$ and their absolute values $|Z_t^{(n)}|$ (assume $\mathbf{X}^{(n)} = (X_0^{(n)}, X_1^{(n)}, \dots, X_n^{(n)})$). If the density (under $\rho = \rho_0$) cannot be assumed symmetric, unsigned ranks and unsigned autocorrelations should be used instead of the signed ones.

The asymptotic relative efficiencies of the signed and unsigned rank procedures described in this section with respect to each other, and with respect to their parametric counterparts, are computed in Section 5.3.

5.2 Locally maximin tests. The local alternatives considered in Section 5.2 are basically one–dimensional, one–sided alternatives: the tests provided by Propositions 5.1 and 5.2 are optimal against a specified $(\boldsymbol{\gamma}, \boldsymbol{\delta})$ direction in the parameter space, along with specified innovation density. In many situations of practical interest, the alternative is inherently multidimensional, and no particular $(\boldsymbol{\gamma}, \boldsymbol{\delta})$ direction is to be privileged. A natural idea then consists in considering tests which are *locally asymptotically maximin*.

Informally speaking, a maximin test, at given probability level α, is a test whose worst performance is best within the class of all tests at level α. More precisely, let $H^{(n)}$ and $K^{(n)}$ denote two sequences of (nonoverlapping) hypotheses. The corresponding sequence of *envelope power functions* is

$$\beta(\alpha; H^{(n)}, K^{(n)}) = \sup_{\phi} \inf_{\boldsymbol{\theta}} E_{\boldsymbol{\theta}}(\phi) , \quad \alpha \in (0,1) ,$$

where the $\sup\limits_{\phi}$ and $\inf\limits_{\boldsymbol{\theta}}$ are taken over all tests ϕ satisfying $E_{\boldsymbol{\theta}}\phi \leq \alpha$, $\boldsymbol{\theta} \in H^{(n)}$, and all values of $\boldsymbol{\theta}$ in $K^{(n)}$, respectively. A sequence $\phi_*^{(n)}$ of tests is called *asymptotically (locally,* if $K^{(n)}$ is a local alternative to $H^{(n)}$) maximin for $H^{(n)}$ against $K^{(n)}$, at (asymptotic) level α, if

(5.7) $$\limsup_{n \to \infty} \left[E_{\boldsymbol{\theta}^{(n)}}(\phi_*^{(n)}) - \alpha \right] \leq 0 , \quad \boldsymbol{\theta}^{(n)} \in H^{(n)}$$

and

(5.8) $$\liminf_{n \to \infty} \left[E_{\boldsymbol{\theta}^{(n)}}(\phi_*^{(n)}) - \beta(\alpha; H^{(n)}, K^{(n)}) \right] \geq 0 , \quad \boldsymbol{\theta}^{(n)} \in K^{(n)} .$$

If however nontrivial maximin tests are to be obtained, $K^{(n)}$ has to be bounded away from $H^{(n)}$. If indeed the L^1–distance between $H^{(n)}$ and $K^{(n)}$ (for fixed n) would be zero, the envelope power function trivially would reduce to $\beta(\alpha; H^{(n)}, K^{(n)}) = \alpha$. The simplest idea, if $H^{(n)}$ and a local alternative $K^{(n)}$ are to be considered, consists in defining a local alternative whose L^1–distance to $H^{(n)}$ remains (asymptotically) bounded from below by some fixed, strictly positive, constant d. In the problem of testing $H^{(n)}(\boldsymbol{\theta})$ (or $H_+^{(n)}(\boldsymbol{\theta})$) —i.e. a specified ARMA (p_1, q_1) model with unspecified (or symmetric unspecified) innovation density —against unspecified ARMA (p, q) alternatives, with $p_1 \leq p \leq p_2$ and $q_1 \leq q \leq q_2$, such local alternatives with bounded L_1–distance from $H^{(n)}(\boldsymbol{\theta})$ (or $H_+^{(n)}(\boldsymbol{\theta})$) are (for fixed f) of the form

(5.9) $$K_f^{(n)}(d) = \bigcup \{ H_f^{(n)}(\boldsymbol{\theta} + n^{-\frac{1}{2}}\boldsymbol{\tau}) \mid \|\mathbf{a} + \mathbf{b}\| \geq d \} ,$$

where the union is taken over all $\boldsymbol{\tau} \in \mathbb{R}^{p_2+q_2}$ such that $\|\mathbf{a} + \mathbf{b}\|$ (with $\mathbf{a}$ and $\mathbf{b}$ as defined in (4.4) is larger than or equal to d. The alternative (5.9) accordingly can be interpreted as the outside of a L^1–hypersphere. Due to the particular L^1 topology of ARMA likelihoods, $K_f^{(n)}(d)$ however resembles a cylinder rather than a sphere in the $\mathbb{R}^{p_2+q_2}$ space. The choice of d, as we shall see, does not affect the final result and the form of locally maximin tests.

PROPOSITION 5.3. *(i) The (sequence of) rank–based test(s)*

(5.10) $$\phi^{(n)} = 1 \text{ iff } n(\mathbf{T}_{\psi;f}^{(n)})' \begin{pmatrix} \mathbf{I} & \mathbf{0} \\ \mathbf{I} & \mathbf{w}_{\psi}^{-1} \end{pmatrix} \mathbf{T}_{\psi;f}^{(n)} > \chi_{1-\alpha}^2$$

(which, from Proposition 4.4, do not depend on the particular fundamental system adopted), where $\chi_{1-\alpha}^2$ denotes the $(1 - \alpha)$–quantile of a chi–square variable with

$\pi + p_1 + q_1 = \max(p_1 + q_2, p_2 + q_1)$ degrees of freedom, is locally asymptotically maximin, at probability level α, against any alternative of the form $K_f^{(n)}(d)$, $d > 0$.

(ii)   The asymptotic power of (5.10) against $H_g^{(n)}(\boldsymbol{\theta} + n^{-\frac{1}{2}}\boldsymbol{\tau})$ is

$$1 - F(\chi^2_{1-\alpha}; \tfrac{1}{2}\|\mathbf{a} + \mathbf{b}\|^2(I(f|g))^2) \, ,$$

where $F(\ ;\lambda)$ denotes the distribution function of a noncentral chi–square variable with $\pi + p_1 + q_1$ degrees of freedom and noncentrality parameter λ.

(iii)   The envelope power function $\beta(\alpha; H^{(n)}(\boldsymbol{\theta}), K_f^{(n)}(d))$ converges to (same notation as above) $1 - F(\chi^2_{1-\alpha}; \tfrac{1}{2}\|\mathbf{a} + \mathbf{b}\|^2 I(f_1))$.

The same result of course still holds if signed ranks are substituted for unsigned ones. Its parametric counterpart involves a parametric version $\mathbf{T}_{\psi;\mathcal{N}}^{(n)}$, say, of $\mathbf{T}_{\psi;f}^{(n)}$, where classical parametric autocorrelations $r_i^{(n)}$ are substituted for the rank-band ones $r_{i,f}^{(n)}$.

PROPOSITION 5.4. (i) Substituting $\mathbf{T}_{\psi;\mathcal{N}}^{(n)}$ for $\mathbf{T}_{\psi;f}^{(n)}$ in (5.10) yields a parametric test which is locally maximin, at probability level α, for $H^{(n)}(\boldsymbol{\theta})$ (for $H_+^{(n)}(\boldsymbol{\theta})$) against any $K_{\mathcal{N}}^{(n)}(d)$, $d > 0$ (where $\mathcal{N}$ stands for a normal density with mean zero and arbitrary variance).

(ii) The asymptotic power of this test against $H_g^{(n)}(\boldsymbol{\theta} + n^{-\frac{1}{2}}\boldsymbol{\tau})$ is (same notation as above) $1 - F(\chi^2_{1-\alpha}; \tfrac{1}{2}\|\mathbf{a} + \mathbf{b}\|^2)$.

Example 5.3. As in Example 5.1, consider the problem of testing the null hypothesis $H^{(n)}(0)$ that $\mathbf{X}^{(n)}$ is white noise, with unspecified (or unspecified symmetric) density. The alternative now is the whole class of possible ARMA (p, q) dependencies, with $0 < \max(p, q) \leq \pi$ and unspecified coefficients, but the subset of it at which optimality is desired is that with innovation density f. The f–rank (signed or unsigned) portmanteau test of order π

$$\phi^{(n)} = 1 \quad \text{if and only if} \quad \sum_{i=1}^{\pi}(n - i)(r_{i;f}^{(n)})^2 > \chi^2_{1-\alpha} \, ,$$

where $\chi^2_{1-\alpha}$ is the $(1-\alpha)$–quantile of a chi–square variable with π degrees of freedom, is then locally asymptotically maximin. The same property holds for the parametric portmanteau test (where the parametric autocorrelations $r_i^{(n)}$ are substituted for the rank–based $r_{i;f}^{(n)}$) against normal alternatives.

Example 5.4. Consider the problem of testing the null hypothesis that an observed series $\mathbf{X}^{(n)}$ was generated by the ARMA $(1,1)$ model $X_t - \dfrac{4}{5}X_{t-1} = \varepsilon_t + \dfrac{1}{2}\varepsilon_{t-1}$, $t \in \mathbf{Z}$, against unspecified ARMA $(2,2)$ alternatives (or, equivalently, against ARMA $(2,1)$ or ARMA $(1,2)$ alternatives). Let $Z_{-1}^{(n)} = Z_0^{(n)} = 0$ and

$$Z_t^{(n)} = X_t^{(n)} - \frac{4}{5} X_{t-1}^{(n)} - \frac{1}{2} Z_{t-1}^{(n)} \ , \ t = 1, \ldots, n \ ; \ \text{denote by } r_{i,f}^{(n)} \text{ the corresponding}$$

f-rank autocorrelations. A fundamental system of solutions of

$$\left(1 - \frac{4}{5} L\right)\left(1 + \frac{1}{2}\right)\psi_t = 0 \ , \ t \in \mathbb{Z}$$

is

$$\left\{\psi_t^{(1)} = \left(\frac{4}{5}\right)^{t-2} ; \psi_t^{(2)} = \left(-\frac{1}{2}\right)^{t-2}\right\};$$

here $\pi = \max(p_2 - p_1, q_2 - q_1) = 1$. The covariance matrix $\mathbf{W}^2$ is of the form

$$\mathbf{W}^2 = \left(\begin{array}{cc} \left[1 - \left(\frac{4}{5}\right)^2\right]^{-1} & \left[1 + \frac{2}{5}\right]^{-1} \\ \left[1 + \frac{2}{5}\right]^{-1} & \left[1 - \left(\frac{1}{2}\right)^2\right]^{-1} \end{array}\right) = \left(\begin{array}{cc} \frac{25}{9} & \frac{5}{7} \\ \frac{5}{7} & \frac{4}{3} \end{array}\right).$$

The test statistic of Proposition 5.3(i) is thus

$$Q_{\psi;f}^{(n)} = n(r_{1;f}^{(n)})^2 +$$

$$n\left(\sum_{i=2}^{n-1}\left(\frac{4}{5}\right)^{i-2} r_{i;f}^{(n)}, \sum_{i=2}^{n-1}\left(-\frac{1}{2}\right)^{i-2} r_{i;f}^{(n)}\right)\left(\begin{array}{cc} \frac{25}{9} & \frac{5}{7} \\ \frac{5}{7} & \frac{4}{3} \end{array}\right)\left(\begin{array}{c} \sum_{i=2}^{n-1}\left(\frac{4}{5}\right)^{i-2} r_{i;f}^{(n)} \\ \sum_{i=2}^{n-1}\left(-\frac{1}{2}\right)^{i-2} r_{i;f}^{(n)} \end{array}\right)$$

$$\simeq n(r_{1;f}^{(n)})^2 + (1.436)n\left[\sum_{i=2}^{n-1}\left(\frac{4}{5}\right)^i r_{i;f}^{(n)}\right]^2 + (19.600)n\left[\sum_{i=2}^{n-1}\left(-\frac{1}{2}\right)^i r_{i;f}^{(n)}\right]^2$$

$$- (3.937)n\left[\sum_{i=2}^{n-1}\left(\frac{4}{5}\right)^i r_{i;f}^{(n)}\right]\left[\sum_{i=2}^{n-1}\left(-\frac{1}{2}\right)^i r_{i;f}^{(n)}\right].$$

If ARMA (3,3) (or ARMA (3,2), ARMA (3,1), ARMA (1,3), ...) alternatives are to be considered, $Q_{\psi;f}^{(n)}$ has to be modified to

$$n(r_{1;f}^{(n)})^2 + n(r_{2;f}^{(n)})^2 + (2.243)n\left[\sum_{i=3}^{n-1}\left(\frac{4}{5}\right)^i r_{i;f}^{(n)}\right]^2 + (78.400)n\left[\sum_{i=3}^{n-1}\left(-\frac{1}{2}\right)^i r_{i;f}^{(n)}\right]^2$$

$$+ (9.844)n\left[\sum_{i=3}^{n-1}\left(\frac{4}{5}\right)^i r_{i;f}^{(n)}\right]\left[\sum_{i=3}^{n-1}\left(-\frac{1}{2}\right)^i r_{i;f}^{(n)}\right].$$

5.3 Asymptotic Relative Efficiencies. Asymptotic, local power comparisons between tests are usually made on the basis of asymptotic relative efficiencies (ARE's). Recall that the ARE of a test ψ_1 with respect to another one ψ_2, both at probability level α, can be defined as the limiting value of the ratio N_2/N_1, where N_1 is the number of observations required for the power of ψ_1 to equal the power of

ψ_2 based on N_2 observations, as $\min(N_1, N_2) \to \infty$ and the alternative converges to the null hypothesis (see e.g. Puri and Sen 1985, section 3.8, for a more rigorous definition).

ARE computations are greatly simplified in situations where ψ_1 and ψ_2 are based on test statistics which are asymptotically normal or chi–square under the null hypothesis and contiguous consequences of alternatives. If ψ_1 and ψ_2 consist in rejecting the null hypothesis $H^{(n)}$ for "large values of" test statistics, S_1 and S_2 respectively, which are asymptotically normal, $\mathcal{N}(0, \sigma_1)$ and $\mathcal{N}(0, \sigma_2)$ respectively, under $H^{(n)}$, and $\mathcal{N}(\mu_1, \sigma_1)$ and $\mathcal{N}(\mu_2, \sigma_2)$ respectively, under the alternative $K^{(n)}$, then the ARE of ψ_1 with respect to ψ_2 is

$$\text{ARE } (\psi_1/\psi_2) = \left(\frac{\mu_1 \sigma_2}{\sigma_1 \mu_2}\right)^2 .$$

If S_1 and S_2 are both asymptotically chi–square with d degrees of freedom under $H^{(n)}$, and asymptotically noncentral chi–square with d degrees of freedom and noncentrality parameters λ_1 and λ_2, respectively, then

$$\text{ARE } (\psi_1/\psi_2) = (\lambda_1/\lambda_2)^2 .$$

The results of Sections 5.1 and 5.2 thus allow for an explicit computation of the mutual ARE's of the various parametric and nonparametric tests proposed.

PROPOSITION 5.5. *The asymptotic efficiency of the optimal procedures (as described in Propositions 5.1 and (5.3) based on (signed or unsigned) f–rank autocorrelations with respect to their counterparts based on (signed or unsigned) g–rank autocorrelations, against alternatives of the form $H_h^{(n)}(\boldsymbol{\theta} + n^{-\frac{1}{2}}\boldsymbol{\tau})$, is*

$$[I(f|h)/I(g|h)]^2 = \left[\frac{\int_0^1 \phi_f(F^{-1}(u))\phi_h(H^{-1}(u))du \ \int_0^1 F^{-1}(u)H^{-1}(u)du}{\int_0^1 \phi_g(G^{-1}(u))\phi_h(H^{-1}(u))du \ \int_0^1 G^{-1}(u)H^{-1}(u)du}\right]^2 ,$$

where notations $\phi_f, \phi_g, \phi_h, F, G, H$ are used in an obvious fashion. Under the same conditions, the ARE of procedures based on f–rank autocorrelations with respect to the optimal Gaussian parametric procedures, based on classical parametric autocorrelation coefficients (as described in Propositions 5.2 and 5.4) is

$$\left[\int_0^1 \phi_f(F^{-1}(u))\phi_h(H^{-1}(u))du \int_0^1 F^{-1}(u)H^{-1}(u)du\right]^2 / I(f_1) ,$$

which reduces to $I(f_1) \geq 1$ for $h = f$ and $I(f_1) = 1$ if and only if f and h are Gaussian.

Other asymptotic efficiencies can be derived from Proposition 4.3. Explicit numerical values are provided in Table 5.1 below.

Table 5.1. *Mutual ARE's for the various parametric and nonparametric tests described in Sections 5.1 and 5.2, under normal, logistic and double exponential densities, respectively.*

		(1)	(2)	(3)	(4)	(5)	
	Classical	1.000	1.000	1.005	1.634	1.096	normal
(1)	parametric	1.000	0.954	0.911	1.232	1.000	logistic
		1.000	0.816	0.675	0.500	0.790	double exp.
	Signed or	1.000	1.000	1.055	1.634	1.096	normal
(2)	unsigned	1.048	1.000	0.954	1.291	1.048	logistic
	van der Waerden	1.226	1.000	0.827	0.613	0.968	double exp.
	Signed or	0.948	0.948	1.000	1.550	1.091	normal
(3)	unsigned	1.098	1.048	1.000	1.352	1.098	logistic
	Wilcoxon	1.482	1.209	1.000	0.741	1.170	double exp.
	Signed or	0.612	0.612	0.646	1.000	0.671	normal
(4)	unsigned	0.812	0.775	0.740	1.000	0.812	logistic
	Laplace	2.000	1.631	1.350	1.000	1.580	double exp.
	Signed or	0.912	0.912	0.917	1.490	1.000	normal
(5)	unsigned	1.000	0.954	0.911	1.232	1.000	logistic
	Spearman	1.266	1.033	0.855	0.633	1.000	double exp.

An inspection of Table 5.1 reveals the excellent asymptotic performances of rank tests: the van der Waerden tests perform uniformly and strictly better than the corresponding normal–theory tests—except of course under normal densities, where they perform equally well. The ARE of optimal normal theory tests with respect to the corresponding Laplace procedure can be as low as 0.500. Spearman tests (which are never optimal, since the Spearman autocorrelation coefficients do not belong to the class of f-rank autocorrelations) are uniformly (though not very significantly) dominated by Wilcoxon tests.

As expected from Proposition 4.3, signed and unsigned optimal tests are asymptotically equivalent. The fact that their mutual ARE's are one however does not imply that the advantage of using signed–rank tests instead of unsigned ones (whenever innovation densities are symmetric) is nil, or negligible, neither for short series lengths, nor even asymptotically. Numerical investigations of the performance of signed–rank tests (Hallin et al. 1990) indicate that, for moderate, fixed n, the power of signed–rank procedures can be substantially larger than that of unsigned ones, an empirical finding that should be confirmed by a theoretical investigation of the corresponding deficiencies.

A systematic investigation of the finite sample behavior of (unsigned) rank based tests for randomness, both under the null hypothesis (tables of exact critical values) and under alternatives of $AR(1)$ serial dependence (Monte Carlo study of the power function) has been conducted in Hallin and Mélard (1988). This study reveals that rank tests often are substantially more powerful than traditional parametric procedures, even for pretty short series ($n = 20$, e.g.); of course, they are much more reliable under the null hypothesis, and theoretically provide exact tests (whereas the probability level of parametric tests is only approximative) even when traditional procedures are not valid — e.g. under Cauchy innovation densities (see Hallin and Puri (1991a) for a numerical illustration). In addition, they also are considerably more robust, and less sensitive to the presence of outliers, atypical startup behavior, etc.

All these properties show how useful a general rank based methodology would be in the identification and validation steps of time series analysis. Such a methodology however requires additional results allowing for the treatment of nuisance parameters. This is the subject of the next section.

6. ALIGNED RANK TESTS

6.1 Ranks and aligned ranks. The test described in Section 5 are valid if and only if the residuals $Z_t^{(n)}$, the ranks of which are used, under the null hypothesis to be tested, are *exact* residuals. These tests thus allow for testing ARMA models with unspecified innovation densities but completely specified coefficients.

Unfortunately, in most practical problems, one has to test ARMA models with *unspecified* coefficients. This is the case, typically, in diagnostic checking and model validation situations. It is also the case in identification problems, where the main tools (partial correlograms, corner method tables, ...) actually are test statistics— even though the identification process does not exactly reduce to any hypothesis testing problem.

More generally, one might like to test the null hypothesis $H^{(n)}(\mathcal{E}_r)$ (or $H_+^{(n)}(\mathcal{E}_r)$, in the case of unspecified, symmetric innovations) that the parameters $A_1, \ldots, A_{p_1}$, $B_1, \ldots, B_{q_1}$ of the ARMA (p_1, q_1) model underlying the observed series $\mathbf{X}^{(n)}$ satisfy some given linear constraints (to be precise, $p_1 + q_1 - r$ linearly independent ones) and thus belong to some r-dimensional linear subspace $\mathcal{E}_r$ of $\mathbb{R}^{p_1+q_1}$. Innovation densities in $H^{(n)}(\mathcal{E}_r)$ and $H_+^{(n)}(\mathcal{E}_r)$ remain unspecified; whenever they need to be specified, the notation $H_f^{(n)}(\mathcal{E}_r)$ will be used. The alternative consists of unrestricted ARMA (p, q) models with $p_1 \leq p \leq p_2$, $q_1 \leq q \leq q_2$. More specific alternatives however might be considered when optimality properties are to be described.

In what follows, we assume the existence of a root n consistent sequence $\widehat{\boldsymbol{\theta}}^{(n)} = (\widehat{A}_1^{(n)}, \ldots, \widehat{A}_{p_1}^{(n)}, 0, \ldots, 0, \widehat{B}_1^{(n)}, \ldots, \widehat{B}_{q_1}^{(n)}, 0, \ldots, 0)'$ of constrained estimates of $\boldsymbol{\theta} = (A_1, \ldots, A_{p_1}, 0, \ldots, 0, B_1, \ldots, B_{q_1}, 0, \ldots, 0)'$, i.e. a sequence of statistics such that $(\widehat{A}_1^{(n)}, \ldots, \widehat{A}_{p_1}^{(n)}, \widehat{B}_1^{(n)}, \ldots, \widehat{B}_{q_1}^{(n)})' \in \mathcal{E}_r$ (in the sequel, we also write $\widehat{\boldsymbol{\theta}}^{(n)} \in \mathcal{E}_r$ or $\boldsymbol{\theta} \in \mathcal{E}_r$) and the distributions of $n^{\frac{1}{2}}(\widehat{\boldsymbol{\theta}}^{(n)} - \boldsymbol{\theta})$ form, under $H^{(n)}(\mathcal{E}_r)$ (or under $H_+^{(n)}(\mathcal{E}_r)$), a relatively compact sequence.

Root n consistency however is not sufficient for establishing the asymptotic results stated below: additional uniformity properties are required, which are satisfied if "smoother" versions of $\widehat{\boldsymbol{\theta}}^{(n)}$ are used (see LeCam 1960, Appendix 1). One of the "smoothed" versions of $\widehat{\boldsymbol{\theta}}^{(n)}$ is as follows (same reference). Denote by $\mathcal{A}^{(n)}$ the σ–field of Borel sets of $\mathbf{R}^{(n)}$, by V an open convex symmetric neighborhood of the origin in $\mathcal{E}_r$, by $\mathcal{A}_*^{(n)}$ the product σ–field of $\mathcal{A}^{(n)}$ by the σ–field of Borel subsets of $\mathcal{E}_r$. On V, letting ν denote a probability measure having a bounded continuous density with respect to the Lebesgue measure, define $\mathbf{v}^{(n)}$ as a random variable having distribution ν on V, and put

$$\boldsymbol{\theta}_*^{(n)} = \widehat{\boldsymbol{\theta}}^{(n)} + n^{-\frac{1}{2}}\mathbf{v}^{(n)} \ .$$

The sequence of estimates $\boldsymbol{\theta}_*^{(n)}$ is still root n consistent, and meets the uniformity requirements that might not hold for $\widehat{\boldsymbol{\theta}}^{(n)}$ and are technically necessary for establishing Propositions 6.1 and 6.2 below.

Other smoothing methods are also described in LeCam (1960).

We do insist however that these smoothing procedures are of little practical relevance, if any at all. First, because they typically have the nature of analytical procedures guaranteeing ad hoc probabilistic limit results, whereas in statistical practice, one is interested in approximation results only: for given V and ν, and for fixed, "reasonably large" n, it makes extremely little difference whether $\boldsymbol{\theta}_*^{(n)}$ or $\widehat{\boldsymbol{\theta}}^{(n)}$ is used. A second reason is that the actual computation of $\widehat{\boldsymbol{\theta}}^{(n)}$, with a finite number of decimal values, *automatically* provides a "smooth", approximate value $\boldsymbol{\theta}_{**}^{(n)}$, say, of $\widehat{\boldsymbol{\theta}}^{(n)}$ (hence of $\boldsymbol{\theta}_*^{(n)}$).

For all these reasons, this smoothing problem should not be given too much attention and, in the sequel, we shall make no notational difference between $\widehat{\boldsymbol{\theta}}^{(n)}, \boldsymbol{\theta}_*^{(n)}$ and $\boldsymbol{\theta}_{**}^{(n)}$: the simple notation $\widehat{\boldsymbol{\theta}}^{(n)}$ will be used throughout, and $\widehat{\boldsymbol{\theta}}^{(n)}$ will be assumed "smooth enough".

Denote by $\widehat{Z}_t^{(n)} = Z_t^{(n)}(\widehat{\boldsymbol{\theta}}^{(n)}) = (\Sigma_i \widehat{A}_i^{(n)} L^i / \Sigma_i \widehat{B}_i^{(n)} L^i) X_t^{(n)}$ the "estimated" residuals associated with $\widehat{\boldsymbol{\theta}}^{(n)}$, by $\widehat{R}_t^{(n)} = R_t^{(n)}(\widehat{\boldsymbol{\theta}}^{(n)})$, $\widehat{R}_{+,t}^{(n)} = R_{+,t}^{(n)}(\widehat{\boldsymbol{\theta}}^{(n)})$ and $\widehat{s}_t = s_t(\widehat{\boldsymbol{\theta}}^{(n)})$ the corresponding ranks and signs (namely, the *aligned* ranks, *aligned* signed ranks and *aligned* signs for the problem considered). The notation $\widehat{r}_{i;f}^{(n)} = r_{i;f}^{(n)}(\widehat{\boldsymbol{\theta}}^{(n)})$, $\widehat{r}_{i;f}^{(n)+} = r_{i;f}^{(n)+}(\widehat{\boldsymbol{\theta}}^{(n)}), \widehat{a}_i, \widehat{b}_i, \widehat{\psi}_t^{(1)}, \dots, \widehat{\psi}_t^{(p_1+q_1)}, \widehat{\mathbf{w}}_{\widetilde{\psi}}^{(n)}, \widehat{\mathbf{W}}_{\widetilde{\psi}}^{(n)}, \widehat{\mathbf{T}}_{\widetilde{\psi};f}^{(n)}, \widehat{\mathbf{T}}_{\widetilde{\psi};f}^{(n)+}, \dots$ will be used in a similar fashion.

Since the f–rank autocorrelations $r_{i;f}^{(n)}$ and the rank–based statistics $\mathbf{T}_{\psi;f}^{(n)}$ (or the signed ones $\mathbf{T}_{\psi;f}^{(n)+}$) have been shown locally asymptotically sufficient for $H^{(n)}(\boldsymbol{\theta})$ (or $H_+^{(n)}(\boldsymbol{\theta})$) against $H_f^{(n)}(\boldsymbol{\theta} + n^{-\frac{1}{2}}\boldsymbol{\tau}), \boldsymbol{\tau} \in \mathbf{R}^{p_2+q_2}$, a natural idea would consist in using the aligned autocorrelations $\widehat{r}_{i;f}^{(n)}$ (or $\widehat{r}_{i;f}^{(n)+}$) and aligned rank statistics $\widehat{\mathbf{T}}_{\widetilde{\psi}}^{(n)}$ (or $\widehat{\mathbf{T}}_{\widetilde{\psi}}^{(n)+}$) for $H^{(n)}(\mathcal{E}_r)$ (or $H_+^{(n)}(\mathcal{E}_r)$) against $\bigcup\{H_f^{(n)}(\boldsymbol{\theta} + n^{-\frac{1}{2}}\boldsymbol{\tau}), \boldsymbol{\theta} \in \mathcal{E}_r, \boldsymbol{\theta} + n^{-\frac{1}{2}}\boldsymbol{\tau} \notin \mathcal{E}_r\}$. A closer look at this apparently simple idea however reveals a number of apparently unredeemable theoretical drawbacks.

First, invariance is lost. Substituting estimated parameters $\widehat{\boldsymbol{\theta}}^{(n)}$ for exact ones $\boldsymbol{\theta}$ induces among the residuals $\widehat{Z}_t^{(n)}$ complex interrelations that destroy their exchangeability features, hence the invariance property of ranks. This is not just a slight defect in view of the fact, stressed in Section 1, that invariance is the cornerstone of rank–based inference: *if invariance is lost, there is not point in using ranks anymore.*

Second, distribution–freeness, and even asymptotic distribution–freeness, is lost as well: it follows indeed from Proposition 6.1 below that the asymptotic mean of aligned rank autocorrelation coefficients under $H_g^{(n)}(\boldsymbol{\theta})$ depends on g. This again is not just a detail without importance, since distribution-freeness (or, at least, asymptotic distribution-freeness) is crucially necessary if a testing procedure valid under unspecified densities g is to be carried out.

Last and not the least, the local optimality properties of Section 5 obviously cannot be expected to hold anymore in the case of aligned rank tests. Even in the parametric, normal-theory context, additional requirements of unbiasedness or similarity have to be invoked, when nuisance parameters are present, if optimality results are to be obtained. Some form of similarity is likely to be necessary here, too.

6.2 Asymptotic invariance. Since strict invariance apparently is too restrictive a requirement here, some weaker, asymptotic form of it might be more appropriate. Let us define an asymptotically invariant statistic as a statistic asymptotically equivalent (under the null hypothesis considered, hence under contiguous alternatives) to an invariant (hence, in the present context, rank–based) one. More precisely, let $S^{(n)}$ denote a sequence of rank–based statistics such that, for some appropriate centering sequence $m^{(n)}$, $n^{-\frac{1}{2}}(S^{(n)} - m^{(n)})$ is relatively compact under a sequence of null hypotheses $H^{(n)}$. The sequence $\widetilde{S}^{(n)}$ obtained on substituting aligned ranks (aligned signs, aligned signed ranks) for the exact ones in $S^{(n)}$ is said to be *asymptotically invariant* under $H^{(n)}$ if $\widetilde{S}^{(n)} - S^{(n)} = o_P(n^{-\frac{1}{2}})$, under $H^{(n)}$, as $n \to \infty$.

This concept of asymptotic invariance here would be helpful if, e.g. $\widehat{r}_{i;f}^{(n)} - r_{i;f}^{(n)}$ or $\widehat{r}_{i;f}^{(n)+} - r_{i;f}^{(n)+}$ would be $o_P(n^{-\frac{1}{2}})$ under $H^{(n)}(\boldsymbol{\theta})$, i.e. if aligned rank autocorrelation coefficients would be asymptotically equivalent to the genuine ones. Unfortunately, this does not hold, since, from Proposition 4.2, it can be shown that

PROPOSITION 6.1. *Under* $H_g^{(n)}(\boldsymbol{\theta})$,

$$(6.1) \qquad n^{\frac{1}{2}}(\widehat{r}_{i;f}^{(n)} - r_{i;f}^{(n)}) = -(\widehat{a}_i^{(n)} + \widehat{b}_i^{(n)})I(f|g) + o_P(1) , \quad n \to \infty ,$$

where $\widehat{a}_i^{(n)}$ *and* $\widehat{b}_i^{(n)}$ *result from substituting* $n^{1/2}(\widehat{\boldsymbol{\theta}}^{(n)} - \boldsymbol{\theta})$ *for* $\boldsymbol{\tau} = (\boldsymbol{\gamma}, \boldsymbol{\delta})'$ *in* (4.4). *The same result also holds for* $\widehat{r}_{i;f}^{(n)+}$.

It follows that $\widehat{r}_{i;f}^{(n)}$ cannot be asymptotically invariant: $I(f|g)$ indeed is not a distribution–free quantity. Nor can $\widehat{r}_{i;f}^{(n)+}$ be.

Though however no aligned–rank autocorrelation individually is asymptotically invariant, some specific linear combinations of them are. More precisely, denote by $\mathbf{T}_{\psi;f}^{(n)}$ the locally (at $\boldsymbol{\theta}$) sufficient statistic associated with the fundamental system $\{\psi_t^{(1)}, \ldots, \psi_t^{(p_1+q_1)}\}$, by $\widehat{\psi}_t^{(j)}$ the value at t of the solution of $(\Sigma_i \widehat{A}_i L^i)(\Sigma_i \widehat{B}_i L^i)\psi_t =$

0 characterized by the same starting values $\widehat{\psi}_s^{(j)} = \psi_s^{(j)}$, $s = 1, 2, \ldots, p_1 + q_1$ as $\psi_t^{(j)}$, and by $\widehat{\mathbf{T}}_{\widehat{\psi};f}^{(n)}$ the locally (at $\boldsymbol{\theta} = \widehat{\boldsymbol{\theta}}^{(n)}$) sufficient statistic associated with the fundamental system $\{\widehat{\psi}_t^{(n)}, \ldots, \widehat{\psi}_t^{(p_1+q_1)}\}$ but computed from the aligned ranks $\widehat{R}_t^{(n)} \cdot \widehat{\mathbf{T}}_{\widehat{\psi};f}^{(n)+}$ can be defined similarly in the signed rank case.

The asymptotic mean under $H_g^{(n)}(\boldsymbol{\theta} + n^{-\frac{1}{2}}\boldsymbol{\tau})$ of $n^{1/2}\mathbf{T}_{\psi;f}^{(n)}$ is $\boldsymbol{\mu}_{\psi;f}(\boldsymbol{\tau})I(f|g)$ (Proposition 4.4) where $\boldsymbol{\mu}_{\psi;f}$ is a linear transform (of rank $\pi + p_1 + q_1$) of $\boldsymbol{\tau}$, depending on $\boldsymbol{\theta}$ and the choice of the fundamental system $\{\psi_t^{(j)}\}$, but not on g. When $\boldsymbol{\tau}$ unrestrictedly takes its values in $\mathbf{R}^{p_2+q_2}$, $\boldsymbol{\mu}_{\psi;f}$ can take any value in $\mathbf{R}^{\pi+p_1+q_1}$. If $\boldsymbol{\tau}$ is of the form $(\gamma_1, \ldots, \gamma_{p_1}, 0, \ldots, 0, \delta_1, \ldots, \delta_{q_1}, 0, \ldots, 0)'$, where $(\gamma_1, \ldots, \gamma_{p_1}, \delta_1, \ldots, \delta_{q_1})'$ belongs to $\mathcal{E}_r$, thus satisfying $p_1 + q_1 - r$ independent linear restrictions, it can be shown that $\boldsymbol{\mu}_{\psi;f}(\boldsymbol{\tau})$ also satisfies $\pi + p_1 + q_1 - r$ independent linear restrictions, of the form

$$(6.2) \qquad\qquad \boldsymbol{\Omega}(\boldsymbol{\theta})\boldsymbol{\mu}_{\psi;f} = \mathbf{0} \; ,$$

where $\boldsymbol{\Omega}(\boldsymbol{\theta})$, a $(\pi + p_1 + q_1 - r) \times (\pi + p_1 + q_1)$ matrix of rank $\pi + p_1 + q_1 - r = \max(p_2 + q_1, p_1 + q_2) - r$, is a continuous function of $\boldsymbol{\theta}$. The vector $\boldsymbol{\mu}_{\psi;f}$ accordingly lies in a r-dimensional linear subspace of $\mathbf{R}^{\pi+p_1+q_1}$.

The following result can then be established.

PROPOSITION 6.2. *For any $\boldsymbol{\theta} \in \mathcal{E}_r$, any fundamental system $\{\psi_t^{(i)}\}$ and any f,*

$$(6.3) \qquad\qquad \boldsymbol{\Omega}(\widehat{\boldsymbol{\theta}}^{(n)}) \left[\widehat{\mathbf{T}}_{\widehat{\psi};f}^{(n)} - \mathbf{T}_{\psi;f}^{(n)} \right] = o_P(n^{-\frac{1}{2}})$$

under $H^{(n)}(\boldsymbol{\theta})$, $\boldsymbol{\theta} \in \mathcal{E}_r$, as $n \to \infty$. Accordingly, $\boldsymbol{\Omega}(\widehat{\boldsymbol{\theta}}^{(n)})\widehat{\mathbf{T}}_{\widehat{\psi};f}^{(n)}$ is a vector of $\pi + p_1 + q_1 - r$ asymptotically invariant statistics (under $H^{(n)}(\boldsymbol{\theta})$).

A similar result can be proved, under $H_+^{(n)}(\boldsymbol{\theta})$, for $\boldsymbol{\Omega}(\widehat{\boldsymbol{\theta}}^{(n)})\widehat{\mathbf{T}}_{\widehat{\psi};f}^{(n)+}$.

6.3 Locally asymptotically similar tests. The following definition of local asymptotic similarity has been proposed by LeCam (1960) (under the terminology *differential asymptotic similarity*). A sequence $\phi^{(n)}$ of tests is said to be locally asymptotically similar under $H_f^{(n)}(\mathcal{E}_r)$, at asymptotic probability level α, if for every $\boldsymbol{\theta} \in \mathcal{E}_r$ and every bounded $B \subset \mathcal{E}_r$

$$(6.4) \qquad\qquad \lim_{n\to\infty} \sup_{\boldsymbol{\tau} \in B} \left| E_{H_f^{(n)}(\boldsymbol{\theta}+n^{-\frac{1}{2}}\boldsymbol{\tau})}(\phi^{(n)}) - \alpha \right| = 0 \; .$$

LeCam (loc. cit.) then shows how locally optimal tests, asymptotically enjoying, within the class of locally asymptotically similar tests, the same optimal properties (stringency or minimaxity) as the usual normal theory tests for Gaussian linear hypotheses, follow on applying the Gaussian likelihood ratio principle at the local asymptotical level.

For the particular problem of testing $H_f^{(n)}(\mathcal{E}_r)$, this leads to a test statistic of the form

(6.5)
$$
\begin{aligned}
Q_f^{(n)} &= \inf_{\widetilde{\boldsymbol{\mu}}:\boldsymbol{\Omega}(\widehat{\boldsymbol{\theta}}^{(n)})\widetilde{\boldsymbol{\mu}}=\mathbf{0}} [(n^{\frac{1}{2}}\widehat{\mathbf{T}}_{\widehat{\psi};f}^{(n)} - \widetilde{\boldsymbol{\mu}})'\widehat{\mathbf{W}}_{\widehat{\psi}}^{-1}(n^{\frac{1}{2}}\widehat{\mathbf{T}}_{\widehat{\psi};f}^{(n)} - \widetilde{\boldsymbol{\mu}})] \\
&= n(\widehat{\mathbf{T}}_{\widehat{\psi};f}^{(n)})'\left[\widehat{\mathbf{W}}_{\widehat{\psi}}^{-1} - \widehat{\mathbf{W}}_{\widehat{\psi}}^{-1}\mathbf{K}(\mathbf{K}'\widehat{\mathbf{W}}_{\widehat{\psi}}^{-1}\mathbf{K})^{-1}\mathbf{K}'\widehat{\mathbf{W}}_{\widehat{\psi}}^{-1}\right]\widehat{\mathbf{T}}_{\widehat{\psi};f}^{(n)} ,
\end{aligned}
$$

where $\mathbf{K} = \mathbf{K}(\widehat{\boldsymbol{\theta}}^{(n)})$ is any $(\pi + p_1 + q_1) \times r$ full rank matrix such that $\boldsymbol{\Omega}(\widehat{\boldsymbol{\theta}}^{(n)})\mathbf{K} = \mathbf{0}$ (so that $\boldsymbol{\Omega}(\widehat{\boldsymbol{\theta}}^{(n)})\widetilde{\boldsymbol{\mu}} = \mathbf{0}$ if and only if $\widetilde{\boldsymbol{\mu}} = \mathbf{K}\boldsymbol{\lambda}$ for some $\boldsymbol{\lambda} \in \mathbf{R}^r$, and $\mathbf{N}\,\mathbf{K} = \mathbf{0}$ if and only if $\mathbf{N} = \mathbf{M}\boldsymbol{\Omega}(\widehat{\boldsymbol{\theta}}^{(n)})$ for some matrix $\mathbf{M}$ of appropriate dimension).

Denoting by $\widehat{\mathbf{W}}_{\widehat{\psi}}^{\frac{1}{2}}$ an arbitrary symmetric square root of $\widehat{\mathbf{W}}_{\widehat{\psi}}$, (6.5) also takes the form

(6.6)
$$
\begin{aligned}
&n(\widehat{\mathbf{T}}_{\widehat{\psi};f}^{(n)})'\widehat{\mathbf{W}}_{\widehat{\psi}}^{-\frac{1}{2}}\left[\mathbf{I} - \widehat{\mathbf{W}}_{\widehat{\psi}}^{-\frac{1}{2}}\mathbf{K}(\mathbf{K}'\widehat{\mathbf{W}}_{\widehat{\psi}}^{-1}\mathbf{K})^{-1}\mathbf{K}'\widehat{\mathbf{W}}_{\widehat{\psi}}^{-\frac{1}{2}}\right]\widehat{\mathbf{W}}_{\widehat{\psi}}^{-\frac{1}{2}}\widehat{\mathbf{T}}_{\widehat{\psi};f}^{(n)} \\
&= n(\widehat{\mathbf{T}}_{\widehat{\psi};f}^{(n)})'\widehat{\mathbf{W}}_{\widehat{\psi}}^{-\frac{1}{2}}\mathbf{P}\widehat{\mathbf{W}}_{\widehat{\psi}}^{-\frac{1}{2}}\widehat{\mathbf{T}}_{\widehat{\psi};f}^{(n)} ,
\end{aligned}
$$

where $\mathbf{P}$ is a symmetric, idempotent matrix of rank $\pi + p_1 + q_1 - r$. $Q_f^{(n)}$ therefore is asymptotically chi–square under $H_f^{(n)}(\boldsymbol{\theta})$, $\boldsymbol{\theta} \in \mathcal{E}_r$. But, since $\mathbf{P}\,\widehat{\mathbf{W}}_{\widehat{\psi}}^{-\frac{1}{2}}\mathbf{K} = \mathbf{0}$, $Q_f^{(n)}$ is also of the form

$$
n(\widehat{\mathbf{T}}_{\widehat{\psi};f}^{(n)})'\boldsymbol{\Omega}'(\widehat{\boldsymbol{\theta}}^{(n)})\mathbf{M}'\,\mathbf{M}\boldsymbol{\Omega}(\widehat{\boldsymbol{\theta}}^{(n)})\widehat{\mathbf{T}}_{\widehat{\psi};f}^{(n)}
$$

(for some matrix $\mathbf{M}$, the explicit form of which is not needed here); hence, in view of Proposition 6.2, $\mathbf{Q}_f^{(n)}$ is asymptotically invariant under $H^{(n)}(\boldsymbol{\theta})$, $\boldsymbol{\theta} \in \mathcal{E}_r$. Accordingly, $Q_f^{(n)}$ is also asymptotically chi–square, with $\pi + p_1 + q_1 - r$ degrees of freedom, under any hypothesis of the form $H_g^{(n)}(\boldsymbol{\theta})$, hence under $H^{(n)}(\boldsymbol{\theta})$.

A similar result holds for the signed version $Q_f^{(n)+}$ of $Q_f^{(n)}$.

6.4 Optimal aligned rank tests. Summing up the findings of Sections 6.2 and 6.3, we may state the following result.

PROPOSITION 6.3. *The sequence of aligned–rank tests rejecting $H^{(n)}(\mathcal{E}_r)$ whenever*

$$
Q_f^{(n)} > \chi^2_{1-\alpha} ,
$$

where $\chi^2_{1-\alpha}$ denotes the $(1-\alpha)$–quantile of a chi–square variable with $\pi + p_1 + q_1 - r$ degrees of freedom and $Q_f^{(n)}$ is given in (6.5)

(i) is asymptotically invariant

(ii) is locally asymptotically similar under $H^{(n)}(\mathcal{E}_r)$, at probability level α

(iii) is asymptotically most stringent against $\bigcup\{H_f^{(n)}(\boldsymbol{\theta} + n^{-\frac{1}{2}}\boldsymbol{\tau})|\boldsymbol{\theta} \in \mathcal{E}_r, \boldsymbol{\tau} \notin \mathcal{E}_r\}$, within the class of all asymptotically similar (under $H_f^{(n)}(\mathcal{E}_r)$, at probability level α) tests.

A similar proposition holds in the signed–rank case.

The test thus proposed possesses all the asymptotic optimality properties one can expect when testing $H_f^{(n)}(\mathcal{E}_r)$ (with specified innovation density f). In addition, it is asymptotically invariant, similar and distribution–free under the much broader null hypothesis $H^{(n)}(\mathcal{E}_r)$ (with unspecified innovation density); it has much better robustness features than its usual Gaussian parametric competitors, with ARE values (with respect to the latter) as shown in Table 5.1. The amount of computation involved is not heavier than in the case e.g. of traditional Lagrange multiplier tests (see also Kreiss (1990 a) for optimal parametric AR procedures).

For all these reasons, they should be very attractive in a variety of practical problems when innovation densities are suspected to be severely non–Gaussian, or contaminated by possible outliers. For instance, it advantageously could be substituted for parametric Lagrange multiplier statistics in Pötscher (1983, 1985)'s recursive identification procedure.

6.5 Example: testing AR (1) versus ARMA (2,1) dependence. As an illustration, consider the problem of testing a null hypothesis of first–order autoregressive dependence

$$X_t - \rho X_{t-1} = \varepsilon_t , \quad 0 < |\rho| < 1 ,$$

where the parameter ρ and density of ε_t remain unspecified. Denote by $\widehat{\rho}^{(n)}$ a root n consistent estimate of ρ (the least square estimate, for instance), and let $\widehat{Z}_t^{(n)} = X_t^{(n)} - \widehat{\rho}^{(n)} X_{t-1}^{(n)}$, $t = 1, \ldots, n$ (for simplicity, assume that $X_0^{(n)}$ is available). Assume that an ARMA (2,1) alternative is considered. Suppose however that, for some reason, one is willing to be particularly powerful against ARMA processes with innovation density f (f specified up to a scale transformation). The (aligned) ranks $\widehat{R}_t^{(n)}$ here are those of the residuals $\widehat{Z}_t^{(n)}$, and the autocorrelation coefficients to be used are the (aligned) f–rank autocorrelations $\widehat{r}_{i;f}^{(n)}$.

A simple fundamental system (a fundamental system here consists of any non-identically zero solution, since the dimension of the solution space is one) of solutions of $(1 - \widehat{\rho}^{(n)} L)\psi_t = 0$ is $\widehat{\psi}_t = (\widehat{\rho}^{(n)})^{t-2}$, and the corresponding (aligned) locally sufficient statistic is ($p_1 = 1$, $q_1 = 0$; hence $\pi = 1$ and $r = 1$)

$$n^{\frac{1}{2}} \widehat{\underset{\sim}{T}}_{\rho;f}^{(n)} = \begin{pmatrix} (n-1)^{\frac{1}{2}} \, \widehat{r}_{1;f}^{(n)} \\ \sum_{j=2}^{n-1} (n-j)^{\frac{1}{2}} (\widehat{\rho}^{(n)})^{j-2} \widehat{r}_{j;f}^{(n)} \end{pmatrix} .$$

The corresponding, *exact* $\mathbf{T}_{\psi;f}^{(n)}$ (not a statistic since ρ is unknown) is

$$n^{\frac{1}{2}} \mathbf{T}_{\rho;f}^{(n)} = \begin{pmatrix} (n-1)^{\frac{1}{2}} \, r_{1;f}^{(n)} \\ \sum_{j=2}^{n-1} (n-j)^{\frac{1}{2}} \, \rho^{j-2} r_{j;f}^{(n)} \end{pmatrix}$$

which is asymptotically bivariate normal, with mean zero and covariance matrix

$$\mathbf{W}_\psi = \begin{pmatrix} 1 & 0 \\ 0 & (1-\rho^2)^{-1} \end{pmatrix}$$

under $H^{(n)}(\rho)$.

Local ARMA $(2,1)$ alternatives $H_g^{(n)}(\rho,\boldsymbol{\gamma},\delta)$ are of the form

$$X_t - (\rho + n^{-\frac{1}{2}}\gamma_1)X_{t-1} - n^{-\frac{1}{2}}\gamma_2 X_{t-2} = \varepsilon_t + n^{-\frac{1}{2}}\delta\,\varepsilon_{t-1}\;,$$

where $\boldsymbol{\gamma} = (\gamma_1,\gamma_2)'$ and the innovation density is g ; $n^{\frac{1}{2}}\mathbf{T}_{\psi;f}^{(n)}$ under such an alternative is asymptotically bivariate normal, still with covariance matrix $\mathbf{W}$, and with mean

$$\boldsymbol{\mu}_{\rho;f}(\boldsymbol{\gamma},\delta)I(f|g) = \begin{pmatrix} \gamma_1 + \delta \\ [\rho(\gamma_1 + \delta) + \gamma_2]/(1 - \rho^2) \end{pmatrix} I(f|g)$$

since $\qquad a_i = \begin{cases} \gamma_1 & \text{if } i = 1 \\ \rho^{i-2}(\rho\gamma_1 + \gamma_2) & \text{if } i > 1 \end{cases}$ and $b_i = \delta\rho^{i-1}$, $i = 1,2,\dots$.

Clearly, if $\gamma_2 = \delta = 0$, $\boldsymbol{\mu}_{\rho;f}(\boldsymbol{\gamma},\delta)$ is of the form $\gamma_1(1 \quad \rho/(1-\rho^2))'$, and accordingly satisfies the (unique, since $\pi + p_1 + q_1 - r = 1 + 1 + 0 - 1 = 1$) linear constraint

$$(-\rho \qquad 1 - \rho^2)\boldsymbol{\mu}_{\rho;f}(\boldsymbol{\gamma},\delta) = 0\;.$$

Here $\boldsymbol{\Omega}(\boldsymbol{\theta}) = \boldsymbol{\Omega}(\rho)$ is the 1×2 row matrix $(-\rho \quad 1 - \rho^2)$. It follows that

$$(-\widehat{\rho}^{(n)} \qquad 1 - (\widehat{\rho}^{(n)})^2)\;\widehat{\mathbf{T}}_{\rho;f}^{(n)}$$

(6.7)
$$= -(n-1)^{\frac{1}{2}}\widehat{\rho}^{(n)}\widehat{r}_{1;f}^{(n)} + [1 - (\widehat{\rho}^{(n)})^2]\sum_{j=2}^{n-1}(n-j)^{\frac{1}{2}}(\widehat{\rho}^{(n)})^{j-2}\widehat{r}_{j;f}^{(n)}$$

is asymptotically invariant, and asymptotically equivalent to its exact, non–aligned counterpart. The optimal, asymptotically similar aligned rank test of Proposition 6.3 consists in rejecting the null hypothesis whenever the quadratic statistic

(6.8)
$$Q_f^{(n)} = \left\{ -(n-1)^{\frac{1}{2}}\widehat{\rho}^{(n)}\widehat{r}_{1;f}^{(n)} + [1 - (\widehat{\rho}^{(n)})^2]\sum_{j=2}^{n-1}(n-j)^{\frac{1}{2}}(\widehat{\rho}^{(n)})^{j-2}\,\widehat{r}_{j;f}^{(n)} \right\}^2$$

exceeds the $(1 - \alpha)$ quantile of a chi–square variable with one degree of freedom (since the asymptotic variance of (6.7) is one).

The form of the test statistic (6.8) is not quite familiar, and its relation with classical, parametric time series procedures does not straightforwardly appear from (6.8). Denote by $\mathcal{Q}_N^{(n)}$ the Gaussian counterpart of $Q_f^{(n)}$

(6.9)
$$Q_N^{(n)} = \left\{ -(n-1)^{\frac{1}{2}}\widehat{\rho}^{(n)}\widehat{r}_1^{(n)} + [1 - (\widehat{\rho}^{(n)})^2]\sum_{j=2}^{n-1}(n-j)^{\frac{1}{2}}(\widehat{\rho}^{(n)})^{j-2}\widehat{r}_j^{(n)} \right\}^2\;,$$

where $\widehat{\rho}^{(n)}$ denotes the Gaussian maximum likelihood estimator of the AR (1) parameter ρ, and $\widehat{r}_j^{(n)}$ stands for the corresponding classical residual autocorrelations.

The Gaussian Lagrange multiplier approach to the problem (Godfrey 1979; Hosking 1980) leads to the test statistic (Hosking 1980, Theorem 1)

$$(6.10) \qquad Q_{\mathcal{L}}^{(n)} = n \left\{ \sum_{j=2}^{n-1} (\hat{\rho}^{(n)})^{j-2} \hat{r}_j^{(n)} \right\}^2 ,$$

to be compared also with the quantiles of a chi–square variable with one degree of freedom.

$Q_{\mathcal{N}}^{(n)}$ and $Q_{\mathcal{L}}^{(n)}$ apparently are distinct test statistics. However, it follows from asymptotic linear relationships among (parametric) estimated residual autocorrelations (McLeod 1978, Theorem 1) that

$$(6.11) \qquad \sum_{j=1}^{n-1} (\hat{\rho}^{(n)})^{j-1} \hat{r}_j^{(n)} = o_P(n^{-\frac{1}{2}}) , \qquad n \to \infty .$$

Accordingly, our locally optimal Gaussian statistic $Q_{\mathcal{N}}^{(n)}$ satisfies

$$(6.12) \qquad Q_{\mathcal{N}}^{(n)} = n(\hat{r}_1^{(n)})^2 + o_P(1) , \qquad n \to \infty .$$

Similarly, for the Lagrange multiplier statistic, (6.11) implies

$$(6.13) \qquad Q_{\mathcal{L}}^{(n)} = n(\hat{r}_1^{(n)})^2 + o_P(1) , \qquad n \to \infty ,$$

which in turn entails

$$Q_{\mathcal{N}}^{(n)} - Q_{\mathcal{L}}^{(n)} = o_P(1) , \qquad n \to \infty .$$

The parametric, Gaussian counterpart of our rank–based statistic (6.8) is thus asymptotically equivalent to the Gaussian Lagrange multiplier test statistic. It follows that the ARE's of rank tests based on (6.8) with respect either to their Gaussian parametric counterpart (based on (6.9)) or the more familiar Lagrange multiplier test (based on (6.10)) are still those provided in Table 5.1.

Whether an asymptotic equivalence of the form (6.11) or (6.12) also holds for aligned rank–based residual autocorrelations is not known. The asymptotic invariance of (6.7) implies that (6.8) is asymptotically equivalent to its exact unaligned version $\overline{Q}_f^{(n)}$, say. The van der Waerden version of the latter, under Gaussian densities, in turn is asymptotically equivalent to the parametric $Q_{\mathcal{N}}^{(n)}$, hence to $n(\hat{r}_1^{(n)})^2$. Note however that, whereas $Q_{vdW}^{(n)}$, $Q_{\mathcal{N}}^{(n)}$ and $Q_{\mathcal{L}}^{(n)}$ are asymptotically equivalent to the corresponding quantities computed from exact residuals, these equivalences do not hold for $n(\hat{r}_1^{(n)})^2$ which is asymptotically distinct from $n(r_1^{(n)})^2$, where $r_1^{(n)}$ denotes the exact first–order residual autocorrelation. Nor is $\hat{r}_{1;vdW}^{(n)}$ asymptotically invariant. The linear relation (6.11) indeed does not hold for exact residual autocorrelations.

Acknowledgement. This work has greatly benefited from the pleasant environment and peaceful landscapes of La Martinie-Sergeac. Marc Hallin would like to thank Mr. and Mrs. Baudoux for their warm hospitality during the summer of 1990, as well as Dominique, Marie, Catherine and Françoise for their enduring patience.

REFERENCES

AIYAR, R.J. (1981), *Asymptotic efficiency of rank tests of randomness against autocorrelation*, Ann. Inst. Statist. Math **33 A**, 255–262.

AIYAR, R.J., GUILLIER, C.L., and ALBERS, W. (1979), *Asymptotic relative efficiencies of rank tests for trend alternatives*, Jour. Amer. Statist. Assoc. **74**, 226–231.

AKRITAS, M.G., and JOHNSON, R.A. (1982), *Efficiencies of tests and estimators for p-order autoregressive processes when the error is non normal*, Annals Inst. Statist. Math. **34**, 579–589.

ALBERS, W. (1978), *One–sample rank tests under autoregressive dependence*, Ann. Statist. **6**, 836–845.

BARTELS, R. (1982), *The rank version of von Neumann's ratio test for randomness*, Jour. Amer. Statist. Assoc. **77**, 40–46.

BENGHABRIT, Y. and HALLIN, M. (1992), *Optimal rank–based tests against first–order super-diagonal bilinear dependence*, Jour. Statist. Plan. and Inf., to appear.

BERAN, R.J. (1972), *Rank spectral processes and tests for serial dependence*, Ann. Math. Statist. **43**, 1749–1766.

BHATTACHARYYA, G.K. (1984), *Tests for randomness against trend or serial correlations*, In Handbook of Statistics, **4**, (P.R. Krishnaiah and P.K. Sen, eds.), 89–111. North–Holland, Amsterdam and New York.

BROCKWELL, P.J., and DAVIS, R.A. (1987), *Time Series: Theory and Applications*, Springer Verlag, New York.

CAMPBELL, P.J., and DUFOUR, J.–M. (1991), *Over-rejections in rational expectations models: a nonparametric approach to the Mankiw–Shapiro problem*, Economics Letters, (35), 285–290.

DAVID, F.N. (1947), *A power function for tests of randomness in a sequence of alternatives*, Biometrika **34**, 335–339.

DAVID, F.N. and FIX, E. (1966), *Randomization and the serial correlation coefficient*, In Research Papers in Statistics, Festschrift for J. Neyman, (F.N. David, ed.), 461–468. J. Wiley, New York.

DAVIES, R.B. (1973), *Asymptotic inference in stationary Gaussian time series*, Adv. Appl. Prob. **5**, 469–497.

DUFOUR, J.–M. (1981), *Rank tests for serial dependence*, Jour. Time Series Anal. **2**, 117–128.

DUFOUR, J.–M. and HALLIN, M. (1987), *Tests non paramétriques optimaux pour le modèle autorégressif d'ordre un*, Ann. Econ. Stat. **5**, 411–434.

DUFOUR, J.–M. and HALLIN, M. (1990a), *Runs tests for ARMA dependence*, Technical Report, Département de Sciences Economiques, Université de Montréal, Montréal.

DUFOUR, J.–M. and HALLIN, M. (1990b), *An exponential bound for the permutational distribution of a first–order autocorrelation coefficient*, Statistique et Analyse des Données **15**, 45–56.

DUFOUR, J.–M., and HALLIN, M., (1991), *Improved Eaton bounds for linear combinations of bounded random variables, with statistical applications*, Technical Report, C.R.D.E., Université de Montréal, Montréal.

DUFOUR, J.–M., LEPAGE, Y. and ZEIDAN, H. (1982), *Nonparametric testing for time series: a bibliography*, Canad. Jour. Statist., **10**, 1–38.

DUFOUR, J.–M., and ROY, R. (1985), *Some robust exact results on sample autocorrelations and tests of randomness*, Journal of Econometrics **29** , 257–273 and **41**, 279–281.

DUFOUR, J.–M., and ROY, R.(1986), *Generalized portmanteau statistics and tests for randomness*, Communications in Statistics A, Theory and Methods **15**, 2953–2972.

DZHAPARIDZE, K.O. (1986), *Parameter Estimation and Hypothesis Testing in Spectral Analysis of Stationary Time Series*, Springer Verlag, New York.

EDGINGTON, E.S. (1961), *Probability table for number of runs of signs of first differences in ordered series*, J. Amer. Statist. Assoc. **56**, 156–159.

FISHER, R.A. (1926), *On the random sequence*, Quart. J. Roy. Meteorol. Soc. **52**, 250.

GAREL, B. (1989), *The asymptotic distribution of the likelihood ratio for MA processes with a regression trend*, Statist. Decisions **7**, 167–184.

GHOSH, M.N. (1954), *Asymptotic distribution of serial statistics and applications to problems of nonparametric tests of hypotheses*, Ann. Math. Statist. **25**, 218–251.

GODFREY, L.G. (1979), *Testing the adequacy of a time series model*, Biometrika **66**, 67–72.

GOODMAN, L.A. (1958), *Simplified runs tests and likelihood ratio tests for Markov chains*, Biometrika **45**, 181–197.

GOODMAN, L.A. and GRUNFELD, Y. (1961), *Some nonparametric tests for comovements between time series*, J. Amer. Statist. Assoc. **56**, 11–26.

GOVINDARAJULU, Z. (1983), *Rank tests for randomness against autocorrelated alternatives*, In Time Series Analysis: Theory and Practice 4. (O.D. Anderson, ed.) 65–73. North–Holland, Amsterdam and New York.

GRANGER, C.W.J. (1963), *A quick test for serial correlation suitable for use with non–stationary time series*, Jour. Amer. Statist. Assoc. **58**, 728–736.

GUPTA, G.D. and GOVINDARAJULU, Z. (1980), *Nonparametric tests of randomness against autocorrelated normal alternatives*, Biometrika **67**, 375–379.

HÁJEK, J. and ŠIDÁK, Z. (1967), *Theory of Rank Tests*, Academic Press, New York.

HALLIN, M. (1986), *Nonstationary q-dependent processes and time-varying moving-average models: invertibility properties and the forecasting problem*, Advances in Applied Probability **18**, 170–210.

HALLIN, M., INGENBLEEK, J.–FR. and PURI, M.L. (1985), *Linear serial rank tests for randomness against ARMA alternatives*, Ann. Statist. **13**, 1156–1181.

HALLIN, M., INGENBLEEK, J.–FR. and PURI, M.L. (1987), *Linear and quadratic serial rank tests for randomness against serial dependence*, Jour. Time Series Anal. **8**, 409–424.

HALLIN, M., INGENBLEEK, J.–FR. and PURI, M.L. (1989), *Asymptotically most powerful rank tests for multivariate randomness against serial dependence*, Jour. Multivariate Analysis **30**, 34–71.

HALLIN, M., LAFORET, A. and MÉLARD, G. (1990), *Distribution–free tests against serial dependence: signed or unsigned ranks?*, Jour. Statist. Plan. Inf. **24**, 151–165.

HALLIN, M., LEFEVRE, C. and PURI, M.L. (1988), *On time–reversibility and the uniqueness of moving average representations for non–Gaussian stationary time series*, Biometrika **75**, 170–171.

HALLIN, M. and MÉLARD, G. (1988), *Rank–based tests for randomness against first order serial dependence*, Jour. Amer. Statist. Assoc. **83**, 1117–1129.

HALLIN, M., MÉLARD, G. and MILHAUD, X. (1992), *Permutational extreme values of autocorrelation coefficients and a Pitman test against serial dependence*, Ann. Statist. **20**, to appear.

HALLIN, M. and PURI, M.L. (1988), *On locally asymptotically maximin tests for ARMA processes*, in Statistical Theory and Data Analysis II, K. Matusita, Ed., North–Holland, Amsterdam and New-York, 495–500.

HALLIN, M. and PURI, M.L. (1988), *Optimal rank–based procedures for time–series analysis: testing an ARMA model against other ARMA models.* Ann. Statist., **16**, 402–432.

HALLIN, M. and PURI, M.L. (1991a), *Time series analysis via rank–order theory: signed–rank tests for ARMA models*, Jour. Multivariate Analysis **39**, 1–29.

HALLIN, M. and PURI, M.L. (1991b), *A multivariate Wald–Wolfowitz rank test against serial dependence*, Technical Report.

HALLIN, M. and PURI, M.L. (1991c), *Aligned rank tests for linear models with autocorrelated errors and time series with a linear trend*, Technical Report.

HAREL, M. (1988), *Weak convergence of multidimensional rank statistics under φ–mixing conditions*, Jour. Statist. Plan. Inf. **20**, 41–63.

HAREL, M. and PURI, M.L. (1989a), *Limiting behavior of U–statistics, V–statistics and one sample rank order statistics for nonstationary absolutely regular processes*, Jour. Multivariate Analysis **30**, 181–204.

HAREL, M. and PURI, M.L. (1989b), *Weak convergence of the U–statistic and weak invariance of the one sample rank order statistic for Markov processes and ARMA models*, Jour. Multivariate Analysis **31**, 258–265.

HAREL, M. and PURI, M.L. (1990a), *Limiting behavior of one sample rank–order statistics with unbounded scores for nonstationary absolutely regular processes*, Jour. Statist. Plan. Inf. **27**, 1–23.

HAREL, M. and PURI, M.L. (1990b), *Weak convergence of serial rank statistics under dependence with applications in time series and Markov processes*, Ann. Prob. **18**, 1361–1387.

HAREL, M. and PURI, M.L.(1990c), *Weak convergence of the serial linear rank statistic with unbounded scores and regression constants under mixing conditions*, Jour. Statist. Plan. Inf. **25**, 163–186.

HAREL, M. and PURI, M.L. (1991), *Weak invariance of the multidimensional rank statistic for nonstationary absolutely regular processes*, Jour. Multivariate Analysis **36**, 204–221.

HAREL, M. and PURI, M.L. (1992), *Weak convergence of the simple linear rank statistic under mixing conditions in the nonstationary case*, Teor. Veroyatnost. i Premenen., to appear.

HOSKING, J.R.M. (1980), *Lagrange multiplier tests of time series models*, Jour. Roy. Statist. Soc. B. **42**, 170–181.

JOGDEO, K. (1968), *Asymptotic normality in nonparametric methods*, Ann. Math. Statist. **39**, 905–922.

KENDALL, M., and STUART, A. (1968), *The advanced Theory of Statistics*, Vol. 3, Griffin, London.

KERMACK, W.O. and MCKENDRICK, A.G. (1937), *Tests for randomness in a series of numerical observations*, Proc. Roy. Soc. Edinburgh **57**, 228–240.

KNOKE, D.J. (1977), *Testing for randomness against autocorrelation: alternative tests*, Biometrika **64**, 523–529.

KREISS, J.–P. (1987), *On adaptive estimation in stationary ARMA processes*, Ann. Statist. **15**, 112–133.

KREISS, J.–P. (1990a), *Testing linear hypotheses in autoregressions*, Annals of Statistics **18**, 1470–1482.

KREISS, J.–P. (1990b), *Local asymptotic normality for autoregression with infinite order*, Jour. of Statist. Plan. and Inf., **26**, 185–219.

LECAM, L. (1960), *Locally asymptotically normal families of distributions*, Univ. Calif. Publ. Statist. **3**, 37–98.

LECAM, L.(1986), *Asymptotic Methods in Statistical Decision Theory*, Springer–Verlag, New York.

LEVENE, H. and WOLFOWITZ, J. (1944), *The covariance matrix of runs up and down*, Ann. Math. Statist. **15**, 58–69.

MANN, H.B. (1945), *Nonparametric test against trend*, Econometrica **13**, 245–259.

MC.LEOD, A.I. (1978), *On the distribution of residual autocorrelations in Box–Jenkins models*, J.R.S.S. ser. B, **40**, 296–302.

MOOD, A.M. (1940), *The distribution theory of runs*, Ann. Math. Statist. **11**, 367–392.

MOORE, G.H. and WALLIS, W.A. (1943), *Time–series significance tests based on signs of differences*, Jour. Amer. Statist. Assoc. **38**, 153–164.

NIEUWENHUIS, G. and RUYMGAART, F. (1990), *Some stochastic inequalities and asymptotic normality of serial rank statistics in general linear processes*, Jour. Statist. Plan. Inf. **25**, 53–80.

PHAM, D.T. (1987), *Exact maximum likelihood estimate and Lagrange multiplier test statistic for ARMA models*, Jour. Time Series Anal. **8**, 61–78.

PÖTSCHER, B.M. (1983), *Order estimation in ARMA models by Lagrangian multiplier tests*, Annals of Statistics, **11**, 872–885.

PÖTSCHER, B.M. (1985), *The behaviour of the Lagrange multiplier test in testing the orders of an ARMA model*, Metrika, **32**, 129–150.

PURI, M.L. and SEN, P.K. (1971), *Nonparametric Methods in Multivariate Analysis*, Wiley, New York.

PURI, M.L. and SEN, P.K. (1985), *Nonparametric Methods in General Linear Models*, Wiley, New York.

SAVAGE, I.R. (1957), *Contributions to the theory of rank order statistics—the "trend" case*, Ann. Math. Statist. **28**, 968–977.

SERFLING, R. (1968), *The Wilcoxon two–sample statistic on strong mixing process*, Ann. Math. Statist. **39**, 1202–1209.

SWENSEN, A.R. (1985), *The asymptotic distribution of the likelihood ratio for autoregressive time series with a regression trend*, Jour. Multivariate Anal. **16**, 54–70.

SWITZER, P. (1984), *Inference for spatial autocorrelation functions*, In Geostatistics for Natural Resources Characterization, (G. Verly et al., eds.). D. Reidel, Dordrecht,127–140.

TRAN, L.T. (1988), *Rank order statistics for time series models*, Annals Inst. Stat. Math. **40**, 247–260.

TRAN, L.T. (1990), *Rank statistics for testing randomness against serial dependence*, Jour. Statist. Plan. and Inf. **24**, 215–232.

WALD, A. (1943), *Tests of statistical hypotheses concerning several parameters when the number of observations is large*, Trans. Amer. Math. Soc. **54**, 426–482.

WALD, A. and WOLFOWITZ, J. (1943), *An exact test for randomness in the nonparametric case based on serial correlation*, Ann. Math. Statist. **14**, 378–388.

WALLIS, W.A. and MOORE, G.H. (1941), *A significance test for time series*, Jour. Amer. Statist. Assoc. **36**, 401–409.

WEISS, G. (1975), *Time-series reversibility of linear stochastic processes*, Jour. of Appl. Probability, **12**, 831–836.

WOLFOWITZ, J. (1944), *Asymptotic distribution of runs up and down*, Ann. Math. Statist. **15**, 163–172.

YANG, M.C.K. and SCHRECKENGOST, J.F. (1981), *Difference sign test for comovements between two time series*, Comm. Statist. Theor. Meth. A10 **4**, 355–369.

SELECTION OF TIME SERIES MODELS AND SPECTRUM ESTIMATES USING A BIAS-CORRECTED GENERALIZATION OF AIC

CLIFFORD M. HURVICH*

Abstract. We address a general selection problem for time series, namely: Given data from a stationary Gaussian process having a spectral density, and given a class C of candidate spectrum estimates, how should one select a candidate from C for use as a description of the process? Special cases of this problem include order selection for autoregressive models and spectrum estimates, as well as bandwidth selection for nonparametric spectrum estimates. In this paper, we discuss AIC_C, a bias-corrected generalization of the Akaike Information Criterion (AIC; Akaike 1973), which provides a unified solution to the problem. For autoregressive order selection, AIC_C is asymptotically efficient, and provides superior selections in small samples, as shown in Hurvich and Tsai (1989). Unfortunately, AIC will often select models of very high dimension if these are included as candidates. Thus, a difficulty in using AIC (and other selectors in common use) is the need to guard against overfitting by keeping the maximum candidate model order small. The use of a maximum model order cutoff seems arbitrary, however, and is particularly problematic in small samples. The AIC_C criterion avoids these difficulties by providing a more nearly unbiased estimate than AIC of the expected Kullback-Leibler information. For nonparametric spectrum estimation, AIC_C produces good bandwidth selections, as shown in Hurvich and Beltrao (1990). Further, if one allows the class C of candidates to simultaneously contain parametric and nonparametric spectrum estimates, then AIC_C allows the data-driven selection of estimate type (e.g., autoregressive or nonparametric) as well as the corresponding smoothness parameter (e.g., model order or bandwidth). Finally, we briefly discuss some recent improvements in AIC_C for selection of autoregressive models, based on numerical tabulation of penalty functions.

1. Introduction. This paper addresses a general selection problem for time series, namely: Given data $x_0, \ldots, x_{n-1}$ from a zero-mean stationary Gaussian stochastic process with a spectral density, and given a class C of candidate spectrum estimates which can be computed from the data, how should one select a candidate from C for use as a description of the process?

If C consists entirely of finite-parameter models (autoregressions, say) then the problem reduces to model selection, e.g., selection of autoregressive model order. On the other hand, one may wish to avoid finite-parameter models altogether, and use a nonparametric spectrum estimate instead. Here again, though, there are undetermined constants, especially bandwidth, which determine the statistical characteristics (smoothness, mean squared error, etc.) of the estimate. The data-driven selection of these constants is a problem quite analogous to model selection, and is obtained as a special case of our general selection problem by taking C to consist solely of nonparametric spectrum estimates.

Finally, there is the issue of deciding whether it seems best to estimate the spectrum nonparametrically or to fit a finite-parameter model, and of then choosing the corresponding smoothness parameter (e.g., model order or bandwidth). To treat this question, we would allow C to simultaneously contain parametric and nonparametric spectrum estimates.

*Department of Statistics and Operations Research, New York University, Tisch Hall, 40 West Fourth Street, New York NY 10003 U.S.A.

In this paper, we discuss AIC_C, a bias-corrected generalization of the Akaike Information Criterion (AIC; Akaike 1973), which provides a unified solution to the selection problem.

In this introduction, we focus primarily on autoregressive order selection, the problem for which AIC was originally designed. Most recently proposed methods for autoregressive order selection, including AIC, yield the selected model as the minimizer of a data-based criterion function. Asymptotically, a selection criterion may be **efficient** or **consistent**, but not both. An efficient method (Shibata 1980) produces an optimal selection in large samples when the true generating mechanism is infinite-dimensional. A consistent method selects the true model with probability approaching 1 in large samples assuming that there indeed exists a finite-order generating mechanism. Of the two properties, efficiency seems the more useful. Indeed, as pointed out by Tukey (1984), Shibata (1980) and others, it seems rare in practice that an exact finite-dimensional model will hold. The point of view taken in this paper is similar to that of Parzen (1978), namely that model selection is useful because it allows us to select a finite-dimensional model to serve as an *approximation* to the infinite-dimensional truth.

The AIC criterion provides an asymptotically efficient solution to the problem of selecting an approximating autoregressive model. For an $AR(p)$ candidate model, the criterion is given by

$$AIC = n + n\log(2\pi\hat{\sigma}^2) + 2(p+1),$$

where $\hat{\sigma}^2$ is the estimated innovation variance. Although the minimum AIC criterion (i.e., choose the model order to minimize AIC) has proven to be a highly successful and widely applicable method, it does suffer some serious drawbacks, especially in small samples, or more generally when p/n differs appreciably from zero. In fact, as documented below, AIC will often select models of very high dimension if these are included as candidates. Furthermore, even if very high dimensional models are not considered, AIC will sometimes select the highest dimension that is under consideration. Examples of the latter phenomenon appear in Shumway (1988, p. 169), and Linhart and Zucchini (1986, pp. 86–88), who comment (p. 78) that "in some cases the criterion simply continues to decrease as the number of parameters in the approximating model is increased". This is clearly a drawback, since AIC is supposed to provide an optimal trade-off between fidelity to the data and parsimony. To understand the problem, we will examine the finite-sample bias of AIC, viewed (as originally intended by Akaike 1973) as an estimator of the expected Kullback-Leibler information. We find that as p is increased with n held fixed, AIC has a tendency to become strongly negatively biased, leading to gross overfitting.

The generalization of AIC to be developed here is denoted by AIC_C, and defined as

(1) $$AIC_C = n\log(2\pi\hat{\sigma}^2) + n\frac{1+p/n}{1-(p+2)/n}.$$

The AIC_C criterion was originally suggested by Sugiura (1978) for use in a linear regression context. We will show that AIC_C provides a unified solution to the general selection problem. For the selection of autoregressive models, AIC_C is asymptotically efficient, is able to strongly outperform AIC in small samples (in terms of both bias and model selection), and removes the need to set arbitrary cutoffs on the candidate model order. AIC_C also works well in the problem of objectively selecting a spectrum estimate, even from a class which simultaneously includes parametric and nonparametric estimates.

The paper is organized as follows. In Section 2, we present two key expressions for the expected Kullback-Leibler information, which provide the foundation for the subsequent development. In Section 3, we demonstrate in a Monte Carlo study that AIC_C can strongly outperform AIC for autoregressive model selection in small samples. Section 4 provides a theoretical derivation of AIC_C for autoregressive models. Section 5 develops a version of AIC_C suitable for the selection of a nonparametric spectrum estimate. Section 6 gives a further improvement of AIC_C for the selection of autoregressive models, based on numerical tabulation of penalty functions. Section 7 contains some concluding remarks.

2. Kullback-Leibler Information and its Connections with AIC. Let $x = (x_0, \ldots, x_{n-1})^T$ represent data from a real-valued zero-mean stationary Gaussian process having true spectral density $f(\omega)$ for $\omega \in [-\pi, \pi]$. For any candidate spectral density $g(\omega)$, let Σ_g denote the $n \times n$ covariance matrix for x induced by g. Thus Σ_f and Σ_g are the true and candidate covariance matrices, respectively. We have

$$-2 \log \text{ likelihood } (g) = n \log 2\pi + \log |\Sigma_g| + x^T \Sigma_g^{-1} x.$$

If E_0 denotes expectation under the true generating mechanism f, then the Kullback-Leibler information is given by

$$d(f, g) = E_0 \{-2 \log \text{ likelihood } (g)\} = n \log 2\pi + \log |\Sigma_g| + Tr \left(\Sigma_f \Sigma_g^{-1} \right).$$

It can be shown that $d(f, f) \leq d(f, g)$ for any g, so that $d(f, g)$ serves as a measure of the discrepancy between the true and candidate spectral densities, f, g. If $\hat{f}(\omega)$ is a parametric or nonparametric spectrum estimate obtained from the data x, then the expected Kullback-Leibler information is

$$(2) \qquad \Delta_1(\hat{f}) = E_0[d(f, \hat{f})] = n \log 2\pi + E_0 \log |\Sigma_{\hat{f}}| + E_0 Tr \left(\Sigma_f \Sigma_{\hat{f}}^{-1} \right).$$

The criterion Δ_1 provides a theoretical basis for comparing various candidates $\hat{f}$: The best candidate is defined to be the minimizer of Δ_1 in the class C of all candidates under consideration. Apparently, Δ_1 cannot be evaluated in practice since it depends on the unknown true spectral density f. It is possible, however, to try to *estimate* Δ_1, ideally with as little bias as possible. Indeed, AIC was originally proposed and motivated by Akaike (1973) as an approximately unbiased estimator of expected Kullback-Leibler information for parametric models. Findley (1985) presented a rigorous proof that AIC is in fact asymptotically unbiased when the true model is $ARMA$, under certain conditions (the most restrictive of which is that the candidate model is capable of exactly describing the true generating mechanism). Bhansali (1986) demonstrated the asymptotic unbiasedness of AIC in the case where the truth is $AR(\infty)$ and the candidate AR model order p tends to infinity simultaneously (but slowly) with n.

It is often convenient to use

$$\delta(f,g) = 2n \log 2\pi + \frac{n}{2\pi} \int_{-\pi}^{\pi} [\log g(\omega) + f(\omega)/g(\omega)] d\omega$$

as an approximation to the Kullback-Leibler information $d(f,g)$. This approximation, first proposed by Whittle (1953) can be derived from the fact (see, e.g., Parzen 1983, p. 235) that the eigenvectors and corresponding eigenvalues of Σ_g may be approximated by $n^{-1/2}\{\exp(-i\omega_j t)\}_{t=0}^{n-1}$ and $2\pi g(\omega_j)$ for $j = 0,\ldots,n-1$, where $\omega_j = 2\pi j/n$. Thus, an alternative measure of the quality of $\hat{f}$ is

$$(3) \qquad \Delta_2(\hat{f}) = E_0[\delta(f,\hat{f})] = 2n \log 2\pi + \frac{n}{2\pi} E_0 \int_{-\pi}^{\pi} [\log \hat{f}(\omega) + f(\omega)/\hat{f}(\omega)] d\omega.$$

It is known that Δ_1 and Δ_2 are asymptotically equivalent, although we note here that there are some differences in small samples.

3. AIC and AIC$_C$ for Autoregressive Model Selection. Suppose that the candidate spectrum estimate (or equivalently, the candidate model) is pth order autoregressive, $AR(p)$. The Akaike Information Criterion is given in this case by

$$AIC = n + n \log(2\pi\hat{\sigma}^2) + 2(p+1),$$

where $\hat{\sigma}^2$ is the estimated innovation variance.

As shown by Findley (1985), under certain conditions, for p held fixed with $n \to \infty$, AIC is asymptotically unbiased for both Δ_1 and Δ_2. One of the main findings we wish to stress here, however, is that *for a given sample size*, as p is increased, AIC becomes increasingly negatively biased, for either Δ_1 or Δ_2. To illustrate the phenomenon, we generated 100 simulated realizations of length $n = 23$ from the Gaussian $AR(2)$ model

$$x_t = .99x_{t-1} - .8x_{t-2} + \epsilon_t,$$

where the ϵ_t are independent standard normal. The AR parameters and innovation variance estimated by the Burg (1978) method. Figure 1 shows the expected Kullback-Leibler information Δ_1 and Δ_2 given by (2) and (3), together with the average of AIC, all functions of model order $p = 1,\ldots,20$. The bias of AIC which is evident in Figure 1 can lead to gross overfitting. Table 1 shows that AIC selected the correct model in only 7 of the 100 cases, and usually selected an exceedingly high dimensional model instead.

The standard solution to the problem is to set a cutoff on the maximum candidate model order. For example, some authors consider a fixed, finite set of candidates as $n \to \infty$. In practice, however, the user would typically want to allow the set of candidate models to grow with the sample size, especially if it is suspected that the true model is of infinite order. In Shibata (1980), the number of candidates is allowed to grow at the rate $o(\sqrt{n})$. Nevertheless, a cutoff such as this is of limited usefulness in practice since it does not actually rule out any models *a priori*. How, for example, are we to decide whether it is sufficient to consider model orders only up to $\sqrt{n}/\log(n)$, or whether it is necessary instead to go all the way to, say, $20\sqrt{n}/\log(n)$? Further, since the above cutoff is based on asymptotic considerations, its application in small samples would be particularly problematic.

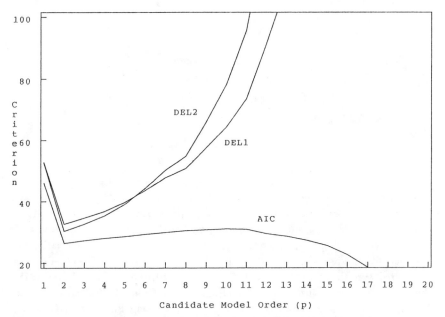

Figure 1. Average AIC and Kullback-Leibler Discrepancy in 100 Realizations from a Gaussian $AR(2)$ with $n = 23$.

Table 1. The Frequency of the Order Selected by AIC in 100
Realizations of a Gaussian $AR(2)$ Model with $n = 23$.

Criterion	Selected Model Order				
	$p = 1$	$p = 2$	$p = 3-5$	$p = 6-10$	$p = 11-20$
AIC	1	7	2	2	88

The AIC_C criterion for a candidate $AR(p)$ model is

$$AIC_C = n\log(2\pi\hat{\sigma}^2) + n\frac{1 + p/n}{1 - (p+2)/n}.$$

Note that

$$AIC_C = AIC + \frac{2(p+1)(p+2)}{n-p-2},$$

so that AIC_C and AIC are asymptotically equivalent for fixed p with $n \to \infty$. It follows from Shibata (1980, p. 160) that AIC_C shares with AIC the property of asymptotic efficiency. Despite the apparent asymptotic similarities, however, AIC_C can strongly outperform AIC in small samples. For the $AR(2)$ process with $n = 23$ described earlier, Figure 2 shows the expected Kullback-Leibler information (3) together with averages of AIC, AIC_C and other selection criteria. The other criteria considered were the efficient methods FPE (Akaike 1970) and CAT (Parzen 1978); the consistent criteria used were HQ (Hannan and Quinn 1979), SIC (Schwarz Information Criterion; Schwarz 1978) and BIC (Bayesian Information Criterion; Akaike 1978, Priestley 1981, p. 375). It is seen that AIC_C is much less biased than AIC, and that all criteria besides AIC_C and BIC have a tendency to favor large model orders. Table 2 summarizes the model orders selected by the various methods, using maximum model order cutoffs of either 10 or 20. It is seen that AIC_C provided excellent model selections, clearly the best among the efficient methods and quite competitive even with the consistent methods. Interestingly, even though AIC_C is asymptotically efficient (and therefore optimal in large samples when the true model is infinite dimensional), it is able to do very well at selecting the correct finite dimensional model in small samples. Further, unlike the other criteria, AIC_C exhibits no sensitivity to the model order cutoff.

161

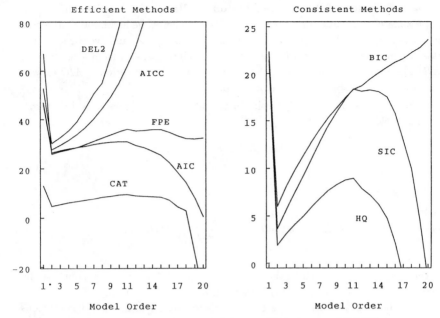

Figure 2. Average of Various Criterion Functions and KL Discrepancy in 100 Realizations from $AR(2)$ with $n = 23$.

Table 2. The Frequency of the Order Selected by Various Criteria in 100 Realizations of a Gaussian $AR(2)$ Model with $n = 23$. Maximum Order Cutoff: 10 (left), 20 (right).

Criterion	Selected Model Order				
	$p = 1$	$p = 2$	$p = 3-5$	$p = 6-10$	$p = 11-20$
AIC_C	6 , 6	80 , 80	10 , 10	4 , 4	0 , 0
AIC	3 , 1	52 , 7	19 , 2	26 , 2	0 , 88
FPE	3 , 2	52 , 19	19 , 5	26 , 7	0 , 67
HQ	4 , 1	56 , 11	19 , 3	21 , 4	0 , 81
SIC	6 , 4	78 , 31	9 , 3	7 , 1	0 , 61
BIC	5 , 4	81 , 77	10 , 10	4 , 3	0 , 6
CAT	3 , 2	54 , 20	19 , 5	24 , 8	0 , 65

4. Derivation of AIC_C for AR Models. Here, we will derive AIC_C as an approximately unbiased estimator of Δ_2. For simplicity, we assume that the true model is $AR(p_0)$ and that $p \geq p_0$. In practice, of course, one can never be sure that this assumption is met, even if the data were really generated from a finite-order AR. The assumption, however, is commonly made in the derivation of AIC as well. See Section 7 for more discussion on this point.

Since $\hat{f}(\omega)$ is the spectral density of an autoregressive process, Kolmogorov's formula yields

$$\frac{1}{2\pi} \int_{-\pi}^{\pi} \log \hat{f}(\omega) d\omega = \log(\hat{\sigma}^2/(2\pi)).$$

Thus,

$$(4) \qquad \delta(f, \hat{f}) = n \log(2\pi\hat{\sigma}^2) + \frac{n}{2\pi} \int_{-\pi}^{\pi} f(\omega)/\hat{f}(\omega) d\omega.$$

Let $\hat{a} = (1, \hat{a}_1, \ldots, \hat{a}_p)^T$ denote the vector of estimated AR parameters, obtained by maximum likelihood, Burg's (1978) method, or some other asymptotically equivalent method. Let R denote the true covariance matrix of any $p + 1$ contiguous observations. Then $1/\hat{f}(\omega) = (2\pi/\hat{\sigma}^2)|1 + \Sigma_{k=1}^p \hat{a}_k \exp(i\omega k)|^2$, and the integral in (4) reduces to

$$\frac{n}{2\pi} \int_{-\pi}^{\pi} f(\omega)/\hat{f}(\omega) d\omega = n\hat{a}^T R \hat{a}/\hat{\sigma}^2 = n[\sigma_0^2 + (\hat{a} - a_0)^T R(\hat{a} - a_0)]/\hat{\sigma}^2,$$

where $a_0 = (1, a_1, \ldots, a_p)^T$ and σ_0^2 denote the true AR parameters and innovation variance. From Brockwell and Davis (1987, p. 184), $(\hat{a} - a_0)^T R(\hat{a} - a_0)$ and $\hat{\sigma}^2$ are approximately independently distributed as $(\sigma_0^2/n)\chi_p^2$ and $(\sigma_0^2/n)\chi_{n-p}^2$, respectively. Thus,

$$\begin{aligned}
\Delta_2(\hat{f}) &= E_0[n \log(2\pi\hat{\sigma}^2)] + nE_0[\sigma_0^2/\hat{\sigma}^2 + (\hat{a} - a_0)^T R(\hat{a} - a_0)/\hat{\sigma}^2] \\
&= E_0[n \log(2\pi\hat{\sigma}^2)] + n^2 E_0[1/\chi_{n-p}^2] + nE_0[\chi_p^2/\chi_{n-p}^2] \\
&= E_0[n \log(2\pi\hat{\sigma}^2)] + n \frac{n + p}{n - p - 2}.
\end{aligned}$$

It follows that

$$AIC_C = n \log(2\pi\hat{\sigma}^2) + n \frac{1 + p/n}{1 - (p + 2)/n}$$

is an approximately unbiased estimator of $\Delta_2(\hat{f})$.

5. AIC_C for Automatic Bandwidth Selection in Nonparametric Spectrum Estimation. Since AIC was originally designed for selection of finite-parameter models, it cannot be immediately applied to the selection of a nonparametric spectrum estimate. A brief attempt made by Hannan and Rissanen (1988) to apply the definition of AIC in a literal fashion failed, as the AIC analog tended to favor the periodogram over all other candidates. Since the periodogram (or more generally, small bandwidth) corresponds to the estimation of a large number of "parameters", there is a strong analogy between the tendency of AIC to select too small a bandwidth in nonparametric spectrum estimation applications and too large a model order in parametric model selection applications. Earlier proposals of automatic bandwidth selection methods for nonparametric spectrum estimation were given by Hurvich (1985) and by Beltrao and Bloomfield (1987).

Here, we obtain AIC_C for automatic selection of nonparametric spectrum estimates $\hat{f}$. The criterion is designed as an estimate of the expected Kullback-Leibler information $\Delta_2(\hat{f})$ given by (3). We consider nonparametric spectrum estimates of form

$$(5) \qquad \hat{f}(\omega_j) = \sum_k g_k I(\omega_j - \omega_k),$$

where $\{g_k\}$ are fixed nonnegative weights with $\sum g_k = 1$, and

$$I(\omega) = \frac{1}{2\pi n} \left| \sum_{t=0}^{n-1} x_t \exp(-i\omega t) \right|^2$$

is the periodogram. Note that the estimator (5) is a discrete average of the periodogram over the grid of Fourier frequencies, $\omega_j = 2\pi j/n$. The problem is to select the form (i.e., bandwidth, and perhaps functional form) of the weight sequence, $\{g_k\}$. The distribution of $\hat{f}(\omega_j)$ may be approximated by $f(\omega_j)(1/\nu)\chi_\nu^2$, where the degrees of freedom are given by $\nu = 2 \left[\sum g_k^2 \right]^{-1}$. Under this approximation, $E_0[f(\omega_j)/\hat{f}(\omega_j)] = E_0[\nu/\chi_\nu^2] = \nu/(\nu - 2)$. Then

$$(6) \qquad AIC_C = 2n \log 2\pi + \sum_{j=0}^{n-1} \log \hat{f}(\omega_j) + n\nu/(\nu - 2)$$

is an approximately unbiased estimator of $\Delta_2(\hat{f})$, since

$$E_0[AIC_C] = 2n \log 2\pi + \frac{n}{2\pi} E_0 \sum_{j=0}^{n-1} \log \hat{f}(\omega_j) \frac{2\pi}{n} + \frac{n}{2\pi} \int_{-\pi}^{\pi} \frac{\nu}{\nu - 2} d\omega$$

$$\approx 2n \log 2\pi + \frac{n}{2\pi} E_0 \int_{-\pi}^{\pi} [\log \hat{f}(\omega) + f(\omega)/\hat{f}(\omega)] d\omega = \Delta_2(\hat{f}).$$

In Hurvich and Beltrao (1990), it is shown that AIC_C provides good bandwidth selections for nonparametric spectrum estimates. Note that the use of the name

AIC_C for both (1) and (6) is warranted, since both are estimates of $\Delta_2(\hat{f})$. In view of this, AIC_C may be used to select the type (e.g., autoregressive or nonparametric) of estimator as well as the corresponding smoothness parameter (e.g., model order or bandwidth). Results on using AIC_C for this purpose will appear in a forthcoming paper. The idea of objectively selecting estimate type using other criteria has been discussed in Hurvich (1985) and Hurvich and Beltrao (1990).

6. An Improvement of AIC_C for AR Model Selection. Here, we describe the AIC_I criterion, designed as an improved version of AIC_C for selection of Gaussian autoregressive models. To gain accuracy, we consider the exact expected Kullback-Leibler information which, ignoring the constant $n \log 2\pi$, is given by

$$(7) \qquad \Delta_1(\hat{f}) = E_0 \log |\Sigma_{\hat{f}}| + E_0 Tr \left(\Sigma_f \Sigma_{\hat{f}}^{-1} \right).$$

The first term in (7) measures the goodness of fit of the model to the observed data and can be estimated without bias by $\log |\Sigma_{\hat{f}}|$. The second term acts as a penalty function to guard against overfitting. It is shown in Hurvich, Shumway and Tsai (1990) that if the process is Gaussian $AR(p_0)$ and $p \geq p_0$ then the asymptotic distribution of $tr \left(\Sigma_f \Sigma_{\hat{f}}^{-1} \right)$ is pivotal, i.e., it does not depend on the true spectrum f. If we treat the distribution as being exactly pivotal for any given sample size, then it is possible to tabulate the penalty term (without knowing f) by numerical methods.

Specifically, a large number of simulated independent realizations can be generated from a zero mean, unit variance Gaussian white noise process. For each realization, parameter estimates $\hat{\theta}^*$, corresponding spectrum estimates $\hat{f}^*$, and $tr \left(\Sigma_{\hat{f}^*}^{-1} \right)$ are computed. Then assuming that the distribution of $tr \left(\Sigma_f \Sigma_{\hat{f}}^{-1} \right)$ does not depend on f, the average of the simulated values of $tr \left(\Sigma_{\hat{f}^*}^{-1} \right)$ will converge to $E_0 \left\{ tr \left(\Sigma_f \Sigma_{\hat{f}}^{-1} \right) \right\}$ as the number of simulated realizations is increased. Thus, an almost exactly unbiased estimator of Δ_1 is given by

$$AIC_I = \log |\Sigma_{\hat{f}}| + E_0 \left\{ tr \left(\Sigma_f \Sigma_{\hat{f}}^{-1} \right) \right\}.$$

Note that in general, $E_0 \left\{ tr \left(\Sigma_f \Sigma_{\hat{f}}^{-1} \right) \right\}$ will depend on the sample size, on the form and dimension of the class of candidate models, and on the form of the estimator $\hat{\theta}$.

Monte Carlo results based on the same 100 simulated realizations from the $AR(2)$ process ($n = 23$) used in Section 3 show that AIC_I outperformed AIC_C (See Table 3), and provided a better estimator of $\Delta_1(\cdot)$ (See Figure 3).

Table 3. The Frequency of the Order Selected by AIC, AIC_C, and AIC_I in 100 Realizations of a Gaussian $AR(2)$ Model with $n = 23$. Maximum Order Cutoff: 10 (left), 20 (right).

Criterion	Selected Model Order				
	$p = 1$	$p = 2$	$p = 3-5$	$p = 6-10$	$p = 11-20$
AIC	3 , 1	52 , 7	19 , 2	26 , 2	0 , 88
AIC_C	6 , 6	80 , 80	10 , 10	4 , 4	0 , 0
AIC_I	6 , 6	85 , 85	8 , 8	1 , 1	0 , 0

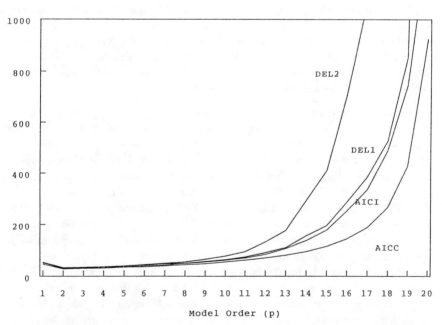

Figure 3. Averages of AICC, AICI and KL for $AR(2)$ Process, $n = 23$.

7. Discussion. The main message of this paper is that simple bias corrections in AIC can yield a large payoff in improved performance. Much work remains to be done, however.

In autoregressive model selection, the main flaw in the derivation of AIC_C is the assumption that the true process is $AR(p_0)$ and that $p \geq p_0$. Clearly, in practice one can never be sure that the true process is $AR(p_0)$, and even if it is, since a variety of candidate models will be tried, it will not always be true that $p \geq p_0$. The derivation of AIC is typically carried out under the same essentially untenable assumptions. Nevertheless, it is interesting that in spite of their derivations, AIC and AIC_C both turn out to be asymptotically efficient criteria, and hence perform optimally in large samples when the true process is infinite dimensional. Monte Carlo evidence presented here suggests that AIC_C can strongly outperform AIC in small samples when the true process is $AR(p_0)$. The question remains as to how AIC_C performs in comparison to AIC when the truth is $AR(\infty)$. In this case, $p_0 = \infty$, so the model is always underfitted, i.e., one will always have $p < p_0$. Preliminary research indicates that even when the true model is $AR(\infty)$ (the most realistic case, in our view), AIC_C is much less biased than AIC for estimating Δ_1, and that AIC_C gives much better model selections than AIC, as measured by Δ_1, whenever the sample size is small, or the maximum candidate model order is an appreciable fraction of the sample size.

For nonparametric spectrum estimation, a careful comparison of the frequency domain versions of AIC (Hannan and Rissanen 1988) and AIC_C remains to be carried out. Preliminary results indicate, however, that unlike AIC_C, the frequency domain AIC has a tendency to favor the periodogram over more highly smoothed estimators, a problem quite analogous to the tendency of the time domain AIC to favor very high order autoregressions to low order ones.

We now present some concluding remarks on the philosophy underlying the use of AIC and AIC_C for autoregressive model selection. We feel that the key contribution of Akaike (1973) is the idea that model selection can be carried out by estimating the expected Kullback-Leibler information. While AIC represents a huge step forward as the first implementation of this idea, it is not the only one possible. To quote Akaike (1974), "Further improvements of definition and use of AIC will be the subjects of further study." We feel that the corrected AIC criterion, AIC_C, originally proposed by Sugiura (1978) in a linear regression context, represents such an improvement. Although AIC and AIC_C are asymptotically equivalent, and hence both asymptotically unbiased under certain conditions for p held fixed as $n \to \infty$, the results reported here show clearly that, viewed globally as functions of p for a given n, AIC_C exhibits much less bias than AIC in estimating the expected Kullback-Leibler information. The AIC criterion tends to become increasingly negatively biased as p increases, a phenomenon which can be linked to the gradual deterioration in the validity of certain Taylor series approximations used in its derivation. This bias can cause AIC to select extremely high dimensional models if these are included as candidates, a problem which has in fact been known for some time (see, e.g., Jones 1976, and Sakamoto, Ishiguro & Kitagawa 1986, pp.

83–85). We have found here that the use of AIC_C corrects this difficulty, without sacrificing the key property of asymptotic efficiency. Further, even if extremely high dimensional models are not considered, AIC_C can often substantially outperform AIC in small samples, or whenever p/n is not vanishingly small, since AIC still exhibits a steadily worsening negative bias as p increases. One can view the maximum model order cutoffs imposed in many theoretical papers on AIC as being necessary so as to exclude from consideration the region where AIC is substantially biased. Since this region is entered gradually as p increases, however, we feel that the cutoff approach will never be completely effective from a practical standpoint. The use of AIC_C removes the need for these cutoffs.

Given a fixed, finite set of candidates, AIC and AIC_C are asymptotically equivalent. Thus, if the series is in fact a finite order AR, then AIC_C will exhibit an asymptotic tendency to sometimes overfit the model, just as AIC does (see Shibata 1976). This problem is unavoidable since, in the Gaussian case, an asymptotically efficient criterion cannot also be consistent. We do not feel that this is a major drawback, however, since it is rare in practice that the true model is an exact finite order autoregression. Nevertheless, the Monte Carlo results presented here show that the efficient AIC_C criterion can outperform even the consistent criteria given short realizations from a finite-order model.

REFERENCES

AKAIKE, H., *Statistical predictor identification*, Ann. Inst. Satist. Math. 22 (1970), pp. 203–217.

AKAIKE, H., *Information theory and an extension of the maximum likelihood principle*, In 2nd International Symposium on Information Theory, Ed. B.N. Petrov and F. Csaki, Budapest: Akademia Kiado (1973), pp. 267–281.

AKAIKE, H., *A new look at the statistical model identification*, IEEE Transactions on Automatic Control AC–19 (1974), pp. 716–723.

AKAIKE, H., *A Bayesian analysis of the minimum AIC procedure*, Ann. Inst. Statist. Math. A 30 (1978), pp. 9–14.

BELTRAO, K. AND BLOOMFIELD, P., *Determining the bandwidth of a kernel spectrum estimate*, J. Time Ser. Anal. 8, (1987), pp. 21–38.

BERK, K., *Consistent Autoregressive Spectral Estimates*, Annals of Statistics 2 (1974), pp. 489–502.

BHANSALI, R.J., *A derivation of the information criteria for selecting autoregressive models*, Adv. Appl. Prob. 18 (1986), pp. 360–387.

BROCKWELL, P.J. & DAVIS, R.A., *Times Series: Theory and Methods*, New York: Springer Verlag (1987).

BURG, J.P., *A new analysis technique for time series data*, In Modern spectrum Analysis, Ed. D.G. Childers, New York: IEEE Press (1978), pp. 42–48.

FINDLEY, D., *On the Unbiasedness Property of AIC for Exact or Approximating Linear Stochastic Time Series Models*, Journal of Time Series Analysis 6, 1985, pp. 229–252.

HANNAN, E.J. & QUINN, B.G., *The determination of the order of autoregression*, J.R. Statist. Soc. B 41 (1979), pp. 190–195.

HANNAN, E.J. & RISSANEN, J., *The width of a spectral window*, In A Celebration of Applied Probability, Special Volume, Journal of Applied Probability 25 A, Ed. J. Gani. Sheffield: Applied Probability Trust (1988).

HURVICH, C.M., *Data-driven choice of a spectrum estimate: extending the applicability of cross-validation methods*, JASA 80 (1985), pp. 933–940.

HURVICH, C.M. & TSAI, C.L., *Regression and time series model selection in small samples*, Biometrika 76 (1989), pp. 297–307.

HURVICH, C.M. & BELTRAO, K.I., *Cross-validatory choice of a spectrum estimate and its connections with AIC*, J. Time Ser. Anal. 11 (1990), pp. 121–137.

HURVICH, C.M., SHUMWAY, R.H. & TSAI, C.L., *Improved estimators of Kullback-Leibler information for autoregressive model selection in small samples*, Biometrika 77 (1990), pp. 709–719.

JONES, R.H., *Autoregression order selection*, In Modern Spectrum Analysis, Ed. D.G. Childers, New York: IEEE Press (1976), pp. 249–251.

LINHART, H. & ZUCCHINI, W., *Model Selection*, New York: Wiley (1986).

PARZEN, E., *Some recent advances in time series modeling*, In Modern Spectrum Analysis, Ed. D.G. Childers, New York: IEEE Press (1978), pp. 226–233.

PARZEN, E., *Autoregressive Spectral Estimation*, In Handbook of Statistics Vol. 3, Eds. D.R. Brillinger and P.R. Krishnaiah, New York: Elsevier (1983), pp. 221–247.

PRIESTLEY, M.B., *Spectral Analysis and Time Series*, New York: Academic Press (1981).

RAO, C.R., *Linear Statistical Inference And Its Applications*, 2nd Ed., New York: Wiley (1973).

SAKAMOTO, Y., ISHIGURO, M. & KITAGAWA, G., *Akaike Information Criterion Statistics*, Dordrecht: D. Reidel (1986).

SCHWARZ, G., *Estimating the dimension of a model*, Ann. Statist. 6 (1978), pp. 461–464.

SHIBATA, R., *Selection of the order of an autoregressive model by Akaike's information criterion*, Biometrika 63 (1976), pp. 117–126.

SHIBATA, R., *Asymptotically efficient selection of the order of the model for estimating parameters of a linear process*, Ann. Statist. 8 (1980), pp. 147–164.

SHUMWAY, R.H., *Applied Statistical Time Series Analysis*, Englewood Cliffs: Prentice Hall (1988).

SUGIURA, N., *Further analysis of the data by Akaike's information criterion and the finite corrections*, Commun. Statist A7 (1978), pp. 13–26.

TUKEY, J.W., *Style of Spectrum Analysis*, In The Collected Works of John W. Tukey, Vol II, Ed. D.R. Brillinger, Monterey, CA: Wadsworth (1984).

WHITTLE, P., *The analysis of multiple stationary time series*, J.R. Statist. Soc. B 15 (1953), pp. 125–139.

CONTRACTION MAPPINGS IN MIXED SPECTRUM ESTIMATION*

BENJAMIN KEDEM†

Abstract. Families and sequences of zero-crossing counts generated by parametric time invariant filters are called higher order crossings or HOC. Because of the close relationship between zero- crossing counts and first order autocorrelations, families of first order autocorrelations are also referred to as HOC. By means of HOC from repeated differencing and repeated summation, it is possible to obtain a complete solution of the problem of hidden periodicities in the purely discrete spectrum case. However, when noise is present, a modification is needed. It is shown how to locate discrete frequencies in the presence of colored noise, using HOC sequences obtained by recursive filtering. By this method, the cosine of each discrete frequency is obtained as a fixed point of a certain contraction mapping. A special feature of this method is that the contraction rate can be enhanced considerably by the iterative reduction in the filter bandwidth.

Key words and phrases: Stationary, non-Gaussian, frequency, contraction coefficient, parametric filter, recursive filter.

AMS subject classification: Primary $62M10$, secondary $62M07$.

1. INTRODUCTION

Consider the mixed spectrum model

$$(1) \qquad Z_t = \sum_{j=1}^{p}(A_j \cos(\omega_j t) + B_j \sin(\omega_j t)) + \zeta_t$$

where, $t = 0, \pm 1, \pm 2, \cdots$, the A's and B's are all uncorrelated, $E(A_j) = E(B_j) = 0$, and $Var(A_j) = Var(B_j) = \sigma_j^2$. Further, suppose $\{\zeta_t\}$ is colored stationary noise with mean 0 and variance σ_ζ^2, independent of the A's and B's. The noise is assumed to possess an absolutely continuous spectral distribution function $F_\zeta(\omega)$ with spectral density $f_\zeta(\omega)$, $\omega \in [-\pi, \pi]$. We do not restrict the process $\{Z_t\}$ to be Gaussian, unless such a restriction is made explicitly. Without loss of generality assume that the frequencies are ordered fixed constants,

$$0 < \omega_1 < \omega_2 < \cdots < \omega_p < \pi$$

In this paper we shall describe a novel iterative scheme for locating the frequencies $\omega_1, \omega_2, \cdots, \omega_p$ from a time series $Z_1, Z_2, \cdots, Z_N$ using HOC. Although in the present work only fixed frequencies are of concern, the scheme can be easily adapted to handle the case of time dependent frequencies. This more general problem will be dealt with elsewhere.

The problem of estimating the frequencies of sinusoidal components in ambient noise is one of the oldest and best known problems in time series analysis (see the review article by Brillinger (1987), and Priestley (1981; Chs. 6,8)). The celebrated solution advanced by Schuster (1898), almost one hundred years ago, in the form of periodogram analysis is still, to a large extent, the prevailing approach to this problem today, and especially so since the popularization of the "fast Fourier transform" (FFT) by Cooley and Tukey (1965). The FFT enables the computation of the periodogram from a data record of length N by requiring only $O(N \log N)$ computational

---

* Work supported by grants AFOSR-89-0049 and ONR -89-J-1051.
† Department of Mathematics, University of Maryland, College Park, Maryland 20742

complexity, a great improvement over the $O(N^2)$ computational complexity needed for direct computation. The question is whether further reduction in computational complexity is possible in the estimation of hidden periodicities.

To gain some insight into this problem, consider a single pure sinusoid in discrete time. It is clear, without getting into the problem of precision, that the number of cycles, or equivalently the number of zero-crossings, can provide an estimate of the period and this arguably requires only $O(N)$ operations versus the $O(N \log N)$ operations needed for the FFT. Thus, this simple example indicates that there are cases where zero-crossing counts have the potential of being computationally competitive with FFT-based methods. The present paper goes far beyond this simple case, by developing a *higher order crossings* (HOC)-based method that together with parallel processing has the potential of requiring only $O(N)$ operations even in the presence of colored noise.

To be more specific, let θ be a parameter taking values in $(-1, 1)$, and let $C(\theta)$ be a function such that $0 < C(\theta) < 1$. In its simplest form, our contraction mapping, denoted by $\rho_1(\theta)$, can be expressed as (He and Kedem (1989), Yakowitz (1989)),

$$(2) \qquad \rho_1(\theta) = r^* + C(\theta)(\theta - r^*)$$

where r^* is the cosine of the frequency to be detected. Then there exists a HOC-sequence $\{\theta_j\}$ such that

$$\theta_{j+1} = \rho_1(\theta_j)$$

and $\theta_j \to r^*$, as $j \to \infty$. Each iteration of the recursion requires $O(N)$ operations, and only very few iterations are needed for a satisfactory level of precision. As we shall see, $\rho_1(\theta)$ is a certain first order autocorrelation indexed by θ.

This paper is a survey of some recent results in discrete spectrum estimation from noisy data, based on HOC concepts and algorithms. The approach that we shall describe is based on a recursive filtering idea first proposed by He and Kedem (1989) and later refined and extended by Yakowitz (1989) and by Kedem and Troendle (1989). By this method, the cosine of each discrete frequency is obtained recursively from a contraction mapping of the form (2). These works led to the general iterative method of Kedem and Yakowitz (1990) in which the convergence is enhanced dramatically by an accelerated iterative reduction in the contraction coefficient $C(\theta)$.

Sections 2 and 3 review some known facts about HOC. The newer results concerning contraction mappings, are considered in Sections 4 and 5. In Section 5.2 we focus on a specific contraction mapping, obtained from a useful parametric family of filters. Concluding remarks are given in Section 6.

In the present paper, by a filter we always mean a *linear* filter.

2. Some Background About HOC

2.1 Parametric families of Zero-Crossing counts. The mathematical formulation of the idea behind HOC can be outlined in five steps as follows.

1. We start with a time series observed in discrete time

$$\{Z_t\}, \ t = 0, \pm1, \pm2, \cdots$$

2. Let
$$\{\mathcal{L}_\theta(\cdot),\ \theta \in \Theta\}$$

be a parametric family of filters indexed by θ. The parameter space Θ may be an interval or a countable set in one or more dimensions. That is, θ may be a vector.

3. Fix $\theta \in \Theta$ and consider the observed time series
$$\mathcal{L}_\theta(Z)_1, \mathcal{L}_\theta(Z)_2, \cdots, \mathcal{L}_\theta(Z)_N$$

of length N from the filtered process
$$\{\mathcal{L}_\theta(Z)_t\},\ t = 0, \pm 1, \pm 2, \cdots$$

4. Define the corresponding binary time series
$$X_t(\theta) = \begin{cases} 1, & \text{if } \mathcal{L}_\theta(Z)_t \geq 0 \\ 0, & \text{if } \mathcal{L}_\theta(Z)_t < 0 \end{cases}$$

$t = 1, 2, \cdots, N$

5. The corresponding *higher order crossings* (HOC) family is given by
$$\{D_\theta, \theta \in \Theta\}$$

where
$$D_\theta = \sum_{t=2}^{N} [X_t(\theta) - X_{t-1}(\theta)]^2$$

Thus for each fixed $\{\theta \in \Theta\}$, D_θ is defined as the number of symbol changes in $X_1(\theta), \cdots, X_N(\theta)$, or the *number of zero-crossings* in discrete time in $\mathcal{L}_\theta(Z)_1, \mathcal{L}_\theta(Z)_2, \cdots, \mathcal{L}_\theta(Z)_N$. When, for some θ, $\mathcal{L}_\theta(\cdot)$ corresponds to the identity filter, then D_θ is the number of zero-crossings in the original unfiltered series $Z_1, \cdots, Z_N$. In addition to dependence on θ, D_θ also depends on N. This dependence is more apparent in the *observed zero-crossing rate* defined by
$$\gamma_N(\theta) \equiv \frac{D_\theta}{N-1}$$

But, clearly, the expected rate
$$\gamma(\theta) \equiv \frac{E[D_\theta]}{N-1}$$

need not depend on N as is the case for a strictly stationary process. In studying HOC families, at times we let θ change while holding N fixed (this amounts to sample reuse), and at other times θ is fixed but N changes.

2.2 Relationship with the autocorrelation. Let $\rho_1(\theta), \theta \in \Theta$, be the first-order autocorrelation of the filtered process $\{\mathcal{L}_\theta(Z)_t\}$. On intuitive grounds, the first-order autocorrelation should be inversely related to the number of zero-crossings. In the Gaussian case this relationship takes the exact form

(3)
$$\rho_1(\theta) = \cos\left(\frac{\pi E[D_\theta]}{N-1}\right)$$

We can see that in the Gaussian case knowledge of $E[D_\theta]$ is equivalent to knowledge of $\rho_1(\theta)$. However, the relationship (3) called the *cosine formula*, is not unique to Gaussian processes, and furthermore, similar but different explicit relationships can also be obtained for non-Gaussian processes. A general method for generating relationships between $E[D_\theta]$ and $\rho_1(\theta)$ by transforming Gaussian processes is described in Barnett and Kedem (1990). As a example consider the integral transformation

$$Z_t = G(Y_t) - \frac{1}{2}$$

where $\{Y_t\}$ is a stationary Gaussian time series with mean 0, and G is the cumulative distribution function of Y_t, for each fixed t. Let θ correspond to the identity filter. Then we obtain a "cosine formula" for a uniform process:

$$(4) \qquad \rho_1(\theta) = \frac{6}{\pi} \sin^{-1} \left(\frac{1}{2} \cos \left(\frac{\pi E[D_\theta]}{N-1} \right) \right)$$

Equation (4) is typical of many other similar formulas.

The point being made here, is that although we do not have exact relationships between $\rho_1(\theta)$ and $E[D_\theta]$ in all cases, quite a few examples as well as experience indicate that in general $E[D_\theta]$ and $\rho_1(\theta)$ are essentially equivalent. Because of this affinity, we find it convenient to refer to the family $\{\rho_1(\theta), \theta \in \Theta\}$ as a (expected) *higher order correlation* family, or simply a HOC family again. Thus, HOC families, observed or expected, can be families of zero-crossing counts or families of first-order correlation coefficients.

The following recent result is of general interest.

THEOREM 0. (Kedem and Li (1990a)). *Let* $\{Z_t\}$, $t = 0, \pm 1, \pm 2, \cdots$, *be a real-valued zero-mean stationary process with autocorrelation* ρ_k. *Let* $|H_\theta(\omega)|$ *be the gain of a linear filter* $\mathcal{L}_\theta(\cdot)$, *and let* $\rho_1(\theta)$ *be the first-order autocorrelation of the filtered process* $\{\mathcal{L}_\theta(Z)_t\}$, $t = 0, \pm 1, \pm 2, \cdots$. *Then we have.*

- *(a)*   *If* $|H_\theta(\omega)|$ *is monotone increasing in* $[0, \pi]$, *then*

$$\rho_1 \geq \rho_1(\theta)$$

If the gain is monotone decreasing the inequality is reversed.

- *(b)*   *Assume that* $|H_\theta(\omega)|$ *is strictly monotone. Then*

$$\rho_1 = \rho_1(\theta)$$

if and only if $\{Z_t\}$ *is a pure sinusoid with probability one.*

Similar results can be obtained for zero-crossing counts. Of interest is a consequence of the cosine formula (3).

Corollary 0. (Kedem and Li (1990a)). *Suppose that the process* $\{Z_t\}$ *is Gaussian with expected zero-crossing rate* γ.

- *(a)   If $|H_\theta(\omega)|$ is monotone increasing in $[0, \pi]$, then*

$$\gamma \le \gamma(\theta)$$

The inequality is reversed if $|H_\theta(\omega)|$ is monotone decreasing.

- *(b)   Assume $|H_\theta(\omega)|$ is strictly monotone. Then*

$$\gamma = \gamma(\theta)$$

if and only if $\{Z_t\}$ is a pure sinusoid with probability one. The frequency of the sinusoid is given by $\pi\gamma$.

2.3 Examples of parametric HOC families

2.3.1 HOC from differences. Let $\bigtriangledown$ be the difference operator defined by

$$\bigtriangledown Z_t \equiv Z_t - Z_{t-1}$$

and define

$$\mathcal{L}_\theta \equiv \bigtriangledown^{\theta-1}, \theta \in \{1, 2, 3, \cdots\}$$

with $\mathcal{L}_1 \equiv \bigtriangledown^0$ being the identity filter. The corresponding HOC sequence

$$D_1, D_2, D_3, \cdots$$

gives the simple HOC. When Z_t is a zero-mean stationary Gaussian process with spectral distribution function $F(\omega)$ then, by virtue of the cosine formula (3), the simple HOC obtained by repeated differencing admit the spectral representation

$$(5) \qquad \rho_1(\theta) = \cos[\pi\gamma(\theta)] = \frac{\int_{-\pi}^{\pi} \cos(\omega)(\sin(\omega/2))^{2(\theta-1)} dF(\omega)}{\int_{-\pi}^{\pi} (\sin(\omega/2))^{2(\theta-1)} dF(\omega)}$$

where $\rho_1(\theta)$ is the first-order correlation in the filtered process $\{\mathcal{L}_\theta(Z)_t\}$. The sequence $\gamma(\theta)$ is bounded and monotone increasing in θ so that

$$(6) \qquad \pi\gamma(\theta) \longrightarrow \omega^*, \ \theta \longrightarrow \infty$$

where ω^* is the highest frequency in the spectral support. We shall provide a proof of this fact in the next section. In the Gaussian case, knowledge of $\gamma(\theta)$ is equivalent to knowledge of the autocorrelation sequence of $\{Z_t\}$ (Kedem (1986), He and Kedem (1989)).

2.3.2 HOC from repeated summation. Let $\mathcal{B}$ be the shift operator defined by

$$\mathcal{B}Z_t = Z_{t-1}$$

and define

$$\mathcal{L}_\theta \equiv (1 + \mathcal{B})^{\theta-1}, \ \theta \in \{1, 2, 3, \cdots\}$$

The corresponding HOC are called HOC from repeated summation and the sequence $\{D_\theta\}$ is monotone decreasing in θ, and in the Gaussian case

$$(7) \qquad \pi\gamma(\theta) \longrightarrow \omega_*, \ \theta \longrightarrow \infty$$

where ω_* is the lowest frequency in the spectral support of $F(\omega)$. This can be shown using the spectral representation

(8) $$\rho_1(\theta) = \cos[\pi\gamma(\theta)] = \frac{\int_{-\pi}^{\pi} \cos(\omega)(1+\cos(\omega))^{(\theta-1)} dF(\omega)}{\int_{-\pi}^{\pi}(1+\cos(\omega))^{(\theta-1)} dF(\omega)}$$

Also here, in the Gaussian case, knowledge of $\gamma(\theta)$ is equivalent to knowledge of the autocorrelation sequence of $\{Z_t\}$.

2.3.3 HOC from exponential smoothing. Consider the the parametric family of filters

(9) $$\mathcal{L}_\alpha \equiv 1 + \alpha\mathcal{B} + \alpha^2\mathcal{B}^2 + \cdots, \alpha \in (-1,1)$$

where $\mathcal{B}$ is the backward shift. We can rewrite this more compactly as follows. Let

$$Z_t(\alpha) \equiv \mathcal{L}_\alpha(Z)_t$$

Then (9) becomes,

(10) $$Z_t(\alpha) = \alpha Z_{t-1}(\alpha) + Z_t, \ \alpha \in (-1,1)$$

Because of this form, (10), or equivalently (9), is referred to as an $AR(1)$ filter. The corresponding HOC family $\{D_\alpha\}$, $\alpha \in (-1,1)$, has been studied by He and Kedem (1989) and by Kedem and Li (1989). When $\{Z_t\}$ is a zero mean stationary Gaussian process then, $\gamma(\alpha)$ is monotone decreasing, and we obtain the spectral representation

(11) $$\rho_1(\alpha) = \cos[\pi\gamma(\alpha)] = \frac{\int_{-\pi}^{\pi} \cos(\omega)|H(\omega;\alpha)|^2 dF(\omega)}{\int_{-\pi}^{\pi} |H(\omega;\alpha)|^2 dF(\omega)}$$

where

(12) $$|H(\omega;\alpha)|^2 = \frac{1}{1 - 2\alpha\cos(\omega) + \alpha^2}, \ \ \alpha \in (-1,1), \ \omega \in [0,\pi]$$

It can be shown that in the Gaussian case $\gamma(\alpha)$ is equivalent to the normalized mixed spectrum. If $\gamma(\alpha) = \gamma(\beta)$ for $\alpha, \beta \in (-1,1)$, then $\{Z_t\}$ is a pure sinusoid with probability 1. Clearly, this is a special case of Theorem 0.

2.3.4 HOC from autoregressive moving average operations. A more general family of parametric filters, and one that contains the previous examples as special cases, is defined by the autoregressive-moving average operation

(13) $$Y_t - a_1 Y_{t-1} - \cdots - a_p Y_{t-p} = Z_t - b_1 Z_{t-1} - \cdots b_q Z_{t-q}$$

where now

$$D_\theta \equiv D_{a_1, a_2, \cdots, b_1, b_2, \cdots}$$

Not much is known at present about the HOC $\{D_\theta\}$ of this general case. However, in the Gaussian case, every statement concerning relationships between the autocorrelation and the parameters, can be translated into a statement regarding the expected HOC $\{E[D_\theta]\}$

3. Detection of the Highest and Lowest Frequencies With an Application to a Purely Discrete Spectrum

Before presenting more general results regarding mixed spectra, it is instructive to consider the case of a purely discrete spectrum, and assume that in (1), $\zeta_t = 0$ for all t. In this case HOC provide a complete solution in locating the frequencies $\omega_1, \cdots, \omega_p$, and determining their number p. For the sake of illustration, $\{Z_t\}$ is assumed to be Gaussian so that the cosine formula (3) can be used. The Gaussian assumption is only a matter of convenience and can in fact be relaxed.

Here and in other places, the algorithms are given in terms of expected HOC. In practice, the expected HOC are replaced by the observed HOC.

Our first result concerns a general spectrum. We need it for the special case at hand.

THEOREM 1. *Suppose $\{Z_t\}$ is Gaussian. Then, regardless of the spectrum type, the simple expected HOC $\{E[D_j]\}$ from repeated differencing are monotone increasing,*

$$(14) \qquad E[D_j] \le E[D_{j+1}], \; j = 1, 2, 3, \cdots.$$

and,

$$(15) \qquad \frac{\pi E[D_j]}{N-1} \longrightarrow \omega^*$$

where $\omega^* \le \pi$ is the highest frequency in the spectral support.

Proof: Because the backward difference operator is a highpass filter, (14) follows from Corollary 0. A more direct proof follows from the combinatorial fact that

$$D_j \le D_{j+1} + 1$$

Therefore, the inequality

$$\gamma(j) \le \gamma(j+1) + \frac{1}{N-1}$$

always holds for any strictly stationary process. By letting $N \to \infty$, (14) follows because $\gamma(j)$ is independent of N. To prove (15), observe that the normalized expected HOC are also bounded:

$$0 \le \frac{\pi E[D_j]}{N-1} \le \pi$$

and therefore must converge. Also note that the spectral measure of $\{(\nabla^j Z)_t\}$, denoted by $\nu_j(\cdot)$, is given by

$$\nu_j(d\omega) = \frac{\sin^{2j}(\omega/2)dF(\omega)}{\int_{-\pi}^{\pi}\sin^{2j}(\lambda/2)dF(\lambda)}$$

where F is the spectral distribution function of $\{Z_t\}$. It follows that (5) can be written more compactly as

$$\cos\left(\frac{\pi E[D_{j+1}]}{N-1}\right) = \int_{-\pi}^{\pi}\cos(\omega)\nu_j(d\omega)$$

But from Kedem and Slud (1982), $\nu_j(\cdot)$ converges weakly,

$$\nu_j \Longrightarrow \frac{1}{2}\delta_{-\omega^*} + \frac{1}{2}\delta_{\omega^*}, \; j \longrightarrow \infty$$

where δ_u is the unit point mass at u. Therefore, as $j \longrightarrow \infty$,

$$\cos\left(\frac{\pi E[D_{j+1}]}{N-1}\right) \longrightarrow \cos(\omega^*)$$

and use the fact that $\cos(x)$, $x \in [0, \pi]$, is monotone. $\qquad\qquad\square$

Denote the HOC from repeated summation by $_jD$. In the same way it can be shown that, regardless of the spectrum type, $E[_jD] \geq E[_{j+1}D]$, and that $\pi E[_jD]/(N-1) \to \omega_*$, $j \to \infty$, where $\omega_* \geq 0$ is the lowest frequency in the spectral support.

We can now apply these results to obtain $\omega_1, \cdots, \omega_p$ and p. The following algorithm portrays an ideal situation by assuming the existence of ideal-pass filters.

Algorithm 1
Determination of $\omega_1, \cdots, \omega_p, \& p$

1. Obtain the lowest frequency $\omega_* \equiv \omega_1$ from repeated summation:

$$\frac{\pi E[_jD]}{N-1} \to \omega_1, \; j \to \infty$$

   Do not filter ω_1 out.

2. Obtain the highest frequency $\omega^* \equiv \omega_p$ from repeated differencing:

$$\frac{\pi E[D_j]}{N-1} \to \omega_p, \; j \to \infty$$

   Filter ω_p out by an ideal lowpass filter with cutoff frequency ω_p-

3. ω_{p-1} is now the highest frequency. Repeat step 2 to determine ω_{p-1}, and filter it out with an ideal lowpass filter with cutoff frequency $\omega_{p-1}-$.

4. Repeat the procedure to obtain $\omega_{p-2}, \omega_{p-3}, \cdots$, until the next highest frequency is equal to ω_1, the lowest frequency that we already know, and stop.

The above serves to illustrate a general principle referred to as the "*dominant frequency principle*". The expected normalized zero-crossing count is a weighted average of the spectral mass, as is apparent from the cosine formula in the Gaussian case. Accordingly, the normalized HOC tend to locate themselves in the dominant spectral region when it exists. Repeated differencing means the sequential application of a highpass filter that puts more and more weight on the highest frequency ω^* rendering it dominant. The corresponding normalized HOC thus move in the direction of ω^* until convergence occurs. Similarly, repeated summation means the sequential application of a lowpass filter that shifts the spectral weight towards the lowest frequency ω_*. As a result, the corresponding normalized HOC are "attracted" to ω_* until convergence occurs. More generally, by appropriate filtering we can shift the spectral weight at will and force the corresponding normalized HOC to move in any desired direction. We shall see examples of this fact in the coming sections where we generate HOC sequences by recursive procedures.

The convergence of the sequence of expected simple HOC, from repeated differencing, to the respective highest frequency, is exponentially fast as is clear from the spectral representation (5). The following computer simulation shows that very few iterations are needed in order to obtain the highest frequency with a satisfactory level of precision. By symmetry, the same can be said of HOC from repeated summation.

Example 1. Convergence to the highest frequency

The fast rate of convergence of $\pi D_j/(N-1)$ with $N = 1000$, as $j \to \infty$, to the highest frequency, is illustrated using (1) with $p = 4$, $\zeta_t \equiv 0$, and uniformly distributed amplitudes. In each of three runs, the amplitudes have the same uniform distribution. In all cases

$$\omega_1 = 0.2, \ \omega_2 = 0.8, \ \omega_3 = 1.2, \ \omega_4 = 2.5$$

The results are given in Table 1.

Table 1. Convergence of $\pi D_j/(N-1)$ towards $\omega_4 = 2.5$ with $N = 1000$. The amplitudes are uniformly distributed as indicated.

	u(-0.5,0.5)	u(-1,1)	u(-2,2)
j	$\pi D_j/999$	$\pi D_j/999$	$\pi D_j/999$
1	1.619540	1.421421	0.905684
2	2.430882	2.141566	2.154145
3	2.500066	2.493777	2.459185
4	2.500066	2.500066	2.500066
5	2.500066	2.500066	2.500066
.	.	.	.
.	.	.	.
.	.	.	.

Example 2. Convergence to the lowest frequency

Using the same type of data as in Example 1, and with the same frequencies, we illustrate the convergence of $\pi E[_j D]/(N-1)$, $N = 1000$, to the lowest frequency $\omega_1 = 0.2$. The results are given in Table 2. Because ω_1 is fairly close to 0, the convergence is slower than in the previous example.

Table 2. Convergence of $\pi_j D/(N-1)$ towards $\omega_1 = 0.2$ with $N = 1000$. The amplitudes are uniformly distributed as indicated.

	u(-0.5,0.5)	u(-1,1)	u(-2,2)
j	$\pi D_j/999$	$\pi D_j/999$	$\pi D_j/999$
1	1.355382	1.534632	1.871119
2	0.801908	0.883671	0.811342
3	0.801908	0.600645	0.597500
4	0.801908	0.603790	0.506303
5	0.801908	2.500066	0.399382
6	0.735869	0.603790	0.399382
7	0.396273	0.603790	0.399382
8	0.396273	0.603790	0.399382
9	0.396237	0.591211	0.399382
10	0.396237	0.402526	0.330197
11	0.393092	0.402526	0.201263
12	0.393092	0.402526	0.201263
13	0.393092	0.402526	0.201263
14	0.393092	0.402526	0.201263
15	0.383658	0.402526	0.201263
16	0.198118	0.383658	0.201263
17	0.198118	0.201263	0.201263
18	0.198118	0.201263	0.201263
.	.	.	.
.	.	.	.
.	.	.	.

4. APPLICATION OF THE $AR(1)$ FILTER IN THE DETECTION OF A SINGLE FREQUENCY IN NOISE

Consider the process (1) with $p = 1$, and assume that $\{\zeta_t\}$ is white noise. The highest frequency is no longer ω_p but π. Cosequently, by Theorem 1, $\pi E[D_j]/(N-1) \to \pi$, and the previous algorithm is not applicable. In order to produce a HOC sequence that converges to ω_1 we must therefore neutralize the noise in some way. The next result, due to He and Kedem (1989), does just that by making use of a HOC sequence from the $AR(1)$ filter (9).

Recall that the $AR(1)$ filter is defined by the operation,

$$(16) \qquad Z_t(\alpha) = \mathcal{L}_\alpha(Z)_t = Z_t + \alpha Z_{t-1} + \alpha^2 Z_{t-2} + \cdots$$

and has a transfer function with squared gain $|H(\omega; \alpha)|^2$ given by (12). Similarly define,

$$\zeta_t(\alpha) = \mathcal{L}_\alpha(\zeta)_t$$

and,

$$(17) \qquad C(\alpha) = \frac{Var(\zeta_t(\alpha))}{Var(Z_t(\alpha))}$$

Then for $\alpha \in (-1, 1)$,

$$0 < C(\alpha) < 1$$

Clearly $C(\alpha)$ also depends on ω_1, but this is not included to keep the notation simple.

THEOREM 2. *Suppose*

$$Z_t = A_1 \cos(\omega_1 t) + B_1 \sin(\omega_1 t) + \zeta_t, \ t = 0, \pm 1, \cdots$$

where $\omega_1 \in (0, \pi)$, A_1, B_1 *are independent* $N(0, \sigma_1^2)$ *random variables, and* $\{\zeta_t\}$ *is Gaussian white noise with mean* 0 *and variance* σ_ζ^2, *independent of* A_1, B_1. *Let* $\{D_\alpha\}$ *be the HOC from (16). Fix* $\alpha_1 \in (-1, 1)$, *and define*

$$(18) \qquad \alpha_{k+1} = \cos\left(\frac{\pi E[D_{\alpha_k}]}{N-1}\right), \ k = 1, 2, \cdots$$

Then, as $k \to \infty$,

$$\alpha_k \to \cos(\omega_1)$$

and

$$(19) \qquad \frac{\pi E[D_{\alpha_k}]}{N-1} \to \omega_1$$

Proof: Note that

$$\int_0^\pi |H(\omega; \alpha)|^2 d\omega = \frac{\pi}{1 - \alpha^2}$$

and

$$\int_0^\pi \cos(\omega) |H(\omega; \alpha)|^2 d\omega = \frac{\pi}{1 - \alpha^2} \times \alpha$$

Therefore, by symmetry,

$$(20) \qquad \int_{-\pi}^\pi \cos(\omega) |H(\omega; \alpha)|^2 d\omega = \alpha \times \int_{-\pi}^\pi |H(\omega; \alpha)|^2 d\omega$$

and so,

$$\rho_1(\alpha) = \cos\left(\frac{\pi E[D_\alpha]}{N-1}\right)$$

$$(21) \qquad = \frac{\sigma_1^2 |H(\omega_1; \alpha)|^2 \times \cos(\omega_1) + \int_{-\pi}^\pi |H(\omega; \alpha)|^2 dF_\zeta(\omega) \times \alpha}{\sigma_1^2 |H(\omega_1; \alpha)|^2 + \int_{-\pi}^\pi |H(\omega; \alpha)|^2 dF_\zeta(\omega)}$$

$$= [1 - C(\alpha)] \times \cos(\omega_1) + C(\alpha) \times \alpha$$

We can see that $\rho_1(\alpha)$ is a convex combination of $\cos(\omega_1)$ and α and that it is a contraction mapping of the form (2),

$$(22) \qquad \rho_1(\alpha) = \alpha^* + C(\alpha)(\alpha - \alpha^*)$$

where $\alpha^* = \cos(\omega_1)$. From the recursion

$$\alpha_{k+1} = \rho_1(\alpha_k)$$

we obtain

$$\rho_1(\alpha_k) = \alpha^* + [\prod_{j=1}^{k} C(\alpha_j)](\alpha_1 - \alpha^*)$$

Because of the convexity in (22), it follows that α_k, $k \geq 1$, is always between α_1 and α^*, and the product goes to 0 with the number of iterations. Therefore $\alpha_k \to \alpha^*$, and α^* is a fixed point of $\rho_1(\cdot)$,

$$\alpha^* = \rho_1(\alpha^*)$$

or

$$\cos(\omega_1) = \cos\left(\frac{\pi E[D_{\alpha^*}]}{N-1}\right)$$

and by the monotonicity of $\cos(x)$, $x \in [0, \pi]$,

$$\omega_1 = \frac{\pi E[D_{\alpha^*}]}{N-1}$$

$\square$

Example 3. Convergence of Observed HOC From the $AR(1)$ Filter

Table 3 shows some examples of the convergence (19) using simulated Gaussian data of a single sinusoid plus noise. As remarked earlier, in practice $\gamma(\alpha)$ is replaced by $\gamma_N(\alpha)$, the observed zero-crossing rate after filtering. Evidently, for reasonable signal to noise ratios, the algorithm converges rather fast. Observe that no Fourier type analysis has been used, and that ω_1 need not be a Fourier frequency of the form $2\pi k/N$.

Table 3. Convergence of $\pi D_{\alpha_j}/(N-1)$ towards $\omega_1 = 0.8$ with $N = 10000$ where the signal to noise ratio (SNR) is given in dB.

1dB	0dB	-1.94dB	-6.02dB	-10dB
$\alpha_0 = -.1$	$\alpha_0 = .9$	$\alpha_0 = .2$	$\alpha_0 = .5$	$\alpha_0 = .1$
0.8848	0.5194	0.9127	0.9291	1.3381
0.8006	0.5904	0.8222	0.8713	1.2294
0.7987	0.6563	0.8015	0.8411	1.1251
0.7987	0.7142	0.7965	0.8191	1.0371
0.7987	0.7600	0.7952	0.8053	0.9617
0.7987	0.7864	0.7952	0.8015	0.9077
0.7987	0.8002	0.7952	0.7990	0.8756
0.7987	0.8065	0.7952	0.7984	0.8511
0.7987	0.8065	0.7952	0.7977	0.8449
0.7987	0.8065	0.7952	0.7971	0.8436
0.7987	0.8065	0.7952	0.7971	0.8423
0.7987	0.8065	0.7952	0.7971	0.8405
0.7987	0.8065	0.7952	0.7971	0.8392
0.7987	0.8065	0.7952	0.7971	0.8379
0.7987	0.8065	0.7952	0.7971	0.8361
0.7987	0.8065	0.7952	0.7971	0.8335
0.7987	0.8065	0.7952	0.7971	0.8323
0.7987	0.8065	0.7952	0.7971	0.8310
0.7987	0.8065	0.7952	0.7971	0.8298
0.7987	0.8065	0.7952	0.7971	0.8310
0.7987	0.8065	0.7952	0.7971	0.8298
.	.	.	.	.
.	.	.	.	.
.	.	.	.	.

Note: $SNR \equiv 20 log_{10} \frac{s.d.signal}{s.d.noise}$ dB

4.1 Yakowitz's extension. An observation made by Yakowitz (1989), extends the above procedure significantly in three ways. First, the Gaussian assumption, that relies so much on the cosine formula, is not really needed. Second, the noise need not be white. Third, many different parametric filters, not just the $AR(1)$ filter, can be used to generate the same contraction mapping as in (22), provided they satisfy a certain parametrization requirement. The key idea of Yakowitz (1989) is to operate on the *noise* so that the parameter of the filter agrees with the first autocorrelation of the filtered noise. In many cases this can be achieved by *reparametrization*.

Consider the process (1) with $p = 1$, and where the noise $\{\zeta_t\}$ is not necessarily white. Let r be a parameter that takes values in $(-1, 1)$, and let $\{H(\omega; r)\}$ be a family of transfer functions indexed by r. Assume that r is the first-order autocorrelation of the filtered noise, where the filter corresponds to $H(\omega; r)$,

$$(23) \qquad r = \frac{\int_{-\pi}^{\pi} \cos(\omega)|H(\omega; r)|^2 dF_\zeta(\omega)}{\int_{-\pi}^{\pi} |H(\omega; r)|^2 dF_\zeta(\omega)}$$

From this we obtain a generalization of (20),

$$(24) \qquad \int_{-\pi}^{\pi} \cos(\omega)|H(\omega;r)|^2 dF_\zeta(\omega) = r \times \int_{-\pi}^{\pi} |H(\omega;r)|^2 dF_\zeta(\omega)$$

As before, let $\rho_1(r)$ be the first-order autocorrelation of the filtered *process* using the same filter with transfer function $H(\omega;r)$. Then from (24),

$$(25) \qquad \rho_1(r) = \frac{\sigma_1^2 |H(\omega_1;r)|^2 \times \cos(\omega_1) + \int_{-\pi}^{\pi} |H(\omega;r)|^2 dF_\zeta(\omega) \times r}{\sigma_1^2 |H(\omega_1;r)|^2 + \int_{-\pi}^{\pi} |H(\omega;r)|^2 dF_\zeta(\omega)}$$

which we recognize to be the same as (21) with r replacing α. Everything now becomes entirely analogous to the previous case discussed in Theorem 2 where the $AR(1)$ filter was used. Thus, if as before we denote by $\{Z_t(r)\}$, and $\{\zeta_t(r)\}$, the filtered process and filtered noise, respectively, and let

$$(26) \qquad C(\alpha) = \frac{Var(\zeta_t(r))}{Var(Z_t(r))}$$

then we obtain a general contraction mapping

$$(27) \qquad \rho_1(r) = r^* + C(r)(r - r^*)$$

where $r^* = \cos(\omega_1)$. Assume that $C(r) < 1$ for all $r \in (-1,1)$. It follows that for any initial point $r_1 \in (-1,1)$, the sequence r_k defined by

$$(28) \qquad r_{k+1} = \rho_1(r_k)$$

converges to r^*,

$$(29) \qquad r_k \to r^*$$

as $k \to \infty$. Thus r^* is a fixed point of $\rho_1(\cdot)$

$$(30) \qquad r^* = \rho_1(r^*)$$

and a limit point of the iterations (28). The desired frequency ω_1 is obtained from the inverse transformation $\omega_1 = \cos^{-1}(r^*)$.

Clearly, under these conditions, Theorem 2 is now a special case, where the family of parametric filters is given by the $AR(1)$ filter with parameter r, and the noise is white.

5. Contraction Mappings from Band-Pass Filters

The previous procedures that treat the single frequency case can be readily extended to the multiple frequency case by applying band-pass filters (Kedem and Troendle (1989), Kedem and Yakowitz (1990)). By controlling the bandwidth by a certain design, it is possible to obtain a greatly enhanced contraction, and hence, an accelerated convergence. This last idea has been considered in Kedem and Yakowitz (1990).

5.1 Ideal band-pass case. For a better understanding of the general procedure described in Kedem and Yakowitz (1990), it is instructive to consider briefly an ideal case first.

Suppose in (1), $\{Z_t\}$ is Gaussian and that the noise $\{\zeta_t\}$ is white Gaussian noise. We apply to $\{Z_t\}$ the ideal band-pass filter defined by the squared gain

$$|H(\omega;\theta)|^2 = \begin{cases} 1, & \omega \in [\theta - \delta, \theta + \delta] \\ 0, & \omega \in [0, \theta - \delta) \cup (\theta + \delta, \pi] \end{cases}$$

where $\delta < C/2$ and

$$C = \min\{|a - b| : a \text{ and } b \text{ are among } \omega_1, \omega_2, \ldots, \omega_p, 0, \pi\}$$

We shall say that the filter *captures* ω_j if ω_j lies in the interval $[\theta - \delta, \theta + \delta]$. This is equivalent to saying "$|H(\omega;\theta)|^2$ captures ω_j".

We have

$$\cos\left(\frac{\pi E[D_\theta]}{N-1}\right) = \frac{\sum_{j=1}^{p} \sigma_j^2 |H(\omega_j;\theta)|^2 \cos(\omega_j) + \sigma_\zeta^2/\pi \int_0^\pi \cos(\lambda)|H(\lambda;\theta)|^2 d\lambda}{\sum_{j=1}^{p} \sigma_j^2 |H(\omega_j;\theta)|^2 + \sigma_\zeta^2/\pi \int_0^\pi |H(\lambda;\theta)|^2 d\lambda}$$

Assume that the filter is applied to $\{Z_t\}$ with an initial value θ_0, and that ω_1 has been captured.

A contraction mapping can be obtained as follows. By the mean value theorem, for any given $\theta \in [\delta, \pi - \delta]$,

$$\int_{\theta-\delta}^{\theta+\delta} \cos(\omega)d\omega = 2\cos(\theta)\sin(\delta) = 2\delta\cos(\omega_\theta)$$

where $\theta - \delta \leq \omega_\theta \leq \theta + \delta$. Therefore, by specifying ω_θ the corresponding θ can be obtained from the equation

$$\theta = \cos^{-1}\left[\frac{\delta\cos(\omega_\theta)}{\sin(\delta)}\right]$$

Taking this into account, the contraction now takes the form,

$$\begin{aligned}
\cos\left(\frac{\pi E[D_{\theta_0}]}{N-1}\right) &= \frac{\sigma_1^2 \cos(\omega_1) + (\sigma_\zeta^2/\pi) \int_{\theta_0-\delta}^{\theta_0+\delta} \cos(\lambda)d\lambda}{\sigma_1^2 + (\sigma_\zeta^2/\pi) \int_{\theta_0-\delta}^{\theta_0+\delta} d\lambda} \\
&= \frac{\sigma_1^2 \cos(\omega_1) + 2\delta(\sigma_\zeta^2/\pi)\cos(\omega_{\theta_0})}{\sigma_1^2 + 2\delta(\sigma_\zeta^2/\pi)} \\
&= r^* + C(\delta)(\cos(\omega_{\theta_0}) - r^*)
\end{aligned}$$

where $r^* = \cos(\omega_1)$, and $C(\delta)$ is the contraction coefficient,

$$C(\delta) = \frac{2\delta(\sigma_\zeta^2/\pi)}{\sigma_1^2 + 2\delta(\sigma_\zeta^2/\pi)}$$

To iterate the contraction, define,

$$\omega_{\theta_{j+1}} = \frac{\pi E D_{\theta_j}}{N-1}$$

(ω_{θ_j} should not be confused with any of the ω_j to be detected) obtain the corresponding θ_{j+1} from,

$$\theta_{j+1} = \cos^{-1}\left[\frac{\delta\cos(\omega_{\theta_{j+1}})}{\sin(\delta)}\right]$$

and observe the resulting $D_{\theta_{j+1}}$. From this follows (Kedem and Troendle (1989)),

THEOREM 3. *Suppose in (1) $\{Z_t\}$ is Gaussian, and $\{\zeta_t\}$ is Gaussian white noise. If ω_i is captured by $|H(\omega; \theta_0)|^2$, and $\{\theta_j\}$ is determined from,*

$$\theta_{j+1} = \cos^{-1}\left[\frac{\delta \cos(\omega_{\theta_{j+1}})}{\sin(\delta)}\right]$$

then

$$\frac{\pi E[D_{\theta_j}]}{N-1} \to \omega_i, \quad j \to \infty$$

We see that once our initial filter captures a frequency, it never looses it, and the expected normalized HOC sequence converges to this frequency. Now suppose that initially, $|H(\omega; \theta_0)|^2$ does not capture any frequency. Then the contraction becomes,

$$\cos\left(\frac{\pi E[D_{\theta_0}]}{N-1}\right) = \frac{\int_{\theta_0-\delta}^{\theta_0+\delta} \cos(\lambda) d\lambda}{\int_{\theta_0-\delta}^{\theta_0+\delta} d\lambda}$$

$$= \frac{\cos(\theta_0) \sin(\delta)}{\delta}$$

and this, along with the definition of ω_{θ_j} and θ_j, gives:

$$\theta_1 = \cos^{-1}\left[\frac{\delta \frac{\cos(\theta_0)\sin(\delta)}{\delta}}{\sin(\delta)}\right]$$

$$= \theta_0$$

Thus, if the initial filter does not hit any frequency, the sequence $\omega_{\theta_{j+1}}$ is constant, and $|H(\omega; \theta_j)|^2$ will not move! This fact can be used to test whether a true frequency has been captured. In the next section, we shall see that this property is shared by non-ideal band-pass filters as well.

Example 4. Contraction using Butterworth filters

To illustrate the HOC convergence of Theorem 3, consider the case of $p = 1$, SNR of $-1.31dB$, and $\omega_1 = 1.2$. The ideal band-pass filter is approximated by a combination of high-pass and low-pass Butterworth filters where the cutoff points are such that the range $[\theta_0 - \delta, \theta_0 + \delta]$ is included in the band-pass. The results are given in Table 4. Note that when ω_1 is captured, the variance of the filtered output after some iterations is much higher than in the case when ω_1 is not captured.

Table 4. Convergence of $\gamma_N(j) = \pi D_{\theta_j}/(N-1)$ towards $\omega_1 = 1.2$ with $N = 2048$, $SNR = -1.31$, $\delta = 0.3$. The variance figure is that of the output after the indicated number of iterations.

	ω_1 captured $\theta_0 = 0.95$	ω_1 not captured $\theta_0 = 1.77$
j	$\gamma_N(j)$	$\gamma_N(j)$
1	1.1487	1.7672
2	1.1919	1.7719
3	1.2056	1.7917
4	1.2102	1.7950
5	1.2067	1.8050
6	1.2064	1.8123
7	1.2078	1.8177
8	1.2092	1.8145
9	1.2019	1.7977
10	1.2087	1.7942
11	1.2084	1.7854
	var=0.64	var=0.22

5.2. A general procedure. One of the things to be noted in the preceding ideal-pass case, is that the gain was shifted *rigidly*. That is, when a frequency is captured, the gain's position changes with a change in the parameter θ (that is before convergence), but the bandwidth does *not* change. However, by narrowing the bandwidth *judiciously*, we can achieve a greater signal-to-noise-ratio (SNR), and as a consequence a speeded up convergence of the corresponding HOC sequence. We emphasize that the reduction in bandwidth must be carried out with care, for otherwise a "captured frequency" may be lost, and no convergence occurs.

In Kedem and Yakowitz (1990), gain shifts, and bandwidth reduction, are carried out *simultaneously* according to some search plans. One of these plans, "Plan B", is described next.

Suppose we have a parametric family of band-pass filters indexed by $r \in (-1,1)$, and by a bandwidth parameter M:

$$\{\mathcal{L}_{r,M}(\cdot), \ r \in (-1,1), \ M = 1,2,\cdots\}$$

Let $h(n;r,M)$ and $H(\omega;r,M)$, be the corresponding complex impulse response and transfer function, respectively. It is required that as $M \to \infty$, $|H(\omega;r,M)|^2$ converges to a Dirac delta function centered at $\theta(r) \equiv \cos^{-1}(r)$. The filtered process and filtered noise are now denoted by $\{Z_t(r,M)\}$, and $\{\zeta_t(r,M)\}$, respectively. Assume that for *any M*

$$(31) \qquad r = \Re\left\{\frac{E[\zeta_t(r,M)\overline{\zeta_{t-1}(r,M)}]}{E|\zeta_t(r,M)|^2}\right\}$$

where the overbar denotes "complex conjugate", and define

$$\rho_1(r,M) = \Re\left\{\frac{E[Z_t(r,M)\overline{Z_{t-1}(r,M)}]}{E|Z_t(r,M)|^2}\right\}$$

If we let

$$C(r, M) = \frac{E|\zeta_t(r, M)|^2}{E|Z_t(r, M)|^2}$$

then the basic contraction (27) has now an extra parameter:

(32) $$\rho_1(r, M) = r^* + C(r, M)(r - r^*)$$

where r^* is the cosine of the frequency to be detected; suppose it is ω_1.

We assume there is an initial search interval Ω in the spectral support containing ω_1:

$$\omega_1 \in \Omega \equiv [\omega_a, \omega_b] \subset (0, \pi)$$

For simplicity we assume that none of the other discrete frequencies is contained in Ω.

The general idea underlying our search plan is the construction of a sequence of nested shrinking intervals that contain ω_1 and whose midpoints converge to ω_1. The corresponding sequence of filters are always centered at these midpoints.

Define

$$\Delta \equiv \omega_b - \omega_a$$

$$r = \cos\left(\frac{\omega_a + \omega_b}{2}\right)$$

and observe that now

$$\theta(r) = \cos^{-1}(r) = \frac{\omega_a + \omega_b}{2}$$

is the midpoint of the initial interval Ω. Center the filter at $\theta(r)$.

Plan B

It makes sense to choose M relative to Δ. We shall argue below that a sensible choice of M is

$$M^+ = O\left(\frac{1}{\Delta}\right)$$

Let $r^+ = \rho_1(r, M^+)$. Suppose $r < r^+$. Then from (32)

$$\cos(\omega_b) < r < r^+ < r^* < \cos(\omega_a)$$

or

$$\omega_a < \omega_1 < \theta(r^+) < \theta(r) < \omega_b$$

The new interval $[\omega_a, \theta(r^+)]$ is shorter than Ω but still contains ω_1:

$$\omega_1 \in [\omega_a, \theta(r^+)] \subset [\omega_a, \omega_b]$$

Similarly, suppose $r^+ < r$. Then,

$$\omega_1 \in [\theta(r^+), \omega_b] \subset [\omega_a, \omega_b]$$

Take $\Omega^+ \equiv [\omega_a^+, \omega_b^+]$ where $[\omega_a^+, \omega_b^+]$ is one of the intervals $[\omega_a, \theta(r^+)]$ or $[\theta(r^+), \omega_b]$, whichever applies. The procedure is now repeated with the new interval $[\omega_a^+, \omega_b^+]$ replacing $[\omega_a, \omega_b]$. □

Thus, by Plan B, ω_1 is contained in a sequence of nested shrinking intervals. As in the previous ideal case, if ω_1 is first captured in Ω, then it is captured by all subsequent nested shrinking intervals whose length converges to 0. The sequence of midpoints of these intervals converges to ω_1.

What happens if ω_1 is not captured in Ω ? To answer this, observe that when the filter passes ω_1,

$$(33) \qquad C(r, M) = \frac{E|\zeta_t(r, M)|^2}{\frac{\sigma_1^2}{2}|H(\omega_1; r, M)|^2 + E|\zeta_t(r, M)|^2}$$

It is important to note from (33) that $C(r, M)$ and the SNR, here relative to the first sinusoid, are inversely related, a point to be considered later. Now, it follows from (33) that as long as the filter passes ω_1, the same analysis as above, only slightly modified, still applies. However, when M is sufficiently large so that $|H(\omega; r, M)|^2$ is narrow enough and only passes an interval Ω that does not contain ω_1, the discrete spectral distribution does not have a jump at Ω and this implies

$$C(r, M) = 1$$

or, from (32),

$$\rho_1(r, M) = r$$

Consequently, r does not change, the filter does not change its location, the signal-to-noise ratio does not increase and no convergence occurs. We refer to this by saying that the "filter does not move". Thus, similarly to the ideal band-pass case, when ω_1 is not contained in Ω, and M is sufficiently large, the filter does not move and no convergence occurs.

5.2.1 An exponential family of filters and the choice of M. We follow Kedem and Yakowitz (1990). Set

$$(34) \qquad h(n; r, M) = \begin{cases} \exp(in\theta(r))/\sqrt{2M+1}, & |n| \le M \\ 0, & |n| > M \end{cases}$$

The corresponding squared gain is

$$(35) \qquad |H(\omega; r, M)|^2 = \frac{1}{2M+1} \frac{\sin^2[\frac{1}{2}(2M+1)(\omega - \theta(r))]}{\sin^2[\frac{1}{2}(\omega - \theta(r))]}, \quad -\pi \le \omega \le \pi$$

Assume the noise process $\{\zeta_t\}$ is white noise. This assumption can be disposed of in practice as we shall argue below. For

$$\theta(r) = \cos^{-1}\left(\frac{2M+1}{2M}r\right)$$

equation (31)

$$r = \Re\left\{\frac{E[\zeta_t(r, M)\overline{\zeta_{t-1}(r, M)}]}{E|\zeta_t(r, M)|^2}\right\}$$

holds with $|r| \le 2M/(2M+1)$. Observe that $|H(\omega; r, M)|^2$ is now centered at $\theta(r)$, the value at the mode is $2M+1$, and that it approaches the Dirac delta function as $M \to \infty$.

Under white noise, the contraction coefficient $C(r, M)$ in (33) becomes,

$$(36) \qquad C(r, M) = \frac{\sigma_\zeta^2}{\frac{\sigma_1^2}{2}|H(\omega_1; r, M)|^2 + \sigma_\zeta^2}$$

When $|\omega_1 - \theta(r)|$ is sufficiently small, and satisfies $|\omega_1 - \theta(r)| < \Delta$, it can be shown that $C(r, M)$ has an upper bound:

$$0 < C(r, M) \leq \tilde{C}(M) < 1$$

where,

$$(37) \qquad \tilde{C}(M) = \frac{\sigma_\zeta^2}{\frac{\sigma_1^2}{2}|H(\Delta; \frac{2M}{2M+1}, M)|^2 + \sigma_\zeta^2}$$

Thus, a sensible choice of M is a value that minimizes the upper bound $\tilde{C}(M)$, and this is the same as maximizing $|H(\Delta; , \frac{2M}{2M+1}, M)|^2$. The optimal value is approximated by,

$$M^+ \approx \frac{2.33}{2\Delta}$$

The point to note here is that M^+ is inversely proportional to Δ as required by Plan B. If M_{opt} is substituted into (37), we see that

$$\tilde{C}(M) = O(\Delta)$$

Upon recalling the contraction equation, the successor interval satisfies

$$\Delta^+ = O(\Delta^2)$$

In other words, application of our rule with the family (34) yields quadratic convergence.

When $\{\zeta_t\}$ is not white noise but merely possesses a sufficiently smooth spectral density $f_\zeta(\omega) > 0$, $\omega \in (-\pi, \pi]$, we have asymptotically as M increases,

$$\Re\left\{ \frac{E[\zeta_t(r, M)\overline{\zeta_{t-1}(r, M)}]}{E|\zeta_t(r, M)|^2} \right\} = \frac{\int_{-\pi}^{\pi} \cos(\omega)|H(\omega; r, M)|^2 f_\zeta(\omega)d\omega}{\int_{-\pi}^{\pi} |H(\omega; r, M)|^2 f_\zeta(\omega)d\omega}$$

$$\approx \cos(\cos^{-1}(r)) = r$$

and so, the above procedure can be reasonably adapted to this case as well. In short, exact knowledge of the noise spectral density is not needed.

5.2.2 A comment about SNR. In reference to the previous exponential parametric family of filters, and with $\{\zeta_t\}$ being white noise, we can explain the important connection between the SNR relative to the $j'th$ sinusoid and the contraction coefficient $C(r, M)$.

Accordingly, from (36), the SNR (in dB) relative to the $j'th$ sinusoid is equal to,

$$(38) \qquad 20\log_{10}\left[\sqrt{\frac{1}{2}}|H(\omega_j; r, M)| \times \frac{\sigma_j}{\sigma_\zeta} \right] = 20\log_{10}\left[\frac{1}{C(r, M)} - 1 \right]$$

Thus, as $\theta(r)$ approaches ω_j while M increases, the gain $|H(\omega_j; r, M)|$ increases as $O(M)$, and simultaneously with this the contraction coefficient $C(r, M)$ goes to 0. It follows that our estimate of $\cos(\omega_j)$, and hence of ω_j, becomes more precise as the bandwidth narrows and the SNR increases, or equivalently, $C(r, M)$ decreases. As remarked earlier, a decrease in $C(r, M)$, also speeds up the convergence of the corresponding HOC sequence. Evidently, equation (38) also gives a rather clear idea about the connection between the SNR and $C(r, M)$ in more general situations when f_ζ is sufficiently smooth.

Example 5. Enhanced contraction

Kedem and Yakowitz (1990) consider the family of filters defined by

$$(39) \qquad \hat{h}(n; r, M) = \cos(n\theta(r))/\sqrt{(M+1)}, \ n = 0, 1, \cdots, M$$

with $\theta(r) = \cos^{-1}(r)$. The enhanced contraction is obtained by taking $M_k = 40 \times k$, k being the iteration number. As in Example 4, when a frequency is "captured", the goodness of detection is judged by the variance of the filtered process. The results after 8 iterations are given in Table 5. The process is (1) with $p = 2, \omega_1 = 0.25, \omega_2 = 0.75$, and $\{\zeta_t\}$ white Gaussian noise.

Table 5. Convergence of $\theta(r_k)$. $N = 2048$, $SNR = 0dB$. **The variance figure is that of the output after the 8 iterations.**

no capture	$\omega_1 = 0.25$ captured	$\omega_2 = 0.75$ captured
$r_1 = 0.5$	$r_1 = 0.1$	$r_1 = 0.8$
$\hat{\omega} = 0.686943$	$\hat{\omega}=0.250267$	$\hat{\omega}=0.750101$
var$=3.7 \times 10^5$	var$=2.2 \times 10^7$	var$=2.1 \times 10^7$

6. CONCLUDING REMARKS

We have shown that there are parametric filters whose application to $\{Z_t\}$ in (1), results in a first-order autocorrelation that follows the form of a contraction mapping. In fact, by suitable reparametrizations, many different families of parametric filters can be used in this construction. By selecting appropriate parametric band-pass filters, the ω's in (1) are obtained as fixed points of such mappings. Putting it differently, with the help of certain contraction mappings, we can construct HOC sequences that converge to the ω's in (1).

In Section 5.2 we presented a specific useful family of band-pass filters that yields a speeded up convergence, once a frequency is located inside a given interval. Imagine now that the interval $[0, \pi]$ is partitioned into m subintervals, $\Omega_1, \cdots, \Omega_m$, such that,

$$[0, \pi] = \Omega_1 \cup \cdots \cup \Omega_m$$

and ω_i is located inside exactly one of these Ω's. Then Plan B can be executed with respect to each of the Ω_j. The analysis over Ω_i is independent of the analysis over Ω_j, $i \neq j$, and the procedure can be applied in *parallel*, simultaneously to all the subintervals.

In the above formulation we considered parametric families of filters in constructing convergent HOC sequences. However, the choice of the filter family is very general and need not at all be parametric in the strict sense. Under some mild assumptions, numerous families of filters can do the same job (Yakowitz (1989), Kedem and Li (1990b)).

Consider again the case $p = 1$ of a single frequency, and let $L(F_\zeta)$ be a family of filters. Suppose $\mathcal{L}(\cdot) \in L(F_\zeta)$ has a transfer function $H(\omega)$, and denote by $\rho(H)$ the first-order autocorrelation of the filtered process $\{\mathcal{L}(Z)_t\}$. Then we require the following.

- $|H(\omega)| = |H(-\omega)|$

- $|H(\omega_1)| > 0$

- Without loss of generality, $\int_{-\pi}^{\pi} |H(\omega)|^2 dF_\zeta(\omega)=1$

- For every filter $\mathcal{L}_{k-1}(\cdot) \in L(F_\zeta)$, there exists a filter $\mathcal{L}_k(\cdot) \in L(F_\zeta)$ such that

$$(40) \qquad \int_{-\pi}^{\pi} |H_k(\omega)|^2 \cos(\omega) dF_\zeta(\omega) = \rho(H_{k-1})$$

Then as before, starting with any filter $\mathcal{L}_0(\cdot) \in L(F_\zeta)$, we immediately obtain the contraction

$$(41) \qquad \rho(H_k) = \cos(\omega_1) + C(H_k)[\rho(H_{k-1}) - \cos(\omega_1)]$$

$k = 1, 2, \cdots$, where

$$(42) \qquad C(H_k) \equiv \frac{1}{1 + \sigma_1^2 |H_k(\omega_1)|^2}$$

is the contraction factor corresponding to $\mathcal{L}_k(\cdot)$. Clearly, $0 < C(H_k) < 1$, and as $k \to \infty$,

$$\rho(H_k) \to \cos(\omega_1)$$

We note that the consistency of the zero-crossing rate in the presence of a mixed spectrum is a challenging problem. This and other consistency issues, related to the convergence of observed HOC sequences generated by our iterative schemes, will be dealt with elsewhere.

Finally, our procedure provides a "local" search in the spectral domain. If an instantaneously changing spectrum is present, some variant of our method can be used in *tracking the instantaneous frequency on-line*. This problem, generally comes under the heading of *spread spectrum*, is currently under investigation. Preliminary results are encouraging.

REFERENCES

1. BARNETT, J. AND B. KEDEM, "Zero-Crossing Rate of Some Non-Gaussian Processes," submitted, 1990.

2. BRILLINGER, D.R.," Fitting cosines: some procedures and some physical examples," in *Applied Probability, Stochastic Processes, and Sampling Theory*, I.B. MacNeill and G.J. Umphry (eds.), Reidel, Dordrecht, Holland, 1987.

3. COOLEY, J. W. AND J. W. TUKEY, "An algorithm for the machine calculation of complex Fourier series", *Math. Comput.*, 19, pp. 297-301, April 1965.

4. HE, S. AND B. KEDEM, "Higher order crossings of an almost periodic random sequence in noise," *IEEE Trans. Infor. Th.*, 35, pp. 360-370, March 1989.

5. KEDEM, B., "Spectral analysis and discrimination by zero-crossings," *PROC. IEEE*, pp. 1477-1493, November 1986.

6. B. KEDEM AND T. LI, "Higher order crossings from a parametric family of linear filters," Tech Report TR-89-47, Math. Dept., Univ. of Maryland, College Park, 1989.

7. B. KEDEM AND T. LI, "Monotone gain, first-order autocorrelation, and zero-crossing rate," to appear in *Annals of Statist.*, 1990a.

8. B. KEDEM AND T. LI, "Adaptive frequency tracking by zero-crossing counts," in preparation, 1990b.

9. KEDEM, B. AND E. SLUD, "Time series discrimination by higher order crossings," *Annals of Stat.*, 10, pp. 786- 794, September 1982.

10. KEDEM, B. AND J. TROENDLE, "Discrete spectrum analysis by filtered zero-crossings counts," submitted, 1989.

11. KEDEM, B. AND S. YAKOWITZ, "A contribution to Frequency Detection," submitted, 1990.

12. PRIESTLEY, M. B., *Spectral Analysis and Time Series*, Vol. 1, Academic Press, London, 1981.

13. SCHUSTER, A., "On the investigation of hidden periodicities with application to a supposed 26-day period of meteorological phenomena," *Terr. Magn. Atmos. Elect.*, 3, 13-41, 1898.

14. YAKOWITZ, S. "Some contributions to a frequency location method due to He and Kedem," submitted, 1989.

ON BOUNDED AND HARMONIZABLE SOLUTIONS
OF INFINITE ORDER ARMA SYSTEMS

A. MAKAGON* AND H. SALEHI†

Abstract. The problem of existence and uniqueness of stationary, harmonizable and bounded solutions to infinite ARMA systems is studied. Specifically it is shown that an infinite ARMA system has a harmonizable solution if and only if it has a stationary solution. Under some regularity conditions on the coefficients of the system (which are automatically satisfied for finite ARMA models) it is proved that if the system has a bounded solution then it has a stationary solution. It is our hope that the consideration of infinite order ARMA systems will contribute to better understanding of the analysis of time series with large sample data.

Key Words and Phrases: Infinite ARMA systems; stationary, harmonizable and bounded solutions; existence and uniqueness of solutions; form of solutions.

AMS(MOS) subject classifications. 60, 40

1. Introduction. In this paper the problem of existence and uniqueness of stationary, harmonizable and bounded solutions to ARMA equations of infinite order is studied. The question of existence and uniqueness of stationary solutions to the classical ARMA equations

$$(1.1) \qquad \sum_{k=0}^{p} \phi_k x_{n-k} = \sum_{k=0}^{q} \theta_k z_{n-k}, \ n \in \mathbf{Z}$$

where (z_n) is a discrete parameter white noise, has been extensively studied in literatures. For example, in [2] it is shown that the equation (1.1) admits a stationary solution if and only if $\dfrac{\theta}{\phi}$ is bounded, where $\theta(t) = \sum_{k=0}^{q} \theta_k e^{itk}, \phi(t) = \sum_{k=0}^{p} \phi_k e^{itk}$. However, to the best of our knowledge, the problem of existence of harmonizable and possibly bounded non-harmonizable solutions have not been studied. We will deal with these questions for finite order as well as infinite order ARMA equations.

The justification for studying infinite order ARMA systems comes from the infinite order autoregressive representation of a stationary sequence whose spectral density satisfies some regularity conditions, see for example [8] and the references therein. It is our hope that the consideration of infinite order ARMA systems will contribute to better understanding of the analysis of time series with large sample data. One such useful study for infinite order autoregressive models is carried out in [9], where asymptotically efficient selection of the order of the model is sought.

---

*Wroclaw Technical University. Presently at the Department of Statistics and Probability, Michigan State University, Wells Halls, East Lansing, Michigan 48824-1027

†Department of Statistics and Probability, Michigan State University, Wells Halls, East Lansing, Michigan 48824-1027

In Section 2 the general equation

(1.2)
$$\sum_{k=-\infty}^{+\infty} \phi_k x_{n-k} = y_n, \; n \in \mathbf{Z} \,,$$

where $\sum |\phi_k| < \infty$ and (y_n) is an arbitrary stationary sequence, is considered. The main thrust of this section is to show that the existence of a stationary solution to equation (1.2) is equivalent to the existence of a harmonizable solution to (1.2); and the latter holds if and only if $\dfrac{1}{\phi}$ is square integrable with respect to (w.r.t.) the spectral measure of the process (y_n).

In Section 3 the sequence (y_n) occurring in equation (1.2) is of the form $y_n = \sum_{k=-\infty}^{+\infty} \phi_k z_{n-k}, \; n \in \mathbf{Z}$, with $\sum |\theta_k|^2 < \infty$, and the search is for a series solution in the form $\sum_{k=-\infty}^{+\infty} \psi_k z_{n-k}, \; n \in \mathbf{Z}$, where (z_n) is a standard white noise sequence.

Section 4 deals with the existence of bounded solutions to infinite order ARMA equations. Under some analytic conditions which are automatically satisfied for finite order ARMA models, it is shown that every bounded solution to the equation

$$\sum_{k=-\infty}^{+\infty} \phi_k x_{n-k} = \sum_{k=-\infty}^{+\infty} \theta_k z_{n-k}, \; n \in \mathbf{Z},$$

where (z_n) is a discrete parameter white noise, is harmonizable. Hence if (1.1) has a bounded solution then it has a stationary solution.

Throughout the paper, $\mathbf{C}, \mathbf{Z}, \mathbf{N}$ and H will stand for complex numbers, integers, positive integers and a complex separable Hilbert space with an inner product $(,)$ and the norm $\| \;\; \|$, respectively; $\mathbf{T} = \{z \in \mathbf{C} : |z| = 1\}$. The set $\mathbf{T}$ will be identified with the interval $(-\pi, \pi]$. For φ defined on $(-\pi, \pi], \tilde{\varphi}$ will stand for its replica on $\mathbf{T}$, i.e., $\varphi(t) = \tilde{\varphi}(e^{it}), \quad t \in (-\pi, \pi]$. All integrals will be over $(-\pi, \pi]$ and all sums and sequences will be indexed by the set of integers, unless it is stated otherwise. If F is a complex Borel measure on $(-\pi, \pi]$, then $L^2(F)$ will denote the space of all square integrable functions w.r.t. the variation of F. dt will denote the Lebesgue measure and we will abbreviate $L^2(dt)$ by L^2. As usual 1_A will denote the indicator of a set A and δ_{nm} will stand for the Kronecker symbol. ℓ^2 will be the class of all complex square summable sequences indexed by $\mathbf{Z}$. Any countably additive set function Z defined on the Borel σ-algebra $\mathcal{B}$ of $(-\pi, \pi]$ with values in a Hilbert space H will be called an (H-valued) measure. An H-valued measure Z is orthogonally scattered (o.s.) if for any disjoint $\Delta_1, \Delta_2 \in \mathcal{B}$, $Z(\Delta_1)$ is orthogonal to $Z(\Delta_2)$. A measure W is absolutely continuous w.r.t. Z ($W \ll Z$) if $Z(\Delta) = 0$ implies $W(\Delta) = 0, \quad \Delta \in \mathcal{B}$.

Let Z be an H-valued measure. Following the notion of integrability introduced in [7], a complex valued function f is said to be Z integrable if

(i) f is (Z, x) integrable for each $x \in H$, and

(ii) for each $\Delta \in \mathcal{B}$ there is an element of H, denoted by $\int_\Delta f dZ$, such that

$$\left(\int_\Delta f dZ, x\right) = \int_\Delta f d(Z, x), \quad x \in H,$$

where (Z, x) denotes the complex valued measure $\Delta \rightarrow (Z(\Delta), x)$, $x \in H$. Below we state couple of basic properties of this integral. Justifications are included in [7].

(A) A function f is Z integrable (in the sense given above) if and only if it is integrable w.r.t. Z in the sense of Dunford–Schwartz ([3], IV.10.7), see [7], Thm. 2.4.

(B1) If f is Z integrable, then the set function defined on $\mathcal{B}$ by $v(\Delta) = \int_\Delta f dZ$ is a measure and the semivariation $|||v|||(E)$ of v on E satisfies

$$|||v|||(E) = \sup_{\|x\| \le 1} \int_E |f| dV(Z, x),$$

where $V(Z, x)$ denotes the total variation measure of the scalar measure (Z, x), (see [7], Thm. 2.2).

(B2) If f is measurable and $\sup_{\|x\| \le 1} \int_{-\pi}^{\pi} |f| dV(Z, x) < \infty$, then f is Z integrable and the conclusion of (B1) holds. Indeed, for every $\Delta \in \mathcal{B}, x \in H$,

$$\left| \int_\Delta f d(Z, x) \right| \le \int_\Delta |f| \, dV(Z, x) \le C\|x\|,$$

with $C = \sup_{\|x\| \le 1} \int |f| \, dV(Z, x)$. Therefore there exists an element $\nu(\Delta)$ in H such that

$$(\nu(\Delta), \ x) = \int_\Delta f d(Z, x), \quad x \in H.$$

(C) The space of all Z integrable functions becomes a Banach space (if two functions which are equal Z almost everywhere are identified) under the norm

$$\|f\|_{L^1(Z)} = ||| \int f dZ ||| \, (-\pi, \pi] = \sup_{\|x\| \le 1} \int |f| \, dV(Z, x), \quad \text{and}$$

the simple functions are dense in $L^1(Z)$. ([1], p. 75).

1.3. LEMMA. *Let Z be an H-valued measure.*

(A) *If Z is orthogonally scattered, then*

$$\mu(\Delta) = \|Z(\Delta)\|^2, \quad \Delta \in \mathcal{B},$$

is a positive measure and $L^1(Z) = L^2(\mu)$.

(B) *If W is an H-valued measure and φ is a bounded Borel measurable function such that for all $\Delta \in \mathcal{B}$*

$$W(\Delta) = \int_\Delta \varphi(t)\, Z(dt)$$

then $\dfrac{1}{\varphi} \in L^1(W)$ and for all $\Delta \in \mathcal{B}$

$$\int_\Delta \frac{1}{\varphi}\, dW = Z(\Delta \cap D^c)$$

where $D = \{t: \varphi(t) = 0\}$.

Proof. (A) If $f = \sum a_k 1_{\Delta_k}$ is a step function and Z is orthogonally scattered then

$$\|f\|^2_{L^1(Z)} = \sup_{\pi = \{D_1, \ldots D_n\}} \|\sum \varepsilon_k \int_{D_k} f\, dZ\|^2 =$$

$$= \sup_\pi \sum_k \|\int_{D_k} \varepsilon_k f dZ\|^2 =$$

$$= \sup_\pi \sum_k \int_{D_k} |f|^2 d\mu = \int |f|^2 d\mu, \quad \text{where } \varepsilon_k = \pm 1.$$

Since step functions are dense in $L^1(Z)$ and $L^2(\mu)$, part (A) is proved.

To prove (B) let $\psi = \dfrac{1}{\varphi} 1_{D^c}$, where $D = \{t: \varphi(t) = 0\}$. Then

$$\int |\psi| dV(W, x) = \int |\psi|\, |\varphi|\, dV(Z, x) = V(Z, x)(D^c) \le C\|x\|,$$

([2], IV 10.4). Therefore $\psi \in L^1(W)$, and because D is a null set of W, $\dfrac{1}{\varphi} \in L^1(W)$. Hence for all $x \in H$ and $\Delta \in \mathcal{B}$,

$$\int_\Delta \frac{1}{\varphi}\, d(W, x) = \int_{\Delta \cap D^c} \frac{1}{\varphi} d(W, x) = (Z(\Delta \cap D^c), x). \quad \square$$

1.4. **Definition.** A sequence $x = (x_n) \subset H$ is called *harmonizable* if there exists an H-valued measure Z_x such that for all $n \in \mathbf{Z}$

$$x_n = \int e^{-int} Z_x(dt).$$

The measure Z_x is uniquely determined by the sequence (x_n) and is called the random spectral measure of (x_n). For a harmonizable sequence $x = (x_n)$ and $A \subset \mathbf{Z}$ we will denote $M_x(A) = \overline{sp}\{x_n : n \in A\}$, $M_x = M_x(\mathbf{Z})$. If Z is an H-valued measure and $\Delta \in \mathcal{B}$ then we define $M(Z, \Delta) = \overline{sp}\{Z(\Delta') : \Delta' \subset \Delta, \Delta' \in \mathcal{B}\}$, $M(Z) = M(Z, (-\pi, \pi])$. For the random spectral measure Z_x of a sequence (x_n) we have $M_x = M(Z_x)$. A harmonizable sequence (x_n) is stationary iff its random spectral measure is orthogonally scattered. If it is so, then $F_x(\Delta) = \|Z_x(\Delta)\|^2, \Delta \in \mathcal{B}$, is a positive measure satisfying

$$(x_n, x_m) = \int e^{-i(n-m)t} F_x(dt), \quad n, m \in \mathbf{Z}.$$

The measure F_x is called the spectral measure of the sequence (x_n). A harmonizable sequence (x_n) is called regular if $\bigcap_n M_x(-\infty, n] = \{0\}$. A stationary sequence (x_n) is regular if and only if F_x is absolutely continuous w.r.t. the Lebesgue measure and its density satisfies $\int \ln \frac{dF_x}{dt} dt > -\infty$ ([6]).

1.5. DEFINITION. Let $\mathbf{A}$ denote the class of all continuous functions f on $(-\pi, \pi]$ such that $\sum_{-\infty}^{+\infty} |\hat{f}(n)| < \infty$, where $\hat{f}(n) = \frac{1}{2\pi} \int e^{-int} f(t) dt, \; n \in \mathbf{Z}$.

2. Existence and Uniqueness. In this section we will discuss the existence of a solution to the equation

$$(2.1) \qquad \sum_{k=-\infty}^{+\infty} \phi_k x_{n-k} = y_n, \quad n \in \mathbf{Z}.$$

Unless otherwise is stated, we will assume that:

$$(2.2) \qquad \sum_{k=-\infty}^{+\infty} |\phi_k| < \infty, \; \phi_k \in \mathbf{C}, \; k \in \mathbf{Z},$$

$(2.3) \qquad \{y_n : n \in \mathbf{Z}\}$ is a stationary sequence with the spectral measure F_y.

Let $\phi(t) = \sum \phi_k e^{ikt}$. If ϕ_k satisfies (2.2) then $\phi(t)$ is a continuous function belonging to the class $\mathbf{A}$ (see Def. (1.5)).

2.4. THEOREM (Existence). *Assume (2.2) and (2.3). Then the following conditions are equivalent:*

(1) *the equation (2.1) has a harmonizable solution,*

(2) *the equation (2.1) has a stationary solution,*

(3) *the equation (2.1) has a unique stationary solution satisfying $M_x \subset M_y$,*

(4) *the equation (2.1) has a unique harmonizable solution satisfying $Z_x \ll Z_y$,*

(5) $\frac{1}{\phi} \in L^2(F_y).$

If (5) is satisfied then the solution in (3) and (4) is given by the formula

$$(2.5) \qquad x_n^0 = \int e^{-int} \frac{1}{\phi(t)} Z_y(dt).$$

Proof. $(1) \Rightarrow (5)$. Suppose that $x_n = \int e^{-int} Z_x(dt)$ is a harmonizable solution of (2.1). Then $\sum \phi_k x_{n-k} = \int \phi(t) e^{-int} Z_x(dt) = \int e^{-int} Z_y(dt)$ for every $n \in \mathbf{Z}$, and from the uniqueness of the Fourier transform, it follows that

$$(2.6) \qquad \phi dZ_x = dZ_y.$$

Therefore, from Lem. 1.3, it follows that

$$\frac{1}{\phi} \in L^1(dZ_y) = L^2(dF_y).$$

$(5) \Rightarrow (4)$. For every $\Delta \in \mathcal{B}$, let

$$Z_0(\Delta) = \int_\Delta \frac{1}{\phi} dZ_y.$$

Then Z_0 is an M_y-valued measure and

$$(2.7) \qquad x_n^0 = \int e^{-int} Z_0(dt), \ n \in \mathbf{Z}$$

is a harmonizable (in fact stationary) sequence. Moreover

$$\sum \phi_k x_{n-k}^0 = \int \phi(t) e^{-int} Z_0(dt) = \int \phi(t) e^{-int} \frac{1}{\phi(t)} dZ_y(t) = y_n, \qquad n \in \mathbf{Z},$$

so (x_n^0) is a solution to (2.1). Clearly $Z_{x^0} = Z_0 \ll Z_y$. Now let (x_n) be any other harmonizable solution to (2.1) satisfying $Z_x \ll Z_y$. From $(1) \Rightarrow (5)$ we conclude that $\phi dZ_x = dZ_y$ and by Lem. 1.3

$$Z_x(\Delta \cap D^c) = \int_\Delta \frac{1}{\phi} d Z_y = Z_0(\Delta), \ \Delta \in \mathcal{B},$$

where $D = \{t : \ \phi(t) = 0\}$. Notice that if $\Delta \subset D, \Delta, D \in \mathcal{B}$, then $Z_y(\Delta) = 0$ and $Z_x(\Delta) = 0$ since $Z_x \ll Z_y$. Therefore for every $\Delta \in \mathcal{B}$

$$Z_x(\Delta) = Z_x(\Delta \cup D^c) = Z_0(\Delta),$$

which proves that

$$x_n = \int e^{-int} Z_x(dt) = x_n^0, \ n \in \mathbf{Z}.$$

$(5) \Rightarrow (3)$. Let (x_n^0) be the stationary solution to (2.1) defined by (2.5). Then $M_{x^0} = M_y$. Assume (x_n) is any other stationary solution satisfying $M_x \subset M_y$. Then from $(1) \Rightarrow (5)$, $\phi dZ_x = dZ_y$ and by Lem. 1.3

$$Z_x(\Delta \cap D^c) = Z_0(\Delta), \ \Delta \in \mathcal{B},$$

where $D = \{t : \phi(t) = 0\}$.

This implies that $M(Z_x, D^c) = M(Z_0, (-\pi, \pi]) = M_{x^0} = M_y$. Since Z_x is orthogonally scattered and by assumption

$$M(Z_x, (-\pi, \pi]) = M(Z_x, D) \oplus M(Z_x, D^c) \subset M_y,$$

$M(Z_x, D) = 0$ and consequently for all $\Delta \in \mathcal{B}$.

$$Z_x(\Delta) = Z_x(\Delta \cap D) \oplus Z_x(\Delta \cap D^c) = Z_0(\Delta).$$

Thus $x_n = x_n^0$, $n \in \mathbf{Z}$. This completes the proof since the implications $(3) \Rightarrow (2) \Rightarrow (1)$ and $(4) \Rightarrow (1)$ are obvious. $\square$

In general the equation (2.1) may have many stationary and harmonizable solutions if we do not require that $Z_x \ll Z_y$ (or $M_x \subset M_y$, which is equivalent to the previous in the stationary case). The following proposition describes all stationary and harmonizable solutions to (2.1). Note that from (2.1) and Thm. 2.4 it follows that if $\dfrac{1}{\phi} \in L^2(F_y)$, then $M_y = M_{x^0}$.

2.8. THEOREM (Form of solutions). *Suppose that* $\dfrac{1}{\phi} \in L^2(F_y)$. *Then*

(1) every harmonizable solution to (2.1) has the form

(2.9)
$$x_n = x_n^0 + x_n^1,$$

where x_n^0 *is the stationary sequence given by (2.5) and* x_n^1 *is a harmonizable sequence with the random spectral measure concentrated on* $D = \{t : \phi(t) = 0\}$,

(2) every stationary solution to (2.1) has the form

(2.10)
$$x_n = x_n^0 \oplus x_n^1$$

where x_n^0 *is given by (2.5),* x_n^1 *is a stationary sequence with random spectral measure concentrated on* D *and* $M_{x^1} \perp M_{x^0}$.

Proof. (1). Let x_n be a harmonizable solution to (2.1). Then $x_n^1 = x_n - x_n^0$, $n \in \mathbf{Z}$ is a harmonizable sequence and $\int e^{-int} \phi(t) dZ_{x^1}(t) = 0$, $n \in \mathbf{Z}$. Therefore Z_{x^1} is concentrated on D.

(2) Now suppose that (x_n) is a stationary solution to (2.1). Then $x_n^1 = x_n - x_n^0$, $n \in \mathbf{Z}$ is harmonizable with the spectral random measure Z_{x^1} concentrated on D. Moreover,

$$x_n = \int e^{-int} \frac{1}{\phi(t)} \, dZ_y + \int e^{-int} dZ_{x^1} = \int_{D^c} e^{-int} \frac{1}{\phi(t)} dZ_y + \int_D e^{-int} dZ_{x^1}$$
$$= \int e^{-int} dZ_x.$$

Hence

$$Z_x(\Delta) = \int_{\Delta \cap D^c} \frac{1}{\phi(t)} dZ_y + \int_{\Delta \cap D} dZ_{x^1}, \quad \Delta \in \mathcal{B}.$$

Since (x_n) is stationary, Z_x is orthogonally scattered and

$$M_{x^1} = \overline{sp}\{Z_x(\Delta \cap D) : \Delta \in \mathcal{B}\} \perp \overline{sp}\{Z_x(\Delta \cap D^c) : \Delta \in \mathcal{B}\} = M_{x^0}.$$

Therefore (x_n^1) is stationary and orthogonal to (x_n^0). $\square$

As an easy consequence of Thm. 2.8 we obtain that

2.11. COROLLARY (Uniqueness). *The following three conditions are equivalent:*

(i) *(2.1) has a unique harmonizable solution,*

(ii) *(2.1) has a unique stationary solution,*

(iii) $\phi(t) \neq 0$ *everywhere.*

Proof. Implications (iii) $\Rightarrow$ (i) $\Rightarrow$ (ii) follow from Thm. 2.8, because ϕ is continuous and $\phi(t) \neq 0$ everywhere implies that $\frac{1}{\phi}$ bounded. If $\phi(t_0) = 0$, then $x_n = x_n^0 + e^{-int_0}v$, $n \in \mathbf{Z}$, where $v \perp M_y = M_{x^0}$, satisfies (2.1). This proves (ii) $\Rightarrow$ (iii). $\square$

Now suppose that the sequence (y_n) in (2.1) has a spectral density (with respect to the Lebesgue measure). Then, as it is shown below, $y_n = \sum \theta_k z_{n-k}$ where $\sum |\theta_k|^2 < \infty$ and (z_n) is an orthonormal system. In this case the equation (2.1) takes the form

$$(2.12) \qquad \sum_{-\infty}^{\infty} \phi_k x_{n-k} = \sum_{-\infty}^{\infty} \theta_k z_{n-k},$$

$$\sum |\phi_k| < \infty, \ \sum |\theta_k|^2 < \infty, (z_n, z_m) = \delta_{nm}.$$

2.13. LEMMA. *A stationary sequence (y_n) admits a representation*

$$y_n = \sum \theta_k z_{n-k}$$

with (z_n) being an orthonormal system if and only if $F_y \ll dt$. Moreover, if it does then

(i) $\sum |\theta_k|^2 < \infty$ *(so $\theta(t) = \sum \theta_k e^{ikt} \in L^2$),*

(ii) $\dfrac{dF_y}{dt} = |\theta(t)|^2 dt$ *a.e.,*

(iii) $M_y = M_z$ *if and only if $\theta(t) \neq 0$ dt a.e..*

Proof. Suppose that $y_n = \sum \theta_k z_{n-k}$. Since (z_n) is an orthonormal system, $\sum |\theta_k|^2 < \infty$ and $\theta(t) = \sum \theta_k e^{ikt}$ converges in L^2; hence (i). Consider the L^2-valued process $y_n^0 = e^{-in\cdot}\theta(\cdot)$. Then

$$(y_n, y_m) = (y_n^0, y_m^0)_{L^2} = \int e^{-i(n-m)t} |\theta(t)|^2 dt$$

Therefore $F_y \ll dt$ and $\dfrac{dF_y}{dt} = |\theta|^2 dt$ a.e. (so we have (ii)). To prove sufficiency

suppose that $F_y \ll dt$. Let $\theta = \sqrt{\dfrac{dF_y}{dt}}$. Then $\theta \in L^2$, $\theta = \sum \theta_k e^{ikt}$, where $\theta_k = \frac{1}{2\pi} \int e^{-ikt}\theta(t)dt$, and the series above converges in L^2. Let

$$y_n^0(\cdot) = e^{in\cdot}\theta(\cdot), \; n \in \mathbf{Z},$$
$$z_n^0(\cdot) = e^{in\cdot}, \; n \in \mathbf{Z}.$$

Then (z_n^0) is a complete orthonormal system in L^2 and

$$(y_n^0, y_m^0)_{L^2} = \int e^{-i(n-m)t} |\theta(t)|^2 dt = (y_n, y_m).$$

Therefore there exists an isometry $V : M_y \to L^2$ such that $V(y_n) = y_n^0$, $n \in \mathbf{Z}$. Note that

$$VM_y = M_{y^0} = \overline{sp}\{e^{-in\cdot}\theta : n \in \mathbf{Z}\}$$
$$= \{f \in L^2 : f = f1_{\{\theta \neq 0\}} dt \text{ a.e.}\}$$

Let $N = L^2 \ominus M_{y^0}$ (N could be equal to $\{0\}$) and let $K = M_y \oplus N$. Define

$$U : L^2 = M_{y^0} \oplus N \to M_y \oplus N$$

by $U = V^{-1} \oplus I$. Then U is unitary, $z_n = U z_n^0$, $n \in \mathbf{Z}$, is an orthonormal system in K and

$$y_n = U y_n^0 = \sum \theta_k z_{n-k}.$$

To see (iii) notice that if

$$y_n = \sum \theta_k z_{n-k}.$$

and $U : M_z \to L^2$ is the unitary operator defined by

$$U\left(\sum a_k z_k\right) = \sum a_k e^{int}, (a_k) \in \ell^2,$$

then $U y_n = e^{-n\cdot}\theta(\cdot) = y_n^0$. Clearly $M_y = M_z$ if and only if $L^2 = M_{y^0} = \{f \in L^2 : f = f1_{\{\theta(t)\neq 0\}}\}$, which holds if and only if $\{t : \theta(t) \neq 0\}$ has the Lebesgue measure zero. $\square$

The following two results follow immediately from Thms. 2.4 and 2.8.

2.14. THEOREM. *Suppose that* $\sum |\phi_k| < \infty$, $\sum |\theta_k|^2 < \infty$. *Let* $\phi(t) = \sum \phi_n e^{-int}$, $\theta(t) = \sum \theta_n e^{-int}$. *Then the following conditions are equivalent:*

(i) *(2.12) has a harmonizable solution,*

(ii) *(2.12) has a stationary solution,*

(iii) *(2.12) has a unique stationary solution in the space* M_y, *where* $y_n = \sum \theta_k z_{n-k}$,

$n \in \mathbf{Z}$,

(iv) $\dfrac{\theta}{\phi} \in L^2$ *(with convention* $\dfrac{0}{0} = 0$*).*

Moreover, if (iv) holds then the unique stationary solution in M_y *is given by*

$$(2.15) \qquad x_n^0 = \int e^{-int} \frac{\theta(t)}{\phi(t)} \, Z_z(dt) = \sum \psi_k \, z_{n-k}, \quad n \in \mathbf{Z},$$

where Z_z *is the random spectral measure of* (z_n) *and* $\psi_k = \dfrac{1}{2\pi} \displaystyle\int e^{-ikt} \frac{\theta(t)}{\phi(t)} dt$,
$k \in \mathbf{Z}$.

Note that in general M_y might be a proper subset of M_z so (2.15) need not be a unique stationary solution in M_z.

2.16. THEOREM. *Let* $\dfrac{\theta}{\phi} \in L^2$. *Then*

(1) every harmonizable solution to (2.12) in M_z *has the form*

$$x_n = x_n^0 + x_n^1$$

where

(i) $x_n^0 = \int e^{-int} \dfrac{\theta(t)}{\theta(t)} \, dZ_z$, $n \in \mathbf{Z}$,

(ii) x_n^1 *is an* M_z-*valued harmonizable sequence with the random spectral measure concentrated on* $D = \{t : \phi(t) = 0\}$,

(2) every stationary solution to (2.12) in M_z *has the form*

$$x_n = x_n^0 \oplus x_n^1, \quad n \in \mathbf{Z}$$

where

(i) $x_n^0 = \int e^{-int} \frac{\theta(t)}{\phi(t)} \, dZ_z$, $n \in \mathbf{Z}$

(ii) $M_{x^0} \perp M_{x^1}$,

(iii) (x_n^1) *is an* M_z-*valued stationary sequence with spectral measure concentrated on* $D = \{t : \phi(t) = 0\}$.

The next proposition provides a necessary and sufficient conditions for the existence of a unique stationary solution to (2.12) in M_z.

2.17. THEOREM. *The equation (2.12) has a unique stationary solution in M_z if and only if either one of the following two conditions is satisfied:*

(1) $\phi(t) \neq 0$ *everywhere, or*

(2) $\theta(t) \neq 0$ *dt a.e., and* $\dfrac{\theta}{\phi} \in L^2$.

Proof. If (1), then by Cor. 2.11, the equation (2.12) has a unique stationary solution. If (2), then from Lem. 2.13 (iii), $M_y = M_z$, where (y_n) is defined in Prop. 2.14. Therefore by Prop. 2.14 (iii), (2.12) has a unique stationary solution in M_z. Conversely, suppose that $\theta(t) = 0$ on a set Δ of positive Lebesgue measure, $\phi(t_0) = 0$ and suppose that (2.12) has a stationary solution. Then $\dfrac{\theta}{\phi} \in L^2$ and the sequence

$$x_n^1 = \int e^{-int} \frac{\theta(t)}{\phi(t)} dZ_z(t) + Z_z(\Delta)e^{-int_0}, \ n \in \mathbf{Z}$$

is a stationary solution to (2.12) with values in M_z different from (x_n^0). $\square$

3. Series Solution. Suppose that $\sum |\phi_k| < \infty, \sum |\theta_k|^2 < \infty$ and consider the equation

(3.1) $$\sum \phi_k \, x_{n-k} = \sum \theta_k z_{n-k}$$

where $(z_n, z_m) = \delta_{nm}$. Let, as before, $\phi(t) = \sum \phi_k e^{ikt}, \theta(t) = \sum \theta_k e^{ikt}$

3.2. DEFINITION. We will say that (3.1) has a series solution if there exists a sequence of scalars $(\varphi_k)_{-\infty}^{\infty}$ such that

$$x_n = \sum_{-\infty}^{+\infty} \varphi_k z_{n-k}, \ n \in \mathbf{Z},$$

is a solution to (3.1).

It is obvious that if (3.1) has a series solution (x_n), then $\sum |\varphi_k|^2 < \infty$ and (x_n) is stationary. Therefore by Thm. 2.14 and the formula (2.15) therein

the equation (3.1) has a series solution iff $\dfrac{\theta}{\phi} \in L^2$.

Theorem below discusses the problem of the uniqueness of a series solution.

3.3. THEOREM. *The equation (3.1) has a unique series solution iff*

(1) $\dfrac{\theta}{\phi} \in L^2$, *and*

(2) $\phi(t) \neq 0$ *dt a.e.*

Proof. We have already noticed that (3.1) has a series solution iff $\dfrac{\theta}{\phi} \in L^2$. Let $D = \{t : \phi(t) = 0\}$. Assume first that (3.1) has a series solution and $dt(D) \neq 0$. Then

$$x_n^0 = \int e^{-int} \left(\frac{\theta(t)}{\phi(t)}\right) dZ_z = \sum_{-\infty}^{+\infty} (\theta/\phi)\widehat{}(k) z_{n-k} \quad \text{and}$$

$$x_n^1 = \int e^{-int} \left[\left(\frac{\theta(t)}{\phi(t)}\right) + 1_D(t)\right] dZ_z = \sum_{-\infty}^{+\infty} \left(\frac{\theta}{\phi} + 1_D\right)\widehat{}(k) z_{n-k}$$

are two different series solutions to (3.1). Conversely, assume (1) and (2). Suppose that (x_n) is a series solution to (3.1). Then $x_n^2 = x_n - x_n^0$, where x_n^0 is as above has the form $x_n^2 = \sum a_k z_{n-k}$, $n \in \mathbf{Z}$, and by Thm. 2.8 its spectral measure is concentrated on D. Since $dt(D) = 0$, from Lem. 2.13 we conclude that $x_n^2 = 0$. $\square$

3.4. REMARK. If (3.1) has a series solution $x_n = \sum \varphi_k z_{n-k}$, $n \in \mathbf{Z}$, then the coefficients (φ_k) satisfy the infinite system of equations

$$(3.5) \qquad\qquad \sum_{-\infty}^{+\infty} \phi_k \varphi_{n-k} = \theta_n, \ n \in \mathbf{Z}.$$

In fact, there is one-to-one correspondence between ℓ^2- solutions of (3.5) and series solutions of (3.1). To see this it is enough to notice that the Fourier transform converts (3.5) into the functional equation

$$\phi(t)\varphi(t) = \theta(t) \quad dt \text{ a.e.,}$$

where $\varphi(t) = \sum \varphi_k e^{ikt}$, $\varphi \in L^2$. On the other hand, if $x_n = \sum \varphi_k z_{n-k}$ is a solution to (3.1) then

$$\int e^{-int} \phi(t)\varphi(t) dZ_z = \int e^{-int} \theta(t) dZ_z, \ n \in \mathbf{Z},$$

and $\phi(t)\varphi(t) = \theta(t) \ dt$ a.e.. This leads to an alternative proof of Thm. 3.3.

If the coefficients ϕ_k and θ_k in (3.1) vanish for negative k the functions ϕ and θ are in the Hardy class H^2 (H^2 consists of all functions in L^2 whose negative Fourier coefficient vanish). Since for every $f \in H^2$, $\log |f| \in L^1$ (e.g. [5]) and in particular $f(t) \neq 0 \ dt$ a.e., we obtain the following theorem.

3.6. THEOREM. *Assume that* $\displaystyle\sum_{k=0}^{\infty} |\phi_k| < \infty, \sum_{k=0}^{\infty} |\theta_k|^2 < \infty$. *Consider the equation*

$$(3.7) \qquad\qquad \sum_{k=0}^{\infty} \phi_k\, x_{n-k} = \sum_{k=0}^{\infty} \theta_k z_{n-k}, \ n \in \mathbf{Z},$$

where $(z_n, z_m) = \delta_{nm}$.

Then the following conditions are equivalent:

(1) *(3.7) has a harmonizable solution,*

(2) *(3.7) has a stationary solution,*

(3) *(3.7) has a unique series solution,*

(4) *(3.7) has a unique regular stationary solution,*

(5) *(3.7) has a unique stationary solution in M_z,*

(6) $\dfrac{\theta}{\phi} \in L^2$.

If (6) holds then the unique regular series solution, whose existence is guaranteed by (4) and (5), is given by

$$(3.8) \qquad x_n^0 = \int e^{-int} \frac{\theta(t)}{\phi(t)} dZ_z = \sum_{-\infty}^{+\infty} \psi_k z_{n-k}, \; n \in \mathbf{Z},$$

where $\psi_n = \dfrac{1}{2\pi} \displaystyle\int e^{-int} \dfrac{\theta(t)}{\phi(t)} dt, \; n \in \mathbf{Z}.$

Moreover, if $\phi(t) \neq 0$ everywhere, then the equation (3.7) has only one harmonizable solution given by (3.8).

Proof. Because $\phi \in H^2$, $\phi(t) \neq 0$ dt a.e. Therefore the equivalence of (1), (2), (3), (5), (6) follows from Thms. 2.14, 2.17 and 3.3. We need to prove only that (6) $\Rightarrow$ (4). Since the spectral density of (x_n^0), $|\frac{\theta}{\phi}|^2$, has integrable logarithm, (x_n^0) is regular. Let (x_n) be any other regular stationary solution to (3.7). Then from Thm. 2.16 it follows that $dF_x = |\frac{\theta}{\phi}|^2 dt + dF_{x^1}$, and dF_{x^1} is singular w.r.t. dt. Since (x_n) is regular, $F_{x^1} = 0$. The part "moreover" follows from Cor. 2.11. $\square$

3.9. REMARK. The solution (3.8) has the form

$$(3.10) \qquad x_n = \sum_{k=0}^{\infty} \psi_k z_{n-k},$$

iff $\dfrac{\theta}{\phi} \in H^2$. In this case in view of Rem. 3.4 the coefficients ψ_k, $k \geq 0$, can be computed by solving the system of equations

$$\sum_{k=0}^{n} \phi_k \psi_{n-k} = \theta_n, \; n \geq 0.$$

4. Bounded Solutions. In this section we discuss the problem of the existence of bounded solutions to the equation (2.12).

4.1.   DEFINITION. Assume that $\sum |\phi_k| < \infty$, $\sum |\theta_k|^2 < \infty$, $(z_k, z_n) = \delta_{kn}$, $n, k \in \mathbf{Z}$. Every norm bounded sequence $x_n \in H$, $n \in \mathbf{Z}$, satisfying the equation

$$(4.2) \qquad \sum_{-\infty}^{\infty} \phi_k x_{n-k} = \sum_{-\infty}^{+\infty} \theta_k\, z_{n-k}, n \in \mathbf{Z}$$

is called a bounded solutions to (4.2).

As before, let $\mathbf{A}$ be the space of all continuous functions $\varphi(t)$, $t \in (-\pi, \pi]$ with absolutely summable Fourier series. It is known ([4], 11.4.17) that $\mathbf{A}$ with the norm $\|\varphi\|_{\mathbf{A}} = \sum_{-\infty}^{+\infty} |\varphi_k| < \infty$ is a Banach algebra under pointwise multiplication. Moreover it is easy to see that the relations

$$(4.3) \qquad \begin{cases} x_n = F(e^{-in\cdot}), & n \in Z \\ F(\varphi) = \sum_{k=-\infty}^{+\infty} \varphi_k\, x_{n-k}, & \varphi \in \mathbf{A}, \end{cases}$$

establishes a one-to-one correspondence between the class of all bounded linear operators from $\mathbf{A}$ to H and the class of all bounded H-valued sequences (x_n). Here and in the sequel (φ_k) will stand for Fourier coefficients of φ.

4.4. LEMMA. Let Z denote the random spectral measure of (z_n). The equation (4.2) has a bounded solution if and only if there exists a bounded linear operator $F : \mathbf{A} \to H$ such that

$$F(\phi\varphi) = \int \theta\varphi dZ$$

for every trig polynomial φ (or equivalently for every $\varphi \in \mathbf{A}$).

Proof. Let (x_n) be a bounded solution of equation (4.2) and let F denote the corresponding bounded operator as described above, that is $x_n = F(e^{-in\cdot})$, $n \in Z$. Then (4.2) takes the form

$$F(\phi e^{-in\cdot}) = \sum \phi_k F(e^{-i(n-k)\cdot}) = \sum \theta_k \int e^{-i(n-k)t} Z(dt) = \int \theta(t)e^{-int} Z(dt), n \in \mathbf{Z}.$$

Let $\varphi \in \mathbf{A}$. Then $\varphi(t) = \sum_{-\infty}^{+\infty} \varphi_k e^{ikt}$ where the convergence is uniform and in $\mathbf{A}$. Since multiplication in $\mathbf{A}$ is continuous and F is bounded $\sum \varphi_k F(\phi e^{ik\cdot}) = F(\phi\varphi)$. On the other hand

$$\sum \phi_k \int \theta(t)F(t)e^{-int} Z(dt) = \int \theta(t)\, \varphi(t)\, Z(dt)$$

because the Lebesgue dominated theorem applies. Consequently we have

$$F(\phi\varphi) = \int \theta\, \varphi dZ, \ \varphi \in \mathbf{A}.$$

The proof of the converse is easy and can be similarly carried out. □

As an immediate consequence of the Lemma above and the Wiener theorem for the algebra $\mathbf{A}$ we obtain the following corollary.

4.5. COROLLARY. *If $\phi(t) \neq 0$ everywhere then the equation (4.2) has a unique bounded solution given by $x_n^0 = \int e^{-int} \dfrac{\theta}{\phi} dZ$, $n \in \mathbf{Z}$.*

Proof. Since $\phi(t)$ is continuous, $\dfrac{\theta}{\phi} \in L^2$. By Thm. 2.14 $x_n^0 = \int e^{-int} \dfrac{\theta(t)}{\phi(t)} dZ$ is a stationary, and hence bounded, solution to (4.2). Let (y_n) be any bounded solution to (4.2). Then $x_n = y_n - x_n^0$ satisfies $\displaystyle\sum_{k=-\infty}^{+\infty} \phi_k x_{n-k} = 0$ and the operator F generated by (x_n) by the formula (4.3) satisfies $F(\phi\varphi) = 0$, $\varphi \in \mathbf{A}$. Since, by the Wiener theorem ([4], 11.4.17), $\dfrac{1}{\phi} \in \mathbf{A}$, we conclude that $F(\varphi) = 0$ for all $\varphi \in \mathbf{A}$. Thus $F = 0$. $\Box$

The problem of describing all bounded solutions to (4.2) for arbitrary $\phi \in \mathbf{A}$ and $\theta \in L^2$ seems to be rather difficult. However, if:

(4.6)   the functions $\tilde{\phi}(z) = \displaystyle\sum_{k=-\infty}^{+\infty} \phi_k z^k$ and $\tilde{\theta}(z) = \displaystyle\sum_{k=\infty}^{\infty} \theta_k z^k$ $z \in \mathbf{C}$.

are analytic in an annulus $1 - \varepsilon < |z| < 1 + \varepsilon$, $\varepsilon > 0$,

then one can prove that every bounded solution to (4.2) is harmonizable, which if combined with previous results, gives a complete description of all bounded solutions to (4.2) under assumptions (4.6). The proof of this statement is divided into three lemmas, which seem to be of independent interest. Note that if $\tilde{\phi}$ and $\tilde{\theta}$ satisfy condition (4.6), then by Laurent expansion the functions $\phi(t) = \tilde{\phi}(e^{it})$ and $\theta(t) = \tilde{\theta}(e^{it})$ belong to class $\mathbf{A}$.

4.7. LEMMA. *If $\phi(t) = \phi_1(t)\phi_2(t)$, $\phi_1, \phi_2 \in \mathbf{A}$ and the equation $\sum \phi_k x_{n-k} = \sum \theta_k z_{n-k}, \sum |\theta_k|^2 < \infty, (z_n, z_m) = \delta_{nm}$ has a bounded solution, say (x_n), then the sequence $y_n = \sum \phi_{2,k} x_{n-k}$ is a bounded solution to the equation*

$$\sum \phi_{1,k} y_{n-k} = \sum \theta_k z_{n-k}$$

where $\phi_j(t) = \displaystyle\sum_{k=-\infty}^{+\infty} \phi_{j,k} e^{ikt}$ $j = 1, 2$.

Proof. Let F be associated with (x_n) by the formula (4.3). Since (x_n) is a bounded solution to $\sum \phi_k x_{n-k} = \sum \theta_k z_{n-k}$, by Lem. 4.4 we have

$$F(\phi_1 \phi_2 \varphi) = \int \varphi \theta dZ$$

for all $\varphi \in \mathbf{A}$. Since $\mathbf{A}$ is a Banach algebra, the mapping $\phi_2 F : \mathbf{A} \to H$ defined by $(\phi_2 F)(\psi) = F(\phi_2 \psi)$ is bounded and satisfies

$$(\phi_2 F)(\phi_1 \varphi) = \int \theta \varphi dZ, \qquad \varphi \in \mathbf{A}.$$

Therefore, using once again Lem. 4.4, it is easy to see that $y_n = (\phi_2 F)(e^{-in\cdot}) = \sum \phi_{2,k} x_{n-k}$, $n \in \mathbf{Z}$, satisfies

$$\sum \phi_{1,k} y_{n-k} = \sum \theta_k z_{n-k}, \quad n \in \mathbf{Z}.$$

4.8. LEMMA. *Assume that* $\tilde{\phi}(z) = \sum\limits_{k=-\infty}^{+\infty} \phi_k z^k$, $z \in \mathbf{Z}$ *is a nonzero analytic function in an annulus* $1 - \varepsilon < |z| < 1 + \varepsilon$, $\varepsilon > 0$. *Let* $t_0, \ldots, t_k$ *be the zeros of the function* $\phi(t) = \tilde{\phi}(e^{it})$, $t \in (-\pi, \pi]$. *Then a sequence* (x_n) *is a bounded solution to the homogeneous equation*

$$(*) \qquad\qquad \sum_{k=-\infty}^{+\infty} \phi_k x_{n-k} = 0$$

if and only if $x_n = \sum\limits_{j=0}^{k} v_j e^{-int_j}$, $v_j \in H$.

Proof. First we note that since ϕ is analytic, $\tilde{\phi}(z)$ has only finitely many zeros in $1 - \varepsilon/2 \le |z| \le 1 + \varepsilon/2$, unless it is the zero function. Therefore $\phi(t) = \tilde{\phi}(e^{it})$ has only finitely many zeros on $(-\pi, \pi]$ and $\phi(t)$ can be written in the form

$$\phi(t) = (e^{it} - e^{it_0})^{j_0}(e^{it} - e^{it_1})^{j_1} \cdots (e^{it} - e^{it_k})^{j_k} \phi_0(t),$$

where $\phi_0(t) \ne 0$ on $(-\pi, \pi]$, $j_r \in \mathbf{N}, 0 \le r \le k$.

Let

$$\phi_0^m(t) = \left[\left[(e^{it} - e^{it_0})^m \right]^{-1} \phi(t), \qquad 1 \le m \le j_0, \right.$$

$$\phi_1^m(t) = \left[(e^{it} - e^{it_0})^{j_0}(e^{it} - e^{it_1})^m \right]^{-1} \phi(t), \qquad 1 \le m \le j_1,$$

$$\vdots$$

$$\phi_k^m(t) = \left[(e^{it} - e^{it_0})^{j_0} \cdots (e^{it} - e^{it_{k-1}})^{j_{k-1}}(e^{it} - e^{it_k})^m \right]^{-1} \phi(t), \quad 1 \le m \le j_k,$$

$$\varphi_{N,j}(t) = \frac{e^{iNt} - e^{iNt_j}}{e^{it} - e^{it_j}},$$

$$\varphi_{-N,j}(t) = \frac{e^{-iNt} - e^{-iNt_j}}{e^{it} - e^{it_j}}.$$

At the points $t_0, \ldots, t_k$ the functions are defined by continuity. In particular

$$\varphi_{N,j}(t_j) = Ne^{-i(N+1)t_j}, \quad \varphi_{N,j}(t_j) = Ne^{i(N-1)t_j}, \ N \in \mathbf{N}.$$

Suppose that (x_n) is a bounded solution to $(*)$ and let the associated F be given by (4.3). Then from Lems. 4.4 and 4.7 it follows that

$$(\phi_0^1 F)((e^{it} - e^{it_0})\varphi) = 0$$

for every polynomial φ. Letting $\varphi = \varphi_{N,0}$ and $\varphi = \varphi_{-N,0}$ we obtain $(\phi_0^1 F)(e^{iNt}) = e^{iNt}(\phi_0^1 F)(1)$, $N \in \mathbf{Z}$. Therefore for every trig polynomial φ, $(\phi_0^1 F)(\varphi) = \varphi(t_0)w_0$, where $w_0 = (\phi_0^1 F)(1)$. If $j_0 = 1$ this shows that $(\phi_0^{j_0} F)(\varphi) = \varphi(t_0)w_0$, $\varphi \in \mathbf{A}$. If $j_0 \ge 2$, let us assume that $m \le j_0 - 1$ and that

$$(\phi_0^m F)(\varphi) = \varphi(t_0)w_0.$$

We will show that $(\phi_0^{m+1}F)(\varphi) = \varphi(t_0)w_0$, for possibly different w_0. From Lem. 4.7

$$(\phi_0^{m+1}F)((e^{it} - e^{it_0})\varphi) = \varphi(t_0)w_0.$$

Letting φ equal $\varphi_{N,0}$ and $\varphi_{-N,0}$ we obtain

$$(\phi_0^{m+1}F)(e^{iNt} - e^{iNt_0}) = \varphi_N(t_0)w_0 = Ne^{i(N-1)t_0}w_0, \quad N \in \mathbf{N},$$

and

$$(\phi_0^{m+1}F)e^{-iNt_0} - e^{-iNt}) = \varphi_{-N}(t_0)w_0 = Ne^{-i(N+1)t_0}w_0, \quad N \in \mathbf{N}.$$

Since the left sides are bounded as functions of N, $w_0 = 0$ and consequently

$$(\phi_0^{m+1}F)(e^{iNt}) = e^{iNt_0}(\phi_0^m F)(1), \quad N \in \mathbf{Z}.$$

Hence $(\phi_0^{m+1}F)(\varphi) = \varphi(t_0)w_0$, for every trig polynomial φ. Repeating the argument we conclude that $(\phi_0^{j_0}F)(\varphi) = \varphi(t_0)w_0$, for every $\varphi \in \mathbf{A}$ and possibly different w_0. Using Lem. 4.7 once again we obtain

$$(\phi_1^1 F)[(e^{it} - e^{it_1})\varphi] = \varphi(t_0)w_0, \quad \varphi \in \mathbf{A}.$$

Setting $\varphi = \varphi_{N,1}$ and $\varphi = \varphi_{-N,1}$ we get,

$$(\phi_1^1 F)(e^{iNt} - e^{iNt_1}) = \varphi_{N,1}(t_0)w_0 = (e^{it_0 N} - e^{it_1 N})\frac{w_0}{e^{it_0} - e^{it_1}}, \quad N \in \mathbf{N},$$

and

$$(\phi_1^1 F)(e^{-iNt_1} - e^{-iNt}) = \varphi_{-N,1}(t_0)w_0 = (e^{-iNt_1} - e^{-iNt_0})\frac{w_0}{e^{it_0} - e^{it_1}}, \quad N \in \mathbf{N}.$$

Hence for all $N \in \mathbf{Z}$, $(\phi_1^1 F)(e^{iNt}) = e^{iNt_1}((\phi_1^1 F)(1) - \frac{w_0}{e^{it_0} - e^{it_1}}) + e^{iNt_0}\frac{w_0}{e^{it_0} - e^{it_1}}$.
From the linearity we obtain

$$(\phi_1^1 F)(\varphi) = \varphi(t_1)w_1 + \varphi(t_0)w_0, \quad \varphi \in \mathbf{A}.$$

where w_0 and w_1 are now the coefficients of e^{iNt_0} and e^{iNt_1}, respectively. If $j_1 = 1$, this step is completed. If $j_1 \geq 2$, let us assume that $m \leq j_1 - 1$ and that

$$(\phi_1^m F)(\varphi) = \varphi(t_1)w_1 + \varphi(t_0)w_0, \quad \varphi \in \mathbf{A}.$$

Then

$$(\phi_1^{m+1}F)[(e^{it} - e^{it_1})\varphi] = \varphi(t_1)w_1 + \varphi(t_0)w_0.$$

Now setting again $\varphi = \varphi_{N,1}$ and $\varphi = \varphi_{-N,1}$ we obtain

$$(\phi_1^{m+1}F)(e^{iNt} - e^{it_1 N}) = Ne^{i(N-1)t_1}w_1 + (e^{iNt_0} - e^{iNt_1})\frac{w_0}{e^{it_0} - e^{it_1}}, \quad N \in \mathbf{N},$$

$$(\phi_1^{m+1}F)(e^{-iNt_1} - e^{-iNt}) = Ne^{-i(N+1)t_1}w_1 + (e^{-iNt_1} - e^{-iNt_1})\frac{w_0}{e^{it_0} - e^{it_1}}, \quad N \in \mathbf{N}.$$

Therefore $w_1 = 0$, since the terms not containing w_1 are bounded in N. Hence

$$(\phi_1^{m+1} F)(e^{iNt}) = e^{iNt_1}\left((\phi_1^{m+1} F)(1) - \frac{w_0}{e^{it_0} - e^{it_1}}\right) + e^{iNt_0}\frac{w_0}{e^{it_0} - e^{it_1}}, \quad N \in \mathbf{Z}.$$

This yields that $(\phi_1^{j_1} F)(\varphi) = \varphi(t_0)w_0 + \varphi(t_1)w_1$, $\varphi \in \mathbf{A}$, for some w_0, w_1 (which may not be the same as earlier w_0, w_1). Repeating the sequence of arguments given above k times, we obtain

$$F(\phi_0\varphi) = \varphi(t_0)w_0 + \cdots + \varphi(t_k)w_k. \quad \varphi \in \mathbf{A},$$

for some $w_0, \ldots, w_k$ in H. Setting $\varphi = \dfrac{1}{\phi_0}\psi, \psi \in \mathbf{A}$, we obtain

$$F(\psi) = \sum_{j=0}^{k} \psi(t_j) \frac{w_j}{\phi_0(t_j)}$$

Therefore $x_n = F(e^{-int}) = \displaystyle\sum_{j=0}^{k} e^{int_j} v_j$, with $v_j = \dfrac{w_j}{\phi_0(t_j)}$, $j = 1, \ldots, k$. The converse implication is trivial since

$$\sum_{m=-\infty}^{+\infty} \phi_m\left(\sum_{j=0}^{k} e^{-int_j} e^{imt_j} v_j\right) = \sum_{j=0}^{k} e^{-int_j} \phi(t_j)v_j = 0. \quad \square$$

4.9. LEMMA. *Assume (4.6). Then the equation*

$$(4.10) \qquad \sum \phi_k x_{n-k} = \sum \theta_k z_{n-k}, \quad (z_n, z_m) = \delta_{nm}$$

has a bounded solution iff $\dfrac{\theta}{\phi}$ *is a bounded function on* $(-\pi, \pi]$,

Proof. Since $\tilde{\phi}$ is analytic in some neighborhood of $\mathbf{T}$, $\phi(t)$ has finitely many zeros, say $t_0, \ldots, t_k$. We examine the behavior of $\dfrac{\theta}{\phi}$ on a neighborhood of each zero of ϕ. Specifically we show that $\frac{\theta}{\phi}$ is bounded on each neighborhood. Since the proofs are the same for all points we will merely give the proof for t_0. Let $z_0 = e^{it_0}$, $\tilde{\phi}(e^{it_0}) = 0$. Then $\tilde{\phi}(t) = (e^{it} - e^{it_0})^m \phi_0(t)$, where $\tilde{\phi}_0(z) = \dfrac{\tilde{\phi}(z)}{(z - z_0)^m}$ is analytic, $\tilde{\phi}_0(z_0) \neq 0$, and m is the order of z_0. Let $\theta(t) = a_0 + (e^{it} - e^{it_0})\theta_1(t)$ with $\tilde{\theta}_1(z) = \dfrac{\tilde{\theta}(z) - a_0}{z - z_0}$, $a_0 = \tilde{\theta}(z_0)$. Suppose first that the equation (4.10) has a bounded solution (x_n). Let F be associated with (x_n) by formula (4.3). Then (4.10) can be written as

$$(\phi_0 F)((e^{it} - e^{it_0})^m \varphi) = \int a_0\varphi dZ + \int (e^{it} - e^{it_0})\theta_1\varphi dZ,$$

$\varphi \in \mathbf{A}$, where Z is the random spectral measure of the sequence (z_n). First we prove that $a_0 = 0$. Let $F_1(\psi) = ((e^{it} - e^{it_0})^{m-1}\phi_0 F)(\psi)$, $F_2(\psi) = \int \theta_1 \psi dZ$, $\psi \in \mathbf{A}$. Then $F_0 = F_1 - F_2$ is a bounded linear operator from $\mathbf{A}$ to H which satisfies

$$F_0((e^{it} - e^{it_0})\varphi) = a_0 \int \varphi dZ, \ \varphi \in \mathbf{A}.$$

Setting

$$\varphi = \varphi_N = \frac{e^{iNt} - e^{iNt_0}}{e^{it} - e^{it_0}} = e^{i(N-1)t_0} \sum_{k=0}^{N-1} e^{itk} e^{it_0 k}$$

we obtain

$$F_0(e^{iNt} - e^{iNt_0}) = a_0 \int \varphi_n dZ, \ N \in \mathbf{N}.$$

Since F_0 is bounded,

$$\|F_0(e^{iNt} - e^{iNt_0})\| \ \leq 2\|F_0\|,$$

while on the other hand

$$\|a_0 \int \varphi_N dZ\|^2 = |a_0|^2 \int | \sum_{k=0}^{N-1} e^{itk}(e^{-it_0 k})|^2 \ dt = N|a_0|^2.$$

Therefore $a_0 = 0$ and $\theta(t) = (e^{it} - e^{it_0})\theta_1(t)$, where $\tilde{\theta}_1(z)$ is analytic in the annulus $1 - \varepsilon < |z| < 1 + \varepsilon$. If $m = 1$ this shows that $\dfrac{\theta}{\phi}$ is bounded in a neighborhood of t_0. Assume that $m > 1$, $1 \leq k \leq m - 1$, and suppose that $\theta(t) = (e^{it} - e^{it_0})^k \theta_k(t)$, where $\tilde{\theta}_k(z)$ is analytic in the annulus $1 - \varepsilon < |z| < 1 + \varepsilon$. We show that $\theta_k(t) = (e^{it} - e^{it_0})\theta_{k+1}(t)$, where $\tilde{\theta}_{k+1}(z)$ is analytic in $1 - \varepsilon < |z| < 1 + \varepsilon$. Write $\theta_k(t) = a_k + (e^{it} - e^{it_0})\theta_{k+1}(t)$, where $\tilde{\theta}_{k+1}(z) = \dfrac{\tilde{\theta}_k(z) - \tilde{\theta}_k(z_0)}{z - z_0}$. Then (4.10) can be written as

$$F((e^{it} - e^{it_0})^m \phi_0 \varphi) = \int (e^{it} - e^{it_0})^k a_k \ \varphi dZ + \int (e^{it} - e^{it_0})^{k+1} \theta_{k+1} \varphi dZ, \ \varphi \in \mathbf{A}.$$

$$\text{Let } F_0(\psi) = (\phi_0 (e^{it} - e^{it_0})^{m-k-1} F)(\psi) - \int \theta_{k+1} \psi dZ, \ \psi \in \mathbf{A}.$$

Then F_0 is bounded and

$$F_0((e^{it} - e^{it_0})^{k+1}\varphi) = a_k \int (e^{it} - e^{it_0})^k \ \varphi dZ, \ \varphi \in \mathbf{A}.$$

Setting $\varphi = (\varphi_N)^{k+1}$, we obtain

$$F_0((e^{iNt} - e^{iNt_0})^{k+1}) = a_k \int (e^{iNt} - e^{iNt_0})^k \ \varphi_N dZ.$$

Since F_0 is bounded, $\|F_0(e^{iNt} - e^{iNt_0})^{k+1}\| \le C_k$ for all N, while

$$\|a_k \int (e^{iNt} - e^{iNt_0})^k (e^{(i(N-1)t_0} \sum_{k=0}^{N-1} e^{itk} e^{-it_0 k}) dZ\|^2 =$$

$$\|a_k e^{i(N-1)t_0} \sum_{m=0}^{k} \sum_{r=0}^{N-1} e^{it(Nm+r)} \binom{k}{m} (-1)^{m-k} e^{i(Nk-nm-r)t_0}\|_{L^2}^2 =$$

$$= |a_k|^2 \, N \sum_{m=0}^{k} \binom{k}{m}^2 \to +\infty.$$

Hence $a_k = 0$. By repetitions of this argument we obtain that $\theta(t) = (e^{it} - e^{it_0})^m \theta_m(t)$, where $\tilde{\theta}_m(z)$ is analytic on the annulus $1 - \varepsilon < |z| < 1 + \varepsilon$. Thus $\dfrac{\theta}{\phi}$ is bounded in a neighborhood of t_0. The converse implication is a trivial consequence of Thm. 2.14. ☐

4.11 THEOREM. *If ϕ and θ satisfy assumptions (4.6) then the following conditions are equivalent:*

(1) *the equation (4.2) has a bounded solution,*

(2) *the equation (4.2) has a stationary solution,*

(3) $\dfrac{\theta}{\phi}$ *is bounded.*

Moreover, if (3) is satisfied then every bounded solution to (4.2) is harmonizable and is given by

$$x_n = \int e^{-int} \frac{\theta(t)}{\phi(t)} \, dZ + \sum_{j=0}^{k} e^{-int_j} v_j,$$

where Z is the random spectral measure of (z_n); $t_0, \ldots, t_k$ are zeros of $\phi(t)$ and $v_0, \ldots, v_k$ are arbitrary vectors in H.

Proof. The equivalence of (1), (2), (3) follows from Lem. 4.9 and Thm. 2.14. Let $x_n^0 = \int e^{-int} \frac{\theta(t)}{\phi(t)} dZ$, $n \in \mathbf{Z}$, and let x_n be an arbitrary bounded solution to (4.2). Then by Lem. 4.8, $x_n^1 = x_n - x_n^0$ has the form

$$x_n^1 = \sum_{j=0}^{k} v_j e^{-int_j} = \int e^{-int} (\sum_{j=0}^{k} \delta_{\{t_j\}}(dt) v_j), \; n \in \mathbf{Z},$$

so it is harmonizable. Therefore x_n^1 is harmonizable and has the required form. ☐

Since a classical ARMA equation is a special case of the system of equations considered in this paper, the results obtained in Thm. 4.11 and those of the earlier sections also hold for ARMA equations. We will not discuss in details the ramifications of our results to the ARMA case. We merely state just one result.

4.12. THEOREM. *Consider the ARMA equation*

$$(4.13) \qquad \sum_{k=0}^{p} \phi_k x_{n-k} = \sum_{k=0}^{q} \theta_k z_{n-k}, \ n \in \mathbf{Z},$$

where (z_n) is an orthonormal system and $\phi_0, \theta_0, \phi_p, \theta_q$ are nonzero. The equation (4.13) has a bounded solution iff $\frac{\theta}{\phi}$ is bounded. When this is so, every bounded solution has the form

$$(4.14) \qquad x_n = \sum_{k=-\infty}^{+\infty} \psi_k z_{n-k} + \sum_{k=1}^{N} e^{-int_k} v_k, \ n \in \mathbf{Z},$$

where $\psi_k = \frac{1}{2\pi} \int e^{-ikt} \frac{\theta(t)}{\phi(t)} \, dt, k \in \mathbf{Z}$, $v_1 \ldots, v_N$ are elements of H and $t_1, \ldots, t_N$ are zeros of $\phi(t)$.

4.15. REMARK. The fact that when $\frac{\theta}{\phi}$ is bounded, every bounded solution to (4.13) has the form (4.14) is well known and can be derived from the fact that in this case every solution to the ARMA equation is given by

$$(4.16) \qquad x_n = x_n^0 + \sum_{m=1}^{M} \left(\sum_{k=0}^{r_m-1} y_{mk} \, |n|^k (\frac{1}{a_m})^n \right), \ n \in \mathbf{Z}$$

where $x_n^0 = \int e^{-int} \frac{\theta}{\phi} dZ = \sum_{-\infty}^{+\infty} \psi_k z_{n-k}, \ n \in \mathbf{Z}$, $\psi_k = \frac{1}{2\pi} \int e^{-ikt} \frac{\theta(t)}{\phi(t)} dt, \ k \in \mathbf{Z}$, $\alpha_1, \ldots, a_M$ are zeros of $\tilde{\phi}(z) = \sum_{k=0}^{p} \phi_k z^k$ and $r_1, \ldots, r_M$ their multiplicities (see [2]). However, even in this case, the fact that all solutions to (4.13) are unbounded, if $\frac{\theta}{\phi}$ is not bounded, seems to be new.

REFERENCES

[1] ABREU, J.L., SALEHI, H., *Schauder basic measures in Banach and Hilbert spaces*, Bol. Soc. Mat. Mexicana 29 (2) (1984), 71–84.

[2] BROCKWELL, P.J., DAVIS, R.A., *Time Series: Theory and methods*, Springer–Verlag (1987).

[3] DUNFORD, N., SCHWARTZ, J.T., *Linear Operators, Part I*, Interscience Publ. (1958).

[4] EDWARDS, R.E., *Fourier Series: a Modern Introduction*, v. II, Holt, Rinehart & Winston Inc. (1967).

[5] HOFFMAN, K., *Banach Spaces of Analytic Functions*, Prentice–Hall Inc. (1962).

[6] KOLMOGOROV, A.N., *Stationary sequences in Hilbert space*, Bull. Math Univ. Moscov 2 (1961), 1–60.

[7] LEWIS, D.R., *Integration with respect to vector measures*, Pacific J. Math. 33 (1) (1970), 157–165.

[8] MIAMEE, A.G., SALEHI, H., *On an explicit representation of the linear predictor of a weakly stationary stochastic sequence*, Bol. Soc. Mat. Mexicana 28 (1) (1983), 81–93.

[9] SHIBATA, R., *Asymptotically efficient selection of the order of the model for estimating parameters of a linear process*, Ann. Stat. 8 (1) (1980), 147–164.

LEAST SQUARES ESTIMATION OF THE LINEAR MODEL WITH AUTOREGRESSIVE ERRORS*

NEERCHAL K. NAGARAJ** AND WAYNE A. FULLER†

Abstract. A Monte Carlo study of the least squares estimator of the regression model with autocorrelated errors is presented. The model contains a stationary explanatory variable and a random walk explanatory variable. The error model is a first order autoregressive model and the unit root case is included in the simulations. The limiting distribution of the regression pivotals for the basic model are normal, while the statistics for the autoregressive coefficient have a distribution that depends on the true parameter. The agreement between the Monte Carlo results and the asymptotic theory depends upon the autoregressive coefficient and on the nature of the explanatory variable.

Key words. Least squares, nonlinear estimation, Monte Carlo, time series.

AMS(MOS) subject classifications. Primary 62M10; secondary 62J02, 62F12.

1. Introduction. The regression model with autocorrelated errors is a natural model to use in many situations where the regression variables are observed over time. The basic model can be written as

$$Y_t = X_t\beta + u_t, \qquad t = 1, 2, \ldots \tag{1.1}$$

$$u_t = \sum_{i=1}^{p} \alpha_i u_{t-i} + e_t, \quad t = 1, 2, \ldots \tag{1.2}$$

where X_t is a q-dimensional row vector of explanatory variables, u_t is the unobserved error and $\{e_t\}$ is a sequence of zero mean random variables that are independent, or that satisfy conditions, such as those of martingale differences, that lead to behavior similar to that of independent random variables. Let

$$m^p - \sum_{i=1}^{p} \alpha_i m^{p-1} = 0 \tag{1.3}$$

be the characteristic equation associated with the autoregressive process.

By substituting the definition of u_t from (1.1) into (1.2), we obtain

$$\begin{aligned} Y_t &= X_t\beta + \sum_{i=1}^{p} \alpha_i(Y_{t-i} - X_{t-i}\beta) + e_t \\ &= f(Z_t, \eta) + e_t, \end{aligned} \tag{1.4}$$

*This research was partly supported by Joint Statistical Agreement JSA 88-1 with the U.S. Bureau of the Census.

**Department of Mathematics, University of Maryland at Baltimore County, Catonsville, Maryland 21228.

†Department of Statistics, Iowa State University, Ames, Iowa 50011.

where $Z_t = (X_t, Y_{t-1}, X_{t-1}, \ldots, Y_{t-p}, X_{t-p})$ and $\eta' = (\beta', \alpha_1, \alpha_2, \ldots, \alpha_p)$. Given a sample of n observations (Y_t, Z_t), $t = 1, 2, \ldots, n$, the problem is to estimate η.

Model (1.4) can also be written as

$$
(1.5) \qquad Y_t = X_t\beta + \sum_{i=1}^{p} X_{t-i}\zeta_i + \sum_{i=1}^{p} \alpha_i Y_{t-i} + e_t,
$$

where $\zeta_i = -\alpha_i\beta$. Model (1.5) is linear in the parameters β, ζ_i, and α_i. However, the matrix of sums of squares and products for n observations on the vector $(X_t, X_{t-1}, \ldots, X_{t-p})$ may be singular. For example, the matrix will be singular if the intercept is included in the model. Let θ be the portion of $(\beta', \zeta_1', \zeta_2', \ldots, \zeta_p', \alpha_1, \alpha_2, \ldots, \alpha_p)$ associated with the part of the vector $(X_t, \ldots, X_{t-p})$ that has a non-singular sum of squares and products matrix. We call the ordinary least squares estimator of θ, denoted by $\hat\theta$, the unrestricted least squares estimator.

If the largest root of the characteristic equation (1.3) is less than one in absolute value, Theorem 1 of Fuller, Hasza and Goebel (1981) can be used to show that the limiting distribution of the regression pivotal for the ordinary least squares estimator of an element of θ is normal for a wide range of explanatory variables.

If the largest root is equal to one, the limiting distribution of the least squares coefficient associated with that root is not normal and the limiting distribution depends on the explanatory variables in the equation. See Dickey and Fuller (1979), Fuller (1984), Fuller, Hasza and Goebel (1981), Phillips and Durlauf (1986), and Chan and Wei (1988).

Define the least squares estimator of η, denoted by $\tilde\eta$ of model (1.4), to be the value of η that minimizes

$$
(1.6) \qquad Q_n(\eta) = \sum_{t=1}^{n} [Y_t - f(Z_t, \eta)]^2.
$$

We sometimes call this estimator the restricted estimator.

The limiting properties of the least squares estimator of η depend on the properties of the sequence $\{X_t\}$, on the properties of the sequence $\{e_t\}$ and on the roots of the characteristic equation. Nagaraj and Fuller (1989) have given a theorem for the estimation of a linear model subject to nonlinear restrictions that is applicable to the model defined by (1.1) and (1.2). The theorem permits the sum of squares of the explanatory variables to grow at different rates. For example, the vector X_t could contain a stationary variable, a time trend and (or) a random walk.

An interesting special case of model (1.1, 1.2) is the model with a random walk explanatory variable. Let

$$
(1.7) \qquad \begin{aligned} Y_t &= \beta_0 + \beta_1 Z_t + u_t, \\ u_t &= \rho u_{t-1} + e_t, \end{aligned}
$$

where $Z_t = \sum_{i=1}^{t} d_i$, and $(e_t, d_t)' \sim NI[0, \mathrm{diag}(\sigma_{ee}, \sigma_{dd})]$. The the unrestricted model can be written as

$$
Y_t = \beta_0(1 - \rho) + (\beta_1 - \rho\beta_1)Z_{t-1} + \beta_1(Z_t - Z_{t-1}) + \rho Y_{t-1} + e_t
$$

or as

$$
\begin{aligned}
Y_t &= \beta_0(1-\rho) + (\beta_1 - \rho\beta_1 + \rho\beta_1^0)Z_{t-1} + \beta_1(Z_t - Z_{t-1}) \\
&\quad + \rho(Y_{t-1} - \beta_1^0 Z_{t-1}) + e_t, \\
&= X_t\gamma + e_t,
\end{aligned}
$$

(1.8)

where β_1^0 is the true value of β_1, $\gamma = [\beta_0(1-\rho), \beta_1 - \rho\beta_1 + \rho\beta^0, \beta_1, \rho]'$, and $X_t = [1, Z_{t-1}, Z_t - Z_{t-1}, Y_t - \beta_1^0 Z_{t-1}]$. The transformed version of the model in (1.8) is only used to identify the limiting distribution. It cannot be used in the actual estimation because β_1^0 is unknown. The transformation is used to define the limiting distribution in some important cases. For example, if $\rho = 1$, then the correlation between Z_t and Z_{t-1} converges to one. Therefore, the limit of the normalized sums of squares and product matrix of the original variables is singular. The transformation to the parameter vector γ removes the singularity.

Let $\rho = 1$ and write the transformed model as

$$
Y_t = \gamma_0 + \gamma_1 Z_{t-1} + \gamma_2(Z_t - Z_{t-1}) + \gamma_3(Y_{t-1} - \beta_1^0 Z_{t-1}) + e_t,
$$

where the true value of γ_0 is zero. Then it can be shown that the regression pivotal for the unrestricted least squares estimator of γ_1 converges in distribution to

(1.9)
$$
\frac{T_{ab} - W_a T_b - 1/2(\Gamma_{bb} - W_b^2)^{-1}(\Gamma_{ba} - W_b W_a)(T_b^2 - 1 - 2T_b W_b)}{[\Gamma_{aa} - W_a^2 - (\Gamma_{bb} - W_b^2)^{-1}(\Gamma_{ba} - W_b W_a)]^{1/2}},
$$

where

$$
(\Gamma_{aa}, \Gamma_{bb}, \Gamma_{ab}) = \sum_{i=1}^{\infty} \zeta_i^2(a_i^2, b_i^2, a_i b_i),
$$

$$
T_{ab} = \sum_{i=1}^{\infty}\sum_{j=1}^{\infty} v_{ij} a_i b_j,
$$

$$
T_b = \sum_{i=1}^{\infty} 2^{1/2} \zeta_i b_i,
$$

$$
(W_a, W_b) = \sum_{i=1}^{\infty} 2^{1/2} \zeta_i^2(a_i, b_i),
$$

$$
v_{ij} = 2[\zeta_j + \zeta_i]^{-1}\zeta_i^2\zeta_j,
$$

$$
(a_i, b_i) \sim NI(0, I) \quad \text{and} \quad \zeta_i = (-1)^{i+1}2[(2i-1)\pi]^{-1}.
$$

The regression pivotal for γ_3 in the unrestricted model with $\rho = 1$ converges in distribution to

(1.10)
$$
\frac{1/2(T_b^2 - 1 - 2T_b W_b) - (\Gamma_{aa} - W_a^2)^{-1}(\Gamma_{ab} - W_a W_b)(T_{ab} - W_a T_b)}{[\Gamma_{bb} - W_b^2 - (\Gamma_{aa} - W_a^2)^{-1}(\Gamma_{ab} - W_b W_a)]^{1/2}}.
$$

The regression pivotal for γ_2 has a limiting normal distribution. The limiting results were obtained using the transformation given in Dickey and Fuller (1979). Also see

Phillips (1986) and Phillips and Durlauf (1986) for a different representation of the limiting distribution.

Using the results of Nagaraj and Fuller (1989), it can be shown that the limiting distribution for the estimator of γ_2 under the restricted model is the same as under the unrestricted model. The limiting distribution of the regression pivotal for the estimator of ρ under the restricted model is the distribution of

$$\hat{\tau}_\mu = \frac{1/2(T_b^2 - 1 - 2T_b W_b)}{(\Gamma_{bb} - W_b^2)^{1/2}},$$

where the distribution of $\hat{\tau}_\mu$ is tabulated in Fuller (1976).

Because the model (1.4) is a restricted version of the model (1.5), it is natural to consider a test of the restrictions. One test is constructed by analogy to the F-test in linear regression. This test is

(1.11) $$F = (d_f - d_r)^{-1}(n - d_f)[Q_n(\hat{\theta})]^{-1}[Q_n(\tilde{\eta}) - Q_n(\hat{\theta})],$$

where d_f is the number of parameters in the full model associated with θ, $d_r = p + q$ is the number of parameters in the restricted model and $Q_n(\hat{\theta})$ is the residual sum of squares from the unrestricted model. Provided the unrestricted estimator has a limiting distribution, Corollary 2 of Nagaraj and Fuller (1989) can be used to obtain the limiting distribution of the test statistic. In model (1.7) with $\rho = 1$, the limiting distribution of the test statistic is the square of (1.9).

We use the Monte Carlo method to study the small sample behavior of the least squares estimator for a regression model with two regressors and an intercept. The error in the regression is a first order autoregression. One of the regressors is the normal random walk and the other is a sequence of independent $N(0,1)$ random variables.

There are some Monte Carlo experiments in the literature related to model (1.1). For example, Rao and Griliches (1969) study the model with a stationary regressor and stationary first order autoregressive error process. A regression model with a regressor which follows a random walk and errors which follow a stationary first order autoregression was considered by Krämer (1986). The Monte Carlo experiment in Krämer (1986) compared the ordinary least squares estimator to the corresponding generalized least squares estimators constructed with the true parameter of the error process. The generalized least squares estimator is the best linear unbiased estimator for the regression parameters. However, the generalized least squares estimator is generally unattainable in practice because it requires knowledge of the variance-covariance matrix of the errors.

2. Monte Carlo Study. The model of our study is an extension of the model considered by Krämer (1986). Our model contains two regressors and we include nonstationary autoregressive errors in our study. The model is

(2.1) $$Y_t = \beta_0 + \beta_1 X_{1t} + \beta_2 X_{2t} + u_t, \qquad t = 1, 2, \ldots,$$

$$u_t = \rho u_{t-1} + e_t, \qquad t = 1, 2, \ldots,$$

(2.2) $$= 0, \qquad t = 0 \quad \text{and} \quad |\rho| = 1,$$

$$= (1 - \rho^2)^{-1/2} e_0, \qquad t = 0 \quad \text{and} \quad |\rho| < 1,$$

where $\beta_0 = 0$, $\beta_1 = 1$, $\beta_2 = 1$, and $e_t \sim NI(0,1)$. The sequence $\{X_{1t}\}$ in the model (2.1) is a random walk generated by the following stochastic difference equation,

$$X_{1t} = X_{1,t-1} + w_t, \quad t = 1, 2, \ldots, n$$
$$= 0, \quad\quad\quad\ t = 0$$

where $w_t \sim NI(0,1)$. The X_{2t} are $NI(0,1)$ random variables independent of w_t. The sequence $\{u_t\}$ is independent of the sequence $\{X_{1t}, X_{2t}\}$. The sum of squares of X_{1t} is order in probability n^2 and the sum of squares of X_{2t} is order in probability n. We rewrite the model as

$$Y_1 = \beta_0 + \beta_1 X_{11} + \beta_2 X_{21} + u_1,$$
$$(2.3) \quad Y_t = \beta_0(1 - \rho) + \beta_1 X_{1t} - \beta_1 \rho X_{1,t-1} + \beta_2 X_{2t} - \beta_2 \rho X_{2,t-1} + \rho Y_{t-1} + e_t,$$
$$t = 2, 3, \ldots, n.$$

Because of the degeneracy of model (2.3) associated with $\rho = 1$, we defined a new parameter $\beta_* = \beta_0(1 - \rho)$. Although the parameter β_* is zero under model (2.1) with $\rho = 1$, we estimated β_* in all cases.

Let $\theta = (\beta_*, \beta_1, -\beta_1 \rho, \beta_2, -\beta_2 \rho, \rho)$ and let $\hat{\theta}$ be the unrestricted least squares estimator of θ. The correlation between X_{1t} and $X_{1,t-1}$ approaches one as the sample size increases. Therefore, it is necessary to transform the independent variables with a transformation such as that described in Fuller, Hasza and Goebel (1981) to define the limiting distribution of the transformed unrestricted least squares estimator. The limiting distribution of the transformed parameter vector depends upon the parameters. In particular, the limiting distribution of the estimator of ρ (the coefficient of Y_{t-1}) depends on the value of ρ. The limiting distribution of the regression pivotal for ρ is given in (1.10). The addition of the stationary process to the model (1.7) does not alter the limiting distribution of the estimator of ρ.

It can be shown that assumptions of Theorem 1 of Nagaraj and Fuller (1989) are satisfied for the model (2.1) and (2.2) for all values of ρ in the interval $[-1, 1]$. Hence, the restricted least squares estimator, properly normalized, has a limiting distribution.

Samples were generated for several values of ρ in the range -1 to 1. Because the values of ρ close to 1 or -1 are more interesting than the values of ρ close to zero, more values of ρ close to the boundary were used in the study.

The nonlinear least squares estimators were obtained using the Gauss-Newton method. Estimation of the model (2.3) by the Gauss-Newton method consists of repeated regressions using derivatives evaluated at the estimates from the previous step as independent variables, and using the residuals $\hat{e}_t$ from the previous step as the dependent variable. See Section 5.5 of Fuller (1976) for a description of the method. The derivatives for $t = 2, 3, \ldots, n$ are given in Table 1 with the parameters at the top of the columns. The first row of the table contains the partial derivatives for the first observation, multiplied by $(1 - \rho^2)^{1/2}$, where $(1 - \rho^2)$ is the ratio of the variance of e_t to the variance of u_1.

Using the last $n-1$ rows of the Table 1, four Gauss-Newton iterations were performed. The first iteration used the coefficients of $(X_{1t}, X_{2t}, Y_{t-1})$ in the regression of Y_t on $(X_{1t}, X_{1,t-1}, X_{2t}, X_{2,t-1}, Y_{t-1})$ as start values for β and ρ, respectively. The estimate of ρ was restricted to the interval $[-1, 1]$ at the end of each iteration. If the estimate of ρ was less than one in absolute value at the end of four iterations then the fifth iteration was performed using all n observations. During the last iteration, the first observation was weighted by the factor $(1 - \rho^2)^{1/2}$ as illustrated in Table 1. If the estimate of ρ was either -1 or 1 at the end of the four iterations, the fifth iteration was based on only the last $n - 1$ observations of Table 1. The nonlinear least squares estimator obtained by this procedure is denoted by $(\tilde{\beta}_*, \tilde{\beta}_1, \tilde{\beta}_2, \tilde{\rho})$.

Table 1. Matrix used in estimation.

t	$\beta_0(1 - \rho) = \beta_*$	β_1	β_2	ρ
1	$(1 - \rho^2)^{1/2}(1 - \rho)^{-1}$	$(1 - \rho^2)^{1/2} X_{11}$	$(1 - \rho^2)^{1/2} X_{21}$	0
2	1	$X_{12} - \rho X_{11}$	$X_{22} - \rho X_{21}$	u_1
3	1	$X_{13} - \rho X_{12}$	$X_{23} - \rho X_{22}$	u_2
$\vdots$	$\vdots$	$\vdots$	$\vdots$	$\vdots$
n	1	$X_{1n} - \rho X_{1n-1}$	$X_{2n} - \rho X_{2n-1}$	u_{n-1}

The usual "t-statistics" are constructed using the estimated standard errors obtained at the last iteration. For example, the t-statistic, $t_{\tilde{\beta}_1}$, for β_1 is

$$t_{\tilde{\beta}_1} = [\tilde{V}\{\tilde{\beta}_1\}]^{-1/2}(\tilde{\beta}_1 - \beta_1^0),$$

where $\tilde{V}\{\tilde{\beta}_1\}$ is the second diagonal element of the inverse of the sum of squares and cross products matrix for the derivatives multiplied by s^2, and s^2 is the residual mean square obtained at the last iteration.

For model (2.1, 2.2), the "t-statistics" corresponding to the regression parameters $(\beta_0, \beta_1, \beta_2)$ are distributed as standard normal random variables, asymptotically, for $|\rho| < 1$. The "t-statistic" corresponding to the parameter ρ also has a limiting normal distribution when the true value of ρ is strictly less than one in absolute value.

If $\rho = 1$, the pivotal statistic corresponding to ρ has the same limiting distribution as the τ_μ statistic characterized in Fuller (1976, Chapter 8). If $\rho = -1$, the limiting distribution of the pivotal is the limiting distribution of $-\tau$. When $|\rho| = 1$, the limiting distributions of the "t-statistics" for $\tilde{\beta}_1$ and $\tilde{\beta}_2$ are independent of $\tilde{\rho}$. However, the distribution of the t-statistic for β_* is not normal and is not independent of $\tilde{\rho}$. See Dickey and Fuller (1981).

Tables 2, 3, and 4 give empirical percentiles of the pivotal statistics corresponding β_1, β_2 and ρ for sample sizes 25 and 100. The results are based on 1000 samples. The standard errors of the estimates in Tables 2 and 3 are about 0.07 and the standard errors of the estimates in Table 4 are about 0.18.

Table 2. Empirical percentiles of $t_{\tilde{\beta}_1}$ for 1000 samples.

ρ	$n = 25$		$n = 100$	
	5%	95%	5%	95%
1.00	−2.29	2.23	−1.77	1.75
0.99	−2.25	2.20	−1.86	1.78
0.95	−2.42	2.35	−1.85	2.24
0.90	−2.29	2.77	−1.96	1.99
0.70	−2.27	2.56	−1.90	2.00
0.50	−2.30	2.14	−1.87	1.89
0.25	−1.97	1.86	−1.82	1.75
0.00	−1.94	1.91	−1.69	1.67
−0.25	−1.80	1.86	−1.69	1.71
−0.50	−1.76	1.71	−1.69	1.65
−0.70	−1.62	1.77	−1.61	1.58
−0.90	−1.59	1.60	−1.54	1.56
−0.95	−1.62	1.65	−1.48	1.66
−0.99	−1.56	1.74	−1.76	1.60
−1.00	−1.68	1.65	−1.75	1.65
$N(0,1)$	−1.65	1.65	−1.65	1.65

Table 3. Empirical percentiles of $t_{\tilde{\beta}_2}$ for 1000 samples.

ρ	$n = 25$		$n = 100$	
	5%	95%	5%	95%
1.00	−1.63	1.68	−1.64	1.66
0.99	−1.66	1.55	−1.53	1.67
0.95	−1.83	1.77	−1.53	1.63
0.90	−1.55	1.63	−1.63	1.68
0.70	−1.65	1.78	−1.58	1.58
0.50	−1.67	1.74	−1.60	1.64
0.25	−1.63	1.91	−1.75	1.55
0.00	−1.73	1.78	−1.80	1.73
−0.25	−1.73	1.76	−1.70	1.71
−0.50	−1.81	1.72	−1.58	1.64
−0.70	−1.60	1.66	−1.75	1.71
−0.90	−1.64	1.70	−1.65	1.71
−0.95	−1.54	1.56	−1.59	1.72
−0.99	−1.56	1.70	−1.60	1.68
−1.00	−1.74	1.53	−1.60	1.62
$N(0,1)$	−1.65	1.65	−1.65	1.65

Table 4. Empirical percentiles of $t_{\hat{\rho}}$ for 1000 samples.

	$n = 25$		$n = 100$	
ρ	5%	95%	5%	95%
1.00	−3.27	0.00	−3.69	0.00
0.99	−9.37	0.11	−5.83	0.13
0.95	−4.06	0.39	−2.87	0.75
0.90	−3.15	0.52	−2.39	0.92
0.70	−2.52	0.96	−2.10	1.15
0.50	−2.38	1.31	−2.01	1.35
0.25	−2.16	1.21	−2.02	1.51
0.00	−2.23	1.33	−1.86	1.32
−0.25	−1.94	1.43	−1.75	1.43
−0.50	−1.80	1.29	−1.81	1.59
−0.70	−1.69	1.30	−1.64	1.65
−0.90	−1.44	1.60	−1.42	1.71
−0.95	−1.10	1.49	−1.28	1.62
−0.99	−0.55	1.54	−1.06	1.85
−1.00	0.00	1.68	0.00	1.87
$N(0,1)$	−1.65	1.65	−1.65	1.65
τ_μ	−3.00	0.00	−2.89	−0.05
$-\tau$	−1.33	1.95	−1.29	1.95

When $n = 25$, the percentiles for the pivotal associated with the random walk explanatory variable are close to those of a $N(0,1)$ variable if ρ is less than -0.70. When $n = 100$, the percentiles differ from those of the normal distribution by a considerable amount for ρ in the range 0.7 to 0.95, but the agreement is quite good for ρ less than zero. The deviation from normality is partly explained by the serious downward bias in $\tilde{\rho}$ for ρ close to 1.00.

The percentiles of $t_{\tilde{\beta}_2}$ for the coefficient of the normal $(0,1)$ explanatory variable are closer to the limiting distribution than are those of the coefficient of the random walk explanatory variable for all parameter configurations, except $n = 25$ and $\rho = -1$.

The regression pivotals for the estimated autocorrelation coefficient are given in Table 4. The small sample bias in the estimator is clear from this table. Because the random walk has a very high positive sample autocorrelation, the estimator of a positive ρ has a large negative bias. Even for $\rho = 0$ and $n = 100$, the negative bias is not trivial.

From Table 4 it is clear that the finite sample distribution of $t_{\tilde{\rho}}$ differs considerably from the limiting distribution when ρ is close to 1. This is well known. See, for example, Ahtola and Tiao (1984), Chan and Wei (1987), and Chan (1988).

The Monte Carlo distribution becomes closer to the normal approximation as the sample size is increased.

The generalized least squares estimator of $(\beta_0, \beta_1, \beta_2)$ was computed for each sample using the true value of ρ. The generalized least squares estimators of the regression parameters are conveniently obtained by a regression where the variables are defined in Table 1. The last $n-1$ observations are $(Y_t - \rho Y_{t-1}, 1, X_{1t} - \rho X_{1t-1}, X_{2t} - \rho X_{2t-1})$. For values of ρ other than -1 and 1 the first observation in the regression is that given in the first row of Table 1. When ρ is either 1 or -1 the first observation is deleted from the sample. We denote the generalized least squares estimator of (β_1, β_2) by $(\hat{\beta}_{1GLS}, \hat{\beta}_{2GLS})$.

For our model with estimated (β_1, β_2, ρ), it can be shown that the limiting distribution of the nonlinear least squares estimator of (β_1, β_2) is the same as the limiting distribution of the generalized least squares estimator constructed with known ρ. Table 5 compares the Monte Carlo variances of the nonlinear least squares estimator of (β_1, β_2) with the infeasible generalized least squares estimator of (β_1, β_2). The ratios of the empirical variance of the nonlinear least squares estimator to the empirical variance of the generalized least squares estimator for various values of ρ are given in the table. The convergence of the nonlinear least squares estimator to the limiting form is faster for the $N(0,1)$ variable than for the random walk explanatory variable. Convergence is slowest for ρ near to, but not too near to, one.

Table 5. Ratios of the variance of the nonlinear least squares estimator to the variance of the generalized least squares estimator.

| | $n = 25$ | | $n = 100$ | |
ρ	$\tilde{\beta}_1$	$\tilde{\beta}_2$	$\tilde{\beta}_1$	$\tilde{\beta}_2$
1.00	1.27	1.04	1.06	1.00
0.99	1.28	1.05	1.09	1.00
0.95	1.35	1.06	1.17	1.00
0.90	1.28	1.09	1.19	1.01
0.70	1.27	1.07	1.12	1.01
0.50	1.40	1.10	1.08	1.03
0.25	1.16	1.15	1.06	1.03
0.00	1.21	1.14	1.03	1.03
−0.25	1.12	1.09	1.00	1.04
−0.50	1.04	1.09	1.00	1.03
−0.70	1.04	1.04	1.00	1.01
−0.90	1.03	1.02	1.00	1.00
−0.95	0.99	1.00	1.00	1.00
−0.99	1.00	1.00	1.00	1.00
−1.00	1.00	1.01	1.00	1.00

It can be shown that the limiting distribution of the test statistic (1.7) is that of a chi-square random variable with two degrees of freedom divided by its degrees of freedom when $|\rho| < 1$. This follows from the fact that the unrestricted estimators standardized by the square root of the sums of squares have limiting normal distributions.

If $\rho = 1$, the limiting distribution of the test statistic is the average of two random variables. The first variable in the test statistic is associated with the stationary variable and is a one-degree-of-freedom chi-square random variable. The second variable in the test statistic is independent of the first and the limiting distribution of the second variable is the distribution of the square of the pivotal given in (1.9). If $\rho = -1$, the limiting distribution of the test statistic is the average of two random variables, where the first is a chi-square random variable and the second is the square of variable like (1.9), but without the mean adjustment parts. The 5% and 95% points of the distribution of the test statistics are given in the last part of Table 6. The statistic denoted by $1/2(\chi^2 + \xi_\mu^2)$ is the statistic for $\rho = 1$. The percentage points are based upon a Monte Carlo run of size 10,000.

Table 6. Empirical percentiles of test statistic.

ρ	$n = 25$		$n = 100$	
	5%	95%	5%	95%
1.00	0.07	4.83	0.08	4.33
0.99	0.09	5.01	0.08	5.48
0.95	0.09	5.33	0.07	4.26
0.90	0.07	4.81	0.08	3.63
0.70	0.06	3.76	0.06	3.05
0.50	0.05	3.61	0.06	3.05
0.25	0.05	3.43	0.05	3.21
0.00	0.05	3.42	0.05	3.31
−0.25	0.05	3.55	0.05	3.02
−0.50	0.05	3.39	0.05	3.00
−0.70	0.05	3.31	0.05	2.90
−0.90	0.06	3.31	0.04	2.85
−0.95	0.04	3.10	0.05	2.87
−0.99	0.06	3.09	0.06	3.46
−1.00	0.05	3.36	0.06	2.95
$F(2, n - 6)$	0.05	3.55	0.05	3.09
$1/2(\chi^2 + \xi_\mu^2)$	0.04	3.79	0.04	3.79
$1/2(\chi^2 + \xi^2)$	0.04	3.45	0.04	3.45

225

Table 6 contains the 5% and 95% points of the Monte Carlo distribution of the test statistics based upon 1000 samples. For $\rho = 0$, the percentiles of the test statistic are close to those of the F-distribution. For ρ close to one, the 95-th percentage points for both sample sizes are beyond that of the F-distribution and beyond the limiting distribution of the test statistic for $\rho = 1$. For ρ close to negative one, the percentage points are close to those of the F-distribution. For $\rho = +1$, and $n = 100$ the observed percentage points are in reasonable agreement with those of the limiting distributions.

AHTOLA, J.A., AND TIAO, G.C., *Parameter inference for a nearly nonstationary first-order autoregressive model*, Biometrika 74 (1984), pp. 263–272.

CHAN, N.H., *The parameter inference for nearly nonstationary time series*, J. Amer. Statist. Assoc. 83 (1988), pp. 857–862.

CHAN, N.H. AND WEI, C.Z., *Asymptotic inference for nearly nonstationary AR(1) processes*, Ann. Statist. 15 (1987), pp. 1050–1063.

CHAN, N.H. AND WEI, C.Z., *Limiting distributions of least squares estimates of unstable autoregressive processes*, Ann. Statist. 16 (1988), pp. 367–401.

DICKEY, D.A., AND FULLER, W.A., *Distribution of the estimators for autoregressive time series with a unit root*, J. Amer. Statist. Assoc. 74 (1979), pp. 427–431.

DICKEY, D.A., AND FULLER, W.A., *Likelihood ratio statistics for autoregressive time series with a unit root*, Econometrica 49 (1981), pp. 1056–1072.

FULLER, W.A., *Introduction to Statistical Time Series*, Wiley, New York (1976).

FULLER, W.A., *Nonstationary autoregressive time series*, In Handbook of Statistics 5, Time Series in the Time Domain. (E.J. Hannan, P.R. Krishnaiah and M.M. Rao, Eds.) North Holland, Amsterdam (1984).

FULLER, W.A., HASZA, D.P., AND GOEBEL, J.J., *Estimation of the parameters of stochastic difference equations*, Ann. Statist. 9 (1981), pp. 531–543.

KRÄMER, W., *Least squares regression when the independent variable follows an ARIMA process*, J. Amer. Statist. Assoc. 81 (1986), pp. 150–154.

NAGARAJ, N.K. AND FULLER, W.A., *Estimation of the parameters of linear time series models subject to nonlinear restrictions*, Ann. Statist. 19 (1991), pp. 1143–1154.

PHILLIPS, P.C.B., *Understanding spurious regressions in econometric*, J. Econometrics 33 (1986), pp. 311–340.

PHILLIPS, P.C.B., AND DURLAUF, S.N., *Multivariate time series with integrated variables*, Review of Economic Studies 53 (1986), pp. 473–496.

RAO, P., AND GRILICHES, Z., *Small sample properties of several two stage regression methods in the context of auto-correlated errors*, J. Amer. Statist. Assoc. 64 (1969), pp. 253–272.

SIMS, C.A., STOCK, J.H. AND WATSON, M.W., *Inference in linear time series models with some unit roots*, Econometrica 58 (1990), pp. 113–144.

IDENTIFICATION OF
NONLINEARITIES AND NON-GAUSSIANITIES
IN TIME SERIES

TOHRU OZAKI*

Abstract. The identification of nonlinearities and non-Gaussianities of time series is discussed based on the observation that many nonlinear and/or non-Gaussian time series are generated from Gaussian white noise. The importance of nonlinear stochastic dynamical systems as their generation mechanisms is pointed out. A statistical identification method for nonlinear stochastic dynamical systems is presented. The identification method employs the maximum likelihood method which is based on Gaussian innovations obtained by applying a nonlinear filtering method to the observation data. The application of the method for the identification of nonlinear random vibration systems is presented with numerical results. Implications of the present method in nonlinear and non-Gaussian time series analysis are discussed.

Key words. Gaussian white noise, System identification, chaos, nonlinear stochastic dynamical systems, maximum likelihood method, local linearization, Markov diffusion process, nonlinear random vibrations, van der Pol oscillations, innovations, nonlinear state-space representation, nonlinear filtering, non-Gaussian distribution, Pearson system.

1. Introduction. In linear Gaussian time series analysis, Gaussian white noise has played a very important role. An ARMA(AutoRegressive Moving Average) time series is understood as the output of a system with rational transfer function driven by a Gaussian white noise. The model identification and diagnostic checking of the Box-Jenkins procedure for the ARMA time series modelling ([3]) are based entirely on this understanding, which suggests finding a linear dynamic mechanism to whiten the given time series data. In nonlinear and/or non-Gaussian time series analysis, however, we do not have any general standard model like the ARMA models in linear case. Instead we have many varieties of nonlinear time series models (see [23]). Some of them, such as Exponential AR models, Threshold AR models, Bilinear models and Random Coefficient models are based on Gaussian white noise, but some are not. Some models are based on non-Gaussian white noise with fat tail distribution like the Cauchy distribution, and some models are based on Exponentially distributed driving noise. How are we supposed to understand these nonlinear and/or non-Gaussian time series models? Isn't Gaussian white noise important in nonlinear and non-Gaussian time series analysis anymore?

In this paper we will see that Gaussian white noise will still play as important a role in nonlinear and/or non-Gaussian time series analysis as in the linear Gaussian case. We will see that very wide class of nonlinear and/or non-Gaussian stochastic processes are in the category of models driven by Gaussian white noise. What is different from the linear Gaussian case is that the dynamics of the system driven by Gaussian white noise are nonlinear. This means the idea of finding the whitening operation systematically used in linear Gaussian case in Box-Jenkins procedure is still valid in nonlinear and/or non-Gaussian time series model identification.

*Institute of Statistical Mathematics, 4-6-7 Minami Azabu Minato-ku, Tokyo Japan-106

What is required in nonlinear and/or non-Gaussian case is to find nonlinear dynamics to whiten the time series data. This data has been brought in through analysing real nonlinear and non-Gaussian time series data in ocean engineering, where the time series data come from the dynamics of a moored ship which is described by a continuous time stochastic dynamical system model. Typical non-linearity and non-Gaussianity observed in real time series of moored vessel motion is shown in section 2. Many dynamic phenomena in scientific fields are modeled by a continuous time stochastic dynamical system driven by a Gaussian white noise $\underline{w}(t)$ as in,

$$\dot{\underline{z}} = \underline{f}(\underline{z} \mid \underline{\theta}) + \underline{w}(t).$$

This is because modelling dynamics in continuous time makes things easier to understand than in discrete time when the dynamics is nonlinear.

Data are mostly sampled at discrete time points. In some cases we can record analog data of some phenomena, but that does not mean we have continuous time observation data of the stochastic dynamical system. It is because the system is driven by a continuous time Gaussian white noise and the continuous time record of such system is supposed to contain infinite information even in a very short microscopic time interval. A machine which records such fast movement of continuous time white noise does not exist. Even though it exists it is impossible to store the exact continuous time data in a computer even before we apply any estimation procedure to the data. To store the data we need to cut off the information of the high frequency fluctuations of the process outside a certain frequency band.

In practice we need to use a discrete time dynamic model to whiten the time series data measured in discrete time. This suggests us to introduce a nonlinear discrete time dynamic model which is obtained from the continuous time model by some time discretization method. A useful discretization method, called local linearization method, is introduced in section 3. The discretization method leads us to the identification method for a nonlinear stochastic dynamical system, which generates the time series data, based on the approximate discrete time model. The identification of nonlinearities in time series is realized by identifying the nonlinear stochastic dynamical system generating the time series.

The dimension r of the observation data $\underline{x}_t$ is sometimes less than the dimension k of the state $\underline{z}(t)$. In this situation the idea of finding a whitening operator for $\underline{x}_t$ is still realized by the above method combined with a nonlinear filtering method which transforms the time series data to Gaussian white innovation sequences. A nonlinear filtering scheme based on the locally linearized model is given in section 4.

The identification method can be used to judge if a deterministic dynamical system, i.e. a chaos model, is really more appropriate than a stochastic model by checking the estimated variance of the system noise. Topics related to the chaos model are discussed in section 5.

The process defined by the continuous time dynamical system driven by a Gaussian white noise is a Markov diffusion process. A class of Markov diffusion processes in general is a very wide class of stochastic processes indeed, including many interesting non-Gaussian distributed processes. In section 6 it is pointed out that any Markov diffusion process has a stochastic dynamical system representation associated with an instantaneous variable transformation. It means that any diffusion process can be whitened into a Gaussian white noise process by a nonlinear dynamic model combined with an instantaneous variable transformation. It is also shown that there can be infinitely many different nonlinear dynamical system models which produce one and the same non-Gaussian marginal distribution. Thus the identification of a non-Gaussianities in time series is very much dependent on the identification of nonlinearities in a dynamical system associated with the process which generates the time series.

2. Time Series for Moored Vessel Motion. It is a common view of applied time series analysts that most non-linear and/or non-Gaussian time series analysis consists of a great deal of complicated mathematics without much contribution to real and general applications. To avoid confusion and uselessness in applications we consider nonlinearities and non-Gaussianities in time series based on some real time series data which have clear and significant nonlinearity and non-Gaussianity.

A typical example of nonlinear and non-Gaussian time series is the measurement of the motion of moored vessel in the sea. Fig. 2.1 shows time series data obtained in an experiment where a ship is moored in a experimental water tank and random waves are artificially generated to excite the moored ship. The data include time series of the height of the waves near the moored ship, the tension of the mooring rope and the dynamics of the moored ship, i.e. surging, pitching, heaving and yawing. The data was sampled every 0.03 seconds and the number of data points for each series is 2000.

Fig. 2.2 shows the histogram of these time series data. The tension data of Fig. 2.1 shows a clear non-Gaussian characteristic which can be noticed without even in looking at the asymmetric shape of its histogram in Fig. 2.2. The histogram of yawing does not contain reliable information since the data length is rather short compared with the length of the period of its main frequency (see Fig. 2.1). Interesting data is the surge data whose non-Gaussian characteristic is not obvious in Fig. 2.1, but is clearly seen in the asymmetric histogram in Fig. 2.2. The tail in the left hand side decays fast and cut off, while its right hand side tail decays more slowly. This happens because when the ship surges to the left direction the mooring rope starts pulling it back. This is confirmed by the diagram of the two dimensional plot of the tension data and other variable in Fig. 2.3 where the tension increases nonlinearly when the surge goes to the left hand side and decreases to zero when the surge goes to the right hand side.

This kind of nonlinear dynamics of the moored vessel is modeled by equation (2.1),

$$(2.1) \qquad \ddot{x}(t) + \dot{x}(t) + bx = n(t) - T(x)$$

where $T(x)$ is some nonlinear function of the surge position x and represents the tension force on the surge motion caused by the mooring rope. A typical example

230

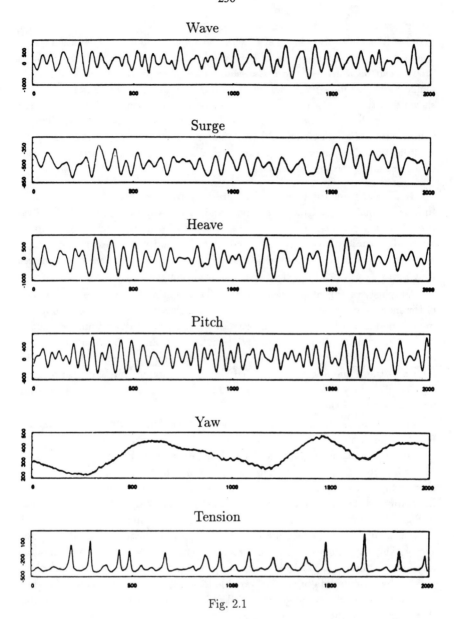

Fig. 2.1

of the function $T(x)$ is

$$
\begin{aligned}
T(x) &= c[(x - x_0)^3]_+ \\
&= c(x - x_0)^3 \qquad \text{for } x > x_0 \\
&\ \ 0 \qquad\qquad\ \ \text{for } x \leq x_0
\end{aligned}
$$

(2.2)

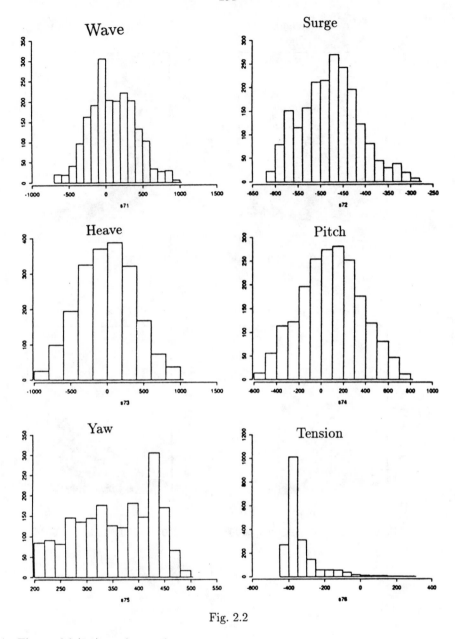

Fig. 2.2

The model (2.1) can be rewritten as

$$(2.3) \qquad \ddot{x}(t) + a\dot{x}(t) + bx + T(x) = n(t).$$

This is a random vibration with restoring force $bx + T(x)$. It means the vibrating system of the moored ship is nonlinear when the restoring force $T(x)$ is nonlinear

232

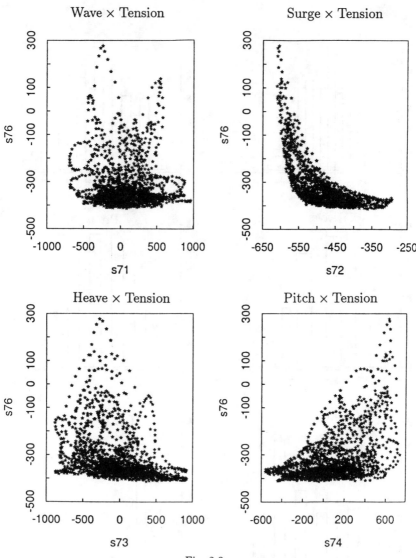

Fig. 2.3

even if the restoring force of the ship is linear. If we can estimate the parameters a and b and identify the function $T(x)$ for this model, we can characterize the nonlinear dynamics and non-Gaussian feature of the surge motion of the moored vessel.

3. How to Identify the Dynamics? In this section we try to identify the model,

$$(3.1) \qquad \ddot{x}(t) + a\dot{x}(t) + bx + T(x) = n(t)$$

from the observation data. If we employ the parametric form (2.2) for $T(x)$, the model (3.1) can be rewritten in a two dimensional stochastic dynamical system representation as

$$(3.2) \qquad \dot{\underline{z}} = \underline{f}(\underline{z} \mid \underline{\theta}) + \underline{w}(t)$$

where

$$
\begin{aligned}
&\underline{\theta} = (a, b, c, x_0)' \\
&\underline{z}(t) = \{x(t), x(t)\}' \\
&\underline{w}(t) = \{n(t), 0\}' \\
&\underline{f}(\underline{z} \mid \underline{\theta}) = \{-a\dot{x} - bx - c[(x - x_0)^3]_+, \dot{x}\}'.
\end{aligned}
$$

$n(t)$ is a Gaussian white noise with variance σ^2. We first consider the identification of the model (3.2) when we have the observation data $\underline{z}_1 = (\dot{x}_1, x_1)', \underline{z}_2 = (\dot{x}_2, x_2)', \ldots, \underline{z}_N = (\dot{x}_N, x_N)'$. The identification problem for the model when we have only observations $x_1, x_2, \ldots, x_N$ without the observation of the velocity of x is discussed in the next section.

Our basic idea for model identification is the same as in ARMA modelling, i.e. to find a whitening mechanism for the data $\underline{z}_1, \underline{z}_2, \ldots, \underline{z}_N$ into Gaussian white noise $\underline{w}_1 \underline{w}_2, \ldots, \underline{w}_N$. For this purpose we introduce a discrete time dynamical system model of the following type from model (3.2),

$$(3.3) \qquad \underline{z}_{t+\Delta t} = A(\underline{z}_t \mid \underline{\theta})\underline{z}_t + B(\underline{z}_t \mid \underline{\theta})\underline{w}_{t+\Delta t}$$

If this discrete time model is a good approximation for the original continuous time model (3.2), we can whiten the data $\underline{z}_1\underline{z}_2, \ldots, \underline{z}_N$ into Gaussian white noise $\underline{w}_1\underline{w}_2, \ldots, \underline{w}_N$, by

$$(3.4) \qquad \underline{w}_{t+\Delta t} = B(\underline{z}_t \mid \underline{\theta})^{-1}\{\underline{z}_{t+\Delta t} - A(\underline{z}_t \mid \underline{\theta})\underline{z}_t\}$$

The estimates of the parameter $\underline{\theta}$ of the model are given by the maximum likelihood estimates $\hat{\underline{\theta}}$ for the model (3.3).

3.1 Time Discretization. For the above purpose, we try to introduce a "good" discrete time approximation to the continuous time stochastic dynamical system model,

$$\dot{\underline{z}} = \underline{f}(\underline{z} \mid \underline{\theta}) + \underline{w}(t)$$

for a nonlinear and non-Gaussian process $\underline{z}(t)$, where $\underline{w}(t)$ is a Gaussian white noise and $\underline{f}(\underline{z} \mid \underline{\theta})$ is a nonlinear vector function of $\underline{z}(t)$. Before we go into the definition of our discretization method, which we call the local linearization method, let us see why conventional and common methods of time discretization methods are not satisfactory and why our new "good" method is needed.

For simplicity of explanation we first confine ourselves to the time discretization of the scalar stochastic dynamical system model,

$$(3.5) \qquad\qquad \dot{z} = f(z \mid \underline{\theta}) + w(t).$$

The most simple and common discrete time model for the scalar stochastic dynamical system (3.5) is

$$(3.6) \qquad\qquad z_{t+\Delta t} = z_t + \Delta t f(z_t \mid \underline{\theta}) + \sqrt{\Delta t} w_{t+\Delta t}.$$

This discretization method is a stochastic analogue of Euler's method for the discretization of deterministic dynamical system,

$$\dot{z} = f(z \mid \underline{\theta}).$$

The time discretization method (3.6) is consistent, i.e.

$$\frac{z_{t+\Delta t} - z_t}{\Delta t} \to f(z_t \mid \underline{\theta}) + w(t) \quad \text{for } \Delta t \to 0.$$

Actually the Ito stochastic differential equation (and the Stratonovich stochastic differential equation as well, since the variance of $w(t)$ is independent of $z(t)$) is defined as a limit of this discrete time model (see [13]). Therefore the distribution of z_t defined by the approximate discrete model is known to converge to the distribution of $z(t)$ in probability. The only disadvantage of this model is that the discrete time process, i.e. the Markov chain in a continuous state space defined by (3.6) is not stationary for most nonlinear function $f(\underline{z} \mid \underline{\theta})$ if the time interval Δt is fixed, no matter how small it is (see [8] and [31]). We may have a non-explosive sample path if we simulate the model with a small Δt for a certain finite time interval. However the probability of computational explosion in a finite time step is always 1, i.e. if we carry on the simulation it always ends up with computational explosion in finite steps. For example, for the process defined by (3.5) with

$$f(z \mid \underline{\theta}) = -z^3,$$

the model (3.6) explodes if z_t starts from an initial value greater than $\sqrt{2/(\Delta t)}$, even though the original continuous time process is stationary and never explodes no matter how large an initial value it starts from.

Some computationally stable time discretization schemes of the deterministic dynamical system,

$$(3.7) \qquad\qquad \dot{z} = f(z \mid \underline{\theta})$$

have been used in the engineering field (see p264 of [28]). The discrete model derived from the scheme applied to the deterministic dynamical system (3.7) is

$$(3.8) \qquad\qquad z_{t+\Delta t} = e^{J_t \Delta t} z_t$$

where J_t is the Jacobian of $f(z \mid \underline{\theta})$ at time point t. The model (3.8) is known to be computationally stable whenever the original nonlinear deterministic dynamical system is stable. When $f(z \mid \underline{\theta})$ is linear the trajectory of (3.8) coincides with the exact trajectory of the original system (3.7) on the discrete time points $t = \Delta t$, $2\Delta t$, $3\Delta t \ldots$,. If we replace the deterministic part of the model (3.6) by (3.8) we have the following discrete time stochastic dynamical system model,

$$(3.9) \qquad\qquad z_{t+\Delta t} = e^{J_t \Delta t} z_t + \sqrt{\Delta t} w_{t+\Delta t}.$$

The discrete time stochastic process z_t of (3.9), which is a Markov chain with a continuous state space, is known to be computationally stable and ergodic. Unfortunately the model is not consistent unless $f(z \mid \underline{\theta})$ is linear. If we let $\Delta t \to 0$, the process defined by the model (3.9) converges to

$$(3.10) \qquad\qquad \dot{z}(s) = J_t z(s) + w(s).$$

Incidentally we note that if we assume the model (3.10) on the interval $[t, t + \Delta t)$, the autocovariance function $\gamma_t(s)$ of the process $z(s)$ of (3.10) is

$$\gamma_t(s) = \left(\frac{\sigma_w^2}{2J_t} \right) e^{J_t s}.$$

Since

$$\frac{\gamma_t(\Delta t)}{\gamma_t(0)} = e^{J_t \Delta t},$$

it is reasonable to have $e^{J_t \Delta t}$ as a coefficient in its discretized model (3.9). This is used later when we introduce our local linearization scheme, which is consistent and computationally stable.

Some more sophisticated time discretization methods for the deterministic dynamical system (3.7) are known in numerical analysis, such as the Heun method and the Runge-Kutta method. It may be natural to substitute the deterministic part of (3.6) by some of the deterministic discrete time models obtained by these more sophisticated methods. We call these discrete time stochastic dynamical system models Heun scheme model and Runge-Kutta scheme model respectively. Actually these models are also used for simulation of stochastic differential equations (see [2] and [11]). Since the difference of these models and model (3.6) is of order $(\Delta t)^r$

with $r > 1$, these models have the same consistency property as the Euler scheme model (3.6).

Unfortunately, however, the disadvantage of model (3.6), i.e. its non-stationarity, is not remedied by these more sophisticated discrete time models. For example the Runge-Kutta scheme, applied to

$$\dot{z} = -z^3 + w(t),$$

gives us a stochastic discrete time dynamic model,

(3.11) $$z_{t+\Delta t} = p_{81}(z_t) + \sqrt{\Delta t} w_{t+\Delta t}$$

where $p_{81}(z_t)$ is a 81-st order polynomial of z_t. Therefore the Markov chain defined by (3.11) is, like the one in (3.6) introduced by the Euler scheme, transient (see [8] and [31]) and is not stationary.

In numerical analysis it is known that the error of the Euler method applied to the deterministic dynamical system (3.7) is of order $O(\Delta t)$, whereas the Heun method is of order $O(\Delta t)^2$, and the Runge-Kutta method is of order $O(\Delta t)^4$. In the stochastic situation one way of evaluating the goodness of the approximations may be by the mean square one step error

$$E_{z_t}[z(t + \Delta t) - z_{t+\Delta t}]^2,$$

where E_{z_t} denotes the conditional expectation with respect to z_t. The Euler scheme model (3.6) is known to have expected mean square error of order $O(\Delta t)^3$([12]). Higher order expected mean square errors are expected for the Heun scheme model and the Runge-Kutta scheme model. However it is shown that the order of the expected mean square error of these models are also $O(\Delta t)^3$. Moreover it was proved that the maximum possible speed of convergence attained by a discrete time approximate model is $O(\Delta t)^3$([26]). This result is, in a sense, natural because even though we approximate the deterministic part of the model (3.5) more accurately by Runge-Kutta method etc., the approximation of the white noise part in these models stays as poor as in the Euler scheme model, dominating the overall performance, and is impossible to improve. Thus other discrete time models with more sophisticated time discretization schemes are no better than the Euler method (3.6).

From the above discussion we now know that we should not expect to have a new time discretization scheme with a higher order convergence speed of the approximation. What we need is a new discretization method which is consistent as the Euler scheme method and at the same time yields a computationally stable and stationary Markov chain as (3.9) for a fixed Δt whenever the original continuous time Markov diffusion process is stationary.

3.2. Local Linearization. A key for the introduction of a new time discretization scheme lies in the scheme (3.9) which is the only one giving a computationally stable scheme out of the conventional methods. The scheme is also the only one whose deterministic part gives a trajectory which exactly coincides with the true trajectory of the continuous deterministic system (3.7) at the discrete time points when $f(z \mid \theta)$ is linear. The scheme (3.9) can be introduced by approximating the original process by the linear process (3.10) on each short interval $[t, t + \Delta t)$, where we assume that the coefficient of the linear function of the dynamical system on the interval is given by the Jacobian J_t of $f(z \mid \theta)$ at time point t. A disadvantage of this scheme is that it is not consistent. This is because J_t is constant on the interval $[t, t + \Delta t)$ and the function $J_t z$ does not converge to $f(z(t) \mid \theta)$ for $\Delta t \to 0$ unless $f(z(t) \mid \theta)$ is linear. This consideration suggests that we may be able to get a computationally stable and consistent scheme if we use some different function L_t to approximate the original continuous model on each short time interval $[t, t + \Delta t)$ by using, instead of (3.9), a linear stochastic dynamical system,

$$(3.12) \qquad \dot{z}(s) = L_t z(s) + w(s),$$

where L_t is some function of z_t and

$$L_t z_t \to f(z(t)) \quad \text{for} \quad \Delta t \to 0.$$

A simple scheme which satisfies this requirement is obtained by a simple assumption, i.e. "The Jacobian of the linear dynamical system,

$$(3.13) \qquad \dot{z}(s) = L_t z(s)$$

is equivalent to the Jacobian,

$$J_t = \left(\frac{\partial f}{\partial z} \right)_{z = z_t}$$

of the original dynamical system on each short time interval $[t, t + \Delta t)$". We call it the **Local Linearization (L.L.) assumption**. From this assumption we have

$$\ddot{z}(s) = J_t \dot{z}(s),$$

on $[t, t + \Delta t)$. If we integrate this on the interval $[t, t + \tau)$, where $0 \le \tau < \Delta t$, we have

$$\dot{z}(t + \tau) = e^{J_t \tau} \dot{z}(t)$$
$$= e^{J_t \tau} f(z(t)).$$

By integrating this again with respect to τ on the interval $[0, \Delta t)$ we have,

$$(3.14) \qquad z(t + \Delta t) = z(t) + J_t^{-1}(e^{J_t \Delta t} - 1) f(z(t))$$

Since J_t is given as a function of $z(t)$, $z(t + \Delta t)$ is explicitly given as a function of $z(t)$ at time point t. Thus the trajectory of the system (3.13) takes the value (3.14) at the discrete time point, $t + \Delta t$. On the other hand the solution $z(t + \Delta t)$ of the linear system (3.13) is given explicitly as a function of L_t and $z(t)$ by integrating $\dot{z}(s)$ of (3.13) from t to $t + \Delta t$ thus,

$$(3.15) \qquad\qquad z(t + \Delta t) = e^{L_t \Delta t} z(t).$$

From (3.14) and (3.15) we can write down L_t explicitly as a function of $z(t)$, as

$$(3.16) \qquad\qquad L_t = \frac{1}{\Delta t} \log\{1 + J_t^{-1}(e^{J_t \Delta t} - 1)F_t\}$$

where

$$F_t = \frac{f(z(t))}{z(t)}.$$

L_t of (3.16) is defined on the region of $z(t)$ where $J_t z_t \neq 0$. When $J_t z_t \neq 0$, it holds, for sufficiently small Δt, that

$$1 + J_t^{-1}(e^{J_t \Delta t} - 1)F_t > 0.$$

Therefore L_t of (3.16) is well defined. When $z(t)$ is on the region where $J_t z_t = 0$ (which is of measure zero), the $z(t + \Delta t)$ is defined separately in a way consistent with the rest of the region (see [20]).

Next we try to use L_t of (3.16) to introduce a stochastic version of the discrete time approximation to the continuous time system. One simple way is, like those conventional discretization schemes for stochastic dynamical systems, to add $\sqrt{\Delta t} w_{t+\Delta t}$ to the deterministic part, as

$$(3.17) \qquad\qquad z_{t+\Delta t} = e^{L_t \Delta t} z_t + \sqrt{\Delta t} w_{t+\Delta t}.$$

However if we stick to the L.L. assumption (3.16) and extend it to stochastic dynamical systems, we have more consistent model than (3.17). In the stochastic situation it will be natural to start from the assumption, "The stochastic dynamical system is locally linear and so it is Gaussian on each short time interval $[t, t + \Delta t)$, i.e. we employ the model (3.12) on the interval to approximate the original process, where L_t is given by (3.16)". Then $\dot{z}(s)$ of (3.12) can be integrated from t to $t + \Delta t$ on the interval giving

$$z_{t+\Delta t} = e^{L_t \Delta t} z_t + a_{t+\Delta t}$$

where

$$a_{t+\Delta t} = \int_t^{t+\Delta t} e^{L_t(t+\Delta t - s)} w(s) ds.$$

The variance of $a_{t+\Delta t}$ is

$$\sigma_a^2 = \int_t^{t+\Delta t} e^{2L_t(t+\Delta t - s)} \sigma_w^2 ds$$

$$= \frac{(e^{2L_t \Delta t} - 1)}{2L_t} \sigma_w^2.$$

where σ_w^2 is the variance of the continuous time white noise $w(t)$. Therefore it will be natural and more consistent if we use, instead of model (3.17), the following discrete time model,

(3.18)
$$z_{t+\Delta t} = A_t z_t + B_t w_{t+\Delta t}$$
$$A_t = e^{L_t \Delta t}$$
$$B_t = \sqrt{\frac{(e^{2L_t \Delta t} - 1)}{2L_t}}$$

where $w_{t+\Delta t}$ is a discrete time Gaussian white noise of variance σ_w^2. Since L_t of (3.16) satisfies

$$L_t z(t) \to f(z(t)) \quad \text{for} \quad \Delta t \to 0,$$

model (3.18) is consistent. The deterministic part of the model (3.18) inherits the zero points and Jacobian of the original continuous time deterministic dynamical system (3.18). Therefore the stability property of the original continuous time stochastic dynamical system is also preserved in scheme (3.18). Ergodicity of the discrete time model is proved by direct application of Tweedie's theorem (see [31]) when a proper $f(\underline{z} \mid \underline{\theta})$ of the original continuous time system is given ([16], [20]). Thus the Markov chain process defined by (3.18) gives us a stationary Markov chain if the original continuous time diffusion process is stationary.

Since

$$\sqrt{\frac{(e^{2L_t \Delta t} - 1)}{2L_t}} \approx \sqrt{\Delta t}$$

we can replace B_t of (3.18) by $\sqrt{\Delta t}$. Then we have the following nonlinear autoregressive model driven by a Gaussian white noise,

(3.19)
$$z_{t+\Delta t} = A_t z_t + \sqrt{\Delta t} w_{t+\Delta t}$$

which is equivalent to (3.17). When $f(z \mid \underline{\theta})$ of (3.5) is nonlinear the autoregressive coefficient A_t of the model (3.19) is a nonlinear function of z_t. Figures of the functions A_t for the following three examples are illustrated in Fig. 3.1 where we set $\Delta t = 0.1$.

Example 3.1. $\dot{z} = -z^3 + w(t)$

Example 3.2. $\dot{z} = -6z + 5.5z^3 - z^5 + w(t)$

Example 3.3. $\dot{z} = -\tanh(\sqrt{2}z) + w(t)$

An interesting common feature seen in the figures of the functions A_t in Fig. 3.1 is that the A_t's are all smooth functions of z_t and tend to a constant when z_t tends to $\pm\infty$.

This suggests that among several well-known conventional nonlinear autoregressive time series models, the exponential autoregressive(ExpAR) model, Ozaki [14], and nonlinear threshold model, Ozaki[15], are more appropriate than others such as the linear threshold autoregressive (TAR) model of Tong and Lim [30] whose autoregressive coefficient is a discontinuous step function.

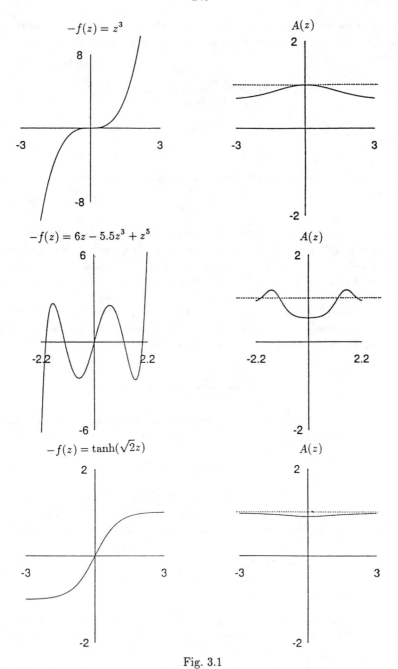

Fig. 3.1

The local linearization can be applied to a k-dimensional stochastic dynamical system,

$$(3.20) \qquad \dot{\underline{z}} = \underline{f}(\underline{z} \mid \underline{\theta}) + \underline{w}(t)$$

without any essential change, where $\underline{w}(t)$ is a k-dimensional Gaussian white noise with variance-covariance matrix $\sigma^2 I$. Since the solution of the linear k-dimensional differential equation,

$$\dot{\underline{z}} = L\underline{z}(t)$$

is

$$\underline{z}(t) = \text{Exp}(Lt)\underline{z}(0),$$

the discrete time model for the multi-dimensional stochastic dynamical system (3.20) is given by

(3.21)
$$\underline{z}_{t+\Delta t} = A_t \underline{z}_t + \underline{a}_{t+\Delta t}$$

$$\underline{a}_{t+\Delta t} = \int_t^{t+\Delta t} \text{Exp}\{L_t(t + \Delta t - s)\}\underline{w}(s)ds$$

$$A_t = \text{Exp}(L_t \Delta t)$$

$$L_t = \frac{1}{\Delta t}\text{Log}\{1 + J_t^{-1}(e^{J_t \Delta t} - 1)F_t\}$$

$$J_t = \{\frac{\partial \underline{f}(\underline{z})}{\partial \underline{z}}\}_{\underline{z}=\underline{z}_t}$$

F_t is such that $F_t \underline{z}_t = \underline{f}(\underline{z}_t)$. The matrix functions $\text{Exp}(.)$ and $\text{Log}(.)$ are defined by

$$\text{Exp}(L) = \sum_{k=0}^{\infty} \frac{1}{k!} L^k$$

and

$$\text{Log}(L) = \sum_{k=0}^{\infty} \frac{(-1)^k}{k!}(L - I)^k$$

respectively. The variance-covariance matrix of $\underline{a}_{t+\Delta t}$ is Σ_t whose elements are given as functions of the eigenvalues of L_t ([18], [22]). Using the unitary matrix U_t and the eigenvalues $\lambda_1, \ldots, \lambda_k$ of the diagonalized representation of Σ_t,

$$\Sigma_t = \sigma^2 U_t \begin{bmatrix} \lambda_1, 0 & \cdots & 0 \\ \cdots & \cdots & \cdots \\ 0 & \cdots & 0, \lambda_k \end{bmatrix} U_t'$$

we have, from (3.21), the following model,

(3.22)
$$\underline{z}_{t+\Delta t} = A_t \underline{z}_t + B_t \underline{w}_{t+\Delta t}$$

where

$$B_t = U_t \begin{bmatrix} \sqrt{\lambda_1}, 0 & \cdots & 0 \\ \cdots & \cdots & \cdots \\ 0 & \cdots & 0, \sqrt{\lambda_k} \end{bmatrix},$$

and $\underline{w}_{t+\Delta t}$ is a discrete time k-variate Gaussian white noise with variance-covariance matrix $\sigma^2 I$. Since the elements of the variance-covariance matrix Σ_t of $\underline{a}_{t+\Delta t}$ are explicitly given as a function of the eigenvalues of the matrix L_t the elements of the unitary matrix U_t and the eigenvalues $\lambda_1, \ldots, \lambda_k$, and so of B_t, are all given functions of $\underline{z}_t$.

An example of a two dimensional stochastic dynamical system is given by the following model for moored vessel motion.

$$(3.23) \qquad \ddot{x}(t) + 0.25\dot{x}(t) + 14.8x = n(t) - 4[x^3]_+.$$

From this example we have, through the two dimensional stochastic dynamical system representation (3.2), a bi-variate time series model representation,

$$(3.24) \qquad \underline{z}_{t+\Delta t} = A_t \underline{z}_t + B_t \underline{w}_{t+\Delta t}.$$

The eigenvalues of A_t, which are given by $\exp(\lambda_1 \Delta t)$ and $\exp(\lambda_2 \Delta t)$ using the eigenvalues λ_1 and λ_2 of L_t, are functions of z_t. If x is negative, the mooring force $4[x^3]_+$ is zero and the system is linear. When the system is linear, the eigenvalues of A_t, i.e. $\exp(\lambda_1 \Delta t)$ and $\exp(\lambda_2 \Delta t)$, are constant and the model (3.24) is known to be equivalent to a linear ARMA(2,1) model. The autoregressive coefficients ϕ_1 and ϕ_2 of the ARMA(2,1) model are given by

$$\phi_1 = \exp(\lambda_1) + \exp(\lambda_2)$$
$$\phi_2 = -\exp(\lambda_1)\exp(\lambda_2).$$

The first autoregressive coefficient of ARMA(2,1) model is known to characterize the proper frequency of the vibrating system and the second coefficient to characterize the damping property of the vibrating system. However for $x > 0$ the system becomes nonlinear because of the presence of $4[x^3]_+$. We can see how ϕ_1 and ϕ_2 change for positive x in Fig. 3.2, where the figures of the function ϕ_1 and ϕ_2 are calculated from (3.24) where $\Delta t = 0.1$ (see Fig. 3.2).

In the figure we notice that ϕ_1 decreases when x increases. This means that the proper frequency of the vibrating system shifts to higher frequency when x is positive and increasing. This frequency shift phenomenon is a common nonlinearity seen in hard spring type random vibrations described by a stochastic Duffing equation,

$$\ddot{x}(t) + a\dot{x}(t) + bx + cx^3 = n(t).$$

3.3. Maximum Likelihood Method. The log-likelihood function of the model (3.24) is given by

$$(3.25)$$
$$\log p(\underline{z}_1, \underline{z}_2, \ldots, \underline{z}_N \mid \underline{\theta}, \sigma^2)$$
$$= \log p(\underline{z}_2, \underline{z}_3, \ldots, \underline{z}_N \mid \underline{z}_1, \underline{\theta}, \sigma^2) + \log p(\underline{z}_1 \mid \underline{\theta}, \sigma^2)$$
$$= \log p(\underline{w}_2, \underline{w}_3, \ldots, \underline{w}_N \mid \underline{z}_1, \underline{\theta}, \sigma^2) + \log |J(\underline{z}, \underline{w})| + \log p(\underline{z}_1, \mid \underline{\theta}, \sigma^2)$$
$$= -\sum_{t=1}^{N-1} \frac{\|B_t^{-1}\{\underline{z}_{t+1} - A_t \underline{z}_t\}\|^2}{2\sigma^2} - \frac{N-1}{2}\log|\sigma^2 I| - \frac{N-1}{2}\log 2\pi$$
$$+ \sum_{t=1}^{N-1} \log \det(B) + \log p(\underline{z}_1 | \underline{\theta}, \sigma^2)$$

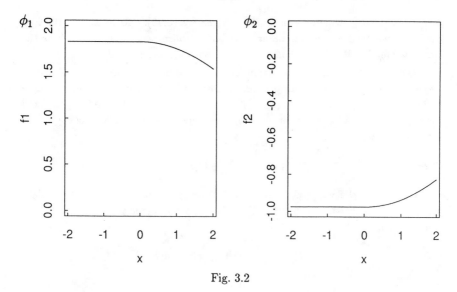

Fig. 3.2

where $J(\underline{z},\underline{n})$ is the Jacobian of the transformation from $\underline{z} = (\underline{z}_2, \underline{z}_3, \ldots, \underline{z}_N)$ to $\underline{w} = (\underline{w}_2, \underline{w}_3, \ldots, \underline{w}_N)$ which is equivalent to

$$\prod_{t=1}^{N-1} \det(B_t)$$

When N, the data length, is large the last term of the log-likelihood (3.25) is negligibly small compared with the rest and we can drop the term from the log-likelihood representation. Since the log-likelihood function satisfies the relation,

$$\left[\frac{\partial \log p(\underline{z}_2, \underline{z}_3, \ldots, \underline{z}_N \mid \underline{z}_1, \underline{\theta}, \sigma^2) + \log p(\underline{z}_1 \mid \underline{\theta}, \sigma^2)}{\partial \sigma^2} \right]_{\sigma^2 = \hat{\sigma}^2} = 0$$

at the maximum likelihood estimate $\hat{\sigma}^2$ of σ^2, σ^2 takes the following form

(3.26) $$\sigma^2 = \frac{1}{N-1} \sum_{t=1}^{N-1} \|B_t^{-1}\{\underline{z}_{t+1} - A_t \underline{z}_t\}\|^2.$$

Then in the log-likelihood representation (3.25) the first term of the right hand side becomes constant at $\sigma^2 = \hat{\sigma}^2$ and the maximum log-likelihood is obtained by maximizing

(3.27) $$-\frac{N-1}{2} \log \sigma^2 + \sum_{t=1}^{N-1} \log \det(B_t)$$

with respect to $\underline{\theta}$, where σ^2 in (3.27) is given by (3.26). To maximize the log-likelihood we need to use a numerical nonlinear optimization method such as the Davidon-Fletcher-Powell method or the Newton-Raphson method.

We can check how the above maximum likelihood method works in the following two examples of typical nonlinear random vibration time series data. One example is the data $\underline{z}_1 = (\dot{x}_1, x_1)', \underline{z}_2 = (\dot{x}_2, x_2)', \ldots, \underline{z}_N = (\dot{x}_N, x_N)'(N = 1000)$ generated from the van der Pol type nonlinear vibration model where the damping coefficient is nonlinear as in,

$$(3.28) \qquad \ddot{x}(t) + (x^2 - 1)\dot{x}(t) + 14.8x = n(t)$$

The variance of the white noise is 1. The time series $x_1, x_2, \ldots, x_N$ of the data $\underline{z}_1, \underline{z}_2, \ldots, \underline{z}_N$ is shown in Fig. 3.3.a. For this case we assumed the following parameteric model,

$$(3.29) \qquad \ddot{x}(t) + g_1(x)\dot{x}(t) + g_2(x)x = n(t)$$

with

$$(3.30) \qquad g_1(x) = a_1 + a_2 x + a_3 x^2$$
$$g_2 = b$$

The estimated parameters obtained by applying the maximum likelihood method are shown in Table 3.1. The second example is the data $\underline{z}_1, \underline{z}_2, \ldots, \underline{z}_N$ $(N = 1000)$ generated from the Duffing type nonlinear vibration model where the restoring force $g_2(x)x$ is nonlinear and so the coefficient $g_2(x)$ is non-constant. The variance of the white noise is 1. The time series $x_1, x_2, \ldots, x_N$ of the data $\underline{z}_1, \underline{z}_2, \ldots, \underline{z}_N$ is shown in Fig. 3.3.b. For this case we assumed the model (3.29) with

$$(3.31) \qquad g_1(x) = a,$$
$$g_2(x) = b_1 + b_2 x + b_3 x^2.$$

The estimation results are in Table 3.1. In both examples the results are satisfactory and convincing. Since we are using a numerical optimization method, standard errors of the estimates are easily obtained numerically from the inverse of the Hessian, the second derivative of the log-likelihood function.

Table 3.1 Maximum likelihood estimates of the model (3.30) and the model (3.31).

	Model (3.30)			Model (3.31)	
	true	estimated		true	estimated
a_1	-1	-0.9742	a	0.25	0.1589
a_2	0	-0.0112	b_1	14.8	14.9385
a_3	1	0.9878	b_2	0	0.3551
b	14.8	14.8520	b_3	4	3.7954
σ_n^2	1	1.0141	σ_n^2	4	4.0270

245

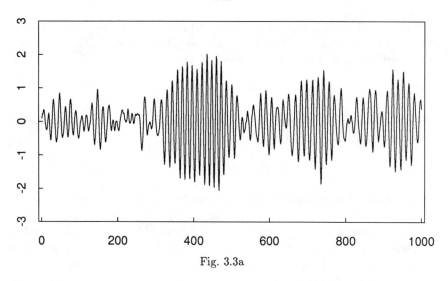

Fig. 3.3a

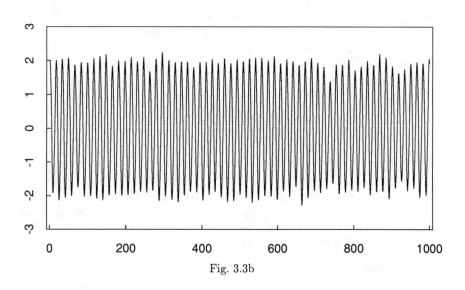

Fig. 3.3b

We note that the maximum likelihood method which we have used is not for the continuous time model (3.20) but for the discrete time model (3.21). If $\underline{\theta}$ is the estimated parameter the model which we obtained is

$$(3.32) \qquad \underline{z}_{t+\Delta t} = A(\underline{z}_t \mid \hat{\underline{\theta}})\underline{z}_t + B(\underline{z}_t \mid \hat{\underline{\theta}})\underline{w}_{t+\Delta t}$$

Then the following continuous time model (3.32) will be a natural estimate for the continuous time model (3.20),

$$(3.33) \qquad \dot{\underline{z}} = \underline{f}(\underline{z} \mid \hat{\underline{\theta}}) + \underline{w}(t)$$

since the local linearization of (3.33) is equivalent to (3.32). However since the local linearization is an approximation this correspondence is not unique and we have some other function $\underline{f}_{\Delta t}(\underline{z} \mid \hat{\underline{\theta}})$ which produces (3.32). A numerical method to obtain $f_{\Delta t}(\underline{z} \mid \hat{\underline{\theta}})$ from $A(\underline{z}_t \mid \hat{\underline{\theta}})$ is given in Ozaki ([17], [19]). The numerical method provides us with a useful method to guess the functional form of $\underline{f}(\underline{z} \mid \underline{\theta})$ from an estimated nonlinear autoregressive coefficient of some general nonlinear autoregressive model with fairly general parameterization such as the ExpAR model (see [17] and [20] for numerical examples).

4. Nonlinear Filtering as a Whitening Operator. The above maximum likelihood method is not valid when we observe, instead of a k-dimensional state vector $\underline{z}_i$, a scalar time series data x_i, which is a linear transformation of the state vector $\underline{z}_i$. In many practical identification problems we have only observations, $\underline{x}_1, \underline{x}_2, \ldots, \underline{x}_N$, whose dimension r is smaller than the system dimension k. Also the observation error is not negligibly small. In these situations what we need is an identification method for the following nonlinear state space representation model,

$$(4.1) \qquad \begin{aligned} \dot{\underline{z}} &= \underline{f}(\underline{z} \mid \underline{\theta}) + \underline{w}(t) \\ \underline{x}_t &= C\underline{z}(t) + \underline{v}_t \end{aligned}$$

where C is an $r \times k$ rectangular observation matrix, and $\underline{v}_t$ is an observation error vector which is an r-dimensional Gaussian white noise. In this situation we can still identify the model (4.1) based on the same idea in the previous section, i.e. to try to find a whitening operator which transforms the r-dimensional data, $\underline{x}_1, \underline{x}_2, \ldots, \underline{x}_N$ into r-dimensional Gaussian white innovations, $\underline{v}_1, \underline{v}_2, \ldots, \underline{v}_N$. This is realized by using a nonlinear filtering method for (4.1).

4.1. Local Linearization Filter. The filtering problem for (4.1) is sometimes called continuous-discrete nonlinear filtering ([7]) since the first equation for the state dynamics is in continuous time and the other observation equation is in discrete time. It is well known that nonlinear filtering is a difficult problem even in the scalar case. Its main difficulty is the computational instability which comes from the nonlinearity of the state dynamics. Our idea of obtaining a computationally stable filtering scheme for (4.1) is to introduce a stable approximate discrete time L.L. model from (4.1) and give a discrete time filtering scheme for the L.L. model, instead of introducing an approximate continuous time filtering scheme as in most nonlinear filtering methods ([10]). In (4.1), $\underline{w}(t)$ is a continuous time k-dimensional Gaussian white noise. We assume that Σ_w, the variance-covariance matrix of $\underline{w}(t)$ is $\sigma^2 I$ for simplicity. The variance-covariance matrix of discrete time r-dimensional Gaussian white noise $\underline{v}_t$ is Σ_v and $\underline{w}(t)$ and $\underline{v}_t$ are independent. Thus from (4.1), using the L.L. method in the previous section, we have,

$$(4.2) \qquad \underline{z}_{t+\Delta t} = A(\underline{z}_t)\underline{z}_t + B(\underline{z}_t)\underline{w}_{t+\Delta t}$$
$$\underline{x}_t = C\underline{z}_t + \underline{v}_t$$

where $A(\underline{z}_t)$ and $B(\underline{z}_t)$ are given by A_t of (3.20) and B_t of (3.21). $\underline{w}_{t+\Delta t}$ is a k-dimensional discrete time Gaussian white noise with variance-covariance matrix Σ_w. Also between i-th component $w^{(i)}$ of $\underline{w}_{t+\Delta t}$ and j-th component $v^{(j)}$ of $\underline{v}_t$ it follows that $E[w^{(i)}v^{(j)}] = 0$ for $1 \le i \le k$ and $1 \le j \le r$. Based on the minimum variance principle ([28]), which is to minimize

$$E[(\underline{z}_t - \underline{z}_{t|t})'(\underline{z}_t - \underline{z}_{t|t})],$$

the filtering equation for (4.2) is obtained as follows,

$$(4.3) \qquad \underline{z}_{t+\Delta t|t} = A(\underline{z}_t)\underline{z}_{t|t}$$
$$\underline{z}_{t|t} = \underline{z}_{t|t-\Delta t} + K_t\underline{\nu}_t$$
$$\underline{\nu}_t = \underline{x}_t - C\underline{z}_{t|t-\Delta t}$$
$$K_t = P_t C'(C P_t C' + \Sigma_v)^{-1}$$
$$P_t = E[(\underline{z}(t) - \underline{z}_{t|t-\Delta t})(\underline{z}(t) - \underline{z}_{t|t-\Delta t})']$$
$$V_t = P_t - K_t C P_t$$
$$\qquad = P_t - P_t C'(C P_t C' + \Sigma_v)^{-1} C P_t$$

where $\underline{\nu}_t$ is the innovation of the filtering model, K_t is the filter gain and P_t is the variance-covariance matrix of the one step ahead prediction error of the state $\underline{z}_t$. The evolution of P_t is obtained from

$$P_{t+\Delta t} = E[(\underline{z}_{t+\Delta t} - \underline{z}_{t+\Delta t|t})(\underline{z}_{t+\Delta t} - \underline{z}_{t+\Delta t|t})']$$
$$= E[\{A(\underline{z}_t)(\underline{z}_t - \underline{z}_{t|t}) + B(\underline{z}_t)\underline{w}_{t+\Delta t}\}\{A(\underline{z}_t)(\underline{z}_t - \underline{z}_{t|t}) + B(\underline{z}_t)\underline{w}_{t+\Delta t}\}'].$$

Since we are assuming $A(\underline{z}_t)$ and $B(\underline{z}_t)$ are constant on $[t, t + \Delta t)$, we have

$$(4.4) \qquad P_{t+\Delta t} = A(\underline{z}_t)V_t A(\underline{z}_t)' + B(\underline{z}_t)\Sigma_w B(\underline{z}_t)'$$

As is mentioned in the previous section, the function $B(\cdot)$ is almost constant for small Δt, while $A(\cdot)$ is not. Since we assume the system (4.2) is linear on $[t, t + \Delta t)$ and the system transition is characterized by $A(\underline{z}_t)$ as in (3.21) on the interval, it will be reasonable to replace $A(\underline{z}_t)$ and $B(\underline{z}_t)$ in (4.3) and (4.4) by $A(\underline{z}_{t|t})$ and $B(\underline{z}_{t|t})$ respectively. Then we have

$$(4.5) \qquad \underline{z}_{t+\Delta t|t} = A(\underline{z}_{t|t})\underline{z}_{t|t},$$

and

$$(4.6) \qquad P_{t+\Delta t} = A(z_{t|t})V_t A(z_{t|t})' + B(z_{t|t})\Sigma_w B(z_{t|t})'$$

The local linearization filter works very well compared with conventional nonlinear filtering method such as the Extended Kalman filter or the Minimum Variance filter. The error reduces to one tenth or less compared with the conventional methods (see [21] for numerical results).

4.2. Innovation Likelihood. Before we go into the maximum likelihood method for the nonlinear state space models, let us see how the linear state space model parameters are estimated by innovation maximum likelihood method. The parameters of a discrete time linear state space model,

$$(4.7) \qquad \begin{aligned} z_{t+\Delta t} &= A(\theta)z_t + B(\theta)w_{t+\Delta t} \\ x_t &= Cz(t) + v_t \\ C &= \begin{pmatrix} 0...0 & 1.....0 \\ ..... & ..... \\ 0...0 & 0.....1 \end{pmatrix} \end{aligned}$$

can be estimated by maximizing the likelihood,

$$p(x_1, x_2, \ldots, x_N \mid \theta)$$

of the model for the given data $x_1, x_2, \ldots, x_N$. If we use the Kalman filter, the data $x_1, x_2, \ldots, x_N$ are transformed to innovations $\nu_1, \nu_2, \ldots, \nu_N$ as

$$(4.8) \qquad \nu_i = x_i - \hat{x}_{i|i-1}$$

where $\hat{x}_{i|i-1}$ is the one-step ahead prediction value of x_i at time point $i-1$ and is a function of z_{i-1}, V_{i-1} and θ. If the initial values z_0 and V_0 are known, then the determinant of the Jacobian of the transformation from $x_1, x_2, \ldots, x_N$ to the innovations $\nu_1, \nu_2, \ldots, \nu_N$ is 1. Usually they are unknown and may be specified by a probability density function $p(z_0, V_0|\theta)$. Then the likelihood is written down in terms of the innovations, obtained by applying the Kalman filter, thus

$$\begin{aligned} p(x_1, x_2, \ldots, x_N \mid \theta) &= \int \int p(x_1, x_2, \ldots, x_N | z_0, V_0, \theta) p(z_0, V_0 \mid \theta) dz_0 dV_0 \\ &= \int \int p(\nu_1, \nu_2, \ldots, \nu_N | z_0, V_0, \theta) p(z_0, V_0 \mid \theta) dz_0 dV_0 \end{aligned}$$

There are many ways of specifying $p(z_0, V_0|\theta)$. One simple and practical way of specification is to use the delta function. If we assume a delta function for z_0 and V_0, the likelihood becomes

$$p(x_1, x_2, \ldots, x_N \mid z_0, V_0, \theta) = p(\nu_1, \nu_2, \ldots, \nu_N \mid z_0, V_0, \theta),$$

where we have two more parameters z_0 and V_0. Since innovations are Gaussian white noise, we have

$$(4.9) \qquad p(\nu_1, \nu_2, \dots, \nu_N \mid z_0, V_0, \theta) = \prod_{i=1}^{N} \frac{1}{\sqrt{2\pi |\Sigma_{\nu_i}|}} \exp(-\frac{1}{2} \nu_i' \Sigma_{\nu_i}^{-1} \nu_i).$$

Then we have

$$(4.10) \quad (-2) \log p(x_1, x_2, \dots, x_N | z_0, V_0, \theta) = \sum_{i=1}^{N} (\log |\Sigma_{\nu_i}| + \nu_i' \Sigma_{\nu_i}^{-1} \nu_i) + N \log 2\pi.$$

Then the maximum likelihood estimates of the parameters z_0, V_0 and θ are obtained by minimizing the (-2)log-likelihood (4.10). The maximum likelihood method based on this innovation likelihood has been used for the estimation of parameters of linear state space models ([27], [6] and [9]).

In the nonlinear case,

$$(4.11) \qquad \qquad \dot{z} = f(z \mid \theta) + w(t)$$
$$x_t = C z(t) + v_t$$
$$C = \begin{pmatrix} 0...0 & 1....0 \\ ..... & ..... \\ 0...0 & 0.....1 \end{pmatrix}$$

we expect to use the L.L. filter to transform the data $x_1, x_2, \dots, x_N$ into Gaussian white noise innovations $\nu_1, \nu_2, \dots, \nu_N$ as we did in the linear case. When $f(z \mid \theta)$ is nonlinear the state $z(t)$ is not Gaussian. However if the sampling interval Δt is sufficiently small, we can assume that the data $x_1, x_2, \dots, x_N$ are generated from the discrete time state space model,

$$(4.12) \qquad \qquad z_{t+\Delta t} = A(z_t) z_t + B(z_t) w_{t+\Delta t}$$
$$x_t = C z(t) + v_t,$$

whose discrete time state space process z_t is driven by w_t. Then we can get discrete time Gaussian white noise innovations, from observations x_t by applying the L.L. filter, since w_t is a discrete time Gaussian white noise. It is also proved by Frost and Kailath [5] that the continuous time innovation process for a non-Gaussian diffusion process is a Gaussian white noise. Since the continuous time innovations are defined as a limit of discrete time innovations as

$$\nu(t) = \lim_{\Delta t \to 0} \nu_t$$
$$= \lim_{\Delta t \to 0} (x_t - C z_{t|t-\Delta t})$$

their result implies that the innovation ν_t obtained by the L.L. filter is very close to a Gaussian white noise. This consideration justifies and guarantees the use of the following representation for the approximation of the (-2) log-likelihood,

$$(4.13) \quad (-2) \log p(x_1, x_2, \dots, x_N \mid z_0, V_0, \theta) = \sum_{i=1}^{N} (\log |\Sigma_{\nu_i}| + \nu_i' \Sigma_{\nu_i}^{-1} \nu_i) + N \log 2\pi$$

where Σ_{ν_i} is the variance-covariance matrix of the innovation ν_t and is given by the L.L. filter (4.3) as

(4.14) $$\Sigma_{\nu_i} = CP_iC' + \Sigma_{\nu}.$$

Thus the maximum likelihood estimates of the parameters z_0, V_0 and θ of (4.7) are obtained by minimizing (4.13).

4.3. Numerical Results Applied to Surge Data. We applied the innovation maximum likelihood method to the surge data in section 2(see Fig.2.1). To estimate the nonlinear restoring function $bx + T(x)$ of the surge motion model (3.1) we used the following parametric model,

(4.15) $$\ddot{x}(t) + a\dot{x}(t) + b_1x + b_2x^2 + b_3x^3 = n(t).$$

With this model we expect to obtain an asymmetric estimated restoring function. Before we apply the maximum likelihood method we calibrate the data x_t to the range between -0.64 and 1.2 so that $-0.64 < x_t < 1.2$. The calibration is often useful to avoid computational difficulties, such as overflowings, in likelihood calculation for nonlinear models. We assumed $\Delta t = 0.03$. We also assumed the observation error is zero because the measurement is quite accurate and the sampled data show quite smooth movements of the surge motion. The estimated parameters are as follows: $a = 51.4807$, $b_1 = 15.7963$, $b_2 = -14.3694$, $b_3 = 10.8584$ and $\sigma_N^2 = 221.9206$.

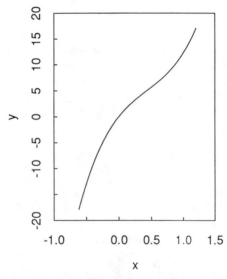

Figure 4.1

Fig. 4.1 shows the estimated restoring force function $b(x) = b_1 x + b_2 x^2 + b_3 x^3$. The function shows that when x is on the left hand side it sharply decreases and pulls the ship back strongly, while if the surge motion goes to the right hand side the restoring force does not increase so sharply. It is known (see [4] and [18]) in Markov diffusion theory that the equilibrium density distribution $p(x)$ of x of (4.15) is

(4.16)
$$p(x) = C \frac{1}{\sqrt{2\pi\sigma^2}} \exp\left(-\frac{2a(\frac{b_1}{2}x^2 + \frac{b_2}{3}x^3 + \frac{b_3}{4}x^4)}{\sigma^2}\right)$$

where C is a normalizing constant. The density function (4.16) explains why the asymmetric non-Gaussian like histogram of surge data is seen in Fig. 2.2 through the nonlinear function $b(x)$ of Fig. 4.5.

5. Deterministic or Stochastic Models? The identification method introduced in the previous section can be used not only for stochastic dynamical system but for deterministic dynamical system models. Recently it has been shown that some non-deterministic process can be generated by a deterministic dynamical system model. There have been introduced many interesting nonlinear deterministic dynamical system models which show stochastic behaviour, called chaos ([25]). Some people seem to be led, from these results, to the belief in determinism, which claims that all the stochastic models should be replaced by deterministic models referring to Albert Einstein's famous phrase (see [29] for example) which sounds misleading for those who know Einstein's innovative contribution to Brownian motion theory in statistical mechanics.

With so much evidence of random fluctuations ([24]) at a microscopic level of quantum mechanics it does not seem to be so promising to revive the determinism of Newton and Laplace. Recently there has been some interesting work reformulating quantum mechanics based on Brownian motion theory. Actually it was Norbert Wiener who initiated the white noise approach to quantum mechanics as an alternative to the conventional quantum mechanics formulation of Niers Bohr (see [32] and [33]). Incidentally he started it after being stimulated by Albert Einstein's hidden variable argument in the famous debate between Niels Bohr over the wave function formulation of quantum mechanics.

The chaotists' (as we might call scientists who believe in determinism) claim implies that $\underline{w}(t)$ and $\underline{v}_t$ of (4.11) should be zero. However, as one cannot deny the uncertainty principle in quantum mechanics, we cannot assume $\underline{v}_t = 0$. If the system noise $\underline{w}(t)$ really was zero, then the maximum log-likelihood of the model should occur when the estimated variance-covariance matrix Σ_w is close to zero. This never happens in numerical studies except for artificially generated time series data. Real data does not come from a simple mathematical model but comes from nature, which is too complicated to be totally explained by a simple deterministic model. Actually what Akaike's Entropy Maximization Principle ([1]) and AIC aim to do is to find a model with Σ_w as small as possible and at the same time with as small a number of parameters as possible. The present identification method will provide chaotists with a clue to understanding the big gap between reality and fantasy in real data analysis.

6. Non-Gaussianities in Time Series. In this section we see how non-Gaussianities of time series are specified by parametric models. In the past decades high order spectra have been studied to characterize non-Gaussianities of time series. However there is not any guarantee that major non-Gaussianity is specified by the bispectrum. The fact is that not only the bispectrum, but also all the higher order spectra from fourth order to infinite order are needed to characterize a single non-Gaussian process. Estimating so many quantities from a finite set of time series data does not sound a sensible thing to do for statisticians. To obtain some suggestions for the direction of the study of non-Gaussian time series, again, a real example of non-Gaussian phenomena in the real world is very useful.

6.1. Non-Gaussianity of Moored Vessel Motion. We have seen that the dynamics of moored vessel motion is characterized by a nonlinear random vibration model,

$$(6.1) \qquad \ddot{x}(t) + a\dot{x}(t) + bx + T(x) = n(t)$$

where

$$
\begin{aligned}
T(x) &= c[(x - x_0)^3]_+ \\
&= c(x - x_0)^3 \qquad \text{for } x > x_0 \\
&= 0 \qquad\qquad \text{for } x \le x_0.
\end{aligned}
$$

From (6.1) we have the following two dimensional stochastic dynamical system representation,

$$(6.2) \qquad \dot{z} = \underline{f}(\underline{z} \mid \underline{\theta}) + \underline{w}(t)$$

where

$$
\begin{aligned}
\underline{\theta} &= (a, b, c, x_0)' \\
\underline{z}(t) &= \{\dot{x}(t), x(t)\}' \\
\underline{w}(t) &= \{n(t), 0\}' \\
\underline{f}(\underline{z} \mid \underline{\theta}) &= \{-a\dot{x} - bx - c[(x - x_0)^3]_+, \dot{x}\}'
\end{aligned}
$$

The process $\underline{z}(t)$ defined by (6.2) is a bivariate Markov diffusion process. The evolution of the transition function $p(\underline{z}, t)$ of the Markov diffusion process is characterized by the Fokker-Planck equation,

$$(6.3) \qquad \frac{\partial p}{\partial t} = -\sum_{i=1}^{2} \frac{\partial}{\partial z_i}[f_i(z)p] + \frac{1}{2}\sum_{i=1}^{2}\sum_{j=1}^{2} \frac{\partial^2}{\partial z_i \partial z_j}[\sigma_{ij}p]$$

where σ_{ij} is (i, j)-th element of the variance-covariance matrix of $\underline{w}(t)$. The marginal distribution of the process $\underline{z}(t)$ is given by $p(\underline{z}, t)$ with $t \to \infty$, which is obtained as

$$(6.4) \qquad p(\underline{z}, \infty) = p_0 \exp(-\frac{a\dot{x}^2}{\sigma^2})\exp\{-\frac{abx^2}{\sigma^2} - \frac{ac(x - x_0)_+^4}{2\sigma^2}\}$$

by solving the Fokker-Planck equation (6.3) with $\partial p / \partial t = 0$ ([4]). p_0 is a normalizing constant. We see from (6.4) that the tail of the distribution of x decreases faster than the Gaussian case for $x > x_0$ because of the presence of the nonlinear term $c(x - x_0)_+^3$ in the original dynamical system of (6.2). This explains the asymmetric shape of the histogram of the surge data of Fig. 2.2 where the left hand side of the tail decreases faster and is cut off compared with the right hand side tail. It is interesting to see that the distribution of the velocity is not affected by the nonlinear term and is Gaussian distributed.

The above consideration suggests that the nonlinear dynamics specified by $c(x - x_0)_+^3$ also specifies the non-Gaussianity of the distribution of $x(t)$. Then a natural question may be "Does the non-Gaussian distribution of $x(t)$ specify the nonlinear dynamics of the process?" The answer is "No" which comes from further analysis of the relationship between distribution systems and diffusion processes.

6.2. Pearson System and Markov Diffusion Process. Wong [34] pointed out an interesting relationship between the Pearson system and Markov diffusion process. He showed that for any distribution $W(x)$ which belongs to the Pearson system,

$$\text{(6.5)} \qquad \frac{dW(x)}{dx} = \frac{c_0 + c_1 x}{d_0 + d_1 x + d_2 x^2} W(x)$$

it is possible to give a diffusion process whose marginal distribution is $W(x)$. Ozaki [16] showed that it can be extended to any distribution $W(x)$ defined by proper analytic functions $c(x)$ and $d(x)$ in a distribution system given by,

$$\text{(6.6)} \qquad \frac{dW(x)}{dx} = \frac{c(x)}{d(x)} W(x)$$

and also showed that the corresponding Markov diffusion process is given, with $c(x)$ and $d(x)$, by the following Fokker-Planck equation,

$$\text{(6.7)} \qquad \frac{\partial p}{\partial t} = -\frac{\partial}{\partial x}[\{c(x) + d'(x)\}p] + \frac{1}{2}\frac{\partial^2}{\partial x^2}[2d(x)p].$$

The class of distributions defined by the system (6.6) is very wide indeed, including the Pearson system and the exponential family of distributions. The distribution system (6.6) is sometimes called the Extended Pearson system.

From Markov diffusion theory ([13]) we know that the diffusion process $x(t)$ defined by (6.7) has a stochastic differential equation representation,

$$\text{(6.8)} \qquad \dot{x} = a(x) + \sqrt{b(x)}n(t)$$

where

$$a(x) = c(x) + d'(x),$$
$$b(x) = 2d(x),$$

in the Ito form of stochastic calculus. In Stratonovich form they are,

$$a(x) = c(x) - \frac{1}{2}d'(x)$$

$$b(x) = 2d(x).$$

From (6.8) we have,

(6.9)
$$\frac{\dot{x}}{\sqrt{b(x)}} = \frac{a(x)}{\sqrt{b(x)}} - \frac{1}{4}\frac{b'(x)}{\sqrt{b(x)}} + n(t).$$

With the variable transformation,

$$y = \int^x \frac{1}{\sqrt{b(\xi)}}d\xi$$

we can replace (6.8) by a dynamical system driven by a Gaussian white noise $n(t)$ as follows.

(6.10)
$$x = h(y)$$

$$\dot{y} = f(y) + n(t)$$

where

$$f(y) = \frac{a(x)}{\sqrt{b(x)}} - \frac{1}{4}\frac{b'(x)}{\sqrt{b(x)}}$$

Since the white noise in (6.10) is free from x, the representation (6.10) is uniquely given and coincides for both Ito calculus and Stratonovich calculus.

The above consideration suggests that we can generate, with the locally linearized time series model,

(6.11)
$$x_t = h(y_t)$$

$$y_{t+\Delta t} = A_t y_t + B_t n_{t+\Delta 1},$$

a time series with a marginal distribution which is close to any density distribution belonging to the Extended Pearson system defined by (6.6). We can confirm this in simulated time series of these diffusion processes and their histograms shown in Fig. 6.1.

The simulated time series in Fig.6.1 are obtained using the local linearization method in section 3 applied to examples shown below.

Example 6.1. $W(x) = W_0 \exp\{-6x^2 + \frac{11}{4}x^4 + \frac{1}{3}x^6\}$

$\dot{x} = -6x + 5.5x^3 + x^5 + n(t)$

Example 6.2. $W(x) = \frac{\Gamma(\alpha+1)}{\Gamma(\frac{1}{2})\Gamma(\alpha)}(1+x^2)^{-(\alpha+1/2)}$

$x = \sinh\sqrt{2}y$

$\dot{y} = -\sqrt{2}\alpha \tanh\sqrt{2}y + n(t)$

Example 6.3. $W(x) = \frac{\Gamma(\alpha+\gamma+2)}{\Gamma(\alpha+1)\Gamma(\gamma+1)}\frac{(1+x)^\alpha(1-x)^\gamma}{2^{\alpha+\gamma+1}}$

$x = \sin\sqrt{2}y$

$\dot{y} = \frac{\alpha-\gamma}{\sqrt{2}}\frac{1}{\cos\sqrt{2}y} - \frac{\alpha+\gamma+1}{\sqrt{2}}\tan\sqrt{2}y + n(t)$

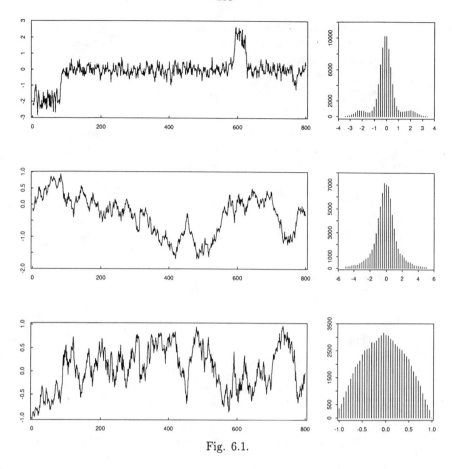

Fig. 6.1.

Example 6.1 is a process with a trimodal marginal distribution. A part of simulated time series (N=800) is shown in Fig.6.1.a. We can see the trimodal histogram from the simulated series (N=80,000) in Fig.6.1.a. Example 6.2 is a process with the stable marginal distribution. The stable distribution is equivalent to the Cauchy distribution when $\alpha = 1/2$. A part of simulated time series (N=800) is shown in Fig.6.1.b, where we used $\alpha = 1/2$. The histogram of the simulated series (N=80,000) shows long tails which are typical for the Cauchy distribution. A part of simulated time series (N=800) is in Fig.6.1.c, where we used $\alpha = 1$ and $\gamma = 1$. The bell shaped histogram is shown in Fig.6.1.c.

The representation (6.10) suggests that any Markov diffusion process has a stochastic dynamical system representation associated with an instantaneous variable transformation. It means that any non-Gaussian time series data which comes from a diffusion process can be whitened into a Gaussian white noise process using (6.11), i.e. by an instantaneous variable transformation, $y_t = h^{-1}(x_t)$, and a nonlinear dynamic model for y_t.

6.3. Different Dynamics with Same Distribution. We note that $c(x)$ and $d(x)$ of the Extended Pearson system (6.6) need not to be mutually irreducible. We can think of a diffusion process which corresponds to

$$(6.12) \qquad \frac{dW(x)}{dx} = \frac{c(x)x}{d(x)x}W(x),$$

from which we have a diffusion process defined by the following Fokker-Planck equation,

$$(6.13) \qquad \frac{\partial p}{\partial t} = -\frac{\partial}{\partial x}[\{c(x)x + d(x) + d'(x)x\}p] + \frac{1}{2}\frac{\partial^2}{\partial x^2}[2d(x)xp]$$

The diffusion process has the same marginal distribution as the diffusion process of (6.7). For example the density $W(x)$ of Gamma distribution is given by

$$W(x) = \frac{x^{\alpha-1}\exp(-\frac{x}{\beta})}{\Gamma(\alpha)\Gamma(\beta)}.$$

Its Pearson system form is

$$(6.14) \qquad \frac{dW(x)}{dx} = \frac{(\alpha-1)\beta - x}{\beta x}W(x).$$

A diffusion process for the Gamma distribution with this Pearson system is given by Wong [34] by

$$(6.15) \qquad \frac{\partial p}{\partial t} = -\frac{\partial}{\partial x}[(\alpha\beta - x)p] + \frac{1}{2}\frac{\partial^2}{\partial x^2}[2\beta xp]$$

If we multiply the numerator and denominator of (6.14) by x we obtain

$$(6.16) \qquad \frac{dW(x)}{dx} = \frac{(\alpha-1)\beta x - x^2}{\beta x^2}W(x)$$

and we have another diffusion process defined by

$$(6.17) \qquad \frac{\partial p}{\partial t} = \frac{\partial}{\partial x}[\{(\alpha+1)\beta x - x^2\}p] + \frac{1}{2}\frac{\partial^2}{\partial x^2}[2\beta x^2 p].$$

We call the process of (6.15) a type-I Gamma distributed process and the process of (6.17) a type-II Gamma distributed process. From (6.15) we have the following stochastic dynamical system representation with an instantaneous variable transformation,

$$(6.18) \qquad x = \frac{\beta y^2}{2}$$

$$\dot{y} = \frac{(\alpha\frac{1}{2})}{y} - \frac{y}{2} + n(t).$$

From (6.17) we have the following model,

$$(6.19) \qquad x = \exp(\sqrt{2\beta}y)$$

$$\dot{y} = \frac{\alpha\beta}{\sqrt{2\beta}} - \exp(\sqrt{2\beta}y) + n(t).$$

Figs.6.2 and Figs.6.3 show time series data simulated by our locally linearized models for Type-I Gamma processes and for Type-II processes for several different shape parameters. Their histograms are also in the figures. We can confirm from these examples that actually many different dynamics can produce the same distribution.

In time series data analysis the model (6.18) suggests taking the square root transformation of the positive valued time series data before fitting a nonlinear time series model, while the model (6.19) suggests taking the log-transformation before fitting a nonlinear time series model. It is interesting to see that the use of these variable transformations, commonly seen in empirical non-Gaussian time series data analysis, is suggested from our theoretical analysis. In some case (Type-I Gamma process with $\alpha = 1/2$) the process is transformed to a Gaussian process by a memoryless transformation, where a linear modelling is sufficient. This means that memoryless variable transformations are important for nonlinear time series models in non-Gaussian time series data analysis.

The above discussion shows that the introduction of common factors in the extended Pearson system (6.6) could lead us to infinitely many different representations of (6.11), a memoryless variable transformation with a nonlinear time series model, which produce one and the same non-Gaussian marginal distribution. A variety of different nonlinear dynamics for the same marginal distribution shows how differently a k-step ahead prediction distribution of each model approaches the marginal distribution for $k \to \infty$. Thus the identification of a non-Gaussian distribution character in time series is very much dependent on the identification of nonlinearities of a dynamical system and a variable transformation associated with the process which generates the time series.

6.4. Second Order Dynamics with a Given Distribution. The two examples of Gamma distributed processes in the previous section were generated by combinations of a memoryless variable transformation and a nonlinear dynamical system. In both cases the dynamical systems used are one dimensional, and cannot have a vibrating mechanism contributing to the existence of a peak in the spectrum of the process. A stochastic process having both vibration character

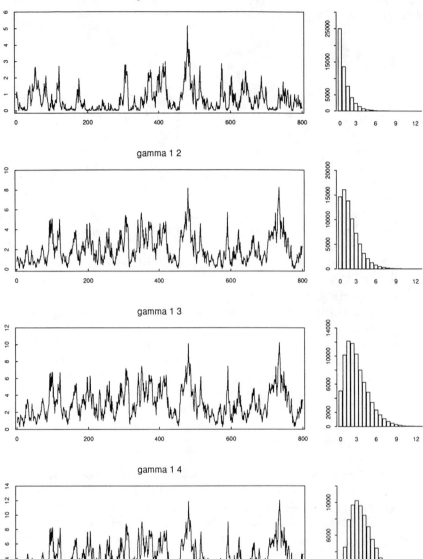

Fig. 6.2.

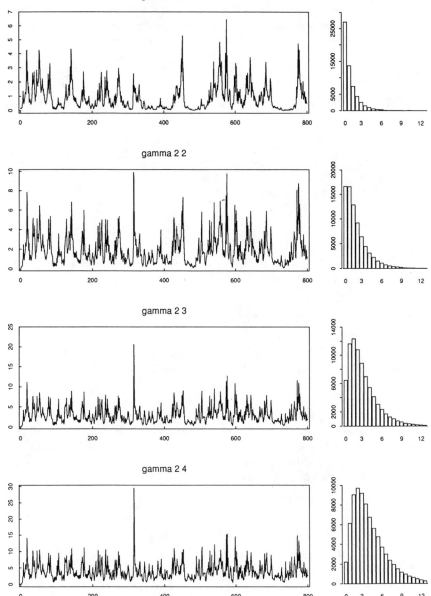

Fig. 6.3.

and non-Gaussian marginal distribution character is easily obtained by combining a stochastic differential equation model of (6.1) type and one of these models of (6.10) type. If a nonlinear random vibration model has a general nonlinear restoring function $b(x)$, instead of $bx + T(x)$ in (6.1) the model is given by

$$(6.20) \qquad \ddot{x}(t) + a\dot{x}(t) + b(x) = n(t),$$

where the variance of $n(t)$ is σ^2. The marginal distribution of x of (6.20) is given, in the same way as for (6.1), by

$$(6.21) \qquad p(\dot{x}, x) = p_0 \exp(-\frac{a\dot{x}^2}{\sigma^2}) \exp\{-\frac{2a \int^x b(\xi)d\xi}{\sigma^2}\}.$$

On the other hand we know that the marginal distribution of y of (6.10) is

$$p(y) = p_0 \exp\{\frac{2 \int^y f(\eta)d\eta}{\sigma^2}\}$$

Then if we replace $b(x)$ in model (6.20) by $f(.)$ of the dynamical system of the Type-II Gamma distributed process (6.19) in model (6.20), i.e. if we take

$$(6.22) \qquad \ddot{y} + a\dot{y} - \frac{\alpha\beta}{\sqrt{2\beta}} + \exp(\sqrt{2\beta}y) = n(t)$$

we have the same distribution $p(y)$ as the associated variable y of the Type-II Gamma process. By combining (6.22) and the variable transformation,

$$(6.23) \qquad x(t) = \exp\{\sqrt{2\beta}y(t)\}$$

we can have another Gamma-distributed process (Type-III Gamma-distributed process) $x(t)$ which has a vibration mechanism. Some of the Type-III Gamma distributed processes simulated by the locally linearized time series models with several shape parameter α's and their histograms are shown in Fig.6.4.

From (6.22) and (6.23) we have

$$(6.24) \qquad \ddot{x} + (a - \frac{\dot{x}}{x})\dot{x} + (\sqrt{2\beta}x - \alpha\beta)x = \sqrt{2\beta}xn(t),$$

which is equivalent to

$$(6.25) \qquad \ddot{x} + (a - \frac{\dot{x}}{x})\dot{x} + \{\sqrt{2\beta} - \alpha\beta - \sqrt{2\beta}n(t)\}x = 0.$$

The representation (6.25) implies not only a dynamical system driven by a Gaussian white noise but also a deterministic dynamical system whose coefficients are disturbed by a Gaussian white noise produce a random process. It implies that with chaos models not only an inexact initial value but also an inexact coefficient of a deterministic dynamical system model could produce to large future uncertainty.

261

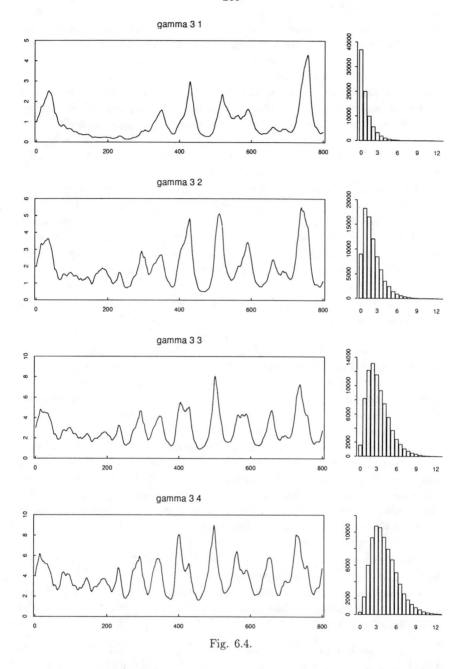

Fig. 6.4.

7. Conclusions. We have seen that Gaussian white noise still plays a very important role in nonlinear and/or non-Gaussian time series analysis as it does in the linear Gaussian case. Any diffusion process is generated from a dynamical system model driven by Gaussian white noise associated with a memoryless variable transformation. We have also seen that any time series generated from (possibly high dimensional) stochastic dynamical system can be transformed into Gaussian white innovation sequences with an identified model by maximizing the likelihood of innovations generated by the local linearization filter. It means any diffusion process is transformed into Gaussian white noise by a memoryless variable transformation and a nonlinear dynamic model. The idea of finding a whitening operation systematically used in the linear Gaussian case in the Box-Jenkins procedure is still valid in many cases of nonlinear and/or non-Gaussian time series analysis.

It has been also shown that there can be infinitely many different nonlinear dynamical system models which produce one and the same non-Gaussian marginal distribution. Each different dynamic produces different multi-step ahead prediction distributions although the infinite-step ahead prediction distributions have the same marginal distribution. Thus the identification of non-Gaussianity in time series reduces to choosing one out of many possible combinations of variable transformations and nonlinear time series models.

The fact that any non-Gaussian distributed diffusion processes are all generated from Gaussian white noise leads to a natural question "Is non-Gaussian white noise unnecessary in non-Gaussian time series analysis?". The answer is "Yes!" if the time series is generated from a process which is Markov and continuous in type, i.e. the transition probability from x to $R - (x - \epsilon, x + \epsilon)$ in Δt time goes to zero for $\Delta t \to 0$ where ϵ is a small value. This is because if the process is Markov and of continuous type then the process is a Markov diffusion process and has the Gaussian white noise representation (6.10). Non-Gaussian distributed time series does not necessarily mean the presence of non-Gaussian white noise. A nonlinear time series model driven by non-Gaussian white noise may look more general than a nonlinear time series model driven by a Gaussian white noise, but actually in many cases it is an unnecessary generalization. However, in real applications, we sometime have time series data which come from a continuous type Markov diffusion process contaminated by a Markov jump process, where we need to consider some non-Gaussian white noise with a fat tail distribution. Thus the non-Gaussian generalization of white noise in nonlinear time series models is justified in the sense of robustification of the whitening method for the identification of the associated nonlinear dynamics and variable transformation.

Our final conclusion is "Nonlinear and non-Gaussian time series analysis can be less confusing and simple if we go back to Wiener's idea, **Explain time series with Gaussian white noise**".

Acknowledgement. The author would like to thank Prof. Y. Yamanouchi and D. H. Oda for useful discussions and for providing him with the data of Fig. 2.1.

REFERENCES

[1]   AKAIKE, H., *On entropy maximization principle*, in *Application of Statistics*, P.R. Krishnaiah,ed., North Holland, 1977, pp. 27–41.

[2] ARNOLD, L., *Stochastic Differential Theory and Applications*, John Wiley & Sons, 1974.

[3] BOX, G.E.P. AND JENKINS, G.M., *Time Series Analysis, Forecasting and Control*, Holden-Day, San Francisco, 1970.

[4] CAUGHEY, T.K., *Derivation and application of the Fokker-Planck equation to discrete non-linear dynamic systems subjected to white random excitation*, J. Acoust. Soc. Amer. 35 (1963), pp. 1683–1692.

[5] FROST, P.A. AND KAILATH, T., *An innovation approach to least squares estimation-part III: nonlinear estimation in white Gaussian noise*, IEEE Trans. Automatic Control, Vol. AC-16 (1971), pp. 217–226.

[6] HARVEY, A.C., *A unified view of statistical forecasting procedures*, Journal of Forecasting, vol. 3 (1984), pp. 245–275.

[7] JAZWINSKI, A.H., *Stochastic Processes and Filtering Theory*, Academic Press, New York, 1970.

[8] JONES, D.A., *Nonlinear autoregressive processes*, Proc. Roy. Soc. London Ser. A 360 (1978), pp. 71–95.

[9] KITAGAWA, G., *A nonstationary time series model and its fitting by a recursive filter*, J. Time Ser. Anal. 2 (1981), pp. 103–116.

[10] LIANG, D.F., *Comparisons of nonlinear recursive filters for systems with nonnegligible non-linearities*, in *Control and Dynamical Systems, Advances in Theory and Applications, Vol 20, Nonlinear and Kalman Filtering Techniques* C.T. Leondis Eds. Academic Press, New York, 1983, pp. 341–401.

[11] MCSHANE, E.J., *Stochastic Calculus and Stochastic Models*, Academic Press, 1974.

[12] MILSHTEIN, G.N., *Approximate integration of stochastic differential equations*, Theory Prob. Appl. 23 (1974), pp. 396–401.

[13] MORTENSEN, R.E., *Mathematical problems of modeling stochastic dynamic systems*, J. Statist. Phys. 1 (1979), pp. 271–296.

[14] OZAKI, T., *Nonlinear time series models for nonlinear random vibrations*, J. Appl. Probab. 17 (1980), pp. 84–93.

[15] ————, *Nonlinear threshold autoregressive models for nonlinear random vibrations*, J. Appl. Probab. 18 (1981), pp. 443–451.

[16] ————, *Nonlinear time series models and dynamical systems*, in *Handbook of Statistics, Vol 5*, Hannan, E.J. et al Ed., North-Holland yr 1985, pp. 25–83.

[17] ————, *Statistical identification of storage models with application to stochastic hydrology*, Water Resources Bulletin, 21 (1985), pp. 663–675.

[18] ————, *Local Gaussian modelling of stochastic dynamical systems in the analysis of nonlinear random vibrations*, in *Essays on Time Series and Allied Processes, Festschrift in honour of Prof. E.J. Hannan*, Applied Probability Trust, 1986, pp. 241–255.

[19] ————, *Statistical identification of nonlinear random vibration systems*, J. of Appl. Mechanics, vol 56 (1989), pp. 186–191.

[20] ————, *A bridge between nonlinear time series models and nonlinear stochastic dynamical systems: a local linearization approach*, Statistica Sinica, vol. 2., 1992 (1990) (to appear).

[21] ————, *A local linearization approach to nonlinear filtering*, Research Memo. No.378, Institute of Statistical Mathematics, Toyko (1990).

[22] OZAKI, T. AND OZAKI, V.H., *Statistical identification of nonlinear dynamics in macroeconomics using nonlinear time series models*, in *Statistical Analysis and Forecasting of Economic Structural Change*, P. Hackl Eds., Springer-Verlag, 1989, pp. 345–365.

[23] PRIESTLY, M.B., *Non-linear and Non-stationary Times Series Analysis*, Academic Press, London, 1988.

[24] PUTHOFF, H., *Everything for nothing*, New Scientist, No 1727, 28, July, (1990), pp. 36–39.

[25] ROSLER, O.E., *Chaos*, in *Structure Stability in Physics*, W. Guttinger and H. Eikemeter eds, Springer, 1979.

[26] RUMELIN, W., *Numerical treatment of stochastic differential equations*, SIAM J. Numer. Anal. Vol 19 (1982), pp. 604–613.

[27] SCHWEPPE, H., *Evaluation of likelihood functions for Gaussian signals*, IEEE Trans. Automatic Control Vol. AC-11 (1965), pp. 61–70.

[28] SORENSON, H.W., *Kalman filtering technique*, Academic Press, in, 1966, pp. 219–292.

[29] STEWART, I., *Does God Play Dice? The Mathematics of Chaos*, Oxford: Basil Blackwell, 1989.

[30] TONG, H. AND LIM, K.S., *Threshold autoregression, limit cycles and cyclical data*, J. Roy. Statist. Soc. Ser. B 42 (1980), pp. 245–292.

[31] TWEEDIE, R.L., *Sufficient conditions for ergodicity and stationarity of Markov chains on a general state space*, Stoch. Proc. Their Appl. 3 (1975), pp. 385–403.

[32] WIENER, N. AND SIEGEL, A., *A new form for the statistical postulate of quantum mechanics*, Phys. Rev. 91 (1953), pp. 1551–1560.

[33] WIENER, N., SIEGEL, A. RANKIN, B. AND MARTIN, W.T., *Differential Space, Quantum Systems, and Prediction*, The MIT Press, 1966.

[34] WONG, E., *The construction of a class of stationary Markoff process*, Proc. Amer. Math. Soc. Symp. Appl. Math. 16, (1963), pp. 264–276.

TIME SERIES, STATISTICS, AND INFORMATION*

EMANUEL PARZEN**

Abstract. This paper is a broad survey of ideas for the future development of statistical methods of time series analysis based on investigating the many levels of relationships between time series analysis, statistical methods unification, and inverse problems with positivity constraints. It is hoped that developing these relations will: help integrate old and new directions of research in time series analysis; provide research tools for applied and theoretical statisticians in the 1990's and coming era of statistical information; make possible unification of statistical methods and the development of statistical culture. New results include a new information divergence between spectral density functions. Topics discussed include

1. Traditional entropy and cross-entropy,
2. Renyi and Chi-square information divergence,
3. Comparison density functions,
4. Approximation of positive functions (density functions) by minimum information divergence (maximum entropy),
5. Equivalence and orthogonality of Normal Time Series,
6. Asymptotic Information of Stationary Normal Time Series,
7. Estimation of Finite Parameter Spectral Densities,
8. Minimum information estimation of spectral densities and power index correlations,
9. Tail classification of probability laws and spectral densities,
10. Sample Brownian Bridge exploratory analysis of time series.

0. Introduction. The general level of the current relation (or non-relation) between statistics and time series analysis is: (1) many applied statisticians are ignorant about the theory of time series analysis, (2) many departments of statistics offer almost no courses in time series analysis (often relying on courses taught in economics or engineering departments), (3) theoretical statisticians traditionally have regarded time series analysis as safe to ignore because it is a "technical" subject in which it is difficult to confront basic issues of statistical inference which are the problem about which they want to do research.

Time series models are becoming of research interest to some theoretical statisticians whose primary research areas involve statistical analysis of data obeying the classical model of independent observations. They would like to investigate the extension of their work to data obeying probability models of dependence. This relation of time series analysis and statistics has a possibility of being superficial because it is completely methods-driven, rather than problem-driven. Consequently it may not handle problems that are of real interest to applied users of time series analysis. Questions of asymptotic rate of convergence of parameter estimators are technical problems which fail to treat basic problems (such as model identification and/or non-regular estimation for long-memory or non-Gaussian time series) that are usually the central problems in a time series analysis.

---

*Presented at Workshop in New Directions in Time Series Analysis at Institute for Mathematics and Its Applications, University of Minnesota on July 16, 1990, a day dedicated to honor John Tukey's contributions to Time Series Analysis. Research supported by U.S. Army Research Office.

**Department of Statistics, Texas A&M University, College Station, TX 77843-3143.

I propose that the narrow reasons why statisticians should learn about the methods of time series analysis (they are important for applications and many potential clients have time series problems) should be supplemented by broad information age reasons; the development of statistical culture requires that statisticians should learn about the theory of time series because it will help them improve their mastery of the basic methods of statistical analysis for traditional data consisting of independent observations. The theory of time series analysis needs to become *exoteric* (belonging to the outer or less initiate circle) as well as *esoteric*.

My concept of Statistical Culture (Parzen (1990)) proposes that a statistical analysis should aim to provide not a single answer but a choice of answers (answers by several methods for the same problem). Therefore a framework for comparing answers is required. A framework should also provide ways of thinking for classical problems that extends to as many modern problems as possible. This paper discusses: the basic ideas of a unification of statistical methods in terms of information concepts, the relation between time series and statistics in terms of their relations with information statistics, a framework for time series methods in terms of information concepts, and suggestions for research problems in time series analysis. The references aim to include many influential papers on information methods in statistics; additional references are warmly solicited.

An aim of this paper is to stimulate discussion of the mind-boggling discovery which appears to be emerging in modern statistical research and which I call I. O. U. (Information, Optimization, Unification). It appears that one can find a common type of optimization and approximation problem that provides a link among almost all classical and modern statistical analysis problems! An inner product $\langle f, g \rangle$ is the integral of the product fg. Let the information about an unknown NON-NEGATIVE function f be the values of linear functionals which are inner products of f with specified score functions J_k: for $k = 1, \ldots, m$,

$$\langle f, J_k \rangle = \tau_k$$

for specified constants τ_k called "moment parameters". Find a NON-NEGATIVE function, denoted

$$f^\wedge \text{ or } f^\wedge(\tau_1, \ldots, \tau_m),$$

which among all functions satisfying the above constraints minimizes an information divergence criterion (which includes as a special case maximum entropy). This problem is called an inverse problem with positivity constraints.

We favor Renyi information criteria which imply that $f^\wedge$ has a representation f_θ: for suitable index λ and parameters θ_k (which we call inverse parameters)

$$(f_\theta^\lambda - 1)/\lambda = \sum_{k=1}^{m} \theta_k J_k.$$

The inverse parameters θ are functions of the moment parameters and obtained by solving $\langle f_\theta, J_k \rangle = \tau_k$.

Uncertainty (probability and statistics) enters the picture because one observes a raw random function (denoted $f^\sim$) or at least its inner products

$$\langle f^\sim, J_k \rangle = \tau_k^\sim, \quad k = 1, \ldots, m.$$

One then seeks $f^\wedge$ which is non-negative and minimizes the specified information criterion among all functions satisfying $\langle f, J_k \rangle = \tau_k^\sim$, $k = 1, \ldots, m$. Among new data analytic tools that are open problems for research are "profile functions", defined as the minimum value of the criterion as a function of the moment parameters.

The problems that need to be solved to apply the foregoing approach to unifying statistical methods include

(1) introducing suitable density function $d(u)$, $0 < u < 1$, whose estimation underlies conventional problems,

(2) determining sufficient statistics $J_k(u)$ whose inner products with $d(u)$ or $d^\lambda(u)$ for some power λ, are regarded as most significantly different from zero and therefore provide the constraints on the unknown d,

(3) determining information measures whose index λ provides a parameter formula for d of the form $d^\lambda(u)$ is a linear combination of known functions $J_k(u)$ with coefficients θ_k to be estimated,

(4) developing and implementing algorithms to compute the solutions of the optimization problems.

Other aspects of unification of time series analysis methods were discussed in Parzen (1958), Parzen (1961), Parzen (1965), Parzen (1971), and Parzen (1974).

1. Traditional Entropy and Cross-Entropy. The (Kullback-Liebler) information divergence between two probability distributions F and G is defined (Kullback (1959)) by (our definitions differ from usual definitions by a factor of 2)

$$I(F;G) = (-2) \int_{-\infty}^{\infty} \log\{g(x)/f(x)\} f(x) dx$$

when F and G are continuous with probability density functions $f(x)$ and $g(x)$; when F and G are discrete, with probability mass functions $p_F(x)$ and $p_G(x)$, information divergence is defined by

$$I(F;G) = (-2) \sum \log\{p_G(x)/p_F(x)\} p_F(x).$$

An information decomposition of information divergence is

$$I(F;G) = H(F;G) - H(F),$$

in terms of entropy $H(F)$ and cross-entropy $H(F;G)$;

$$H(F) = (-2) \int_{-\infty}^{\infty} \{\log f(x)\} f(x) dx,$$

$$H(F;G) = (-2) \int_{-\infty}^{\infty} \{\log g(x)\} f(x) dx.$$

2. Renyi and Chi Square Information. Adapting the fundamental work of Renyi (1961) this section offers a new definition of Renyi information of index λ. For continuous F and G: for $\lambda \neq 0, -1$

$$IR_\lambda(F;G) = \frac{2}{\lambda(1+\lambda)} \log \int \left\{ \frac{g(y)}{f(y)} \right\}^{1+\lambda} f(y) dy$$

$$= \frac{2}{\lambda(1+\lambda)} \log \int \left\{ \frac{g(y)}{f(y)} \right\}^{\lambda} g(y) dy$$

$$IR_0(F;G) = 2 \int \left\{ \frac{g(y)}{f(y)} \log \frac{g(y)}{f(y)} \right\} f(y) dy$$

$$= 2 \int \left\{ \frac{g(y)}{f(y)} \log \frac{g(y)}{f(y)} - \frac{g(y)}{f(y)} + 1 \right\} f(y) dy$$

$$IR_{-1}(F;G) = -2 \int \left\{ \log \frac{g(y)}{f(y)} \right\} f(y) dy$$

$$= -2 \int \left\{ \log \frac{g(y)}{f(y)} - \frac{g(y)}{f(y)} + 1 \right\} f(y) dy$$

An analogous definition holds for discrete F and G.

The second definition provides: (1) extensions to non-negative functions which are not densities, and also (2) a non-negative integrand which can provide diagnostic measures at each value of y.

Renyi information, for $-1 < \lambda < 0$, is equivalent to Bhattacharyya distance (Bhattacharyya (1943)).

In addition to Renyi information divergence (an extension of information statistics) one uses as information divergence between two non-negative functions an extension of chi-square statistics which has been developed by Read and Cressie (1988). For $\lambda \neq 0, -1$, Chi-square divergence of index λ is defined for continuous F and G by

$$C_\lambda(F;G) = \int B_\lambda \left(\frac{g(y)}{f(y)} \right) f(y) dy$$

where

$$B_\lambda(d) = \frac{2}{(1+\lambda)} \left\{ d \left(\frac{d^\lambda - 1}{\lambda} \right) - d + 1 \right\}$$

$$B_0(d) = 2\{d \log d - d + 1\}$$

$$B_{-1} = -2\{\log d - d + 1\}$$

Important properties of $B_\lambda(d)$ are:

$$B_\lambda(d) \geq 0, B_\lambda(1) = B'_\lambda(1) = 0,$$

$$B'_\lambda(d) = \frac{2}{\lambda}(d^\lambda - 1), B''_\lambda(d) = 2d^{\lambda-1}$$

$$B_1(d) = (d-1)^2$$

$$B_0(d) = 2(d \log d - d + 1)$$

$$B_{-.5}(d) = 4(d^{.5} - 1)^2$$

$$B_{-1}(d) = -2(\log d - d + 1)$$

$$B_{-2}(d) = d(d^{-1} - 1)^2$$

An analogous definition holds for discrete F and G. Axiomatic derivations of information measures similar to C_λ are given by Jones and Byrne (1990).

The Renyi information and chi-square divergence measures are related:

$$IR_0(F;G) = C_0(F;G)$$
$$IR_{-1}(F;G) = C_{-1}(F;G)$$

For $\lambda \neq 0, -1$,

$$IR_\lambda(F;G) = \frac{2}{\lambda(1+\lambda0)} \log\left\{1 + \left(\frac{\lambda(1+\lambda)}{2}\right) C_\lambda(F;G)\right\}$$

Interchange of F and G is provided by the Lemma:

$$C_\lambda(F;G) = C_{-(1+\lambda)}(G;F)$$
$$IR_\lambda(F;G) = IR_{-(1+\lambda)}(G;F)$$

Our survey in this paper suggests (in section 6) a new class of information measures:

$$A_\lambda(F;G) \int A_\lambda\left(\frac{g(y)}{f(y)}\right) f(y)dy$$

For $\lambda \neq 0, -1$, perhaps most usefully $-1 < \lambda < 0$,

$$A_\lambda(d) = \{2/\lambda(1+\lambda)\}\{(1+\lambda)\log d - \log\{1 + (1+\lambda)(d-1)\}_+\}$$

Note $A_{-1}(F;G) = IR_{-1}(F;G)$,

$$A_0(d) = 2\{\log d - 1 + (1/d)\} = (1/d)B_0(d).$$

3. Comparison Density Functions. Information divergence $I(F;G)$ is a concept that works for both multivariate and univariate distributions. This section shows that the univariate case is distinguished by the fact that we are able to relate $I(F;G)$ to the concept of comparison density $d(u;F,G)$.

Quantile domain concepts introduced in Parzen (1979) play a central role; $Q(u) = F^{-1}(u)$ is the quantile function. When F is continuous, we define the density quantile function $fQ(u) = f(Q(u))$, score function $J(u) = -(fQ(u))'$, and quantile density function

$$q(u) = 1/fQ(u) = Q'(u).$$

When F is discrete, we define $fQ(u) = p_F(Q(u))$, $q(u) = 1/fQ(u)$.

The comparison density $d(u;F,G)$ is defined as follows: when F and G are both continuous,

$$d(u;F,G) = g(F^{-1}(u))/f(F^{-1}(u));$$

when F and G are both discrete

$$d(u : F, G) = p_G(F^{-1}(u))/p_F(F^{-1}(u)).$$

In the continuous case $d(u; F, G)$ is the derivative of

$$D(u; F, G) = G(F^{-1}(u));$$

in the discrete case we define the comparison distribution function

$$D(u; F, G) = \int_0^u d(t; F, G)dt.$$

Let F denote the true distribution function of a continuous random variable Y. To test the goodness of fit hypothesis $H_0 : F = G$, one transforms to $W = G(Y)$ whose distribution function is $F(G^{-1}(u))$ and whose quantile function is $G(F^{-1}(u))$. The comparison density $d(u; F, G)$ and $d(u; G, F)$ are respectively the quantile density and the probability density of W.

For a density $d(u)$, $0 < u < 1$, Renyi information (of index λ), denoted $IR_\lambda(d)$, is nonnegative and measures the divergence of $d(u)$ from uniform density $d_0(u) = 1, 0 < u < 1$. It is defined

$$IR_0(d) = 2 \int_0^1 \{d(u) \log d(u)\} du = 2 \int_0^1 \{d(u) \log d(u) - d(u) + 1\} du$$

$$IR_{-1}(d) = -2 \int_0^1 \{\log d(u)\} du \quad = -2 \int_0^1 \{\log d(u) - d(u) + 1\} du$$

for $\lambda \neq 0$ or -1

$$IR_\lambda(d) = \{2/\lambda(1 + \lambda)\} \log \int_0^1 \{d(u)\}^{1+\lambda} du$$

$$= \{2/\lambda(1 + \lambda)\} \log \int_0^1 \left(\{d(u)\}^{1+\lambda} - (1 + \lambda)\{d(u) - 1\}\right) du.$$

To relate comparison density to information divergence we use the concept of Renyi information IR_λ which yields the important identity (and interpretation of $I(F; G)$!)

$$I(F; G) = (-2) \int_0^1 \log d(u; F, G) du$$

$$= IR_{-1}(d(u; F, G)) = IR_0(d(u; G, F)).$$

For a density $d(u)$, $0 < u < 1$, define

$$C_\lambda(d) = \int_0^1 B_\lambda(d(u))du.$$

The comparison density again unifies the continuous and discrete cases. One can show that for univariate F and G

$$C_\lambda(F, G) = C_\lambda(d(u; F, G))$$

4. Approximation of Positive Functions (Density Functions) by Minimum Information Divergence (Maximum Entropy). This section discusses how approximation theory provides models for comparison density functions. To a density $d(u)$, $0 < u < 1$, approximating functions are defined by constraining (specifying) the inner product between $d(u)$ and a specified function $J(u)$, called a score function. We often assume that the integral over $(0,1)$ of $J(u)$ is zero, and the integral of $J^2(u)$ is finite. A score function $J(u)$, $0 < u < 1$ is always defined to have the property that its inner product with $d(u)$, denoted

$$[J, d] = [J(u), d(u)] = \int_0^1 J(u)d(u)du,$$

is finite. The inner product is called a *component* or *linear detector*; its value is a measure of the difference between $d(u)$ and 1.

The question of which distributions to choose as F and G is often resolved by the following formula which evaluates the inner product between $J(u)$ and $D(u; F, G)$ as a moment with respect to G if $J(u) = \varphi(F^{-1}(u))$:

$$[\varphi(F^{-1}(u)), d(u; F, G)] = \int_{-\infty}^\infty \varphi(y)dG(y) = E_G[\varphi(Y)]$$

Often G is a raw sample distribution and F is a smooth distribution which is a model for G according to the hypothesis being tested.

We propose that non-parametric statistical inference and density estimation can be based on the same criterion functions used for parametric inference if one uses the minimum Renyi information approach to density estimation (which extends the maximum entropy approach); form functions $d^\wedge_{\lambda,m}(u)$ which minimize $IR_\lambda(d^\wedge(u))$ among all functions $d^\wedge(u)$ satisfying the constraints

$$[J_k, d^\wedge] = [J_k, d] \text{ for } k = 1, \ldots, m$$

where $J_k(u)$ are specified score functions. One expects $d^\wedge_{\lambda,m}(u)$ to converge to $d(u)$ as m tends to ∞, and $IR_\lambda(d^\wedge_{\lambda,m})$ to non-decreasingly converge to $IR_\lambda(d)$.

The case $\lambda = 1$ provides approximations in L_2 norm which are based on a sequence $J_k(u)$, $k = 1, 2, \ldots$, which is a complete orthonormal set of functions. If $d(u)$, $0 < u < 1$, is square integrable (equivalently, $IR_1(d)$ is finite) one can represent $d(u)$ as the limit of

$$d_m(u) = 1 + \sum_{k=1}^{m} [J_k, d] J_k(u), m = 1, 2, \ldots$$

When $\varphi_k(y)$, $k = 1, 2, \ldots$, is complete orthonormal set for $L_2(F)$, a density $g(y)$ can be approximated by

$$g_m(y) = f(y) \left\{ 1 + \sum_{k=1}^{m} E_G[\varphi_k(Y)] \varphi_k(y) \right\}$$

We call $d_m(u)$ a truncated orthogonal function (generalized Fourier) series.

An important general method of density approximation, called a weighted orthogonal function approximation, is to use suitable weights w_k to form approximations

$$d^*(u) = 1 + \sum_{k=1}^{\infty} w_k [J_k, d] J_k(u).$$

to $d(u)$. Often w_k depends on a "truncation point" m, and $w_k \to 1$ as $m \to \infty$.

Quadratic Detectors. To test $H_0 : d(u) = 1$, $0 < u < 1$, many traditional goodness of fit test statistics (such as Cramer-von Mises and Anderson-Darling) can be expressed as quadratic detectors

$$\sum_{k=1}^{\infty} \{w_k [J_k, d]\}^2 = \int_0^1 \{d^*(u) - 1\}^2 du = C_1(d^*) = -1 + \exp IR_1(d^*).$$

We propose that these nonadaptive test statistics should be expressed as information measures and compared with minimum Renyi information detectors $IR_\lambda(d^\wedge_{\lambda,m})$; in this way information can provide unification of statistical methods.

Maximum entropy approximators correspond to $\lambda = 0$; $d^\wedge_{0,m}(u)$ satisfies an exponential model (whose parameters are denoted $\theta_1, \ldots, \theta_m$)

$$\log d^\wedge_{0,m}(u) = \sum_{k=1}^{m} \theta_k J_k(u) - \Psi(\theta_1, \ldots, \theta_m)$$

where Ψ is the integrating factor that guarantees that $d^\wedge_{0,m}(u)$, $0 < u < 1$, integrates to 1:

$$\Psi(\theta_1, \ldots, \theta_m) = \log \int \exp \left\{ \sum_{k=1}^{m} \theta_k J_k(u) \right\} du$$

The approximating functions formed in practice are not computed from the true components $[J_k, d]$ but from raw estimators $[J_k, d^\sim]$ for suitable raw estimators $d^\sim(u)$. The approximating functions are interpreted as estimators of a true density. Methods proposed for unification and generalization of statistical methods use minimum Renyi information estimation techniques. Different applications of these methods differ mainly in how they define the raw density $d^\sim(u)$ which is the starting point of the data analysis.

5. Equivalence and Orthogonality of Normal Time Series. This section formulates in terms of Renyi information some classic results of the theory of time series that should be part of the education of Ph.D.'s in statistics.

To apply Neyman Pearson statistical inference to a time series $\{Y(t), t \in T\}$ with abstract index set T, we must first define the probability density functional, denoted $p(Y(\cdot); \theta)$. We assume a family of probability models (for the time series) parametrized by a possibly infinite dimensional parameter θ.

The model "$Y(\cdot)$ is normal with known covariance kernel $K(s,t) = \text{cov}\ \{Y(s), Y(t)\}$ and unknown mean value function $m(t) = E[Y(t)]$" is equivalent to a probability measure P_m on the function space R_T of all functions on T. We define $p(Y(\cdot); m)$ to be the Radon-Nikodym derivative of P_m with respect to P_0, the probability measure corresponding to $m(t) = 0$ for all t.

THEOREM (Parzen (1958)). *In order that P_m be absolutely continuous with respect to P_0 it is necessary and sufficient that m is in $H(K)$, the reproducing kernel Hilbert space of functions on T with reproducing kernel K and inner product between functions f and g in $H(K)$ denoted $\langle f, g \rangle_{H(K)}$ and satisfying*

$$\langle f, K(\cdot, t) \rangle_{H(K)} = f(t)$$

for every f in $H(K)$ and t in T.

The probability measures P_m and P_0 of normal time series are either equivalent or orthogonal; when they have the same covariance kernel K and mean value function differing by m they are orthogonal if and only if $\|m\|_{H(K)} = \infty$, and they are equivalent if and only if $\|m\|_{H(K)}$ is finite [which is equivalent to m is a member of $H(K)$] and

$$\log p(Y(\cdot); m) = \langle Y, m \rangle_{\sim H(K)} - .5\|m\|_{H(K)}^2$$

The random variable $\langle Y, m \rangle_{\sim H(K)}$ is a "congruence inner product" (Parzen (1970); it is the linear combination of $\{Y(t), t \text{ in } T\}$ corresponding to m under the congruence which maps $K(\cdot, t)$ into $Y(t)$.

To compute the Renyi information of index λ let $\sigma = \|m\|_{H(K)}$, $\langle Y, m \rangle_{\sim H(K)} = \sigma Z$, where Z denotes a Normal $(0,1)$ random variable. Then

$$IR_\lambda(P_0, P_m) = \{2/\lambda(1 + \lambda)\} \log \text{ Int },$$

where

$$
\begin{aligned}
\text{Int } &= \int_{R_T} \{p(Y(\cdot); m)\}^{1+\lambda} dP_0 \\
&= E[\exp\{(1 + \lambda)(\sigma Z - .5\sigma^2)\}] \\
&= \exp\{.5(1 + \lambda)^2\sigma^2 - .5(1 + \lambda)\sigma^2\} \\
&= \exp\{.5\lambda(1 + \lambda)\sigma^2\}
\end{aligned}
$$

THEOREM. *Renyi information of two common covariance normal time series:*

$$IR_\lambda(P_0; P_m) = \sigma^2 = \|m\|_{H(K)}^2$$

This beautiful formula illustrates that our definition of Renyi information has been adjusted to be equivalent to a chi-squared statistic.

Method of proof uses limits of information numbers: To prove results about equivalence and orthogonality one studies the limit of information measures of finite dimensional restrictions $P_m^{(n)}$ representing probability measures of $Y(t)$, t in a finite set $T^{(n)}$ of points in T converging monotonely to a set $T^{(\infty)}$ dense in T. The norm of the restriction of m to $T^{(n)}$, denoted $\|m\|_{H(K,T^{(n)})}$, may be shown to converge to $\|m\|_{H(K)}$. For a time series $\{Y(t), t$ in $T\}$ with abstract index set T, the finite dimensional Renyi information is denoted

$$IR_\lambda^{(n)} = IR_\lambda \left(P_0^{(n)}; P_m^{(n)} \right)$$

THEOREM. *General martingale theory can be used to show that*
(1) P_0 and P_m are orthogonal if and only if $IR_{-.5}^{(n)}$ converges to ∞ as n increases,
(2) P_m is absolutely continuous with respect to P_0 if $IR_1^{(n)}$ has a finite limit as n increases.

PROPOSITION. *Renyi information divergence of two zero mean univariate normal distributions. Let P_j be the distribution on the real line corresponding to Normal $(0, K_j)$ with variance K_j. Let $p(y)$ denote the probability density of P_1 with respect to P_2. Let $\kappa = \frac{K_2}{K_1}$. Then*

$$p(y) = \kappa^{.5} \exp \left\{ -.5(\kappa - 1) \frac{y^2}{K_2} \right\}$$
$$IR_{-1}(P_2; P_1) = \kappa - 1 - \log \kappa,$$
$$IR_\lambda(P_2; P_1) = (1/\lambda)\{\log \kappa - (1 + \lambda)^{-1} \log\{1 + (1 + \lambda)(\kappa - 1)\}_+\}$$
$$C_\lambda(P_2; P_1) = \{2/\lambda(1 + \lambda)\}\kappa^{.5(1+\lambda)}\{1 + (1 + \lambda)(\kappa - 1)\}_+^{-.5}$$

Asymptotic information can be computed from the fact (compare Hannan (1970), p. 429)

$$\lim_{n\to\infty} \|m\|_{H(K,T^{(n)})}^2 / \sum_{t=1}^n m^2(t) = \int_0^1 \frac{1}{R(0)f(\omega)} dM(\omega)$$

where $M(\omega)$ is the asymptotic spectral distribution function of $M(\cdot)$ assuming that $m(\cdot)$ obeys Grenander's conditions, or is "persistently exciting" in the language of control engineers (Bohlin (1971)):

$$\rho_m(v) = \int_0^1 e^{2\pi i v\omega} dM(\omega),$$

where $\rho_m(v)$ is limit of sample autocorrelations of $m(\cdot)$.

6. Asymptotic Information of Stationary Normal Time Series. This section discusses unification of information measures of stationary normal time series and information measures of non-negative functions which are spectral density functions.

When a time series $\{Y(t), t = 1, 2, \dots\}$ is modeled by probability measures which are orthogonal over the infinite sequence but equivalent for any finite sample, we define asymptotic information divergence (or rate of information divergence)

$$\text{Asym } IR_\lambda(P_2; P_1) = \lim_{n \to \infty} (1/n) IR_\lambda(P_2^{(n)}; P_1^{(n)}).$$

Let $Y(\cdot)$ be zero mean normal stationary with covariance function

$$R(v) = E[Y(t)Y(t-v)].$$

The correlation function is defined

$$\rho(v) = R(v)/R(0).$$

We prefer to analyze the time series after first subtracting sample mean and dividing by its sample standard deviation; its covariance function asymptotically equals $\rho(v)$.

An important classification of time series is by memory type: no memory, short memory, long memory according as $I_\infty = 0$, $0 < I_\infty < \infty$, $I_\infty = \infty$ where

$$I_\infty = I(Y|Y_{-1}, Y_{-2}, \dots) = E_{Y_{-1}Y_{-2}, \dots} I\left(f_{Y|Y_{-1}, Y_{-2}, \dots}; f_Y\right)$$

is the information about $Y(t)$ in $Y(t-1)$, $Y(t-2)$, $\dots$, its infinite past (see Parzen (1981), (1983)).

Assume that $Y(\cdot)$ is short memory and satisfies

$$\sum_{v=-\infty}^{\infty} |R(v)| \text{ finite.}$$

The spectral density function $f(\omega)$, $0 \leq \omega < 1$, is defined as the Fourier transform of the correlation function:

$$f(\omega) = \sum_{v=-\infty}^{\infty} \exp(-2\pi i v \omega) \rho(v)$$

We call a time series *bounded* memory if the spectral density is bounded above and below:

$$0 < c_1 \leq f(\omega) \leq c_2 < \infty.$$

Let P_f denote the probability measure on the space of infinite sequences R_∞ corresponding to a normal zero mean stationary time series with spectral density function $f(\omega)$.

A result of Pinsker [(1964), p. 196] can be interpreted as providing a formula for asymptotic information divergence between two zero mean stationary time series with respective rational spectral density functions $f(\omega)$ and $g(\omega)$. Write Asym $IR_\lambda(f, g)$ for Asym $IR_\lambda(P_f; P_g)$. Adapting Pinsker (1964) one can prove that

$$\text{Asym } IR_{-1}(f, g) = \int_0^1 \{(f(\omega)/g(\omega)) - 1 - \log(f(\omega)/g(\omega))\} d\omega$$

The definition of Renyi information can be extended to non-negative functions $d(u)$ which do not necessarily integrate to 1. Because spectral densities are even functions we take the integral to be over $0 \leq \omega < .5$. One obtains the following important theorem.

THEOREM. *Unification of information measures of Pinsker (1964) and Itakura-Saito (1970).*

$$\text{Asym } IR_{-1}(f, g) = IR_{-1}(f(\omega)/g(\omega))_{0, .5}$$

The validity of this information measure can be extended to non-normal asymptotically stationary time series (Ephraim, Lev-Ari, Gray (1988)).

One can heuristically motivate Pinsker's information theoretic justification of the Itakura-Saito distortion measure by the formula (at the end of section 5) for the information divergence between two univariate normal distributions with zero means and different variances. Motivated by this formula we propose a formula for bounded memory time series (whose proof is given by Kazakos and Kazakos (1980)) which motivates a *new* distortion measure: Asym $IR_\lambda(f, g) =$

$$(1/\lambda) \int_0^1 \left\{ \log(f(\omega)/g(\omega)) - (1/(1+\lambda)) \log \left\{ 1 + (1+\lambda)((f(\omega)/g(\omega)) - 1\}_+ \right\} dw \right.$$

The properties of the integrand are the same as those of $B_\lambda(d)$.

Kazakos and Kazakos (1980) also give formulas for asymptotic information of multiple stationary time series.

7. Estimation of Finite Parameter Spectral Densities. This section formulates in terms of Renyi information the classic asymptotic maximum likelihood Whittle theory of spectral estimation.

For a random sample of a random variable with unknown probability density f, maximum likelihood estimates $\theta^\wedge$ of the parameters of a finite parameter model f_θ of the probability density f can be shown to be equivalent to minimizing

$$IR_{-1}(f^\sim, f_\theta)$$

where $f^\sim$ is a raw estimator of f (initially, a symbolic sample probability density formed from the sample distribution function $F^\sim$). A similar result, called Whittle's estimator (Whittle (1953)), holds for estimation of spectral densities of a bounded memory zero mean stationary time series for which one assumes a finite parametric model $f_\theta(\omega)$ for the true unknown spectral density $f(\omega)$.

A raw fully nonparametric estimator of $f(\omega)$ from a time series sample $Y(t)$, $t = 1, \ldots, n$, is the sample spectral density (or periodogram)

$$f^\sim(\omega) = \left| \sum_{t=1}^{n} Y(t) \exp(-2\pi i \omega t) \right|^2 \div \sum_{t=1}^{n} |Y(t)|^2$$

Note that $f^\sim(\omega)$ is not a consistent estimator of $f(\omega)$; nevertheless,

$$E[f^\sim(\omega)] \text{ converges to } f(\omega),$$

a fact which can be taken as the definition of $f(\omega)$.

An estimator $\theta^\wedge$ which is asymptotically equivalent to maximum likelihood estimator is obtained by minimizing Asym $IR_{-1}(f^\sim; f_\theta) = IR_{-1}(f^\sim, f_\theta)_{0,.5} =$

$$\int_0^1 \{(f^\sim(\omega)/f_\theta(\omega)) - 1 - \log(f^\sim(\omega)/f_\theta(\omega))\} \, d\omega$$

which can be interpreted as choosing θ to make $\frac{f^\sim(\omega)}{f_\theta(\Omega)}$ as flat or constant as possible.

We usually use the representation

$$f_\theta(\omega) = \sigma^2 / \gamma_\theta(\omega)$$

where $\gamma_\theta(\omega)$ is the square modulus of the transfer function of the whitening filter represented by the spectral density model f_θ, constructed so that

$$\int_0^1 \log f_\theta(\omega) d\omega = \log \sigma^2.$$

Minimizing Asym $IR_{-1}(f^\sim, f_\theta)$ is equivalent to minimizing

$$(1/\sigma^2) \int_0^1 \{f^\sim(\omega)\gamma_\theta(\omega)\} \, d\omega + \log(\sigma^2)$$

which is equivalent to minimizing over θ

$$\sigma_\theta^2 = \int_0^1 \gamma_\theta(\omega) f^\sim(\omega) d\omega$$

and setting

$$\sigma^{\wedge 2} = \int_0^1 \gamma_{\theta^\wedge}(\omega) f^\wedge(\omega) d\omega = \sigma_{\theta^\wedge}^2$$

The information divergence between the data and the fitted model is given

$$IR_{-1}(f^{\sim}, f_{\theta^{\wedge}}) = \log \sigma_{\theta^{\wedge}}^2 - \log \sigma^{\sim 2} = I_{\infty}^{\sim} - I_{\infty}^{\wedge}$$

defining $-I_{\infty}^{\wedge} = \log \sigma^{\wedge 2}$,

$$-I_{\infty}^{\sim} = \log \sigma^{\sim 2} = \int_0^1 \log f^{\sim}(\omega) d\omega$$

This criterion (however, corrected for bias in $I_{\infty}^{\sim}$) arises from information approaches to model identification (Parzen (1983)). A model fitting criterion (but not a parameter estimation criterion) is provided by the information increment

$$I(Y| \text{ all past } Y; Y \text{ values in model } \theta)$$

$$= \int_0^1 -\log\{f^{\sim}(\omega)/f_{\theta^{\wedge}}(\omega)\} d\omega = IR_{-1}(f^{\sim}/f_{\theta^{\wedge}})_{0,-.5}$$

One can regard it as a measure of the distance of the whitening spectral density

$$f^*(\omega) = f^{\sim}(\omega)/f_{\theta^{\wedge}}(\omega)$$

from a constant function; note that $f^*(\omega)$ is constructed to integrate to 1. When one accepts that the optimal smoother of $f^*(\omega)$ is a constant, a "parameter-free" non-parametric estimator of the spectral density $f(\omega)$ by a smoother of $f^{\sim}(\omega)$ is given by the parametric estimator f_{θ^*}. By "parameter-free" we mean that we are free to choose the parameters to make the data (raw estimator) shape up to a smooth estimator. The parameters are not regarded as having any significance or interpretation; they are merely coefficients of a representation of $f(\omega)$.

Portmanteau statistics to test goodness of fit of a model to the time series use sums of squares of correlations of residuals; an analogous statistic is

$$IR_1(f^{\sim}/f_{\theta^{\wedge}})_{0,.5} = \log \int_0^{.5} \{f^{\sim}(\omega)/f_{\theta^{\wedge}}(\omega)\}^2 d\omega$$

Goodness of fit of the model to the data (as measured by how close $f^*(\omega)$ is to the spectral density of white noise) is the ultimate model identification criterion to decide between competing parametric models.

8. Minimum Information Estimation of Spectral Densities. This section provides a perspective on maximum entropy spectral estimation from the point of view of minimum Renyi information approximation.

The maximum entropy approach to the problem of spectral estimation of a stationary time series was originated by Burg (1967). It derives a parametric model

f_θ for the true f by imposing constraints on linear functionals of f of the form: for $v = 0, 1, \ldots, m$

$$\rho(v) = \langle \exp(2\pi i v w), f(\omega) \rangle_{L_2(0,1)} = \rho^{\sim}(v)$$

where $\rho^{\sim}(v)$ are estimators (from a sample) of autocorrelations. Note that the remarkable properties of Burg's estimators derive from the fact that he first estimates in a novel way the partial correlations and should not be interpreted as proof of the superiority of maximum entropy philosophy.

Let $f_m^\wedge$ denote the function (among all functions f satisfying these constraints) which minimizes the neg-entropy (of order -1)

$$IR_{-1}(f)_{0,.5} = \int\limits_0^1 -\{\log f(\omega)\}d\omega$$

The solution $f_m^\wedge$ has the following parametric form:

$\{1/f_m^\wedge\}$ is linear combination of $\exp(2\pi i v \omega)$, $v = -m, \ldots, m$

The non-negativity of $f^\wedge$ then guarantees that $f^\wedge$ is an autoregressive spectral density:

$$f_m^\wedge(\omega) = \sigma_m^2 \left| \sum_{j=0}^m a(j) \exp(2\pi i j \omega) \right|^{-2}$$

A negative opinion about applying autoregressive spectral estimates is expressed by Diggle (1990), p. 112: "A final method, which we mention only briefly, is to fit an $AR(p)$ process to the data $\{y_t\}$ and to use the fitted autoregressive spectrum as the estimate of $f(\omega)$. The motivation for this is threefold: fitting an autoregressive process is computationally easy, autoregressive spectra can assume a wide variety of shapes, and automatic criteria are available for choosing the value of p. Nevertheless, the method seems to fit uneasily into a discussion of what is essentially a nonparametric estimation problem. It is analogous to the use of polynomial regression for data smoothing, and is open to the same basic objection, namely that it imposes global assumptions which can lead to artefacts in the estimated spectrum."

If one's criterion is to minimize

$$IR_0(f)_{0,.5} = \int\limits_0^1 \{f(\omega) \log f(\omega)\}d\omega,$$

the neg-entropy of order 0, the solution $f^\wedge$ obeys an exponential model (Bloomfield (1973)):

$\{\log f^\wedge\}$ is linear combination of $\exp(2\pi i v \omega)$, $v = -m, \ldots, m$

These optimization problems are related to the problem: subject to the constraints, with specified score functions $J_k(\omega)$,

$$\langle f, J_k \rangle = \text{specified constant for } k = 1, \ldots, m$$

find a density $f(\omega), 0 \leq \omega < 1$, minimizing the L_p norm, with $p = 1 + \lambda$,

$$\int_0^1 \{f(\omega)\}^{1+\lambda} d\omega$$

Theorems about this problem are given by Chui, Deutsch, and Ward (1990).

Power correlations, inverse and cepstral correlations: Parametric models for the spectral density f can be obtained from various maximum entropy criteria. To check which model is parsimonious, one requires goodness of fit procedures which check the significant difference from zero of the Fourier transforms of various functions of f such as $1/f$, $\log f$, or f^λ.

Let $e_v(\omega) = \exp(2\pi ij\omega)$, and interpret inner products as $L_2(0,1)$. Define: inverse correlations

$$\rho^{(-1)}(v) = \langle e_v, 1/f \rangle$$

cepstral correlations

$$\rho^{(0)}(v) = \langle e_v, \log f \rangle$$

ordinary correlations

$$\rho^{(1)}(v) = \langle e_v, f \rangle$$

power correlations of power λ

$$\rho^{(\lambda)}(v) = \langle e_v, f^\lambda \rangle$$

Identification of a parametric model for f should include routine estimation and interpretation of these various correlations.

In general if one expands $f^\lambda(\omega)$ as a linear combination of orthogonal functions $J_k(u)$, $0 \leq \omega < 1$, one forms the transform (called power orthogonal series coefficients)

$$\rho^{(J,\lambda)}(k) = \langle J_k, f^\lambda \rangle$$

An open research problem is identification of appropriate orthogonal functions $J_k(u)$, $0 \leq \omega < 1$.

9. Tail Classification of Probability Laws and Spectral Densities.
This section discusses models for spectral density functions which are based on their analogy with quantile density functions.

From extreme value theory, statisticians have long realized that it is useful to classify distributions according to their tail behavior (behavior of $F(x)$ as x tends to $\pm\infty$). It is usual to distinguish three main types of distributions, called (1) limited, (2) exponential, and (3) algebraic. Parzen (1979) proposes that this classification be expressed in terms of the density quantile function $fQ(u)$; we call the types short, medium, and long tail.

A reasonable assumption about the distributions that occur in practice is that their density-quantile functions are *regularly varying* in the sense that there exist tail exponents α_0 and α_1 such that, as $u \to 0$,

$$fQ(u) = u^{\alpha_0} L_0(u), \quad fQ(1-u) = u^{\alpha_1} L_1(u)$$

where $L_j(u)$ for $j = 0, 1$ is a slowly varying function.

A function $L(u)$, $0 < u < 1$ is usually defined to be *slowly varying* as $u \to 0$ if, for every y in $0 < y < 1$, $L(yu)/L(u) \to 1$ or $\log L(yu) - \log L(u) \to 0$. For estimation of tail exponents we will require further that, as $u \to 0$,

$$\int_0^1 \{\log L(uy) - \log L(u)\} dy \to 0$$

which we call *integrally slowly varying*. An example of a slowly varying function is $L(u) = \{\log u^{-1}\}^\beta$.

Classification of tail behavior of probability laws. A probability law has a left tail type and a right tail type depending on the value of α_0 and α_1. If α is the tail exponent, we define:

$$\alpha < 0 \quad \text{super short tail}$$
$$0 \le \alpha < 1 \quad \text{short tail}$$
$$\alpha = 1 \quad \text{medium tail}$$
$$\alpha > 1 \quad \text{long tail}$$

Medium tailed distributions are further classified by the value of $J^* = \lim J(u)$:

$$\alpha = 1, J^* = 0 \quad \text{medium-long tail}$$
$$\alpha = 1, 0 < J^* < \infty \quad \text{medium-medium tail}$$
$$\alpha = 1, J^* = 0 \quad \text{medium-short tail}$$

One immediate insight into the meaning of tail behavior is provided by the hazard function $h(x) = f(x) \div \{1 - F(x)\}$ with hazard quantile function $hQ(u) = fQ(u) \div 1 - u$. The convergence behavior of $h(x)$ as $x \to \infty$ is the same as that of $hQ(u)$ as $u \to 1$. From the definitions one sees that $h^* = \lim_{x \to \infty} h(x)$ satisfies

$$h^* = \infty \text{ (increasing hazard rate) Short or medium-short tail}$$
$$0 < h^* < \infty \text{ (constant hazard rate) Medium-medium tail}$$
$$h^* = 0 \text{ (decreasing hazard rate) Long or medium-long tail}$$

Formulas for computing tail exponents. The representation of $fQ(u)$ suggests a formula for computation of tail exponents α_0 and α_1 (which may be adapted to provide estimators from data):

$$-\alpha_0 = \lim_{u \to 0} \int_0^1 \{\log fQ(uy) - \log fQ(u)\} dy$$

$$\alpha_1 = \lim_{u \to 0} \int_0^1 \{\log fQ(1 - yu) - \log fQ(1 - u)\} dy$$

Memory classification of spectral densities: Spectral densities with no poles or zeroes represent time series with bounded memory. We regard spectral density functions as analogous to quantile density functions. A model for a spectral density with a pole or zero at zero frequency (a similar representation holds for an arbitrary frequency ω_0) is (Parzen (1986))

$$f(\omega) = \omega^{-\delta} L(\omega)$$

where L is a slowly varying function at $\omega = 0$ and $L(0) > 0$. An important role is played by $f(1/n) = n^\delta L(1/n)$.

The spectral density is integrable if $\delta < 1$, which is the condition for stationarity. The spectral density of a non-stationary time series needs careful definition. The case $\delta = 1$ is of particular interest; it corresponds to "$1/f$" noise. The case $\delta > 1$ could be called "fractal noise". A time series whose first difference is stationary has $\delta = 2$. Heuristically, δ is interpreted for a zero mean time series $Y(\cdot)$ by

$$E\left[\frac{1}{n}\left|\sum_{t=1}^n Y(t)\right|^2\right] \text{ grows as } n^\delta L(1/n).$$

The index δ associated with frequency ω is interpreted:

$$E\left[\frac{1}{n}\left|\sum_{t=1}^n \exp(2\pi i\omega t)Y(t)\right|^2\right] \text{ grows as } n^\delta L(1/n),$$

and, when $\delta = 0$, converges to $R(0)f(\omega)$ if it is finite.

This approach provides definitions of spectral density for asymptotically stationary time series (Parzen (1962)).

Note that a finite dynamic range (bounded memory) spectral density has $\delta = 0$, but $\delta = 0$ does not imply finite dynamic range since $f(\omega)$ can tend to ∞ as $\omega \to 0$; an example is $f(\omega) \sim (\log \omega)^2$, $\rho(v) \sim (\log v)/v$ as $v \to \infty$.

A traditional parametrization of stationary long memory time series is $\delta = 2H - 1$, where H is the Hurst index satisfying $.5 < H < 1$; Hurst estimated $H = .7$ for the Nile water level time series. The covariance function has the asymptotic representation

$$R(v) \text{ decays slowly like } v^{2H-2} = v^{\delta-1}.$$

The memory index delta plays an important theoretical role. In many time series theorems the asymptotic behavior of a statistic is expressed in terms of $f(0)$, the value at zero frequency of the spectral density function. These results often have analogies for long memory time series if one replaces $f(0)$ by $f(1/n) = n^\delta$ asymptotically; compare Samarov and Taqqu (1988).

Estimation of delta can be considered estimating a "fractal dimension", the exponent of the rate of growth of the mean of the sample spectral density. Values of delta are used to describe music and how the brain works! *U. S. News and World Report*, June 11, 1990, p. 62 writes: "Surprisingly, the same mathematical formula that characterizes the ebb and flow of music has been discovered to exist widely in nature, from the flow of the Nile to the bea ting of the human heart to the wobbling of the earth's axis. Remarkably, this equation is closely related to other mathematical formulas used by computer experts to generate amazingly realistic pictures of coastlines, clouds and mountain ranges and other natural scenery."

Estimating delta from data has many of the same difficulties as estimating the tail index of a probability distribution. Since delta is a property of a long memory time series, it undoubtedly can not be estimated with great accuracy from relative short lengths of observed time series.

10. Sample Brownian Bridge Exploratory Analysis of Time Series. To a time series sample $\{Y(t), t = 1, 2, \ldots, n\}$ one can associate functions $d^\sim(u)$, $0 \le u \le 1$, and

$$D^\sim(u) = \int_0^u d^\sim(t)dt$$

satisfying $D^\sim(1) = 0$. Let μ_n and σ_n denote respectively the sample mean and sample standard deviation. Define, for $j = 1, \ldots, n$,

$$d^\sim(u) = \sqrt{n}(Y(j) - \mu_n)/\sigma_n, \quad \frac{j-1}{n} < u < \frac{j}{n}$$

Note that for $k = 1, \ldots, n$

$$D^\sim\left(\frac{k}{n}\right) = \frac{1}{\sqrt{n}} \sum_{t=1}^{k} (Y(t) - \mu_n)/\sigma_n$$

We call $D^\sim(\cdot)$ the *sample Brownian Bridge* of an observed time series. We propose that a time series analysis should routinely examine the graph of $D^\sim(u)$, $0 \le u < 1$; one can show by examples that it provides graphical tools of identification of various types of long memory time series.

THEOREM. *For a stationary time series with bounded memory*

$$\{D^\sim(u), 0 \le u \le 1\} \text{ converges in distribution to } \{f^{.5}(0)B(u), 0 \le u \le 1\},$$

where $B(u)$, $0 < u < 1$, is a Brownian Bridge stochastic process and $f(0)$ is the spectral density at zero frequency.

For a long memory time series we would like to apply the theory of the asymptotic behavior of

$$\{f(1/n)\}^{-.5}D^{\sim}(u), \quad 0 \leq u \leq 1.$$

Simulation and Time Series: Note that similar processes are studied by researchers (Schruben, Iglehart) in operations research departments who study simulation methods of forming confidence intervals for μ, the true mean of $Y(\cdot)$; they standardize $D^{\sim}(u)$ by its maximum minus its minimum.

Quality Control and Time Series: The process $D^{\sim}(\cdot)$ also has applications to quality control problems of identifying departures from the null hypothesis that $Y(\cdot)$ is white noise. Components (linear functionals) of $d^{\sim}(\cdot)$ are related to accumulation analysis methods of Taguchi.

REFERENCES

AKAIKE, H., *Information theory and an extension of the maximum likelihood principle*, Proc. of the Second International Symposium on Information Theory, B. N. Petrov and F. Csaki, Akademiai Kiado, Budapest (1973), pp. 267–281.

AKAIKE, H., *A new look at the statistical model identification*, IEEE Trans. Autom. Contr., AC-19 (1974), pp. 716–723.

AKAIKE, H., *On entropy maximization principle*, Application of Statistics (Krishnaiah ed.), North-Holland (1977), pp. 27–41.

AKAIKE, H., *Canonical correlation analysis of time series and the use of an information criterion*, System Identification: Advances and Case Studies (R.K. Mehra and D.G. Lainiotics, eds.), Academic Press (1978), pp. 27–96.

AKAIKE, H., *Likelihood and the Bayes procedures*, Bayesian Statistics (Bernardo, J.M., De Groot, M.H., Lindley, D.U. and Smith, A.F.M. eds.). University Press: Valencia, Spain (1980).

AKAIKE, H., *Information measures and model selection*, Proc. International Statistical Institute (44th Session, Madrid) 1, (1983), pp. 227–291.

AKAIKE, H., *Statistical inference and measurement of entropy*, in Scientific Inference, Data Analysis, and Robustness, ed. G.E.P. Box and C.J. Wu, Academic Press: New York (1983).

ALI, S. M., SILVEY, S. D., *A general class of coefficients of divergence of one distribution from another*, J. Roy. Stat., 28 (1966), pp. 131–142.

ARIMOTO, S., *Informational-theoretical considerations on estimation problems*, Inf. and Control, 19 (1971), pp. 181–194.

BHATTACHARYYA, A., *On a measure of divergence between two statistical populations defined by their probability distributions*, Bull. Calcutta Math. Soc., Vol. 35 (1943), pp. 99–109.

BLOOMFIELD, P., *An exponential model for the spectrum of a scalar time series*, Biometrika, 60, No. 2 (1973), pp. 217–226.

BOHLIN, T., *On the problem of ambiguities in maximum likelihood identification*, Auotmatica, Vol. 7 (1971), pp. 199–210.

BOX, G. E. P. AND JENKINS, G. M., *Time Series Analysis, Forecasting and Control*, 2nd edition. Holden Day, San Francisco (1976).

BRILLINGER, D. R., *Time Series*, Holden Day, San Francisco (1981).

BROCKETT, P., CHARNES, A. AND COOPER, W. W., *MDI estimation via unconstrained convex programming*, Comm. Statist. B-Simulation Comput., 9 (1980), pp. 223–234.

BROCKWELL, P. J. AND DAVIS, R. A., *Time Series: Theory and Methods*, Springer-Verlag, New York (1987).

BURG, J. P., *Maximum entropy spectral analysis*, in Proc. 37th Meeting Soc. of Exploration Geophysicists. Oklahoma City, OK (1967).

CAINES, PETER E., *Linear Stochastic Systems*, Wiley, New York (1988).

CHARNES, A., COOPER, W. W. AND SEIFORD, L., *Extremal principles and optimization qualities for Khinchin-Kullback-Leibler estimation*, Math Operationsforsch, Statist., Vol. 9 (1978), pp. 21–29.

CHARNES, A., COOPER, W. W., AND TYSSEDAL, J., *Khinchin-Kullback-Leibler estimation with inequality constraints*, Math. Operationsforsch. Statis. Ser. Optim., 14 (1983), pp. 1–4.

CHUI, C. K., DEUTSCH, F., WARD, J. D., *Constrained best approximation in Hilbert space*, Constructive approximation, 67 (1990), pp. 35–64.

CSISZÁR, I., *Information-type measures of divergence of probability distributions and indirect observations*, Studia Sci. Math. Hung. 2 (1967), pp. 299–318.

CSISZÁR, I., *I-divergence geometry of probability distributions and minimization problems*, Ann. Probab., Vol. 3 (1975), pp. 146–168.

DAVID, H. T. AND KIM, GEUNG-HO, *Pragmatic optimization of information functionals*, in Optimizing Methods in Statistics, ed. Jagdish S. Rustagi, Academic Press: New York (1979), pp. 167–181.

DIGGLE, P. J., *Time Series: A Biostatistical Introduction*, Clarendon Press: Oxford (1990).

EPHRAIM, Y., HANOCH, L., AND GRAY, R., *Asymptotic minimum discrimination information measure for asymptotically weakly stationary processes*, IEEE Transactions on Information Theory, Vol. 34, No. 5 (1988), pp. 1033–1040.

GERSCH, W., et al., *Automatic classification of electroencephalograms*, Science, ol. 205 (1979), pp. 193–195.

GUIASU, S., *Information Theory with Application*, Great Britain, McGraw-Hill Book Co., Inc. (1977).

HANNAN, E. J., *Multiple Time Series*, New York: Wiley (1970).

ITAKURA, F. AND SAITO, S., *Analysis synthesis telephony based on the maximum likelihood method*, in Proc. 6th. Int. Conf. Acoustics., Tokyo, Japan. (1968), pp. C17–C20.

ITAKURA, F. AND SAITO, S., *A statistical method for estimation of speech spectral density and format frequencies*, Electron. Commun. Japan, 53-A (1970), pp. 36–43.

JONES, L. K., *Approximation theoretic derivation of logarithmic entropy principles for inverse problems and u nique extension of the maximum entropy method to incorporate prior knowledge*, SIAM J. Appl. Math., vol. 49 (1989), pp. 650–661.

JONES, L. K. AND BYRNE, C. L., *General entropy criteria for inverse problems, with applications to data compression, pattern classification, and cluster analysis*, IEEE Transactions on Information Theory, vol. 36, no. 1 (1990), pp. 23–30.

JONES, L. K. AND TRUTZER, V., *Computationally feasible high resolution minimum distance procedures which extend the maximum entropy method*, Inverse Problems, vol. 5, (1989), pp. 749–766.

KALLIANPUR, G., *On the amount of information in a sigma field*, in Ingram Olkin et al, editors, Contributions to Probability and Statistics in Honor of H. Hotelling, Stanford University Press (1960).

KAZAKOS, D. AND PAPANTONI-KAZAKOS, P., *Spectral distance measures between Gaussian processes*, IEEE Trans. Automat. Contr., vol. AC-25, no. 5 (1980), pp. 950–959.

KULLBACK, S., *Information Theory and Statistics*, New York: Wiley (1959).

MCCLELLAN, J., *Multidimensional spectral estimation*, Proc. IEEE, vol. 70 (1982), pp. 1029–1039.

MANDELBROT, B. B. AND TAQQU, M. S., *Robust R/S analysis of long run serial correlation*, Proceedings of the 42nd session of the International Statistical Institute, Manila. Bull. I. S. I. 48 (Book 2) (1979), pp. 69–104.

NEWTON, H. J., *TIMESLAB: A Time Series Analysis Laboratory*, Wadsworth: Pacific Grove, California (1988).

PARZEN, E., *Time Series Analysis Papers*, Holden-Day, San Francisco, California (1967).

PARZEN, E., *On asymptotically efficient consistent estimates of the spectral density function of a stationary time series*, J. Royal Statist. Soc., Ser. B., 20 (1958), pp. 303–322.

PARZEN, E., *Statistical Inference on Time Series by Hilbert Space Methods, I*, Technical Report 23, January 2, 1959, Statistics Department, Stanford University. Reprinted in Time Series Analysis Papers (1959).

PARZEN, E., *An approach to time series analysis*, Ann. Math. Statist., 32 (1961), pp. 951–989.

PARZEN, E., *Spectral analysis of asymptotically stationary time series*, Bull. Inst. Internat. Statist., 39 (1962), pp. 87–103.

PARZEN, E., *Multiple time series modeling*, Multivariate Analysis – II, edited by P. Krishnaiah, Academic Press: New York (289–409).

PARZEN, E., *Statistical inference on time series by RKHS methods, II*, Proceedings of the 12th Bienniel Seminar of the Canadian Mathematical Congress, edited by R. Pyke, Canadian Mathematical Congress, Montreal (1970), pp. 1–37.

PARZEN, E., *On the equivalence among time series parameter estimation, approximation theory and control theory*, Proceedings of the Fifth Princeton Conference on Information Science, Princeton (1971), pp. 1–5.

PARZEN, E., *Some Recent Advances in Time Series Modeling*, 723–730 (1974).

PARZEN, E., *Multiple Time Series: Determining the Order of Approximating Autoregressive Schemes*, Multivariate Analysis – IV, Edited by P. Krishnaiah, North Holland: Amsterdam (1977), pp. 283–295.

PARZEN, E., *Nonparametric Statistical Data Modeling*, Journal of the American Statistical Association, (with discussion), 74 (1979), pp. 105–131.

PARZEN, E., *Time Series Model Identification and Prediction Variance Horizon*, Academic Press: New York (1981), pp. 425–447.

PARZEN, E., *Maximum Entropy Interpretation of Autoregressive Spectral Densities*, statistics and Probability Letters, 1 (1982), pp. 2–6.

PARZEN, E., *Time Series Model Identification by Estimating Information*, Studies in Econometrics, Time Series, and Multivariate Statistics in Honor of T. W. Anderson, ed. S. Karlin, T. Amemiya, L. Goodman, Academic Press: New York (1983), pp. 279–298.

PARZEN, E., *Time Series ARMA Model Identification by Estimating Information*, Proceedings of the 15th Annual Symposium on the Interface of Computer Science and Statistics, Amsterdam: North Holland (1983).

PARZEN, E., *Time Series Model Identification by Estimating Information, Memory, and Quantiles*, Questo, 7 (1983), pp. 531–562.

PARZEN, E., *Quantile Spectral Analysis and Long Memory Time Series*, Journal of Applied Probability, Vol. 23A (1986), pp. 41–55.

PARZEN, E., *Statistical Culture*, University Faculty Lecture, Texas A&M University (1990).

PARZEN, E., *Unification of statistical methods for continuous and discrete data*, (Preliminary report), Technical report no. 105, Department of Statistics, Texas A&M University (1990).

PINSKER, M. S., *Information and Information Stability of Random Variables and Processes*, San Francisco, CA: Holden-Day (1964).

PRIESTLY, M. B., *Spectral Analysis and Time Series*, Academic Press: London (1981).

READ, T. R. C. AND CRESSIE, N. A. C., *Goodness of Fit Statistics for Discrete Multivariate Data*, Springer-Verlag, New York (1988).

RENYI, A., *On measures of entropy and information*, Proc. 4th Berkeley Symp. Math. Statist. Probability, 1960, 1, University of California Press: Berkeley (1961), pp. 547–561.

RENYI, A., *On some basic problems of statistics from the point of view of information theory*, Proc. 5th Berkeley Symp. on Math., Stat. and Probability (1967), pp. 531–543.

RENYI, A., *Statists and information theory*, Studia Sci. Math. Hungarica 2 (1967), pp. 249–256.

RENYI, A., *Probability Theory*, North-Holland Publ. Amsterdam (1970).

ROSENBLATT, M., *Stationary Sequences and Random Fields*, Birkhauser: Boston (1985).

SAMAROV, A., TAQQU, M. S., *On the efficiency of the sample mean in long-memory noise*, Journal of Time Series Analysis, 9 (1988), pp. 191–200.

SHORE, J., *Minimum cross-entropy spectral analysis*, IEEE Trans Accoust. Speech, Signal Processing, vol. ASSP-29, No. 2 (1981), pp. 230–237.

SHORE, J. AND GRAY, R., *Minimum cross-entropy pattern classification and cluster analysis*, IEEE Trans. Pattern. Anal. Machine Intell., PAMI-4 (1982), pp. 11–17.

SHORE, J. AND JOHNSON, R., *Axiomatic derivation of the principle of maximum entropy and the principle of minimum cross-entropy*, IEEE Trans. Inform. Theory. IT-26 (1980), pp. 26–37.

SAKAMOTO, Y., ISHIGURO, M., AND KITAGAWA, G., *Akaike Information Criterion Statistics*, D. Reidel: Boston (1983).

SMITH, C. R. AND ERICKSON, G. J., *Maximum Entropy and Bayesian Spectral Analysis and Estimation Problems*, D. Reidel: Boston (1987).

THEIL, H. AND FIEBIG, D. C., *Exploiting Continuity: Maximum Entropy Estimation of Continuous Distributions*, Ballinger, Cambridge, MA (1984).

TUKEY, J. W., *Collected Works on Time Series*, Vol. I: 1949–1964, Vol. II: 1965–1984. Ed. D. R. Brillinger, Wadsworth: Belmont, California (1984).

WHITTLE, P., *Estimating and information in stationary time series*, Ark. Math. 2 (1953), pp. 423–434.

WHITTLE, P., *The analysis of multiple stationary time series*, J. Royl Statist. Soc, B. 15 (1953), pp. 125–139.

WOODS, J. W., *Two-dimensional Markov spectral estimation*, IEEE Trans. Inform. Theory, Vol. IT-22, no. 5 (1976).

ZHANG, J. AND BROCKETT, P., *Quadratically constrained information theoretic analysis*, SIAM J. Appl. Math., Vol. 47, no. 4 (1987), pp. 871–885.

FUNDAMENTAL ROLES OF THE IDEA OF REGRESSION AND WOLD DECOMPOSITION IN TIME SERIES

MOHSEN POURAHMADI*

Abstract. The phenomenal theoretical and methodological developments and widespread applications of time series are centered around the probabilistic notion of stationary processes. The spectral representation and the Wold decomposition theorems are the two most powerful tools for understanding structures and introducing parameters of a stationary time series, and solving many prediction problems that are prototypes of problems of interest in time series analysis. We show that the idea of regression, used originally by Wold (38) to obtain the Wold decomposition, plays a more profound role in revealing structural information about a time series than the Wold decomposition itself. We employ the idea of regression and propose a unified and mathematically rigorous method for introducing the AR, MA and $ARMA$ parameters, approximations and representations of a stationary time series without adhering to its spectral representation. A finite Wold decomposition of the time series $\{X_t\}$ is used to compute predictors of $X_1, \ldots, X_m$ based on the infinite past $\{X_t; t \leq 0\}$. We show that the multi-step ahead prediction errors, $X_k - \hat{X}_k, 1 \leq k \leq m$, and their covariance matrix G_m are the most common ingredients in solving various problems such as the missing value problem, identification problem, intervention analysis and computing the cananical correlations, arising in time series analysis. The paper seeks to reconcile and provide a bridge between the Kolmogorov-Wiener prediction theory and the structural aspects of the Box-Jenkins approach to time series analysis.

1. Introduction. The three major and interrelated goals of this paper are: (1) to provide a possible blueprint for developing a *mathematical foundation* for time series analysis, (2) to identify and bring to the fore the minimal and common *mathematical ingredients and ideas* that are necessary to understand the structure of a stationary time series, (3) to identify generic prediction problems of interest in time series analysis.

The need for a *mathematical foundation* for time series analysis is manifested by its widespread use and truly interdisciplinary nature. These aspects are summarized best in the first page of the IMA Newsletter No. 164, announcing the summer program, New Directions in Time Series Analysis: "The theory and methods of time series analysis lie at the intersection of the mathematical, statistical, computational, and system sciences, and provide an elegant interplay among these disciplines. They provide the means of applying advanced mathematical ideas and theorems to contribute towards the solutions of very practical problems. Time series analysis is truly an *interdisciplinary* field, because developments of its theory and methods requires *interaction* between the diverse disciplines in which it is being applied. To harness its great potential there must develop a community of statistical and other scientists who are educated and motivated to have a background in theory and methods of time series analysis adequate to handle the problems of time series analysis in *all* the fields in which they occur."

*Division of Statistics, Department of Mathematical Sciences, Northern Illinois University, DeKalb, IL 60115. Research supported by AFOSR-88-0284. This paper is an expanded version of an invited talk given at the Workshop in New Direction in Time Series Analysis at Institute for Mathematics and Its Applications, University of Minnesota, July 1-29, 1990.

It goes without saying that even a moderate success in achieving these goals should provide an appealing vantage point from which to view the mathematical structure of a time series and attack the associated statistical problems in a systematic manner. Also, the pedagogical implications of these goals are rather self-evident; for the status and issues related to research and teaching time series within the statistics programs see Parzen [32].

Almost all themes in time series analysis have their origin in the probabilistic notion of *stationary stochastic processes*. The Kolmogorov-Wiener *prediction theory* of such processes have for long been the driving force behind the developments of the spectral-domain analysis of time series, cf. [23,51,52]. Prediction theory does provide a viable, though somewhat inaccessible, mathematical foundation for time series analysis in the spectral domain. However, its role and influence on the time-domain techniques are dubious and less clear, possibly because of its mathematical complexity and the difficulty in interpreting the spectrum, there is a propensity for schism in time series analysis with the proponents of the time-domain approach developing various techniques without much regard for the results and ideas from prediction theory. Although on the surface such development seems to be successful due to its reliance on certain *formal manipulations* involving the backshift operator B, it should be noted that this approach is not without pitfalls [30,50]. The rigorous mathematical justification of such formal manipulations does require techniques and ideas related to Fourier transform and spectral analysis, very much like that for the Heaviside calculus in the communication engineering which took place in the 1930's, see [51, p.7] for an illuminating discussion on this topic.

From the dawn of modern time series analysis, the celebrated Wold decomposition theorem [53,18,9], in spite of being a time domain result, has been the guiding light and unifying force in various approaches to time series analysis. These points are made amply clear by Wold himself [15,p.192]: *My main incentive for the decomposition of a general stationary time series was to cover as special cases the three types of stationary time series that were known at the time. These were (i) the scheme of hidden periodicities, (ii) the finite moving average model, (iii) the finite autoregression model. When establishing the decomposition I was only vaguely aware of its close affinity to the spectral analysis of time series, and even less aware of the ensuing affinity to operator theory. It so happened that the decomposition gave a clue to overcoming an earlier impasse in spectral analysis and operator theory, as was shown by Andre Kolmogorov in 1939, and Norbert Wiener in 1942. Thanks to this profound aspect, my decomposition is referred to in the Encyclopedia Britannica, 13th ed., in the article on "Automata": see also Helson [20].*

In this paper we show that the simple geometric (non-analytic) idea of linear regression (approximation in a linear space) that were used originally by Wold [53], to obtain his versatile decomposition, has a more profound and sweeping role to play in time series analysis than the Wold decomposition itself. In fact, we show that an abstract version of the idea of regression in a linear space is the *only mathematical tool* that is needed for a rigorous understanding of the linear structure of a stationary time series and solving various problems related to prediction, interpolation, canonical correlations and intervention analysis, etc., all in the time-domain. We isolate certain difficult mathematical problems that are of peripheral interest in time series analysis, and indicate how their rigorous solutions may require analytic concepts that are pertinent to spectral analysis of a stationary time series.

From the foundational and pedagogical standpoints and as far as prediction of future values is concerned, it is advantageous to divide the body of knowledge commonly known as time series analysis into the following three non-overlapping and increasingly practical categories: (i) prediction based on the infinite past and the knowledge of the covariance sequence $\{\gamma_k\}$, (ii) prediction based on the finite past and the knowledge of the corresponding segment of $\{\gamma_k\}$, (iii) prediction based on the actual data. In this paper we are concerned mostly with the first two categories, results and formulae obtained here are used in actual forecasting by replacing the unknown covariances by their estimated values and truncating the infinite series whenever necessary. This order of coverage is common in prediction theory, cf. Masani [26,p.284], and textbooks on time series, cf. [9].

The outline of the paper is as follows: A brief history of the use of regression methods in time series is discussed in Sec. 2. Sections 3 and 4 are concerned with preliminary results and concepts related to a stationary time series and prediction theory. An abstract version of the idea of regression in a Hilbert space is stated in Sec. 5. Sec. 6 is the heart of the paper, by using the idea of regression we propose a unified time-domain method for introducing the AR, MA and $ARMA$ parameters and representations of a nondeterministic stationary time series, these representations on the one hand can be viewed as Taylor-like expansions of the time series in terms of its past, and on the other hand as an attempt to find various basis for H_0, the closed linear span of the past values $\{X_t : t \leq 0\}$. The classical recurssions between the AR and MA parameters are obtained, for the first time, by relying on the geometry of H_0 instead of analytic properties of certain auxiliary functions. However, the characterization of the MA and AR parameters, postponed to Sec. 6.4, is stated in terms of location of zeros of these functions. It seems use of analytic concepts at this stage is indispensible.

In Sec. 7, by using a finite MA approximation of the underlying process we give a mathematical solution of the problem of predicting future values of a time series (*the first prediction problem*). This solution does not use the full force of the Wold decomposition, yet it reveals a close and unexplored structural affinity between Γ, the covariance matrix of the infinite past $\{X_t : t \leq 0\}$, and G_∞ the covariance matrix of the prediction errors $\{X_k - \hat{X}_k;\ k \geq 1\}$. The finite prediction problem is studied in Sec. 8, and it is shown that all prediction - theoretic quantities of interest converge to their counterparts for the prediction from the infinite past. The main focus of Sec. 8 is on the theoretical roles of the sequence of partial correlations in characterizing nondeterministic processes, AR (p) models and processes with maximum entropy.

Sec. 9 is concerned with the interpolation of a missing value (*the second prediction problem*) and Sec. 10 is concerned with computing the canonical correlation between past and future (*the third prediction problem*) of a time series. In sharp contrast with the first prediction problem, these two problems involve both the infinite past and the infinite future. In spite of this, the mathematical solution of the first prediction problem or more specifically the fact that the multi-step ahead prediction errors are orthogonal to the past, is the only idea that we use to solve these problems. This shows the *fundamental roles of the first prediction problem and the covariance matrix of the multi-step ahead prediction errors in the time-domain approach to prediction theory and time series analysis.*

In Sec. 11, a nonparametric procedure is developed for estimating impacts of interventions on a given response variable in the presence of dependent noise. It is based on the simple and intuitively appealing idea that the multi-step ahead prediction errors of the post-intervention measurements are unbiased estimators of impacts of interventions. When parametric forms are postulated for impacts of interventions, then the results of this section reduce to those in Box and Tiao [8].

In this paper we have restricted our attention to the important class of univariate, second-order, stationary time series, but most of the ideas and techniques of Sections 5 and 6 apply to the classes of nonstationary, multivariate, and infinite variance time series. For the first class the relevant parameters will be time-dependent, and for the second they are matrices and their uniqueness is linked to the rank of the time series. The full force of the idea of regression was used in [27] to study structure of stationary processes with infinite variance. For such processes the traditional Kolmogorov-Wiener prediction theory does not apply, since the notions of covariance and spectral density are not defined. The approach of this paper, however, does not seem to be applicable to continuous-time stochastic processes.

2. A brief history of the role of regression. G. Udny Yule [55,56] guided by the methods of regression analysis was the first to use the AR (2) model

$$X_t = a_1 X_{t-1} + a_2 X_{t-2} + \epsilon_t ,$$

in searching for the periodicity of the well-known sunspots numbers. By this he not only started time-domain time series analysis, but also proposed methods of time series analysis that were seemingly of wider scope than the classical method of periodogram analysis proposed earlier by Schuster [44]. Moving average models were also introduced by Slutsky [46].

The real force of the *idea of regression* was brought to bear on the structure of a stationary time series by H. Wold in 1938. It seems the development of the Kolmogorov-Wiener prediction theory in the early 1940's and its reliance on more powerful ideas and techniques from functional and harmonic analyses had put a temporary halt on the use of the idea of regression in developing time-domain techniques. The book by Wold contains an excellent historical account of the research in time series analysis before 1938, see also Robinson [42].

The use of regression and time-domain methods did resurface again in the 1960's through the work of Durbin [14], Box and Jenkins [7], Kalman and Bucy [22]. A paper of Tukey [49] emphasizing the connection between analysis of variance and spectrum analysis, and subsequent work of Brillinger [6] in this direction have also popularized the use of regression in the spectral domain. An accessible exposition of various techniques of time series analysis based on regression can be found in Chapters 3 and 4 of Shumway [45].

3. Stationary time series: Time-domain structure. Although in practice the data from economics and many other areas employing time series methods are nonstationary, the classical paradigm of time series analysis assumes that data are generated by a weakly stationary stochastic process $\{X_t\}$. This impractical paradigm has survived mainly because of success of techniques such as differencing, detrending, etc. that renders stationarity to data.

Throughout this paper, unless otherwise stated, we assume that $\{X_t\}$ is a real-valued, discrete-time, mean zero, weakly stationary process with covariance sequence $\{\gamma_k\}$:

$$\gamma_k \;=\; \mathrm{cov}(X_{t-k}, X_t)\, , t, k \;=\; 0, \pm 1\,, ...,\, .$$

It is known that the covariance sequence provides a *global* picture of dependence among various values of the time series. However, an attempt to predict the future values of $\{X_t\}$ based on its infinite past, $\{X_t;\; t \leq 0\}$, and the knowledge of the covariance sequence $\{\gamma_k\}$, one of the simplest and yet most important task demands *local* structure and alternative parametrization of $\{X_t\}$, cf. [17,52]. Thus the problem of prediction of future values is of central importance in any approach to time series analysis.

4. The first prediction problem: Formulation and related concepts. For a time series $\{X_t\}$, its time-domain is the Hilbert space $H \;=\; H(X)$ defined as the closed linear span of all elements of $\{X_t\}$ with respect to the norm, $\|X\| = \sqrt{var(X)}$. More concisely,

$$H(X) \;=\; \overline{sp}\,\{X_t; t \;=\; 0, \pm 1, ...\}\ .$$

The span of its infinite past up to time t is defined and denoted by

(4.1) $$H_t \;=\; H_t(X) \;=\; \overline{sp}\{X_s; s \leq t\}\ .$$

The problem of predicting X_t based on the infinite past $\{X_s; s \leq t-1\}$, and the knowledge of its covariance sequence $\{\gamma_k\}$ is that of finding an element $\hat{X}_t \epsilon H_{t-1}$ such that its distance to X_t is as small as possible. More precisely, the *first prediction problem* [21] is concerned with,

(i) finding $\hat{X}_t \epsilon H_{t-1}$ such that

$$var(X_t \;-\; \hat{X}_t) \;=\; \inf_{Y \in H_{t-1}}\; var(X_t \;-\; Y)\, ,$$

i.e., the variance of prediction error is as small as possible,

(ii) expressing $\sigma^2 \;=\; var(X_t - \hat{X}_t)$ in terms of $\{\gamma_k\}$.

Since H_{t-1} is a closed subspace of H, it is well-known [9,p.51] that there exists a unique $\hat{X}_t \epsilon H_{t-1}$ satisfying (i), and it is in fact the orthogonal projection of X_t onto H_{t-1}, that is,

$$(4.2) \qquad \hat{X}_t = P^{X_t}_{H_{t-1}} .$$

The co-projection

$$(4.3) \qquad \epsilon_t = X_t - \hat{X}_t ,$$

is the one-step ahead prediction error and its variance σ^2 in (ii) is known as the *prediction error variance*. The new time series $\{\epsilon_t\}$ is referred to as the *innovation process* of $\{X_t\}$. Of particular interest are processes for which

$$(4.4) \qquad \sigma^2 > 0 ,$$

or equivalently,

$$(4.5) \qquad X_t \notin H_{t-1},$$

such processes are called *nondeterministic*, as their future values cannot be determined precisely via linear functions of past values. Since $\hat{X}_t$ is the orthogonal projection of X_t onto H_{t-1}, the innovation process $\{\epsilon_t\}$ is a sequence of uncorrelated random variables whenever $\{X_t\}$ is nondeterministic. We assume from here on that $\{X_t\}$ is nondeterministic. From (4.3), we have

$$(4.6) \qquad X_t = \hat{X}_t + \epsilon_t ,$$

this provides a useful decomposition of X_t into its predicted value and prediction error. Furthermore, from (4.6), (4.5) and (4.1), we observe that

$$(4.7) \qquad \hat{X}_t \epsilon H_{t-1} = \overline{sp}\{X_{t-1} , H_{t-2}\} = \overline{sp}\{\epsilon_{t-1} , H_{t-2}\} .$$

These two simultaneous representations of $H_t - 1$ is the key to revealing the (local) structural mystery of a stationary time series, see Sec. 6.

4.1 Characterization of a stationary time series: The shift operator. The following fundamental theorem due to Kolmogorov [23] establishes a one-to-one correspondence between the class of stationary time series and that of unitary operators on a Hilbert space.

4.1 THEOREM. *A second-order stochastic process $\{X_t\}$ with time-domain $H(X)$ is stationary, if and only if there exist a unitary operator $U : H(X) \rightarrow H(X)$ and an element $X_0 \epsilon H(X)$ such that*

$$(4.8) \qquad X_t = U^t X_0, \ t = 0, \pm 1, \dots .$$

If such a U exists, it is unique and is referred to as the shift operator of the stationary time series $\{X_t\}$. *Note that from(4.8) it follows that*

$$X_t \;=\; U X_{t-1}\,,$$

indicating the close relation between this shift operator and the so-called backshift operator B used frequently in the context of Box-Jenkins time series analysis [7,9].

This shift operator is essential in both approaches to time series analysis. The spectral representation of a unitary operator [41] can be used to obtain a slick proof of the spectral representation of a stationary time series see [23,52,16,24,12]. In the time-domain, this operator is essential in showing that certain parameters of a stationary time series are not time-dependent, see Sec. 6.

5. A regression principle. The generic problem of regression in statistics and more generally the problem of filtering in engineering can be stated as follows: *There is a variable X that we are interested in, but we observe Y a contaminated version of X with error or noise coming from a source E. Usually the first goal is to relate Y to X and E, namely, postulate a model, and the second goal is to use the postulated model and the available data to make appropriate inferences.*

To state the first goal mathematically, let H be a Hilbert space with $X \epsilon H, E$ a subspace of H. Then, the goal is to represent $Y \epsilon \overline{sp}\{X, E\}$ in terms of X and elements of E. For relevant examples see (4.7). The following lemma in [27,p.154] provides precise information regarding the relation among Y, X and elements of E.

5.1 LEMMA (REGRESSION PRINCIPLE). *Let H be a Hilbert space with* $X \epsilon H, E$ *a subspace of H such that* $X \notin E$. *Then, for any* $Y \epsilon \overline{sp}\{X, E\}$, *there are unique scalar a and vector* $e \epsilon E$ *such that*

$$Y \;=\; aX \,+\, e\,.$$

The uniqueness of a and e is crucial for the applications in Sec. 6 and it is a consequence of $X \notin E$, *the linear independence of X and elements of E.*

6. Structure of a stationary time series: AR, MA and ARMA approximations. The problem of modeling time series data by an appropriate finite-parameter model lies at the heart of time-domain time series analysis [7]. This problem is usually solved by fitting an $ARMA\ (p, q)$ model of the form

$$(6.1) \qquad X_t \;=\; \sum_{i=1}^{p} \alpha_i X_{t-i} \,+\, \epsilon_t \,+\, \sum_{j=1}^{q} \beta_j \epsilon_{t-j}\,,$$

to the data $X_1, ..., X_n$. The operational details for fitting such models are explained in the classic book by Box and Jenkins [7], where it is demonstrated quite successfully as far as (empirical) data analysis is concerned that these models are parsimonious and appropriate. Of course, asymptotic statistical theory is needed to back up any empirical evidence and claim, particularly those related to the estimators of the orders p, q and parameters of the model cf. [18,19]. In case of time series data, inevitably as time passes ($n \rightarrow \infty$) the amount of data gathered increases and therefore a more complex model can be fitted. Starting with the work of Berk [4], see also Hannan [18] and references therein, it is common to assume that the underlying process $\{X_t\}$ has a representation similar to (6.1) with p and/or $q = \infty$.

In this section with the help of the regression principle and (4.7), we obtain Taylor-like expansions for a nondeterministic stationary time series which justify the use of models like (6.1) as approximants for such time series. For arbitrary p and q, we provide precise information about the remainder term in the expansion. As by-products of this approach we introduce unified and rigorous methods for introducing the AR, MA, and $ARMA$ parameters and representations of a time series, and develop the relationship among these parameters, by relying solely on time-domain techniques. It is instructive to compare this approach with its spectral-domain counterpart where approximation of a stationary time series by $ARMA$ models is motivated by the fact that a smooth (density) function can be approximated by rational functions, cf. [9,Sec. 4.4].

6.1 AR representation and approximation. A traditional course in time series begins with the MA (∞) representation of a stationary time series [7], then it proceeds with *formal* inversion of this representation from which the formal AR (∞) representation of the time series and the relationship between the AR and MA parameters are deduced, this format was used first in [53]. It is a rather lengthy and tedious procedure to make these steps mathematically rigorous, see [1,52,25,35,36] for details.

In this section we introduce the essentials of the AR representation of a stationary time series in the time domain rather directly without relying on the MA (∞) representation of such processes. In fact, we develop the results so that the mathematical equivalence between AR and MA representations of a time series becomes rather evident.

First, we shall work on the AR representation. From the first equality in (4.7) and invoking the Regression Principle with

$$Y = \hat{X}_t, \ X = X_{t-1}, E = H_{t-2},$$

it follows that there exist a unique scalar a_1 and a vector $e_{t,1} \epsilon H_{t-2}$ such that

$$(6.2) \qquad \hat{X}_t = a_1 X_{t-1} + e_{t,2}.$$

But, since

$$(6.3) \qquad e_{t,1} \epsilon H_{t-2} = \overline{sp}\{X_{t-2}, H_{t-3}\},$$

by invoking the Regression Principle again, it follows that

$$(6.4) \qquad e_{t,1} = a_2 X_{t-2} + e_{t,2},$$

for unique scalar a_2 and vector $e_{t,2} \epsilon H_{t-3}$. By arguing in this fashion, and combining (4.6), (6.2) - (6.4), it follows that *for any integer $p \geq 1$, there exist unique scalars $a_1, ..., a_p$ and vector $e_{t,p} \epsilon H_{t-p-1}$ such that*

$$(6.5) \qquad X_t = \sum_{k=1}^{p} a_k X_{t-k} + \epsilon_t + e_{t,p}.$$

It should be noted that for any $p \geq 1$, (6.5) provides a "Taylor expansion" or an AR (p) approximant for $\{X_t\}$ with $e_{t,p}$ as its remainder. It is natural to call the *unique* sequence of scalars $\{a_k\}_{k=1}^{\infty}$ the AR parameters of $\{X_t\}$. If $e_{t,p} \equiv 0$, for $p = p_0$, then $\{X_t\}$ is called an AR (p_0) process.

By using the regression principle and the first equality in (4.7) we were able to introduce the AR parameters $\{a_k\}$, and clarify the relation between AR (p) models and a nondeterministic stationary time series rather early in the development. This should be compared with the classical development in the spectral domain where AR parameters and models are introduced rather late in the theory, cf. [52,25]. The AR (∞) representation, in view of (6.5) hinges on the study of l. i. m.$_{p \to \infty}$ $e_{t,p}$ or l. i. m.$_{p \to \infty}$ $\sum_{k=1}^{p} a_k X_{t-k}$, which is postponed to Sec. 7.3.

6.2 MA representation and approximation. In this section by following exactly the same steps as those in 6.1 we obtain MA representation and approximation for a stationary time series. Perhaps, to some extent this explains the mathematical equivalence between AR and MA representations. However, since the MA representation involves elements of $\{\epsilon_t\}$ that are uncorrelated random variables, one is in a better position to obtain more information about the MA parameters and approximations. It should be noted that, in case of infinite variance processes [27], since the notion of orthogonality is not present, such superficial distinction between the MA and AR representations will disappear.

From the second equality in (4.7) with

$$Y = \hat{X}_t , \quad X = \epsilon_{t-1} , \quad E = H_{t-2} ,$$

and invoking the Regression Principle it follows that there exist a unique scalar b_1 and vector $V_{t,1} \epsilon H_{t-2}$ such that

$$\hat{X}_t = b_1 \epsilon_{t-1} + V_{t,2} .$$

Continuing as in 6.1, we get that *for any integer $q \geq 1$, there exist unique scalars $b_1, ..., b_q$ and vector $V_{t,q} \epsilon H_{t-q-1}$ such that*

$$(6.6) \qquad X_t = \sum_{k=0}^{q} b_k \epsilon_{t-k} + V_{t,q} , \quad b_0 = 1 .$$

Note that for any integer $q \geq 1$, (6.6) provides an MA (q) approximant for $\{X_t\}$ with $V_{t,q}$ as its approximation error. The *unique* sequence of scalars $\{b_k\}_{k=1}^{\infty}$

is naturally called the MA parameters of $\{X_t\}$. If $V_{t,q} \equiv 0$, for $q = q_0$, then $\{X_t\}$ is called an $MA\,(q_0)$ process.

As mentioned earlier a unique feature of the MA representation is its use of the orthogonal sequence $\{\epsilon_t\}$. Thus, for any $q \geq 1$, the random variables $\sum_{k=0}^{q} b_k \epsilon_{t-k}$ and $V_{t,q}$ are orthogonal, and consequently

$$\sigma^2 \sum_{k=0}^{q} b_k^2 = Var\left(\sum_{k=0}^{q} b_k \epsilon_{t-k}\right) \leq Var(X_t) = \gamma_0 < \infty ,$$

for all q, which implies that

$$\text{(6.7)} \qquad \sum_{k=o}^{\infty} b_k^2 < \infty .$$

Of course, this condition is necessary and sufficient for the mean-convergence of the sequences $\{\sum_{k=0}^{q} b_k \epsilon_{t-k}\}_q$ and $\{V_{t,q}\}_q$. For later use, we define $\{V_t\}$ by

$$\text{(6.8)} \qquad V_t = \underset{q \to \infty}{\text{l.i.m.}} \; V_{t,q},$$

and note that, for each t,

$$V_t \epsilon \bigcap_q H_{t-q-1} = H_{-\infty}(X) ,$$

that is, V_t belongs to the *remote past* of the time series $\{X_t\}$.

6.3 Relationship between AR and MA parameters. Thus far, to each nondeterministic stationary time series we have associated two unique sequences $\{a_k\}$ and $\{b_k\}$. It is natural to ask as whether these two sets of parameters are related to each other. It turns out that, by taking $p = q$ and substituting (6.6) into (6.5) and doing some algebra, see [27, Corollary 4.7], we get the well-known set of recursions between $\{a_k\}$ and $\{b_k\}$, cf. Wold [53],

$$\text{(6.9)} \qquad \begin{cases} a_0 = b_0 = 1, \\ \\ b_l = \displaystyle\sum_{k=0}^{l-1} b_k a_{l-k}, \; l = 1, 2, \dots . \end{cases}$$

Thus, if the sequence of MA parameters $\{b_k\}$ is known, one may write (6.9) as an infinite (Toeplitz) system of equations with $\{a_k\}$ as the unknowns:

$$\text{(6.10)} \qquad \begin{bmatrix} 1 & 0 & 0 & \cdots & 0 \\ b_1 & 1 & 0 & \cdots & 0 \\ b_2 & b_1 & 1 & \cdots & 0 \\ \cdot & \cdot & \cdot & & \cdot \\ \cdot & \cdot & \cdot & & \cdot \end{bmatrix} \begin{bmatrix} a_1 \\ a_2 \\ \cdot \\ \cdot \end{bmatrix} = \begin{bmatrix} b_1 \\ b_2 \\ \cdot \\ \cdot \end{bmatrix} .$$

We denote the (infinite) lower triangular matrix of the system (6.10) by T, and for any integer $m \geq 1$, its $m \times m$ leading principal submatrix will be denoted by T_m. It is evident that T_m is invertible, and it follows from (6.9) that its inverse is given in terms of the AR parameters of $\{X_t\}$:

$$(6.11) \qquad T_m^{-1} = \begin{bmatrix} 1 & 0 & 0 & \cdots & 0 \\ -a_1 & 1 & 0 & \cdots & 0 \\ -a_2 & -a_1 & 1 & \cdots & 0 \\ \vdots & \vdots & \vdots & \ddots & \vdots \\ -a_{m-1} & & & \cdots & 1 \end{bmatrix}.$$

This form of T_m^{-1} will be used repeatedly in the later sections of this paper, and it plays a crucial role in explaining the duality between MA and AR representations, as well as unifying solutions of several diverse problems in prediction theory and time series analysis, see Sections 9 and 11.

6.4 Characterization of the MA parameters and the need for spectral analysis.
From the analysis so far, we know that the sequence of MA parameters of a nondeterministic stationary time series is square-summable, cf. (6.7). In view of this, it seems natural to ask the following question: *Can every square-summable sequence be the MA parameters of a nondeterministic stationary time series?* Since the answer to this question is negative, we need to study the characterizing properties of the MA parameters. Such properties are crucial in computing the MA parameters from the knowledge of the covariance sequence $\{\gamma_k\}$, and estimating such parameters from the data.

It is part of the time series folklore that the characterizing property of the MA parameters can be sought in the location of the zeros of the analytic function

$$(6.12) \qquad \psi(z) = 1 + b_1 z + b_2 z^2 + \dots, |z| \leq 1 .$$

The origin of this can be traced to a fundamental theorem in harmonic analysis, cf. [13, p.160]: *A square-summable sequence $\{b_k\}_1^\infty$ is the MA parameters of a nondeterministic stationary time series, if and only if all the zeros of $\psi(z)$ lie outside the unit disk in the complex plane.*

Perhaps it is surprising that all the results up to (6.12) had a clear geometric-algebraic flavor, but the characterizing property of the MA parameters is *analytic in nature*. At this point, it seems there is a *mathematical necessity* to bring in techniques of harmonic analysis to bear on this and other related problems whose solutions are analytic in nature. Even though the sequence of AR parameters is not square-summable in general, still the zeros of the (formally defined) function

$$(6.13) \qquad \phi(z) = 1 - a_1 z - a_2 z^2 - \dots, |z| \leq 1,$$

must lie outside the unit disk. From (6.9) it follows that ψ and ϕ satisfy the identity

$$(6.14) \qquad \psi(z)\phi(z) \equiv 1 .$$

In the spectral domain the functions ψ and ϕ are directly related to the spectral density of the time series and have physical interpretation.

6.5 ARMA representation and approximation. The AR and MA representations in Sections 6.1 and 6.2 seem to suggest that the infinite pasts of the process and its innovation form bases for H_0. In this section we show that one can alternate between the two equalities in (4.7) and obtain $ARMA$ representations and approximations for a nondeterministic stationary time series, thus indicating that appropriate mixture of the pasts of the process and its innovation would, in a sense, form a basis for H_0 as well.

For arbitrary integers $0 \leq p, q < \infty$, let $\{P, Q\}$ be a partition of the set of integers $\{1, 2, \ldots, p + q\}$ into two sets P and Q with p and q elements, respectively. By applying the Regression Principle to the first or second equality in (4.7), depending on whether $1 \epsilon P$ or Q, and continuing, it follows that for a nondeterministic stationary times series $\{X_t\}$ there exist unique scalars α_i, $i \epsilon P$, β_j, $j \epsilon Q$ and a unique random variable $e_{t,p,q} \epsilon H_{t-(p+q+1)}$ such that

$$(6.15) \qquad X_t = \sum_{i \epsilon P} \alpha_i X_{t-i} + \epsilon_t + \sum_{j \epsilon Q} \beta_j \epsilon_{t-j} + e_{t,p,q} .$$

It should be noted that when Q or P is the empty set, then (6.15) and other results related to it will reduce to those in Sections 6.1 and 6.2, respectively, as such most of the results of this section subsumes those stated in the previous sections.

When $\{P, Q\}$ is a partition of $N = \{1, 2, \ldots\}$, the set of natural numbers, the corresponding set of scalars $\{\alpha_i, i \epsilon P\} \cup \{\beta_j, j \epsilon Q\}$ appearing in (6.15), which is unique for a fixed partition, is referred to as the $ARMA$ parameters of $\{X_t\}$. It should be noted that a different partition of N does produce a different set of $ARMA$ parameters for the same time series.

A nondeterministic stationary time series $\{X_t\}$ is said to be an $ARMA$ $(p_0, p_0 + q_0)$, $0 \leq p_0, q_0 < \infty$, if in the representation (6.15),

$$e_{t, p_0, q_0} \equiv 0 .$$

In this case

$$(6.16) \qquad X_t = \sum_{i \epsilon P} \alpha_i X_{t-i} + \epsilon_t + \sum_{j \epsilon Q} \beta_j \epsilon_{t-j} ,$$

where $\{P, Q\}$ is any partition of $\{1, 2, \ldots, p_0 + q_0\}$ with p_0 elements in P. Note that an $ARMA$ $(p, p + q)$ model as introduced in (6.16) has only $p + q$ parameters, this is in sharp contrast with the standard $ARMA$ $(p, p+q)$ model, see (6.1), which requires $2p + q$ parameters. *Thus, the class of models in (6.16) is a subclass of those in (6.1), and the former provides more parsimonious approximants for a stationary time series.*

7. Solutions of the first prediction problem. In this section we take the infinite past to be X_t, $t \leq 0$. Our focus will be on predicting the future values X_l, $l \geq 1$, and their prediction errors. For a fixed integer $l \geq 1$, *the l-step ahead best linear predictor* of X_l based on $\{X_t, t \leq 0\}$ is denoted by $\hat{X}_l$ and is obtained as the orthogonal projection of X_l on H_0. The problem of expressing $\hat{X}_l$, and $var(X_l - \hat{X}_l)$ in terms of the observable past and the distribution of $\{X_t\}$ is important in prediction theory and time series analysis. There are several solutions to this problem, however, and we begin with a mathematical solution.

7.1 A mathematical solution. In this section we express $\hat{X}_l$ and $X_l - \hat{X}_l$ in terms of $\{\epsilon_t\}$, which is not observable, and $\{b_k\}$ as yet to be related to the covariance $\{\gamma_k\}$. Of course, this is not a practical solution of the prediction problem, but its reliance on the orthogonality of $\{\epsilon_t\}$ provides some results and insights that are crucial in solving other prediction problems. From (6.6) with $t = l$, and $q \geq l$, we can easily compute $\hat{X}_l$ and the prediction error $X_l - \hat{X}_l$ as

(7.1)
$$\begin{cases} \hat{X}_l = \sum_{k=l}^{q} b_k \epsilon_{l-k} + V_{l,q}, \\ X_l - \hat{X}_l = \sum_{k=0}^{l-1} b_k \epsilon_{l-k} \end{cases}$$

Note that derivation of these formulae does not require the full force of the Wold decomposition theorem.

It is instructive and for future use, we write the prediction errors for $l = 1, 2, ...,$ $m, m \geq 1$, in matrix form:

(7.2)
$$\begin{bmatrix} X_1 - \hat{X}_1 \\ X_2 - \hat{X}_2 \\ \vdots \\ X_m - \hat{X}_m \end{bmatrix} = \begin{bmatrix} 1 & 0 & 0 & \cdots & 0 \\ b_1 & 1 & 0 & \cdots & 0 \\ b_2 & b_1 & 1 & \cdots & 0 \\ \vdots & \vdots & \vdots & & \vdots \\ b_{m-1} & b_{m-2} & \cdot & \cdot & 1 \end{bmatrix} \begin{bmatrix} \epsilon_1 \\ \cdot \\ \cdot \\ \cdot \\ \epsilon_m \end{bmatrix}.$$

Thus, G_m the covariance matrix of the first m prediction errors is given by

(7.3)
$$G_m = \sigma^2 T_m T_m',$$

where T_m is the matrix defined following (6.10). The matrix G_m has an important unifying role in solving several prediction problems, see sections 9-11.

It is perhaps useful to view the prediction errors $\{X_l - \hat{X}_l\}_l$ as a stochastic process with covariance matrix G_∞. In the next section, we show that there is a close link between the matrices G_∞ and $\Gamma = (\gamma_{i-j})$, the covariance matrix of $\{X_t, t \leq 0\}$.

7.2 Wold decomposition. In this section we derive the celebrated Wold decomposition of a nondeterministic stationary time series as the limit of (6.6) when $q \to \infty$. We need it here only to obtain a factorization of Γ similar to that for G_∞ in (7.3).

Wold Decomposition Theorem. Let $\{X_t\}$ be a nondeterministic stationary time series with innovation process $\{\epsilon_t\}$. Then for each t,

(7.4)   (a)
$$X_t = \sum_{k=0}^{\infty} b_k \epsilon_{t-k} + V_t, \quad \sum_{k=0}^{\infty} b_k^2 < \infty$$

for a unique sequence $\{b_k\}$ and deterministic stationary time series $\{V_t\}$ which is orthogonal to $\{\epsilon_t\}$.

$$(b) \qquad H_t(X) \;=\; H_t\,(\epsilon)\; \oplus\; H_t\,(V)\,.$$

$$(c) \qquad H_t(V) \;=\; H_{-\infty}\,(X)\,.$$

Proof of part (a) is immediate from (6.6) - (6.8). Proofs of (b) and (c) are based on (6.6) and we omit the details.

An important consequence of (7.4) is the following factorization of Γ, the (infinite) covariance matrix of $\{X_t;\ t\ \leq\ 0\}$:

$$(7.5) \qquad\qquad \Gamma \;=\; \sigma^2 T'T + \Gamma_V\,,$$

where Γ_V is the covariance matrix of the deterministic part of $\{X_t\}$. When $\Gamma_V\ =\ 0$, or equivalently when $\{X_t\}$ is *purely nondeterministic*, then (7.5) reduces to

$$(7.6) \qquad\qquad \Gamma \;=\; \sigma^2 T'T\,,$$

which gives the Cholesky factorization of the matrix Γ with factor T as in (6.10). This shows that the MA parameters can be obtained from $\{\gamma_k\}$, furthermore through this factorization one can see the structural similarities between Γ and G_∞:

$$(7.7) \qquad\qquad G_\infty \;=\; \sigma^2 TT'\,.$$

7.3 A practical solution. As mentioned in Sec. 7.1 prediction based on the MA representation of $\{X_t\}$ is not practical, since the predictors depend on $\{\epsilon_t\}$ which is not observable. Thus, it seems natural to turn attention to the AR representation of $\{X_t\}$, cf. (6.5). To fully appreciate the implications of such representation, as in Sec. 7.2, we need to allow $p\ \to\ \infty$.

A nondeterministic stationary time series $\{X_t\}$ with AR parameters $\{a_k\}$ is said to have the *autoregression property (ARP)*, if the infinite series

$$\sum_{k=1}^{\infty} a_k X_{t-k}\,,$$

is mean-convergent. It is not hard to show that ARP is satisfied for $\{X_t\}$, if and only if

$$(7.8) \qquad\qquad \sum_{k=1}^{\infty}\sum_{l=1}^{\infty} a_k a_l\, \gamma_{k-l} \;<\; \infty\,.$$

It follows from (6.5) that if $\{X_t\}$ has the ARP, then

$$X_t \;=\; \sum_{k=1}^{\infty} a_k X_{t-k}\; +\; \epsilon_t + e_t,$$

where $e_t = \mathrm{l.i.m.}_{p\to\infty}\, e_t, p \,\epsilon\, H_{-\infty}(X)$. In particular under ARP, if $\{X_t\}$ is *purely nondeterministic*, that is, if $H_{-\infty}(X) = \{0\}$ or the remote past is the trivial subspace then $\{X_t\}$ has *an autoregressive representation in the time-domain*, namely,

$$(7.9) \qquad X_t = \sum_{k=1}^{\infty} a_k X_{t-k} + \epsilon_t .$$

From the AR representation of $\{X_t\}$ or its predictor, it follows that the parameters $\{a_k\}$ and σ^2 can be expressed directly in terms of the covariance sequence $\{\gamma_k\}$. Indeed, from (7.4) we have

$$\hat{X}_1 = \sum_{k=1}^{\infty} a_k X_{1-k} ,$$

but since $X_1 - \hat{X}_1$ is orthogonal to all X_t, $t \leq 0$, it follows that $\{a_k\}$ satisfies the normal equations,

$$\sum_{k=1}^{\infty} a_k \gamma_{j-k} = \gamma_j, \ j \geq 1 ,$$

or in matrix form

$$(7.10) \qquad \begin{bmatrix} \gamma_0 & \gamma_1 & \cdots \\ \gamma_1 & \gamma_0 & \cdots \\ \vdots & \vdots & \vdots \\ \cdot & \cdot & \end{bmatrix} \begin{bmatrix} a_1 \\ a_2 \\ \vdots \\ \cdot \end{bmatrix} = \begin{bmatrix} \gamma_1 \\ \gamma_2 \\ \vdots \\ \cdot \end{bmatrix}$$

Furthermore,

$$(7.11) \qquad \begin{aligned} \sigma^2 &= var(X_1 - \hat{X}_1) = cov(X_1 - \hat{X}_1, X_1) = \\ var(X_1) &- cov(\hat{X}_1, X_1) = \gamma_0 - \sum_{k=1}^{\infty} a_k \gamma_k . \end{aligned}$$

It is instructive to compare and notice the similarities between (7.10) and (6.10). While both give procedures for finding $\{a_k\}$, in the former, one needs to invert larger and larger Toeplitz matrices, and in the latter, inversion of the triangular matrix T is not a problem, but it requires triangular (Cholesky) factorization of the covariance matrix Γ. In practice, these procedures are implemented via recursive algorithms such as the *Durbin-Levinson algorithm* for solving (7.10) and the *innovation algorithm* for solving (6.10), cf. [9,Chap. 5]. Such recursive algorithms are important when predicting future values from the finite past.

8. Prediction from finite past: Partial correlations. Computational aspect of prediction of future values of a time series based on the finite past $\{X_t; \ -n+1 \leq t \leq 0\}$, $0 < n < \infty$, is covered extensively in textbooks on time series. In this section we focus on some theoretical aspects and implications of solving this problem, particularly those emanating from the Durbin-Levinson and Innovation algorithms, and the concept of partial correlations. We shall discuss extensively the role of partial correlations in characterizing, (i) the $AR(p)$ models, (ii) the nondeterministic stationary times series, and (iii) stationary time series with maximum entropy.

In the following $\hat{X}_{t,n}$ denotes the best linear predictor of X_t based on $\{X_s \; ; \; t - n \leq s \leq t-1\}$. It is evident that

$$(8.1) \qquad \hat{X}_{t,n} = \sum_{k=1}^{n} a_{k,n} X_{t-k} ,$$

for some $a_{k,n}$'s satisfying the *normal equations*

$$(8.2) \qquad \begin{bmatrix} \gamma_0 & \gamma_1 & \cdots & \gamma_{n-1} \\ \gamma_1 & \gamma_0 & \cdots & \cdot \\ \vdots & \vdots & \vdots & \vdots \\ \gamma_{n-1} & \cdot & \cdots & \gamma_0 \end{bmatrix} \begin{bmatrix} a_{1,n} \\ a_{2,n} \\ \vdots \\ a_{n,n} \end{bmatrix} = \begin{bmatrix} \gamma_1 \\ \gamma_2 \\ \vdots \\ \gamma_n \end{bmatrix}$$

The prediction error and its variance are given by

$$(8.3) \qquad \epsilon_{t,n} = X_t - \hat{X}_{t,n} = X_t - \sum_{k=1}^{n} a_{k,n} X_{t-k} ,$$

$$(8.4) \qquad \sigma_n^2 = var\,(\epsilon_{t,n}) = \gamma_0 - \sum_{k=1}^{n} a_{k,n} \gamma_k .$$

From (8.3) with $t = n$ we have

$$(8.5) \qquad \epsilon_n, n = X_n - \sum_{k=1}^{n} a_{k,n} X_{n-k}, \; n = 0, 1, 2, \dots ,$$

where $\epsilon_{0,0} = X_0$. Let us define the following two column vectors for any $n \geq 1$.

$$\tilde{\epsilon}_n = (\epsilon_{0,0} , \dots, \epsilon_{n-1,n-1})' ,$$
$$\tilde{X}_n = (X_0, X_1, \dots, X_{n-1})' .$$

Then, (8.5) can be written as

$$(8.6) \qquad \tilde{\epsilon}_n = A_n \tilde{X}_n ,$$

where A_n is a lower triangular matrix with 1 as its diagonal elements, and $-a_{k,n}$ the below diagonal elements of A_n are calculated recursively using the Durbin-Levinson algorithm, cf. [9, Proposition 5.2.1]. Due to the special form of (8.5), these equations can be solved to express the entries of $\tilde{X}_n$ in terms of those of $\tilde{\epsilon}_n$. In fact,

$$(8.7) \qquad X_n = \sum_{k=0}^{n} b_{k,n} \epsilon_{n-k,n-k} , n = 0, 1, 2.. ,$$

and in matrix form

(8.8)
$$\tilde{X}_n = B_n \tilde{\epsilon}_n ,$$

where B_n is also a lower triangular matrix with 1 as its diagonal elements, the subdiagonal elements of the matrix B_n, are obtained using the following recursions (which is a consequence of $B_n A_n = I_n$ and comparable to (6.9)):

(8.9)
$$\begin{cases} b_{n,0} = 1, \quad n = 0, 1, 2, \dots \\ b_{l,n} = \sum_{k=1}^{l} a_{k,n} b_{l-k,n-k} , n \geq 1, \ l = 1, \dots, n \end{cases}$$

Since $\{\epsilon_{n,n}\}_{n=0}^{\infty}$ is an orthogonal sequence, with $\sum_n = diag\,(\sigma_0^2, \sigma_1^2, \dots, \sigma_{n-1}^2)$, (8.8) and (8.6) give the following (modified) Cholesky factorizations of Γ_n and Γ_n^{-1}:

(8.10)
$$\Gamma_n = B_n \Sigma_n B_n', \ \Gamma_n^{-1} = A_n' \Sigma_n^{-1} A_n, n = 0, 1, 2 \dots,$$

these play crucial roles in the numerical evaluation of Gaussian likelihood in time series analysis.

8.1 Partial correlations. The coefficient $a_{n,n}$ in (8.1) can be shown to be the correlation between X_t and $X_{t-(n+1)}$ after removing the linear effect of the intermediate variables $\{X_s ; \ t - n \leq s \leq t - 1\}$, the sequence $\{a_{n,n}\}_1^{\infty}$ is known as the *partial correlation (sequence)* of the time series $\{X_t\}$. Its role as a tool in identification of time series models is well-known, cf. [7]. *Through the Durbin-Levinson algorithm it can be shown that there is, indeed, a one-to-one correspondence between covariance function and partial correlation function of a stationary time series,* for more interesting results concerning the latter see [39]. Also this recursive algorithm gives a useful formula for σ_n^2, defined in (8.4), in terms of the partial correlations:

(8.11)
$$\sigma_n^2 = \gamma_0 \prod_{k=1}^{n} (1 - a_{k,k}^2) .$$

The first important application of (8.11) is the following characterization of $AR(p)$ models, for the proof see [39].

8.1 THEOREM. *A stationary time series $\{X_t\}$ has an $AR(p)$ representation, if and only if*

$$a_{n,n} = 0, \ \text{for} \ n > p .$$

The next theorem, which is also a consequence of (8.11), provides a time-domain characterization of nondeterministic processes, *and a formula for σ^2, the one-step ahead prediction variance, in terms of the partial correlations.* These should be compared with the corresponding results in the spectral-domain, cf. [13,17].

8.2 THEOREM. *(a) A stationary time series with partial correlations $\{a_{n,n}\}$ is nondeterministic, if and only if*

(8.12)
$$\sum_{n=1}^{\infty} a_{n,n}^2 < \infty .$$

(b) The one-step ahead prediction variance of a nondeterministic time series with partial correlations $\{a_{n,n}\}$ is given by

(8.13)
$$\sigma^2 = \gamma_0 \prod_{n=1}^{\infty} (1 - a_{n,n}^2) .$$

This formula is crucial in connecting $AR(p)$ models and the idea of maximum entropy as in Burg [10] and Parzen [32].

8.3 THEOREM. *Among all stationary processes with the same given covariances $\gamma_0, \gamma_1, ..., \gamma_p$, for a fixed $p \geq 1$, the $AR(p)$ processes have the largest one-step ahead prediction variance.*

This theorem follows from (8.13) and Theorem 8.1, by noting that the knowledge of $\{\gamma_0, \gamma_1, ..., \gamma_p\}$ is equivalent to that of $\{\gamma_0, a_{1,1}, ..., a_{p,p}\}$, and for any given p,

$$\sigma^2 = \gamma_0 \prod_{k=1}^{\infty} (1 - a_{k_1 k}^2) \leq \gamma_0 \prod_{k=1}^{p} (1 - a_{k,k}^2),$$

with equality, if and only if

$$a_{k,k} = 0, \quad \text{for all } k > p .$$

8.2 Convergence of finite predictors and their coefficients. Historically, Wold was the first to use (8.1) as the starting point of the development of a structural theory for stationary time series. Noting the similarities of (8.1), (8.7), (8.9) with (6.5), (6.6) and (6.9) it can be seen that Wold had the right starting point. However, some difficulties encountered in studying $\lim_{n\to\infty} \sum_{k=1}^{n} a_{k,n} X_{t-k}$ had made him to change his direction and instead focus on the convergence of (8.7) and its components, from which the Wold decomposition theorem was derived.

In this section we show the convergence of $\hat{X}_{t,n}, \epsilon_{t,n}, \sigma_n^2, a_{k,n}$ and $b_{k,n}$ to their theoretical counterparts defined in Sections 6 and 7. Modes and rates of convergence of these quantities are of critical importance in prediction theory and statistical analysis of time series, see [18,19].

Proofs of the following theorems can be found in S. Degerine (1982): Partial autocorrelation function for a scalar stationary discrete-time process. Proc. 3rd Franco-Belgian Meeting of Statisticians. Bruxelles: Publications des Facultes Universitaires Sait-Louis.pp.79-94.

8.4 THEOREM. *Let* $\{X_t\}$ *be a nondeterministic stationary time series. With notations as before, we have*

(a) $\lim_{n\to\infty} \sigma_n^2 = \sigma^2$.

(b) $var\,(\hat{X}_t - \hat{X}_{t,n}) = \sigma_n^2 - \sigma^2$, l.i.m.$_{n\to\infty} \hat{X}_t, n = \hat{X}_t$ and
   l.i.m.$_{n\to\infty}\ \epsilon_{t,n} = \epsilon_t$.

(c) $\lim_{n\to\infty}\ b_{k,n} = b_k,\ lim_{n\to\infty} a_{k,n} = a_k, k = 1, 2, \cdots$.

The next theorem gives a time-domain characterization of a purely nondeterministic time series in terms of the norm (l^2 − norm) convergence of the sequence $\tilde{b}_n = (1, b_{1,n}, ..., b_{n,n} 0, ..., 0)'$ to $b = (1, b_1, b_2, ...)'$.

8.5 THEOREM. *A nondeterministic stationary time series* $\{X_t\}$ *is purely nondeterministic, if and only if*

$$\lim_{n\to\infty} ||\tilde{b}_n - b||_{l^2} = \lim_{n\to\infty} \sum_{k=0}^{\infty} (b_k - b_{k,n})^2 = 0.$$

A result similar to Theorem 8.5 can be proved for the l^2 − convergence of $\{a_{k,n}\}_{k=1}^n$ to $\{a_k\}_{k=1}^{\infty}$ by imposing a boundedness condition on the spectral density of $\{X_t\}$. A theorem of Baxter [3] shows that $a_{k,n}$ converges to a_k exponentially, as $n \to \infty$, if and only if

$$\sum_{k=1}^{\infty} |a_{k,k}| < \infty\,,$$

it is instructive to compare this condition with (8.12).

9. The second prediction problem: Inverse correlations. The *second prediction problem* [21] or the *interpolation problem* is concerned with estimating X_0 based on the infinite past and future of $\{X_t\}$. More precisely, with $H_0' = \overline{sp}\,\{X_t;\ t \neq 0\}$, the issues of interest are:

(i) finding $\hat{X}_0' \epsilon H_0'$ such that

$$var\,(X_0 - \hat{X}_0') = \inf_{Y \epsilon H_0'}\ var\,(X_0 - Y)\,,$$

i.e., the variance of the interpolation error is as small as possible,

(ii) expressing $\sigma'^2 = var\,(X_0 - \hat{X}_0')$ in terms of γ_k.

This problem is the prototype of several important problems arising in time series analysis. For example, it is related to the missing value problem, outlier detection, influential observations, inverse correlations and processes. It was solved first by Kolmogorov [23] in the spectral domain. In this section we provide a time-domain solution for this problem by relying on the concept and properties of multi-step ahead predictors. This approach provides a link between the two approaches to time series, in the sense that, certain intrinsically spectral-domain concepts such as the inverse correlations and spectrum are given physical meaning in the time-domain, see Sec. 9.2 and [37,5]. More general interpolation problems when several values are missing can be solved by using this approach, cf. [37].

9.1 Semi-finite interpolation problem. For a fixed integer $m \geq 0$, the problem of estimating X_0 based on $\{X_t \; ; \; t \leq m, \; t \neq 0\}$ subsumes the first two prediction problems, and is a more challenging problem than either one, cf. [29,37]. In the following we refer to it as the *semi- finite interpolation problem*, and show that in the time-domain it can be solved rather conveniently by using the fact that the multi-step ahead prediction errors are orthogonal to the past values of the time series, cf. (7.1). Throughout this section, unless stated otherwise, we assume that $\{X_t\}$ is a nondeterministic time series.

Let $H'_{0,m} = \overline{sp}\,\{X_t \; ; \; t \leq m, \; t \neq 0\}$ and $\hat{X}_k$, $0 \leq k \leq m$, be the best linear predictors of X_k based on $\{X_t \; ; \; t \leq -1\}$. Of course, $X_k - \hat{X}_k$, $0 \leq k \leq m$, are orthogonal to $H_{-1} = \overline{sp}\,\{X_t \; ; \; t \leq -1\}$, and it can be shown that [37],

$$(9.1) \qquad H'_{0,m} = H_{-1} \oplus sp\,\{X_k - \hat{X}_k \; ; \; 1 \leq k \leq m\}\,.$$

This orthogonal decomposition of $H_{0,m'}$ is the key in reducing the semi-finite interpolation problem to that of finding the orthogonal projection of X_0 onto the finite-dimensional space

$$(9.2) \qquad S_m = sp\,\{X_k - \hat{X}_k \; ; \; 1 \leq k \leq m\}\,.$$

Let $\hat{X}'_{0,m}$ and σ'^2_m be the best linear interpolator and interpolation error of X_0 based on $H'_{0,m}$. Since $H'_{0,m} \to H'_0$, as $m \to \infty$, it follows from [9,Problem 2.1.8] that

$$(9.3) \qquad \begin{cases} \underset{m \to \infty}{\mathrm{l.i.m.}}\; \hat{X}'_{0,m} = \hat{X}'_0\,, \\ \underset{m \to \infty}{\lim}\; \sigma'^2_m = \sigma'^2\,, \end{cases}$$

providing a solution of the second prediction problem once $\hat{X}_{0,m'}$ and σ'^2_m are computed.

From (9.1), since $\hat{X}'_{0,m}$ is the orthogonal projection of X_0 onto $H'_{0,m}$, we have

$$(9.4) \qquad \hat{X}'_{0,m} = \hat{X}_0 + \sum_{k=1}^{m} \beta_{k,m}\,(X_k - \hat{X}_k)\,,$$

where the regression coefficients $\beta_m = (\beta_{1,m}, ..., \beta_{m,m})'$ satisfy the normal equations, cf. (7.2) - (7.3),

$$(9.5) \qquad (T_m T'_m + bb')\,\beta_m = b\,,$$

with $b = (b_1, ..., b_m)'$ consisting of the first m MA parameters of $\{X_t\}$. Since the coefficient matrix in (9.5) is a rank-1 perturbation of $T_m T'_m$, it can be inverted explicitly and its inverse involves the AR parameters of $\{X_t\}$, cf. (6.11) and [37, Theorem 3.1], thus

$$(9.6) \qquad \begin{cases} \beta_{k,m} = (\sum_{i=0}^{m} a_i^2)^{-1}\,(a_k - \sum_{i=1}^{m-k} a_i a_{i+k}),\; k = 1, 2, ..., m\,, \\ \sigma'^2_m = \sigma^2(\sum_{i=0}^{m} a_i^2)^{-1}\,. \end{cases}$$

These rather simple formulae expressing $\beta_{k,m}$ and $\sigma_m'^2$ in terms of the AR parameters of $\{X_t\}$ are crucial in finding $\hat{X}_0'$ and σ'^2 in (9.3) and solving the second prediction problem. Before we embark on this, it is helpful to define the notion of *m-innovation process for* $\{X_t\}$. For an integer $m \geq 0$, the m-innovation process of $\{X_t\}$ is denoted by $\{\epsilon_{t,m}'\}$ and defined by

$$\epsilon_{t,m}' = X_t - \hat{X}_{t,m}' ,$$

where $\hat{X}_{t,m}'$ is the best linear interpolator of X_t based on $\{X_s : s \leq t+m, s \neq t\}$. From (7.1), (9.4) and (9.6) it follows that the m-innovation process of $\{X_t\}$ can be represented as

$$(9.7) \qquad \epsilon_{t,m}' = (\sum_{i=0}^{m} a_i^2)^{-1} (\epsilon_t - \sum_{i=1}^{m} a_i \epsilon_{t+i}) ,$$

which is an $MA(m)$ in terms of the future values of the innovation process $\{\epsilon_t\}$ of $\{X_t\}$. Thus, when $\{X_t\}$ is a Gaussian process, its m-innovation process is m-dependent. Various applications of the m-innovations in detecting outliers in time series data are discussed in [38] and references therein.

9.2 Solution of the second prediction problem. In view of (9.6) and (9.3) the condition

$$(9.8) \qquad \sum_{i=1}^{\infty} a_i^2 < \infty ,$$

seems to be a natural requirement for the second prediction problem to have a nontrivial solution. Under this condition, it follows that

$$(9.9) \qquad \sigma'^2 = \sigma^2 (\sum_{i=0}^{\infty} a_i^2)^{-1} \neq 0 ,$$

expressing σ'^2 in terms of components of $\{\gamma_k\}$, and

$$(9.10) \qquad \rho_k' := \lim_{m \to \infty} \beta_{k,m} = (\sum_{i=0}^{\infty} a_i^2)^{-1} (a_k - \sum_{i=1}^{\infty} a_i a_{i+k}), \ k \geq 1 ,$$

giving the coefficients of $\hat{X}_0'$ in its formal series representation:

$$(9.11) \qquad \hat{X}_0' \sim \hat{X}_0 + \sum_{k=1}^{\infty} \rho_k' (X_{t+k} - \hat{X}_{t+k}) .$$

Note that $\{\rho_k'\}$ is precisely the *inverse autocorrelations* of $\{X_t\}$, cf. [11,5]. The problem of mean-convergence of (9.11) is closely related to the $AR (\infty)$ representation of $\{X_t\}$, see Sec. 7.3. However, a forward $MA (\infty)$ representation of $\hat{X}_0'$ follows from (9.7) when the condition (9.8) is satisfied.

The condition (9.8) has a geometric interpretation very similar to the notion of nondeterminism, cf. (4.5), (8.12): A stationary time series $\{X_t\}$ is said to be *minimal*, if

$$X_0 \notin H_0' .$$

A characterization of minimality in the spectral domain was given by Kolmogorov [23]. Here, using (9.6) it can be shown that [37]: *A nondeterministic stationary time series is minimal, if and only if (9.8) is satisfied.* The requirement of minimality is closely related to the invertibility of finite-parameter models, see [9,p. 157].

When $\{X_t\}$ is a minimal stationary time series, the AR and MA parameters are both square-summable and have identical properties, see Sec. 6.4. This suggests a duality between $MA(\infty)$ and $AR(\infty)$ representations and models for a time series. Although for finite-parameter models it is easier to notice and exploit such duality, cf. [21,34], for a general stationary time series such duality can be established through (9.10) and $\{\epsilon_t'\}$ the *two-sided innovation process* of $\{X_t\}$, c.f. [25], defined by

$$\epsilon_t' = \underset{m \to \infty}{\text{l.i.m.}} \; \epsilon_{t,m}' = (\sum_{i=0}^{\infty} a_i \epsilon_{t+i}),$$

which is known as the inverse (dual) process of $\{X_t\}$, its correlation function $\{\rho_k'\}$ in (9.10) is known as the *inverse correlations* of $\{X_t\}$, for more information see [11,5,37].

10. The third prediction problem: Canonical correlations. The *third prediction problem* [21] is concerned with computing ρ the *canonical or maximal correlation* between the past $\{X_t \; ; \; t \leq 0\}$ and the future $\{X_t \; ; \; t \geq 1\}$. More precisely, with F denoting the closed linear span of the future values, the problem is to compute

(10.1) $$\rho = \sup_{\substack{X \epsilon H_0 \\ Y \epsilon F}} \; | \; corr \; (X, Y) \; | \; ,$$

and express it in terms of $\{\gamma_k\}$, where $corr(X, Y)$ stands for the correlation coefficient between X and Y. It is evident that $\rho \leq 1$; the past and future of $\{X_t\}$ is said to be at *positive angle* [21], if $\rho < 1$.

The number ρ plays important roles in several theoretical and practical problems arising in prediction theory and time series analysis. For example, when $\rho < 1$ the time series $\{X_t\}$ has AR representation [35], see Sec. 7.3; is minimal and consequently (9.8) is satisfied [37]. Canonical correlations are also used in the state-space representation of a time series [2], and model identification for time series data [47].

For a general stationary time series, Jewell et al. [31] have given an algorithm in the spectral domain for computing the canonical correlations as the eigenvalues of an infinite-dimensional Hankel operator. It is known [21,54] that for $ARMA$ processes the canonical correlations and components can be computed by solving linear systems of algebraic and determinantal equations.

In this section we give a time domain algorithm for computing canonical correlations and components of a time series by solving linear systems and determinantal equations. First, we solve the "semi-finite" version of (10.1), that is for $m \geq 1$, we compute ρ_m the canonical correlation between the infinite past and finite future $\{X_k : 1 \leq k \leq m\}$. *Even though this is an infinite-dimensional problem, in the setup of multivariate analysis [40,28] we are able to relate ρ_m to the generalized eigenvalues of two $m \times m$ matrices G_m and Γ_m.* This is achieved by relying on the orthogonality of the multi-step ahead prediction errors, and reducing the problem of computing ρ_m to the familiar problem of predicting a linear functional of future values *subject to a quadratic constraint on its coefficients*, which is of interest on its own right and related to the first prediction problem. Here we focus only on the computation of the first or largest canonical correlation.

10.1 Best predictable aspect of future. For a fixed integer $m \geq 1$, and given $c = (c_1, ..., c_m)'$ consider the problem of predicting the random variable $X = \sum_{k=1}^{m} c_k X_k$ based on the infinite past $\{X_t ; t \leq 0\}$ of a time series $\{X_t\}$. It follows from the linearity of the projection operator that

$$\hat{X} = \sum_{k=1}^{m} c_k \hat{X}_k , X - \hat{X} = \sum_{k=1}^{m} c_k (X_k - \hat{X}_k) ,$$

and from (7.2) - (7.3) that

(10.2) $$var (X - \hat{X}) = c' G_m c .$$

The measure of (linear) predictability of X is defined in [31], as

$$\lambda(X) = 1 - \frac{var (X - \hat{X})}{var (X)} ,$$

it follows from (10.2) that

(10.3) $$\lambda(X) = 1 - \frac{c' G_m c}{c' \Gamma_m c} .$$

For a given m, whenever the c_k's are unknown, the best predictable aspect of the future $\{X_k ; 1 \leq k \leq m\}$ is obtained by choosing c so that $\lambda(X)$ in (10.3) is maximized. This is equivalent to minimizing $c' G_m c$ subject to $c' \Gamma_m c = 1$, and its solution is given in Rao [40,p. 583]: Let $\lambda_1 < ... < \lambda_k$ $(k \leq m)$ be the distinct roots of the determinantal equation

(10.4) $$det (G_m - \lambda \Gamma_m) = 0 ,$$

and $c_{(1)}, ..., c_{(m)}$ be the corresponding orthonormalized eigenvectors;

$$(G_m - \lambda_j \Gamma_m) c_{(j)} = 0, c'_{(i)} \Gamma_m c_{(j)} = \delta_{i,j}, 1 \leq i, j \leq m .$$

Then, with $F_m = (X_1, ..., X_m)'$, $X_{(i)} = c'_{(i)} F_m$ is the ith best predictable aspect of future with the measure of predictability $1 - \lambda_i$, $i = 1, 2, .., k$.

10.2 Canonical correlation. It is well-known that there is a close connection between prediction (regression) problems and the concept of correlation. The root of this phenomenon can be traced to the following simple property of the geometry of Hilbert spaces: For N a subspace of a Hilbert space H, and X an element of H with P_N^X denoting its orthogonal projection onto N, we have

$$(X, P_N^X) = (X, X - X + P_N^X) = \|X\|^2 - (X, X - P_N^X)$$
$$= \|X\|^2 - \|X - P_N^X\|^2 .$$

Starting from this identity it can be shown that for any two subspaces M and N of H, we have

$$(10.5) \qquad \sup_{\substack{X \in M \\ Y \in N}} |corr\,(X, Y)| = \sqrt{1 - \inf_{\substack{X \in M \\ \|X\|=1}} \|X - P_N^X\|^2}.$$

From this it can be seen that computation of the canonical correlation on the left is reduced to a constrained prediction problem on the right. Applying this identity with $N = H_0$ and $M = sp\,\{X_1, ..., X_m\}$, the left-hand side of (10.5) is ρ_m, and it follows from (10.3) - (10.4) that

$$(10.6) \qquad \rho_m = \sqrt{1 - \lambda_{1,m}} ,$$

where $\lambda_{1,m}$ is the smallest root of (10.4).

The final step is to show that

$$(10.7) \qquad \rho = \lim_{m \to \infty} \rho_m ,$$

which follows from the fact that ρ_m is increasing in m and the definition of ρ, c.f. [28, Lemma 3.2].

Thus we have shown that for m large the roots of the determinantal equation (10.4) can be used to approximate canonical correlations of a stationary time series. For the sunspots numbers even for $m = 4$ the approximation in (10.7) is quite satisfactory [28].

10.3 The role of $\mathbf{G_m}$ in model identification. Perhaps the single most challenging problem in time series analysis is that of estimating the orders of p and q or identifying an appropriate $ARMA\,(p, q)$ model for a given time series data $X_1, ..., X_n$. In this section we indicate the potential role of the matrices G_m, $m \geq 1$, in unifying various existing approaches. We recall that there are two general and rather distinct approaches to the model identification problem in time series.

The first approach is correlation-based in that it involves the sample auto-correlations or some functionals of them. There are many seemingly different methods which belong to this approach, the most notable are the *corner method*, $R-$ and $S-$ array, canonical correlations, cf. [33]. In [33] a strong case is made that most of these methods have a common basis, and that a unified approach may be found in the analysis of the matrix of the covariance of forecast values $\hat{X}_k$, $1 \le k \le m$, where m is a large integer. We note that this matrix is closely related to the matrix G_m in (7.3), thus the latter matrix can play similar roles in unifying these methods.

The second approach is information-based, almost invariably it is concerned with minimizing a criterion which is a linear function of an estimate of $log\ \sigma^2$ and the number of parameters in the candidate model. The most notable examples of these are the AIC, BIC and CAT [9]. Since the criterion involves σ^2, the one-step prediction variance, the identified model may possess certain optimality properties for one-step ahead predictions. However, if the goal is multi-step ahead predictions, then it is plausible that the criterion must be modified to reflect this goal. Thus, to identify an $ARMA(p,q)$ model for predicting the m future values, the criterion to be minimized should possibly contain $log\ c'G_m c$, where the m-vector c is as in Sec. 10.1.

11. Intervention analysis. Time series are frequently affected by policy changes and other external events usually referred to as *interventions*. Intervention analysis as developed by Box and Tiao [8] has been used extensively in various areas of applications, a slight modification of it is used in detecting various outliers and structural disturbances in time series, cf. [48]. In the Box-Tiao approach, difference equation models, similar to (6.1), are used to represent the possible dynamic characteristic of the interventions and the noise. In this section we propose a nonparametric approach to intervention analysis, it is motivated by *the simple and intuitively appealing fact that the multi-step ahead forecast errors of post-intervention measurements contain all the relevant information regarding the interventions and their impacts.* As a result, here too, the matrix G_m and its inverse play a major role in relating the problem of intervention analysis and its solution to the first prediction problem.

Suppose that the first intervention occurs at a known time period, say $t = 1$. Therefore the post-intervention measurements $X_1, ..., X_m$ cannot be observed directly, instead one observes $Y_t = f_t + X_t$, for $t \ge 1$, where f_t is the impact of possibly many interventions on the level of the time series at time t. Let $f = (f_1, ..., f_m)'$ and $\hat{X}_l$, $1 \le l \le m$, be the l-step ahead predictor of X_l based on the pre-intervention data $\{X_t; t \le 0\}$. The *l-step ahead forecast errors* of post-intervention measurements are defined by

$$(11.1) \qquad \hat{f}_l = Y_l - \hat{X}_l, \ 1 \le l \le m.$$

In what follows we take $\hat{f} = (\hat{f}_1, ..., \hat{f}_m)'$ as an estimator of f. To study statistical properties of $\hat{f}$ as an estimator of f, it is advantageous to express f in terms of the familiar components of the series $\{X_t\}$ and the measurements. From (11.1) and (7.2) it follows that

$$(11.2) \qquad \hat{f} = f + T_m \epsilon,$$

where $\epsilon = (\epsilon_1, ..., \epsilon_m)'$. Since $E(\epsilon) = 0$, we have

$$E\hat{f} = f,$$

i.e. $\hat{f}$ is an *unbiased* estimator of f, with covariance matrix

$$(11.3) \qquad\qquad cov\,(\hat{f})\ =\ G_m$$

We note that $\hat{f}$ is a *nonparameteric* estimator of f in that it is not based on a parametric model for f as in [8]. Therefore, the time series plot of the entries of $\hat{f}$ could provide valuable information about the nature and possibly the number of interventions and their impacts.

If parametric models for the impacts of intervention is postulated as in [8], it can be shown that this is equivalent to writing

$$(11.4) \qquad\qquad f_t\ =\ \beta_1 x_{t,1}\ +...+\ \beta_p x_{t,p}\ ,\ 1\leq t\leq m\ ,$$

for some known variables $x_{t,i},\ 1\leq i\leq p$. In matrix form, using (11.2) with obvious notation we have

$$(11.5) \qquad\qquad \hat{f}\ =\ X\beta\ +\ T_m\epsilon\ .$$

Since the covariance matrix of $\hat{f}$ in the linear model (11.5) is not a constant multiple of the identity, the generalized least square estimator of β is given by

$$(11.6) \qquad \hat{\beta}\ =\ (X'V^{-1}X)^{-1}X'V^{-1}\hat{f}\ ,\ \text{where}\ V\ =\ T_m T'_m\ .$$

Next, we provide an alternative representation for $\hat{\beta}$ which not only makes its statistical properties transparent, it also facilitates interpretation and comparison of this estimator with the one proposed in [8]. Defining the auxiliary matrix U by

$$(11.7) \qquad\qquad U\ =\ T_m^{-1}X\ ,$$

and using (11.5), $\hat{\beta}$ can be rewritten as

$$(11.8) \qquad \hat{\beta}\ =\ \beta\ +\ (X'V^{-1}X)^{-1}X'V^{-1}T_m\epsilon\ =\ \beta\ +\ (U'U)^{-1}U'\epsilon\ .$$

Now it is evident that $\hat{\beta}$ is an unbiased estimator of β, and the error of estimating β by $\hat{\beta}$, namely $(U'U)^{-1}U'\epsilon$, has an obvious interpretation in the framework of linear regression, also for some specific cases this error is related to interpolation of missing values and the m-innovation process introduced in Sec. 9. The covariance matrix of $\hat{\beta}$ is given by

$$(11.9) \qquad\qquad cov\,(\hat{\beta})\ =\ \sigma^2(U'U)^{-1}\ .$$

Through examples one can demonstrate the versatility of the intervention analysis, and clarify the relationship between the formula (11.6) with that obtained in [37] and interpolation of missing values and m-innovations, for relevant examples see [48,p. 4].

REFERENCES

[1]   AKUTOWICZ, E.J., *On an explicit formula in least squares prediction*, Math. Scan., 5 (1957), pp. 261-266.

[2]   AKAIKE, H., *Markovian representation of stochastic processes by canonical variables*, SIAM J. Control, 13 (1975), pp. 162-173.

[3]   BAXTER, G., *An asymptotic result for the finite predictor*, Math. Scand., 10 (1962), pp. 137-144.

[4]   BERK, K.N., *Consistent autoregressive spectral estimates*, Ann. Statist. 2 (1974), pp. 489-502.

[5]   BHANSALLI, R.J., *On a relationship between the inverse of a stationary covariance matrix and the linear interpolator*, J. of Applied Probability,, 27 (1990), pp. 156-170.

[6]   BRILLINGER, D.R., *Time Series: Data Analysis and Theory*, Holt, New York, 1981.

[7]   BOX, G.E.P. AND JENKINS, G.M., *Time Series Analysis: Forecasting and Control. Holden-Day*, San Francisco, 1976.

[8]   BOX, G.E.P., AND TIAO, G.C., *Intervention analysis with applications to environmental and economic problems*, J. of the American Statistical Association, 70 (1975), pp. 70-79.

[9]   BROCKWELL, P.J. AND DAVIS, R.A., *Time Series: Theory and Methods*, Springer- Verlag, New York, 1987.

[10] BURG, J.P., *Maximum entropy spectral analysis*, in *Proc. 37th Meeting Soc. of Exploration Geophysicists*, Oklahoma City, OK, 1967.

[11] CLEVELAND, W.S., *The inverse autocorrelations of time series and their applications*, Technometrics, 14 (1972), pp. 277–283.

[12] CRAMER, H., *On the theory of stationary random processes*, Ann. of Math., 41 (1940), pp. 214–230.

[13] DOOB, J.L., *Stochastic Processes*, John Wiley, New York, 1953.

[14] DURBIN, J., *The fitting of time series models*, Rev. Int. Inst. Stat., 1960, pp. 233-244.

[15] GANI, J., *The Making of Statisticians*, Springer-Verlag, New York, 1982.

[16] GRENANDER, U., *Abstract Inference*, Wiley, New York, 1981.

[17] HANNAN, E.J., *Multiple Time Series*, John Wiley, New York, 1970.

[18] HANNAN, E.J., *Rational transfer function approximation*, Statist. Sci., 2 (1987), pp. 135-161.

[19] HANNAN, E.J. AND RISSANEN J., *Recursive estimation of ARMA order*, Biometrika, 69 (1982), pp. 81-94.

[20] HELSON, H., *Lectures on Invariant Subspaces*, Academic Press, New York, 1964.

[21] HELSON, H. AND SZEGO, G., *A problem in prediction theory*, Ann. Mat. Pura Appl., 51 (1960), pp. 107-138.

[22] KALMAN, R.E. AND BUCY, R.S., *New results in linear filtering and prediction theory. J. of Basic Engineering*, Transactions *ASME* Series D, 83 (1961), pp. 95-108.

[23] KOLMOGOROV, A.N., *Stationary sequences in a Hilbert space*, Bull. Moscow State University, 2 (1941), pp. 1-40.

[24] LAMBERT, P.J. AND POSKITT, D.S., *Stationary Processes in Time Series Analysis: The Mathematical Foundations*, Vandenhoeck & Ruprecht, Gottingen, 1983.

[25] MASANI, P.R., *The prediction theory of multivariate stochastic processes, III.*, Acta Math. 104 (1960), pp. 141-162.

[26] MASANI, P.R., *Commentary on the prediction-theoretic papers Norbert Wiener: Collected Works III*, P.R. Masani (ed.), MIT Press, Cambridge, 1981, pp. 276-306.

[27] MIAMEE, A.G. AND POURAHMADI, M., *Wold decomposition, prediction and parameterization of stationary processes with infinite variance*, Probab. Theory and Related Fields, 79 (1988), pp. 145-164.

[28] MIAMEE, A.G. AND POURAHMADI, M., *Computation of canonical correlation between past and future of a time series*, J. of Time Series Analysis (1992).

[29] NAKAZI, T., *Two problems in prediction theory*, Studia Math., 78 (1984), pp. 7-14.

[30] NIEMI, H., *Theoretical properties of linear predictors for ARIMA models*, Scand. J. Statist., II (1984), pp. 113–122.

[31] JEWELL, N.P. AND BLOOMFIELD, P., *Canonical correlations of past and future for time series: Definitions and theory*, Ann. Statist., 11 (1983), pp. 837–847.

[32] PARZEN, E., *Time series, statistics, and information*, IMA preprints series, # 663 (1990).

[33] PICCOLO, D. AND WILSON, T.G., *A unified approach to ARMA model identification and preliminary estimation*, J. of Time Series Analysis, 5 (1984), pp. 183–204.

[34] PIERCE, D.A., *A duality between autoregressive and moving average processes concerning their least squares estimates*, Ann. Math. Statist., 41 (1970), pp. 422–426.

[35] POURAHMADI, M., *A matricial extension of the Helson-Szego theorem and its applications in multivariate prediction*, J. of Multivariate Analysis, 5 (1985), pp. 265-275.

[36] POURAHMADI, M., *Autoregressive representations of multivariate stationary stochastic processes*, Probab. Theory and Rel. Fields, 80 (1988), pp. 315-322.

[37] POURAHMADI, M., *Estimation and interpolation of missing values of a stationary time series*, J. of Time Series Analysis, 10 (1989), pp. 149-169.

[38] POURAHMADI, M., *m-innovation, leave-k-out residuals and outliers in time series*, Submitted (1990).

[39] RAMSEY, F.L., *Characterization of the partial autocorrelation function*, Ann. of Statist., 2 (1974), pp. 1296-1301.

[40] RAO, C.R., *Linear Statistical Inference and its Applications*, Wiley, New York, 1973.

[41] RIESZ, F. AND NAGY, B., *Functional Analysis*, Frederick Ungar Co., New York, 1955.

[42] ROBINSON, E., *Infinitely Many Variates, Griffin's Statistical Monographs & Courses*, London.

[43] ROSENBLATT, M., *Stationary Sequences and Random Fields*, Birkhauser, Boston,, 1985.

[44] SCHUSTER, A., *On the investigation of hidden periodicities with application to a supposed 26 day period of meteorological phenomena*, Terr. Magn., 3 (1898), pp. 13-41.

[45] SHUMWAY, R.H., *Applied Statistical Time Series Analysis*, Prentice-Hall, New Jersey, 1988.

[46] SLUTSKY, E., *The summation of random causes as the source of cyclic processes*, (Russian) Problem of Economic Conditions 3; Engl. transl. Econometrica 5 (1937), 105-106.

[47] TIAO, G.C. AND TSAY, R.S., *Consistent estimates of autoregressive parameters and extended sample autocorrelations functions for stationary and non-stationary ARIMA models*, J. of Amer. Stat. Assoc., 79 (1984), pp. 84-96.

[48] TSAY, R.S., *Outliers, level shifts and variance changes in time series*, J. of Forecasting, 7 (1988), pp. 1-20.

[49] TUKEY, J.W., *Discussion emphasizing the connection between analysis of variance and spectrum analysis*, Technometrics, 3 (1961), pp. 1-29.

[50] WEGMAN, E.J., *Another look at Box-Jenkins forecasting procedures*, Comm. in Statist., B 15 (1986), pp. 523-530.

[51] WIENER, N., *Time Series*, The M.I.T. Press, Cambridge, 1949.

[52] WIENER, N. AND MASANI, P.R., *The prediction theory of multivariate stationary processes*, I. Acta Math., 1957, pp. 111-150; II. Acta Math. 99 (1958), 93-137.

[53] WOLD, H., *A Study in the Analysis of Stationary Time Series*, Stockholm, Almquist & Wiksell, 1954.

[54] YAGLOM, A.M., *Stationary Gaussian processes satisfying the strong mixing condition and best predictable functional*, Proc. Int. Research Seminar of the Statistical Laboratory, University of California, Berkeley (1963), pp. 241-252.

[55] YULE, G.U., *On the time correlation problem*, J. of Roy. Statist. Soc., 84 (1921), pp. 497-526.

[56] YULE, G.U., *On a model of investigating periodicities in disturbed series with special reference to Wolfer's sunspot numbers*, Phil. Trans. A, 226 (1927), pp. 267-298.

SEMIPARAMETRIC METHODS FOR TIME SERIES*

PETER M. ROBINSON†

Abstract. Various methods of semiparametric inference in time series are discussed, each involving some form of smoothed estimation of a nonparametric component. Four main topics are covered. The first is efficient or robust inference on regression-type models in the presence of disturbance autocorrelation of unknown form, with extension to multiple systems in which only some equations are parameterised and the full system has a nonparametric frequency response function. Nonparametric spectral and cross-spectral estimation is involved here, and some discussion of automatic bandwidth determination is included. In the second topic the spectral density itself has a semiparametric character. In a time series exhibiting long-memory behaviour, the logged spectrum is dominated near the origin by a linear component, with unknown slope, but nonparametric effects can be significant at other frequencies, as when no attempt is made to fully parameterise the smooth spectrum of a fractionally differenced process. We consider inference based on a simple estimates of regression models when the errors have this behaviour. The third topic concerns a class of U- or V-statistics which is useful in semiparametric models for time series in which even getting root-N-consistent parameter estimates is challenging; and in specification of nonparametric predictors. The statistics are functionals of nonparametric regression and derivative-of-probability-density estimates. The final topic is concerned with developing tests with good consistency properties, based on an approximation of the Kullback–Leibler information criterion which employs nonparametric probability density estimates. The main application is to testing for independence in time series with marginal density of unknown form; another is to testing for reversibility in time series.

Key words. time series, semiparametric methods, smoothed nonparametric estimation.

AMS(MOS) subject classifications. 62M10

1. Introduction. This paper discusses recent research on semiparametric inference in time series analysis. A common feature of the methods is that they all employ smoothed estimates of a nonparametric function, where these estimates are not computed for their own sake but to fill the gap left by only a partial parameterization of a statistical model, or to test a nonparametric hypothesis with reasonable efficiency. The paper touches on aspects of the main themes of all four weeks of the IMA Summer Program on Time Series Analysis. Its general focus is statistical inference in time series, but the problems included entail variously non-linearity and non-Gaussianity in time series, long-range dependence, and time series models arising in economic and engineering.

We discuss four topics. The first concerns efficient inference on semiparametric regression or dynamic regression models; our theory allows the degree of smoothing in the nonparametric estimation to depend on the data, and we include a discussion of automatic smoothing, both these aspects being of importance also in the other topics. The second topic concerns estimation of semiparametric time series models allowing long-range dependence, and its applications to robust inference in

---

*Invited Paper presented on July 16, 1990 in honour of Professor J.W. Tukey, at the IMA Summer Program on New Directions in Time Series Analysis. This article is based on research funded by the Economic and Social Research Council (ESRC) reference number: R000231441.

†Department of Economics, London School of Economics, Houghton Street, London, WC2A 2AE, England

regression models. The third topic concerns a class of U- or V-statistics which can be useful in testing a variety of semiparametric models for time series, such as regression models including both a parametric and a nonparametric regression component, and index models (such as probit regression) with error distribution of unknown form, as well as in specifying predictors for nonparametric nonlinear time series. The fourth topic concerns a class of tests of nonparametric hypotheses based on nonparametric estimation of entropy; tests of serial independence and time series reversibility are mentioned. The four topics are discussed in much greater detail in the papers Robinson [19], [20], [21], [22].

The nonparametric functions which we have to estimate by smoothing include power and cross-spectral densities, and probability and regression functions and their derivatives. Professor J.W. Tukey's early work lead to the sophisticated development of power spectrum estimation, which has to some extent preceded that of probability density estimation. Additionally, the fast Fourier transform, developed by Cooley and Tukey [6] for a power spectrum estimation, was suggested for probability density estimation at a much later date. Tukey [25] is an early direct contribution to nonparametric regression estimation, while his many writings on data analysis and robust estimation have influenced developments in both time series and nonparametric inference. Our combination of spectrum and probability density estimation is also a reminder of the important work of Parzen, who wrote almost simultaneously on both topics, e.g. [14], [15].

2. Efficient inference on semiparametric linear systems, and automatic bandwidth selection. This section is based on work of Robinson [20]. We consider a vector covariance stationary time series

$$\begin{matrix} Z_t \\ r \times 1 \end{matrix} = \begin{bmatrix} X_t \\ Y_t \end{bmatrix} \begin{matrix} p \times 1 \\ q \times 1 \end{matrix} , \qquad t = 0, \pm 1, \dots ,$$

where $p \geq 1$, $q \geq 1$. We observe Z_t for $t = 1, \dots, N$. It is assumed that there exists a sequence $\{\alpha_j\}$ of $m \times r$ vectors, $m \leq q$, such that

(2.1) $$U_t \triangleq \sum_{j=-\infty}^{\infty} \alpha_j Z_{t-j}$$

is covariance stationary and satisfies

$$E(X_0 U_t') = 0, \qquad t = 0, \pm 1, \dots .$$

This model is a quite general version of a linear system widely employed in econometrics and engineering, and includes regression and dynamic regression models. The X_t represent "inputs" to the system and the Y_t represent "outputs". The α_j are in general unknown, and the unobservable U_t represent "errors" or "disturbances". The number m of equations in the system may equal the number q of outputs Y_t, when the system can be termed "incomplete"; incomplete systems are common in econometrics.

It is assumed that Z_t has a positive definite spectral density (spectrum),

$$(2.2) \qquad S_Z(\lambda) = \frac{1}{2\pi} \sum_{j=-\infty}^{\infty} E(Z_0 Z_j') e^{-ij\lambda}, \quad -\pi < \lambda \leq \pi .$$

Although $S_Z(\lambda)$ is assumed nonparametric, (2.1) implies that

$$\alpha(\lambda) S_{ZX}(\lambda) = 0 ,$$

where S_{ZX} consists of the first p columns of S_Z and $\alpha(\lambda)$ satisfies

$$\alpha_j = \int_{-\pi}^{\pi} \alpha(\lambda) e^{ij\lambda} d\lambda .$$

It is assumed that $\alpha(\lambda)$ is parametric over a subset Λ of $(-\pi, \pi)$, that is we know a function $\alpha(.;.)$ such that for some unknown d-vector θ,

$$\alpha(\lambda) = \alpha(\lambda; \theta), \qquad \lambda \in \Lambda .$$

In view of (2.1), (2.2) the spectrum of U_t is

$$(2.3) \qquad S_U(\lambda) = \alpha(\lambda) S_Z(\lambda) \alpha(\lambda)^* .$$

$S_U(\lambda)$ is nonparametric even on Λ, because S_Z is nonparametric.

The complex regression function of Y_t on X_t is

$$(2.4) \qquad \beta(\lambda) = S_{YX}(\lambda) S_X^{-1}(\lambda)$$

where $S_{ZX}(\lambda) = [S_{YX}'(\lambda), \ S_{XX}'(\lambda)]'$. When the system is complete, $\beta(\lambda)$ is parametric on Λ. When the system is incomplete, $\beta(\lambda)$ is nonparametric even on Λ.

The model is semiparametric in the following sense. It includes a parametric function $\alpha(\lambda; \theta)$, $\lambda \in \Lambda$. It includes the nonparametric functions $S_U(\lambda)$ and $\beta(\lambda)$ for $\lambda \in (-\pi, \pi)$, as well as $\alpha(\lambda)$ for $\lambda \in (-\pi, \pi) - \Lambda$. The semiparametric aspect is relevant to the efficient estimation of θ.

In Robinson [20] a general class of $\sqrt{N}$-consistent and asymptotically normal estimates of θ is given; these can be termed instrumental variables estimates, and there exists a choice of instruments which produces an estimate with the minimum asymptotic variance in this class. These instruments depend on $S_U(\lambda)$ and $\beta(\lambda)$, $\lambda \in \Lambda$. Because these are unknown, a practically feasible efficient instrument entails consistent estimates of S_U and β; because S_U, and possibly β also, are nonparametric, these must be estimated by smoothed nonparametric estimates. In view of (2.3) and (2.4), consistent estimates of S_U and β will depend on consistent estimates of S_Z. (A preliminary $\sqrt{N}$-consistent estimate, based on a fairly arbitrary instrument choice, can be inserted for θ in (2.3) with no loss of efficiency.) In case of regression models the efficient estimates reduce to generalized least squares (GLS) ones of Hannan [9].

One class of smoothed nonparametric estimates of S_Z is the weighted autocovariance form

$$(2.5) \qquad \widehat{S}_Z(\lambda) = \frac{1}{2\pi} \sum_{j=-M}^{M} k\left(\frac{j}{M}\right) C_Z(j) e^{-ij\lambda} .$$

Here $C_Z(j) = N^{-1} \sum_{t=1}^{N-j} Z_t Z'_{t+j}$, $C_Z(-j) = C_Z(j)'$, $j \geq 0$; $k(x)$, the "kernel" or "lag window" is a real-valued, even function on $(-\infty, \infty)$, with $k(0) = 1$; M the "bandwidth", "smoothing number" or "lag number" is positive. See Parzen [14]. A similar type of estimate consists of a weighted average of periodogram ordinates (see Brillinger [3]).

The choice of M is particularly important. Optimality theory suggests that M should vary inversely with the smoothness of S_Z. Unfortunately S_Z is unknown, but we can nevertheless gain information from the data about S_Z that helps in choosing M. Much previous work on statistical theory of estimates of S_Z, and all previous work on the efficient estimates of θ described above which depend on S_Z, assumes M does not depend on the data; the theory requires M to tend to infinity, but not too quickly. In Robinson [20] it is shown that $\widehat{S}_Z$ (2.5) converges uniformly in probability to S_Z, and the efficient estimates of θ are asymptotically normal with a consistent covariance matrix estimate, under conditions which allow M to depend on the data in a very general way; for example the condition on M is of the form

$$(2.6) \qquad \frac{M^\nu}{N} + \frac{1}{M} \to_p 0$$

for suitable $\nu \geq 1$; (2.6) is a stochastic version of the usual sort of deterministic condition on M and can be checked for a variety of data-dependent methods of bandwidth choice.

Some of these methods appear to have optimality properties, such as asymptotically minimizing some objective function measuring discrepancy between $\widehat{S}_Z$ and S_Z. One method of choosing M is motivated by Robinson [20] in connection with efficient estimates of θ in some semiparametric models. It turns out to be a form of cross-validation based on the frequency domain log likelihood. This sort of method was earlier considered by Hurvich [10], Beltrao and Bloomfield [1]. The latter authors showed that for a scalar Gaussian time series the cross-validated log likelihood approximates asymptotic integrated mean squared error, weighted inversely by the spectrum. The M minimizing the latter function is, under regularity conditions, of form $M = \nu N^{1/5}$, where ν depends on S_Z and k. in Robinson [20] we show that $\widehat{\nu} = \widehat{M} N^{-1/5}$, where $\widehat{M}$ is the cross-validated M, converges in probability to the optimal ν. This sort of result is substantially harder to establish than Beltrao and Bloomfield's [1], and we do not assume Gaussianity of Z_t, but rather a linear process with finite fourth moments.

Monte Carlo simulations of Robinson [20] confirm that $\widehat{\nu}$ is very variable, but that it can produce better estimates of θ in semiparametric models than simple 'rule-of-thumb' choices of M.

3. Semiparametric modelling of long-range dependence, with application to robust inference. This section is based on work of Robinson [22]. All the asymptotic statistical theory referred to in the previous section requires the errors U_t to have a spectrum that is at least continuous; we have thus been restricted to weakly dependent U_t. There is increasing interest in time series that exhibit long-range dependence, and have an unbounded spectral density. Such a time series U_t is often assumed to have the property

$$(3.1) \qquad S_X(\lambda) \sim c\lambda^{1-2H} \qquad \text{as} \quad \lambda \downarrow 0$$

or the closely related (see e.g. Yong [28]) property

$$(3.2) \qquad Cov(U_0, U_j) \sim C_j^{2(1-H)} , \qquad \text{as} \quad j \to \infty,$$

where c and C are positive constants and $\frac{1}{2} < H < 1$. Because (3.1) and (3.2) do not model the short-range characteristics of U_t, we might call them "semiparametric" models.

Extensions of the results of §2 to process satisfying (3.1) or (3.2) is a challenging open problem. Instead we discuss robust inference based on much simpler estimates, specifically ordinary least squares (OLS) estimates of the following special case of the general semiparametric model discussed in §1,

$$(3.3) \qquad Y_t = \mu + \theta'(X_t - \overline{X}) + U_t$$

where $\overline{X} = N^{-1} \sum_{t=1}^{N} X_t$, and U_t, μ are scalars.

The OLS estimate of μ is $\widehat{\mu} = \overline{Y} = N^{-1} \sum_{t=1}^{N} Y_t$. Were U_t weakly dependent, under suitable regularity conditions we expect that $\overline{Y}$ is asymptotically as efficient as GLS based on a parametric or nonparametric model for S_U (see e.g. Grenander [8]); also that

$$(3.4) \qquad N^{\frac{1}{2}}(\widehat{\mu} - \mu) \to_d N(0, 2\pi S_u(0))$$

when $0 < S_u(0) < \infty$, an assumption that is violated by (3.2). We can construct robust inferences in the presence of nonparametric autocorrelation by computing a smoothed nonparametric estimate of $S_U(0)$, and remarks in §2 covering data-dependent and automatic smoothing are relevant here.

If U_t has long-range dependence, $\widehat{\mu}$ can be nearly as efficient as GLS (see e.g. Samarov and Taqqu [22]). Also, a result of form

$$(3.5) \qquad \{V(\widehat{\mu})\}^{-\frac{1}{2}}(\widehat{\mu} - \mu) \to_d N(0, 1)$$

may be available. Some sufficient conditions are given by Ibragimov and Linnik [11], Taqqu [24]. Under (3.1) and (3.2) we expect that $V(\widehat{\mu})$ decays like $N^{2(H-1)}$, so the norming rate in (3.5) is slower than the $N^{\frac{1}{2}}$ of (3.4) for $\frac{1}{2} < H < 1$. Under some conditions $\widehat{\mu}$ can have a non-normal limit distribution.

The statistical properties of the OLS estimate $\widehat{\mu}$ of the slope θ can likewise be treated. It is known that under weak dependence of U_t, $\widehat{\theta}$ is asymptotically efficient in case X_t consists of polynomials or trigonometric polynomials in t, Grenander [8]. When X_t is a stationary stochastic process, then OLS is generally inefficient. In either case we may also have a central limit theorem. For example, if X_t is a scalar stationary process, with spectrum S_X and independent of U_t up to fourth moments, then under regularity conditions

$$(3.6) \qquad N^{\frac{1}{2}}(\widehat{\theta} - \theta) \to_d N\left(0,\ 2\pi \frac{\int_{-\pi}^{\pi} S_U(\lambda)S_X(\lambda)d\lambda}{V(X_t)}\right).$$

Remarks in §2 are relevant to the consistent estimation of the limiting variance in the presence of nonparametric U_t, with data-dependent and automatic smoothing.

Under long-range dependence in U_t, Yajima [26,27] has discussed the asymptotic efficiency of OLS in case of deterministic X_t, such as polynomial X_t (when efficiency is apt to be good). The case of stochastic X_t is discussed by Robinson [22]. It is interesting, because even if U_t or X_t or both has long-range dependence, $U_t X_t$ may possess such weak dependence that (3.6) may again hold. On the other hand, if both U_t and X_t have long range dependence such that the sum of the H coefficients exceeds $1\frac{1}{2}$, then we expect a less standard result, such as an analogue of (3.5).

In the latter case, and in case of $\widehat{\mu}$ for long-range dependent U_t, robust rules of inference require a different covariance matrix estimate than under weak dependence. For $\widehat{\mu}$, under regularity conditions

$$V(\widehat{\mu}) \sim \frac{CN^{2(H-1)}}{H(2H-1)}$$

where C is the same as in (3.2). Feasible, consistent interval estimates and asymptotically valid test statistics based on (3.5) require a consistent estimate of C and a log N-consistent estimate of H. (The situation for $\widehat{\mu}$ is similar but more complicated.)

The problem of estimating H and C (or c in (3.1)) is a challenging one. Some estimates have been proposed though asymptotic results claimed by Janacek [12], Geweke and Porter–Hudak [7] can only be regarded as heuristic indications. These and some new proposals are discussed by Robinson [22]. The estimates tend to be based either on low-frequency periodogram ordinates or on sample autocovariances at long lags, or else they entail a parametric model in which the number of parameters tends slowly to infinity with N. The rate of convergence of the semiparametric estimates will be less than $\sqrt{N}$.

4. Kernel-based U– and V–statistics with application to semipara-metric partly linear regression and index models, and nonparametric prediction. This section is based on work of Robinson [19]. The methods described in §§2 and 3 are most suitable for Gaussian environments because they use only second moments; the semiparametric aspect arises due to nonparametric autocorrelation, at least over the short range. The methods of this and the final section are motivated by the possibility of non-Gaussianity. In a variety of problems the extent of knowledge of distributional form of key observable or unobservable variates influences the form of asymptotically valid and reasonably efficient rules of inference. We assume such distributions are nonparametric. This may lead to a semiparametric model, where we wish to conduct statistical inferences on the parametric part with efficiency that is non-negligible relative to ones based on a fully parametric model. The statistics we employ average smoothed nonparametric probability density, derivative-of-density and regression estimates, and achieve the usual "parametric" $\sqrt{N}$ rate of convergence, despite the slow pointwise convergence of the nonparametric estimates. These statistics are also useful in some purely nonparametric problems, such as specification of nonparametric predictors; the averaging of nonparametric estimates again speeds convergence.

Robinson [19] considers a statistic of the following form:

$$(4.1) \qquad \widehat{\tau} = \frac{1}{N^2} \sum_{s,t=1}^{N} G(Y_s; Y_t) K\left(\frac{Z_s - Z_t}{h}\right).$$

This is a V-statistic, or a U-statistic if $G(Y_t; Y_t) \equiv 0$. In (4.1), Y_t represents an observable vector time series, possibly including lagged values, that is Y_t might include scalars y_t and y_{t-1}, for example; Z_t is a sub-vector of Y_t; K is a vector of kernel functions (which are even, real valued and integrate to one) and their partial derivatives; h is a positive bandwidth smoothing number; G is a given, finite-dimensional vector function, defined by the problem at hand.

The statistic (4.1) is used in testing a parametric hypothesis of the form

$$(4.2) \qquad H_0 \; : \; \tau = 0.$$

The falsity of H_0 implies the falsity of the underlying hypothesis of interest. Thus, if H_0 is rejected, we can reject the underlying hypothesis of interest. The worker constructs τ by a suitable choice of G; in a given problem there may be many such G, though computational considerations and simplicity can make some more suitable than others.

To perform the test, an approximation to the null distribution of $\widehat{\tau}$ under H_0 is needed. Under H_0 and a condition of absolute regularity on Y_t (which is intermediate between strong and uniform mixing), and other conditions, Robinson [19] showed that

$$N^{\frac{1}{2}}\widehat{\tau} \to_d \quad N(0, \Omega)$$

for a finite, positive definite Ω. Being a U- or V-statistic, $\widehat{\tau}$ approximates the sample mean of a certain process; thus, Ω can be interpreted as proportional to a certain spectral density matrix. The nonparametric, nonlinear character of the problem suggests that the spectral density will be nonparametric, but it can be consistently estimated by a smoothed estimate $\widehat{\Omega}$, much as indicated in §2. Then an approximate large-sample test consists of rejecting H_0 when $N\widehat{\tau}'\widehat{\Omega}^{-1}\widehat{\tau}$ is significantly large relative to the appropriate χ^2 distribution. In relation to §3, note that absolute regularity is a condition of weak dependence; a partial extension to allow for some long range dependence is considered by Cheng and Robinson [5].

We briefly mention some problems in which this sort of test is useful. Much fuller details are provided by Robinson [19].

(i) Semiparametric partly linear model. For vector-valued series X_t, Z_t, and a scalar series Y_t, let

$$E(Y_t \mid X_t, Z_t) = \theta' X_t + \lambda(Z_t),$$

where θ is an unknown vector and λ is a function of unknown form. X_t and Z_t do not overlap. This model can arise in various ways. In general, use of an incorrect parametric form for λ produces asymptotically invalid inferences on θ. Robinson [18] established $\sqrt{N}$-consistency and asymptotic normality of a semiparametric estimate of θ in case of independent observations. This estimate is technically somewhat harder to handle than (4.1), which suffices if only a test that θ takes given value is desired.

(ii) Semiparametric index model. For a vector-valued series X_t and a scalar series Y_t,

$$(4.3) \qquad E(Y_t \mid X_t) = \lambda(\theta' X_t),$$

where θ and λ are as before. This model can arise in various ways. In particular, in a censored linear model, with slope vector θ, maximum likelihood estimates of θ based on a given parametric model for the errors are generally inconsistent when this error distribution is incorrect (unlike in the ordinary uncensored linear regression (3.3)). It is thus desirable to regard the error distribution as nonparametric, leading to (4.3) with nonparametric λ. Powell et al. [16] employ a special case of $\widehat{\tau}$ to analyze (4.3) in case of independent observations.

(iii) Nonparametric prediction. If a stationary scalar time series X_t has an unknown, possibly non-Gaussian distribution, the least squares predictor $\widehat{X}_{N+1} \overset{\Delta}{=} E(X_{N+1} \mid X_N, \ldots, X_{N-p})$ is of unknown form, and can be non-linear. We can however, estimate $\widehat{X}_{N+1}$ by nonparametric regression, see e.g. Robinson [17]. But due to the curse of dimensionality, we wish to choose p as small as possible. We can do a series of tests, for example we can test whether $p = 1$ by testing

$$(4.4) \qquad E(X_t \mid X_{t-1}, X_{t-2}) = E(X_t \mid X_{t-1}), \qquad \text{a.s.}$$

Due to the slow convergence of nonparametric estimates, the conditional moment restriction (4.1) is difficult to investigate directly, but it implies moment restrictions

$$(4.5) \qquad E\{(X_t - E(X_t \mid X_{t-1}))a(X_{t-1}, X_{t-2})\} = 0$$

for suitable finite-dimensional vectors a. But (4.5) is of form (4.2), and the left side of (4.5) can be estimated by a suitable $\hat{\tau}$. Another $\hat{\tau}$ arises by noting that (4.4) also implies

$$(4.6) \qquad E\left\{\frac{\partial}{\partial X_{t-2}}\ E(X_t \mid X_{t-1}, X_{t-2})a(X_{t-1}, X_{t-2})\right\} = 0.$$

5. Consistent nonparametric entropy-based tests of serial independence and time reversibility. This section is based on work of Robinson [21]. A difficulty with the approach in §4 is illustrated in the nonparametric prediction example just discussed. (4.5) and/or (4.6) can be true even when (4.4) is not, indicating inconsistent directions of our test. In some problems a more sensitive approach is available.

In general setting, let X be a continuous vector variable, and introduce the rival hypotheses

$$H_1 \ : \ pdf(X) \ \text{is} \ f(x)$$

$$H_2 \ : \ pdf(X) \ \text{is} \ g(x)$$

The Kullback–Leibler [13] mean information for discrimination between H_1 and H_2 per observation from f is

$$I(f,g) = \int \left\{\log \frac{f(x)}{g(x)}\right\} f(x)dx.$$

We record the following properties:

(i) I cannot be increased by transforming X.

(ii) I is unchanged if X is transformed to a sufficient statistic.

(iii) $I \geq 0$.

(iv) $I(f,g) = 0$ if and only if $f = g$ a.e.

It follows that I has valuable invariance properties with respect to linear and nonlinear transformation of X, especially bearing in mind that an observable variate may not directly measure the phenomenon of interest. It also follows that I is highly sensitive to discrepancies between f and g, which never cancel, and always lead to a positive, never a negative I. These properties make I a desirable basis for hypotheses testing.

We consider only nested hypotheses, that is $H_2 \subset H_1$, and g is the null density, f the alternative density. We assume that both f and g are nonparametric. In previous uses of I, both f and g have been parametric, or g has been parametric, f nonparametric.

A classical time series problem which is of this form is testing for serial independence in a stationary scalar time series Y_t with marginal *pdf* of unknown, nonparametric form. Put $X_t = (Y_t, Y_{t+1})'$. We write $pdf(Y_t)$ as $h(u)$, and $pdf(Y_{t,t+1})$ as $f(u,v)$. Because h is nonparametric, so is f. Consider the null hypothesis

$$H_0 \; : \; f(u,v) = h(u)h(v), \quad \text{for all} \quad u, v.$$

This is equivalent to the hypothesis that Y_t and Y_{t+1} are independent, for all t. A test that rejects H_0 also rejects

$$H_0' : Y_t, \; Y_s \quad \text{are independent for all} \quad t, s.$$

It is well known that a sequence of uncorrelated variates need not be independent, when tests for serial correlation will prove inadequate.

We write

$$I(f, hh) \stackrel{\Delta}{=} \int f(u,v) \, \log \, \{f(u,v)/h(u)h(v)\}dudv$$

$$= \int f(u,v) \, \log \, f(u,v)dudv - 2 \int h(u) \, \log \, h(u)du.$$

We can estimate this nonparametrically by substituting smoothed nonparametric estimates $\widehat{f}, \widehat{h}$ and averaging over data points, thus

$$\widehat{I} = \frac{1}{N} \sum_{t=1}^{N} c_t \, \log \, \{\widehat{f_t}/\widehat{h_t^2}\}$$

where $\widehat{f_t} = \widehat{f}(Y_t, Y_{t+1}), \widehat{h_t} = \widehat{h}(Y_t)$ and the c_t are given weights.

In Robinson [21] it is shown that, for a sequence $d_N \uparrow \infty$ and any $C < \infty$, the test:

(5.1) $$\text{Reject} \quad H_0 \quad \text{when} \quad d_N\widehat{I} > C$$

is consistent against all alternatives to H_0 satisfying certain regularity conditions. (The consistency and one-sidedness of the test are due to properties (iv) and (iii) respectively.)

In this test, we would like to choose d_N to increase as fast as possible. We would also like to choose C so as to approximate a desired type I error, which is achieved by identifying a nondegenerate random variable X such that $d_N\widehat{I} \rightarrow_d X$ under H_0. It is found by Robinson [21] that under suitable conditions, we may take $d_N = N^{\frac{1}{2}}$ and

(5.2) $$N^{\frac{1}{2}}\widehat{I} \quad \rightarrow_d \quad N(0, V) ,$$

where a consistent estimate $\widehat{V}$ of V can be justified. An asymptotically valid level-α test of H_0 thus consists of rejecting H_0 when

$$N^{\frac{1}{2}}\widehat{I}/\widehat{V}^{\frac{1}{2}} > C_\alpha$$

where $P(X > C_\alpha) = \alpha$ for $X \sim N(0,1)$. It is important to stress that (5.2), though not the consistency of the tests, relies on an appropriate choice of weights which rules out equal weights, $c_t \equiv 1$. An example of an acceptable choice of weights, which also turns out to be computationally attractive, is $c_t = 1$ for t odd, $c_t = 0$ for t even. The choice of c_t affects the power of the test.

In Robinson [21] the consistency and asymptotic normality properties described above are checked in case of kernel density estimates, under a strong mixing alternative. An application to testing the random walk hypothesis in exchange rate series is also reported. A number of possible extensions are described. One important extension would allow for nuisance parameters, as when Y_t is a regression error; here we are again back to a semiparametric model of type (3.3), where interest centers now on the nonparametric, rather than the parametric part.

Our test for serial independence can be compared with numerous others in the literature; it has both advantages and disadvantages relative to these. For example, it has consistency properties and simple distribution theory not shared by the test of Blum et al. [2], but requires choice of a smoothing number and has no obvious optimality properties. Its consistency properties are not shared by the test of Brock et al. [4], which also requires smoothing and a similar level of computation, but our statistic $\widehat{I}$ is harder to handle mathematically.

Some other applications of our entropy test are listed by Robinson [21]. One that is not included there is a test for time series reversibility. A stationary time series Y_t is said to be reversible if the joint distribution of $(Y_t, Y_{t+1}, \ldots Y_{t+p})$ is identical to that of $(Y_{t+p}, \ldots, Y_{t+1}, Y_t)$. Although a Gaussian autoregressive moving average process is reversible, most time series are not reversible. An entropy-based test for reversibility may be constructed after taking $X_t = (Y_t, Y_{t+1})$, denoting $pdf(X_t)$ by $f(u,v)$ and introducing

$$(5.3) \qquad\qquad H_0 : f(u,v) = f(v,u), \quad \text{for all} \quad u, v,$$

which is implied by reversibility, given that X_t is a continuous random variable. A consistent test of (5.3) may be constructed much as before, except that the variance in the null normal limit distribution is more complicated to estimate because Y_t is serially dependent under (5.3). The test is not consistent against full time reversibility, just as (5.1) is not consistent against full serial independence H_0', but the consistent directions can be increased by extending H_0 and (5.3) to multivariate distributions.

REFERENCES

[1]   K.I. BELTRAO AND P. BLOOMFIELD, *Determining the bandwidth of a kernel spectrum estimate*, Journal of Time Series Analysis, 8 (1987), pp. 21–38.

[2]   J.R. BLUM, J. KIEFER AND M. ROSENBLATT, *Distribution free tests of independence based on the sample distribution function*, Annals of Mathematical Statistics, 32 (1961), pp. 485–498.

[3]   D.R. BRILLINGER, *Time Series, Data Analysis and Theory*, Holden–Day, San Francisco (1975).

[4]   W. BROCK, D. DECHERT AND J. SCHEINKMAN, *A test for independence based on the correlation dimension*, preprint, University of Wisconsin.

[5] B. CHENG AND P.M. ROBINSON, *Semiparametric models estimation from time series with long-range dependence*, preprint.

[6] J.W. COOLEY AND J.W. TUKEY, *An algorithm for the machine calculation of Fourier series*, Mathematics of Computation, 19 (1965), pp. 297–301.

[7] J. GEWEKE AND S. PORTER-HUDAK, *The estimation and application of long memory time series models*, Journal of Time Series Analysis, 4 (1984), pp. 221–238.

[8] U. GRENANDER, *On the estimation of regression coefficients in the case of an autocorrelated disturbance*, Annals of Mathematical Statistics, 25 (1954), pp. 252–272.

[9] E.J. HANNAN, *Regression for time series*, in Time Series Analysis, ed. by M. Rosenblatt, Wiley, New York (1963).

[10] C.M. HURVICH, *Data driven choice of a spectrum estimate: extending the applicability of cross-validation methods*, Journal of the American Statistical Association, 80 (1985), pp. 933–940.

[11] I.A. IBRAGIMOV AND Y.V. LINNIK, *Independent and Stationary Sequences of Random Variables*, Wolters–Noordhoff, Groningen (1971).

[12] G.J. JANACEK, *Determining the degree of differences for time series via the log spectrum*, Journal of Time Series Analysis, 3 (1982), pp. 177–183.

[13] S. KULLBACK AND R.A. LEIBLER, *On information and sufficiency*, Annals of Mathematical Statistics, 22 (1961), pp. 79–86.

[14] E. PARZEN, *On consistent estimates of the spectrum of a stationary time series*, Annals of Mathematical Statistics, 28 (1957), pp. 329–348.

[15] —————, *On estimation of a probability density and mode*, Annals of Mathematical Statistics, 38 (1962), pp. 1065–1076.

[16] J.L. POWELL, J.H. STOCK AND T.M. STOKER, *Semiparametric estimation of weighted average derivatives*, Econometrica, 57 (1989), pp. 1403–1430.

[17] P.M. ROBINSON, *Nonparametric estimators for time series*, Journal of Time Series Analysis, 4 (1983), pp. 185–207.

[18] —————, *Root–N–consistent semiparametric regression*, Econometrica, 56 (1988), pp. 931–954.

[19] —————, *Hypothesis testing in nonparametric and semiparametric models for econometric time series*, Review of Economic Studies, 56 (1989), pp. 511–534.

[20] —————, *Automatic frequency-domain inference on semiparametric and nonparametric models*, Econometrica, 59 (1991), pp. 1329–1363.

[21] —————, *Consistent nonparametric entropy-based testing*, Review of Economic Studies, 58 (1991), pp. 437–453.

[22] —————, *Time Series with strong dependence*, Invited Paper, 1990 World Congress of the Econometric Society.

[23] A. SAMAROV AND M.S. TAQQU, *On the efficiency of the sample mean in long-memory noise*, Journal of Time Series Analysis, 9 (1988), pp. 191–200.

[24] M.S. TAQQU, *Weak convergence to fractional Brownian motion and to the Rosenblatt process*, Z. Wahrscheinlichkeitstheorie verw. Geb., 40 (1975), pp. 287–302.

[25] J.W. TUKEY, *Curves as parameters and touch estimation*, Proceedings of the 4th Berkeley Symposium Math. Stat. and Probab., pp. 681–694.

[26] Y. YAJIMA, *On estimation of a regression model with long-memory stationary errors*, Annals of Statistics, 16 (1988), pp. 791–807.

[27] —————, *Asymptotic properties of the LSE in a regression model with long-memory stationary errors*, preprint.

[28] C.H. YONG, *Asymptotic behaviour of trigonometric series with modified monotone coefficients*, Chinese University of Hong Kong, 1972.

GAUSSIAN AND NONGAUSSIAN LINEAR SEQUENCES*

MURRAY ROSENBLATT†

Abstract. A brief discussion of Gaussian linear processes and nonGaussian possibly nonminimum phase linear processes is given. The nonreversibility of many nonGaussian linear processes is noted. The nonlinearity of the best predictor in mean square for a variety of nonGaussian nonminimum phase linear processes is remarked on. Recent results on nonparametric and parametric estimation procedures are examined. The existence of noncontiguous statistical models for autoregressive schemes driven by discrete random variables is specified.

Introduction. Most of the discussion of linear processes in the literature has focused on the Gaussian assumption. A linear process x_t has the form

$$(1) \qquad x_t = \sum_{j=-\infty}^{\infty} \alpha_j \varepsilon_{t-j}, \qquad t = \ldots, -1, 0, 1, \ldots,$$

with the α_j's real constants satisfying

$$(2) \qquad \sum \alpha_j^2 < \infty$$

and the ε_t's independent, identically distributed with mean zero and variance one. If x_t is a Gaussian process, the ε_t's are of course Gaussian. Let

$$(3) \qquad \alpha(e^{-i\lambda}) = \sum \alpha_j e^{-ij\lambda}.$$

The spectral density $f(\lambda)$ of the sequence x_t is

$$(4) \qquad f(\lambda) = \frac{1}{2\pi} |\alpha(e^{-i\lambda})|^2.$$

In the linear prediction problem, the variance of the one step error of predicting x_1 given the past $x_0, x_{-1}, x_{-2}, \ldots$, is

$$(5) \qquad 2\pi \exp\left\{ \frac{1}{2\pi} \int_{-\pi}^{\pi} \log f(\lambda) d\lambda \right\}.$$

The sequence is purely linearly deterministic if $\log f(\lambda) \notin L$ and purely linearly nondeterministic if $\log f(\lambda) \in L$. Gaussian sequences have the property of reversibility, that is, they have the same probability structure with time reversal as they have without. This is a property that is no longer true for most nonGaussian linear processes.

---

*Research supported in part by the Office of Naval Research grant N00014-90-J1371
†Mathematics Department, University of California, San Diego, La Jolla, California 92093

It is of some interest to look at stationary autoregressive moving average processes, that is, stationary solutions x_t of the system of equations

(6)
$$\sum_{j=0}^{p} \beta_j x_{t-j} = \sum_{k=0}^{q} \alpha_k \varepsilon_{t-k}$$

where the ε_t's are independent and identically distributed with mean zero and variance $\sigma^2 > 0$ and $\beta_0 = \alpha_0 = 1$. Consider the polynomials

(7)
$$\alpha(z) = \sum_{k=0}^{q} \alpha_k z^k,$$
$$\beta(z) = \sum_{j=0}^{p} \beta_j z^k.$$

Assume that $\alpha(z)$, $\beta(z)$ have no roots in common. A necessary and sufficient condition for the existence of a stationary solution x_t is that $\beta(z)$ have no roots of absolute value one. If $\alpha(z)$ has all its roots of absolute value greater than or equal to one, $\varepsilon_t \in \mathcal{M}_t(x)$ where $\mathcal{M}_t(x)$ is the closed (in mean square) linear space of random variables generated by x_s, $s \leq t$. If $\beta(z)$ has all its roots of absolute value greater than one then $x_t \in \mathcal{M}_t(\varepsilon)$. The case in which both $\alpha(z)$, $\beta(z)$ have roots greater than one is called minimum phase and in that case the best predictor of x_t in terms of the past x_{t-1}, $x_{t-2}, \ldots$ in mean square is the linear predictor

(8)
$$x_t^* = -\sum_{j=1}^{p} \beta_j x_{t-j} + \sum_{k=1}^{q} \alpha_k \varepsilon_{t-k}$$

with prediction error ε_t. As already suggested, in the nonminimum phase non-Gaussian case the best predictor in mean square is commonly nonlinear. In the Gaussian linear case one cannot distinguish between minimum and nonminimum phase because of the reversibility.

Nonreversibility. A simple example illustrating nonreversibility of some non-Gaussian linear processes is given by the stationary autoregressive scheme

(9)
$$x_n = \frac{1}{2} x_{n-1} + \varepsilon_n$$

with the ε_n's independent and identically distributed

$$\varepsilon_n = \begin{cases} 0 & \text{with probability } 1/2 \\ 1 & \text{with probability } 1/2. \end{cases}$$

The best predictor of x_n in mean square in the forwards direction is

$$x_n^* = \frac{1}{2} x_{n-1} + \frac{1}{2}.$$

In the backwards direction x_n is perfectly predictable

$$(10) \qquad\qquad x_n = 2x_{n+1} \text{ modulo } 1$$

since x_n has the binary expansion $x_n = .\varepsilon_n \varepsilon_{n-1} \cdots$. The invariant probability measures for the nonlinear transformation $f(x) = 2x$ modulo one correspond precisely to the stationary zero one sequences $\{y_n, n = \cdots, -1, 0, 1, \cdots\}$ and are given by the measures of the one sided expansions $.y_n y_{n-1} y_{n-2} \cdots = z_n$. If f is considered a transformation on the circle (the interval $[0, 1]$ where 0 is identified with 1) into itself it is continuous. Given an initial value $w_0 \in (0, 1]$ consider the sequence $f^{(n)}(w) = w_n$, $n = 1, 2, \ldots$ where $w_1 = f(w)$ and $f(f^{(n)}(w)) = f^{(n+1)}(w)$. Let $G_n(z; w)$ be the distribution function with mass $1/n$ at each of the points $f^{(k)}(w)$, $k = 1, \ldots, n$. Consider the family of weak limits of the sequence of distribution functions $G_n(z; w)$. There is at least one weak limit due to the continuity of f and the compactness of the circle. Each of these limit point distribution functions provides an invariant probability measure for the transformation f. It is clear that f is a transformation whose trajectories can be unstable and extremely sensitive to the initial point in the sense of the usual discussions of chaos. Given an initial number w, it is of considerable interest to find the set of limit measures corresponding to the trajectory generated from it by f.

Nonparametric and Parametric Methods. If the linear process (1) satisfies (2) and the random variables ε_t have an sth order cumulant $\mu_s \neq 0 (s > 2)$ the sth order spectral density of the process exists and has the form

$$(11) \qquad (2\pi)^{-s+1} \alpha(e^{-i\lambda_1}) \ldots \alpha(e^{-i\lambda_{s-1}}) \alpha(e^{i\{\lambda_s + \cdots + \lambda_{s-1}\}})$$
$$= b_s(\lambda_1, \ldots, \lambda_{s-1}).$$

All cumulants of order $s > 2$ of a Gaussian process are zero. In the case of a Gaussian process for this reason from data, one can determine at most $|\alpha(e^{-1\lambda})|$. However, with a nonGaussian linear process having $\mu_s \neq 0$ for some $s > 2$ almost all the phase information can be recovered, $\alpha(e^{-i\lambda})$ can be recovered up to a factor $e^{ik\lambda}$ (with k integral) and a global sign ± 1. This can be based on estimation of the second order spectral density (to estimate $|\alpha(e^{-i\lambda})|$) and the estimation of the sth order cumulant spectral density (to estimate $h(\lambda) = \arg\{\alpha(e^{-i\lambda})\alpha(1)/|\alpha(1)|\}$). The identity

$$h(\lambda_1) + \cdots + h(\lambda_{s-1}) - h(\lambda_1 + \cdots + \lambda_{s-1})$$
$$(12) \qquad = \arg \left[\left\{ \frac{\alpha(1)}{|\alpha(1)|} \right\}^2 \mu_s b_s(\lambda_1, \ldots, \lambda_{s-1}) \right]$$

is satisfied by $h(\lambda)$. A discussion of such a procedure that is effective under fairly broad conditions can be found in [3,4]. These methods have a nonparametric character.

Recently, there has been interest in determining methods for the efficient estimation of parameters in finite parameter nonGaussian possibly nonminimum phase models. Let x_t be stationary autoregressive of order p

$$(13) \qquad x_t - \phi_1 x_{t-1} - \cdots - \phi_p x_{t-p} = \varepsilon_t$$

with $\phi(z) = 1 - \phi_1 z - \cdots - \phi_p z^p \neq 0$ for $|z| = 1$ and ε_t independent, identically distributed with mean 0, variance σ^2 and density $f_\sigma(x) = \sigma^{-1} f(x/\sigma)$. Let the ε_t's be nonGaussian. Factor

$$(14) \qquad \phi(z) = \phi^+(z)\phi^*(z)$$

with

$$(15) \qquad \begin{aligned} \phi^+(z) &= 1 - \theta_1 z - \cdots - \theta_r z^r \neq 0 \text{ for } |z| \leq 1, \\ \phi^*(z) &= 1 - \theta_{r+1} z - \cdots - \theta_p z^s \neq 0 \text{ for } |z| \geq 1 \end{aligned}$$

where $r + s = p$. Let

$$U_t = \phi^*(B)\varepsilon_t, \; V_t = \phi^+(B)\varepsilon_t$$

where B is the backshift operator. Since U_t is independent of V_{t-s+1} then the probability density of $(U_1, \ldots, U_r, V_{n-s+1}, \ldots, V_n)'$ can be written as

$$(16) \qquad h_U(U_1, \ldots, U_r) \left\{ \prod_{t=r+1}^{n} f_\sigma(U_t - \theta_1 U_{t-1} - \cdots - \theta_r U_{t-r}) \right\}$$
$$h_V(V_{n-s+1}, \ldots, V_n)$$

where h_U and h_V are probability densities of $(U_1, \ldots, U_r)'$ and $(V_{n-s+1}, \ldots, V_n)'$. The joint probability density of $(U_1, \ldots, U_s, x_1, \ldots, x_n)'$ can be shown to be

$$h_U(U_1, \ldots, U_r) \left\{ \prod_{t=r+1}^{n} f_\sigma(x_t - \phi_1 x_{t-1} - \cdots - \phi_p x_{t-p}) \right\}$$
$$h_v(\phi^+(B)x_{n-s+1}, \ldots, \phi^+(B)x_n) |\det T|$$

with T an $(n+s) \times (n+s)$ matrix. For $s > 0$, $\ln|\det T| \sim \ln|\theta_p|^{n-p}$ and this suggests approximating the ln-likelihood by

$$(17)$$
$$L(\theta_1, \ldots, \theta_p, \sigma) = \sum_{t=p+1}^{n} \{\ln f_\sigma(U_t - \theta_1 U_{t-1} - \cdots - \theta_r U_{t-r}) + \log|\theta_p|\}$$
$$= \sum_{t=p+1}^{n} g_t(\underline{\theta}), \underline{\theta} = (\theta_1, \ldots, \theta_{p+1})', \theta_{p+1} = \sigma,$$

with

$$g_t(\underline{\theta}) = \ln f_\sigma(U_t - \theta_1 U_{t-1} - \cdots - \theta_r U_{t-r}) + \ln|\theta_p|$$
$$= \ln f_\sigma(V_t - \theta_{r+1} V_{t-1} - \cdots - \theta_p V_{t-s}) + \ln|\theta_p|.$$

If f is a nonnormal probability density satisfying the following conditions among others

$$\underline{A1} \quad f(x) > 0 \text{ for all } x, \quad \underline{A2} \quad f \in C^2, \quad \underline{A3} \quad f' \in L, \int f'(x)dx = 0,$$

$$\underline{A4} \quad \int xf'(x)dx = -1, \quad \underline{A5} \quad \int f''(x)dx = 0, \quad \underline{A6} \quad \int xf''(x)dx = 0,$$

$$\underline{A7} \quad \int x^2 f''(x)dx = -2, \quad \underline{A8} \quad \int (1+x^2)(f'(x))^2/f(x)dx < \infty$$

there is a sequence of solutions $\hat{\theta}_n$ to the approximate likelihood equations

$$(18) \qquad \frac{\partial L(\underline{\theta})}{\partial \theta_j} = 0, \; j = 1,\ldots,p+1$$

that is asymptotically normal with mean $\underline{\theta}_0$ and asymptotic covariance matrix $n^{-1}\sum^{-1}$, $\sum = (\sigma_{ij})$, with

$$\sigma_{ij} = \begin{cases} \tilde{I}\gamma_U(i-j) & 1 \le i \le j \le r \\ \tilde{I}\gamma_V(i-j) & r < i \le j \le p, i \ne p \\ \tilde{I}\gamma_V(0) + \beta_s^2\sigma^2(\tilde{J} - \tilde{I}) & i = j = p \\ \displaystyle\sum_{k=0}^{\infty} \alpha_k\beta_{k+i+j-r} & 1 \le i \le r < j \le p \\ -\theta_p^{-1}\sigma\tilde{J} & i = p, j = p+1 \\ \tilde{J} & i = j = p+1 \\ 0 & \text{otherwise} \end{cases}$$

if $i \le j$, $s > 0$ while if $s = 0$

$$\sigma_{ij} = \begin{cases} \tilde{I}\gamma_U(i-j) & 1 \le i \le j \le p \\ \tilde{J} & i = j = p+1 \\ 0 & \text{otherwise.} \end{cases}$$

Here

$$\tilde{I} = \sigma^{-2}\int (f'(x))^2/f(x)dx,$$

$$\tilde{J} = \sigma^{-2}(\int x^2(f'(x))^2/f(x)dx - 1),$$

$\gamma_U(\cdot)$ and $\gamma_V(\cdot)$ are the autocovariance functions of $\{U_t\}$ and $\{V_t\}$ while

$$\phi^+(z)^{-1} = \sum_{j=0}^{\infty} \alpha_j z^j$$

and

$$\phi^*(z)^{-1} = \sum_{j=s}^{\infty} \beta_j z^{-j}.$$

A detailed derivation of this result and related detailed remarks is given in [1].

Models without Contiguity. It is interesting to note that if one has autoregressive schemes x_t with the random variables ε_t having discrete distributions, contiguity of the x_t induced distributions is often not valid. As an example consider the autoregressive process x_t

$$x_t = \alpha x_{t-1} + \varepsilon_t$$

with independent ε_t's

$$\varepsilon_t = \begin{cases} 0 & \text{with probability } 1/2 \\ 1 & \text{with probability } 1/2 \end{cases}$$

and $0 < \alpha < \frac{1}{2}$. Assuming x_t stationary, $y_t = (1-\alpha)x_t$ takes values in $[0,1]$ and the marginal distribution of y_t is concentrated on a perfect Cantor set and is obtained by deleting $(\alpha, 1-\alpha)$ and assigning mass $1/2$ to each of the remaining intervals $[0,\alpha]$, $[1-\alpha, 1]$ etc. The Hausdorff dimension of the support set of the distribution of Y_t is $-\ln 2/\ln \alpha$ and the distributions corresponding to distinct $\alpha, 0 < \alpha < \frac{1}{2}$ are mutually singular. This, of course, implies that the joint distributions of $y_1, \ldots, y_n$ for different α, $0 < \alpha < \frac{1}{2}$, are mutually singular and hence the set of these distributions is not contiguous.

Consider the autoregressive scheme (13) where the ε_t's distribution has at least one discrete mass point c

$$P[\varepsilon_t = c] = q, \quad 0 < q < 1.$$

Introduce the $p+2$ dimensional vectors

(19)
$$\underline{1} = (1, \ldots, 1)'$$
$$\underline{x}_j = (x_j, x_{j+1}, \ldots, x_{j+p+1})'.$$

Conditional on the event $\{\varepsilon_{p+1} = \varepsilon_{p+2} = \cdots = \varepsilon_{2p+2} = c\}$

(20)
$$\text{rank}([\underline{x}_{p+1}, \ldots, \underline{x}_1, \underline{1}]) = p+1$$

with probability 1 where the last matrix is $(p+2) \times (p+2)$. $\phi_1, \ldots, \phi_p$ and a value c that is a discrete mass point for the ε_t distribution can be identified exactly for a large enough sample. One searches until one finds a $j \geq p+1$ (given a sample $x_1, \ldots, x_n$) for which there is a linear relationship between the vectors

(21)
$$\begin{bmatrix} x_j \\ \vdots \\ x_{j+p-1} \end{bmatrix}, \begin{bmatrix} x_{j-1} \\ \vdots \\ x_{j+p} \end{bmatrix}, \ldots, \begin{bmatrix} x_{j-p} \\ \vdots \\ x_{j+1} \end{bmatrix}, \begin{bmatrix} 1 \\ \vdots \\ 1 \end{bmatrix}.$$

If there is a linear relationship then $\varepsilon_j = \varepsilon_{j+1} = \cdots = \varepsilon_{j+p+1} = c$. Such an event occurs infinitely often with probability one. The detailed discussion and derivation of these results can be found in [2]. Notice that here conventional maximum likelihood procedures are not applicable and the behavior of the estimation procedure described is radically different from conventional maximum likelihood asymptotics which involve asymptotic normality and $n^{-1/2}$ convergence.

REFERENCES

[1] BREIDT, F.J., DAVIS, R.A., LII, K.S. AND ROSENBLATT, M., *Maximum likelihood estimation for noncausal autoregressive processes*, Journal of Multivariate Analysis (to appear).

[2] DAVIS, R.A. AND ROSENBLATT, M., *Parameter estimation for some time series models without contiguity*, Statistics and Probability Letters (to appear).

[3] LII, K.S. AND ROSENBLATT, *Deconvolution and estimation of transfer function phase and coefficients for non-Gaussian linear process*, Ann. Statist., 10 (1982), pp. 1195–1208.

[4] ROSENBLATT, M., *Stationary Sequences and Random Fields*, Birkhäuser, Boston, 1985.

PREDICTIVE DECONVOLUTION OF
CHAOTIC AND RANDOM PROCESSES

JEFFREY D. SCARGLE*

Abstract. Four simple yet characteristic nonlinear systems – the the doubling, logistic, cat, and Hénon maps – illustrate the defining properties of chaos and connections between abstract dynamical systems and stochastic processes. Moving average representations of these stationary processes exist by the Wold Theorem. Predictive deconvolution using a special cost function defined in the embedding space provides estimates for the models. This procedure yields exact, symbolic dynamics representations of processes generated by piecewise linear maps, and partially symbolic ones for processes from quadratic maps.

1. What is Chaos? Linear dynamical systems are usually orderly in two ways: the physical variables are smooth functions of time, and sufficiently close initial states evolve along similar paths. Even very simple nonlinear systems are not orderly in this sense – for at least some values of their parameters they are chaotic.

All chaotic systems have a fundamental instability, called sensitivity to initial conditions (often abbreviated SIC): evolutionary paths diverge exponentially from each other, no matter how close their initial states. Strictly identical states evolve identically, but this determinism is irrelevant in practical situations. External disturbances, no matter how small, are amplified by chaotic sensitivity and eventually become significant.

The time history of a variable in a chaotic system frequently has a disordered, random appearance. (This pseudorandomness is exploited in computer "random" number generators.) Nevertheless, some systems remain in regular, orderly states for extended periods, but are chaotic because of erratic jumps between states. Sensitivity to initial conditions is then exhibited only on time scales long enough for such jumps to occur.

A few elements of the dynamical systems view of discrete-time stochastic processes are important to the main topic of this paper – analysis of chaotic time series data. A dynamical system is an abstract space plus a transformation T which maps the space into itself. Repeated transformation of an element X_0 of the space,

$$(1) \qquad\qquad X_{n+1} = TX_n \qquad n = 0, 1, 2, \ldots,$$

yields an orbit X_n in the space. Further, a measure is defined on the space and T is assumed to be *measure preserving*. The physical meaning is as follows: The space itself is the state space of the physical system (usually called its "phase space"). X_0 is an initial state. The transformation T expresses the dynamics, and corresponds to the advance of time by one unit. The initial time is $n = 0$. The measure assigned to a set is the probability that the point representing the state of the system is in

*Theoretical Studies Branch, Space Science Division, National Aeronautics and Space Administration, Ames Research Center

the set at a random time. Finally, measure preservation under T means that the system's statistics do not change with time.

A dynamical system can in turn be associated with a stochastic process, namely the special case where successive random variables are not arbitrary but connected by

$$(2) \qquad X_{n+1} = f(X_n) \qquad n = 0, 1, 2, \ldots,$$

where f is a bounded nonlinear function that maps some region of the real axis into itself. The stochastic-process interpretation is as follows: The space itself is the usual probability space for the process, the Cartesian product of individual random-variable probability spaces. To my knowledge there is no term for the special variable X_0, but it is a kind of starting or generating random variable. The transformation T is called the *shift* (Doob 1953). Measure has the usual probability interpretation, with measure preservation of T equating to stationarity of the stochastic process.

What is unusual about a chaotic process is the fact that all of the randomness is contained in the one random variable X_0, which determines fully the behavior of the others according to the deterministic map. This will be discussed in Section 2.

We now consider four instructive examples of dynamical systems. In order, the dimension and type of nonlinearity of the defining maps are 1-D piecewise linear, 1-D quadratic, 2-D piecewise linear, and 2-D quadratic.

1.1. Example: the Doubling Map (One-Sided Bernoulli Shift). Most of the characteristics of chaos are evident in the extremely simple *doubling map*,

$$(3) \qquad X_{n+1} = 2X_n \bmod 1 \quad 0 \le X_n \le 1.$$

Iterative application of this map, starting from an initial value in $[0, 1]$, generates an infinite sequence $\{X_n\}$ in the same interval. In some cases it is convenient to use the equivalent zero-mean form

$$(4) \qquad X_{n+1} = 2X_n - Sign(X_n) \qquad -1 \le X_n \le 1.$$

The behavior of the resultant time series depends on the initial value, but for almost all X_0 the sequence looks random. To see this, note that each iteration of the map is equivalent to discarding the most significant binary digit of the previous iterate and shifting the remaining digits one place to the left. This map is a one-sided version of the *Bernoulli shift* (Ornstein 1989), the world's most random process, in which bi-infinite digit sequences are shifted in time. For convenience we call the process generated by the doubling map the *Bernoulli process*, but it should not be confused with the two-sided shift.

The digit-shift-and-discard interpretation makes it obvious that any irrational initial value will produce a disordered, nonrepeating sequence. The first panel of Figure 1 shows a sample time series. For convenience we take $X_0 = C_* =$

337

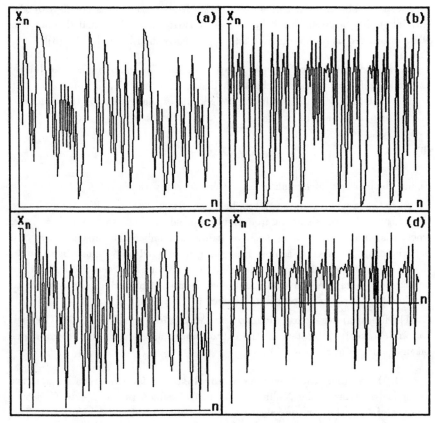

Fig. 1. Chaotic time series generated by (a) the one-sided
Bernoulli shift (doubling map), (b) the logistic map, (c) the
cat map, and (d) the Hénon map.

0.123456789. The binary representation of this constant is rather long, thus delaying
considerably the eventual truncation to zero that befalls any initial value with a
finite representation.

The disorder of this time series arises because the random digits of X_0 produce
the leading digit of each successive iterate. The X_n are nevertheless correlated
with each other because they have an infinitely long digit string in common. The
autocovariance function of the zero-mean process, Equation (4), is (Scargle 1989a)

$$\rho_X(m) = \frac{1}{12}\left(\frac{1}{2}\right)^m \quad m = 0,1,2,\ldots \tag{5}$$

The relative shift of the common digit string by one place does not destroy the correlation, it just diminishes it by a factor of two. We conclude that this process, while disordered, is both correlated and dependently distributed [cf. Eq.(3)].

1.2. Example: the Logistic Map. The doubling map is piecewise linear, and it is useful to consider a chaotic 1-dimensional map that is continuous but nonlinear. Of many that have been studied, one of the simplest is the quadratic *logistic map*, also on [0,1]:

$$(6) \qquad\qquad X_{n+1} = 4X_n(1 - X_n).$$

Figure 1b shows a typical realization of this process, with $X_0 = C*$.

As in the previous case the system behavior depends on the initial value. To see this consider the orbits starting from the two initial values $X_0 = \frac{3}{4}$ and $X_0 = \frac{3}{4} + \epsilon$, where ϵ is an arbitrarily small, positive, irrational number. The first is a very dull periodic orbit, the second a more interesting disordered one – much like that in the figure. As before almost all initial conditions produce disordered series that are nevertheless highly dependent, by virtue of the map. Furthermore, in contrast to the doubling map this process is uncorrelated.

The literature on the logistic map (e.g. May 1976, Feigenbaum 1978, Grebogi, Ott and Yorke 1982) centers on the extraordinary dependence on the parameter taken equal to 4 in Equation (6).

1.3. Example: the Toral Automorphism. A favorite arena for abstract dynamical systems is the N-dimensional torus, defined as the N-fold Cartesian product of N circles (or equivalently unit intervals with 0 and 1 identified). Tori are useful for angles or other periodic coordinates. The study of one-to-one measure preserving maps (i.e., *automorphisms*) defined on the N-torus has been prominent in the history of dynamical systems (Adler and Weiss 1970). Consider the following mapping of this 2-dimensional space onto itself:

$$(7) \qquad\qquad X_{n+1} = (aX_n + bY_n)mod\ 1,$$

$$(8) \qquad\qquad Y_{n+1} = (cX_n + dY_n)mod\ 1,$$

where a, b, c and d are integers and we assume $ad - bc = \pm 1$, which makes the map measure preserving. Of particular fame is the special case $a = b = c = 1, d = 2$ known as the "cat map" (Arnold and Avez 1989).

We use the toral automorphism as a machine to generate sequences of numbers. Going to a two-dimensional map opens up interesting possibilities, but here the only difference is that the initial state needed to start up the process consists of two values, (X_0, Y_0). Figure 1c shows the X time series generated by the cat map, starting from $(C*, C*)$.

1.4. Example: the Hénon Process. The toral automorphism is piecewise linear, and it is useful to consider a chaotic 2-dimensional map that is continuous but nonlinear. A famous example is the map introduced by Hénon (1976):

$$(9) \qquad X_{n+1} = 1 - aX_n^2 + Y_n$$

$$(10) \qquad Y_{n+1} = bX_n.$$

With the commonly used parameters, $a = 1.4, b = 0.3$, this map has a chaotic attractor often used to illustrate the delicate fractal structures of such objects. If the starting point is in the basin of attraction of the attractor, the map generates a bounded vector time series. The X-component of a realization starting from the initial state $(C*, C*)$ is plotted in Figure 1d.

The reader may readily verify with computer experiments that the processes in all the above examples have exponential sensitivity to slightly differing initial values. The property which causes this and other aspects of chaotic behavior, is the "stretch-and-fold" character of map (think of kneading bread dough or taffy). In our examples the stretch comes from the form of the map. For the logistic and Hénon maps the fold also comes from the form of the map – specifically its quadratic term. In the toral automorphism and Bernoulli shift the fold is topological, in that the underlying toroidal space is folded. In short, chaos requires a nonlinear map.

Chaotic processes are also somewhat paradoxical: they generate time series that are disordered yet deterministic. One aspect of this paradox is resolved in Section 2.

1.5. Notes on History. Modern dynamics research began a century ago with Poincaré's (1892) interest in the global structure of solutions in general, as opposed to detailed knowledge of individual solutions. This notion grew into the current theory of chaotic dynamical systems (e.g. Devaney 1986, Wiggins 1990).

The mathematics supporting these physical discoveries has developed into a body of literature far too vast to survey here. (The references here are samples only.) Many fields have contributed, including dynamical systems (Smale 1980, Guckenheimer and Holmes 1983, Wiggins 1988, 1990), ergodic theory (Billingsley 1965, Sinai 1977, Cornfeld, Fomin, and Sinai 1982, Petersen 1983, Ornstein 1989) information and measure theory (Shannon and Weaver 1964, Billingsley 1986), complexity (Crutchfield and Young 1989) and symbolic dynamics (Alekseev and Yakobson 1981, Bai-lin 1989). There are now many general introductions to chaos (e.g., Lichtenberg and Lieberman 1983, Berge, Pomeau, and Vidal 1984, Thompson and Stewart 1986, Moon 1987, Schuster 1988 and Jackson 1989). Compendia of reprints include (Bai-Lin 1984, 1988, Cvitanovic 1984), the first of which has 44 pages of references. A recent tutorial of broad scope, with emphasis on the connections between theory and practice, is Eubank and Farmer (1990).

There are reviews of nonlinear time series analysis theory (Ruelle 1989) and practice (Tong 1990). Mandelbrot (1983) and Feder (1988) discuss interesting connections between fractals, time series, and random walks. A unified theory of chaotic and random ergodic processes, consistent with the modern theory of stochastic processes, is probably not far off.

2. Statistical Aspects of Chaotic Processes. As they stand, none of the above examples quite defines a stochastic process, since each realization with a given initial value is identical. Disordered appearance of the time series does not guarantee randomness of the process. We must define X_0 as a random variable, i.e. specify its distribution, before X is defined as a stochastic process.

Stationarity and Equation (2) impose rigid restrictions on the distribution functions of a chaotic process. This is very different from ordinary stochastic processes, for which joint distributions of different orders must be consistent which each other (Kolmogorov's theorem; e.g., Brockwell and Davis 1987) but are otherwise arbitrary.

To be specific, consider the doubling map. By Equations (3) or (4) the distribution of X_0 determines that of X_1, which in turn determines that of X_2, and so on. Unless all these distributions are the same the process is not stationary. We must therefore address the question: Is there a (unique?) function $g(X)$ on $[0, 1]$ that is the distribution function of all the random variables $X_n, n = 0, 1, 2, ...$?

The answer is that there are many such functions. One such g is a delta-function at $x = 0$; for $X_0 = 0$ implies $X_n = 0$ for all n. Another one is a delta-function at 1. These singular cases do not have the ergodic disorder that arises from most values X_0 randomly chosen from the interval (0,1). Is there a distribution function that describes these more typical cases? The answer is yes, the function $g(X) = 1$ on $(0, 1)$ and 0 elsewhere. The same holds for the other examples. For the logistic map the function $g(x) = [\pi\sqrt{(x - x^2)}]^{-1}$ describes the distribution for almost all initial values.

There have been many studies of conditions for the existence of distribution functions for stationary random variables related as in Equation (2). Such distributions are known as *invariant measures*, because of the requirement that the transformation T not alter the probability distribution. In many cases there is a unique distribution (called the *physical measure*) corresponding to almost all initial values, but there are a number of other invariant measures, corresponding to the few initial values that give periodic orbits (Eckmann and Ruelle 1985, Ruelle 1989).

Because we are interested in data from discrete-time processes, we ignore continuous systems described by differential, difference, or differential-difference (Bellman and Cooke 1963) equations. Yet much can be learned from numerical studies of chaotic differential equations. The instability which characterizes deterministic chaos has many implications for integrations of equations of motion. Integrations of a chaotic system starting from the same initial state, merely specified with different precisions, diverge from each other. So do orbits calculated with different algorithms, time-steps, precisions, or round-off procedures (Quinn and Tremaine 1990). In each case slight differences are amplified by sensitivity to initial conditions. This feature dooms to failure, or at least greatly limits, computation verifications such as convergence of the results with diminishing step size. Further, all bounded numerical solutions are periodic – a computer can represent only a finite set of numbers, so some state must recur – whereas almost all true orbits are not. For these and other reasons it is average, or statistical, behavior of chaotic systems that is important, not details of specific orbits.

We have found that physical systems can exhibit the disordered behavior normally associated with indeterminism, and yet be deterministic. This has several implications for time series analysis. Data that seem random may, in reality, be chaotic. Such processes can be better understood physically if one can "unveil the order hidden in chaos" – e.g. find the map. Chaos and true randomness can be mixed together in the same physical system. We need models and analysis tools to detect chaos and randomness, to study their structure, and to separate them if both are present. In the next sections we introduce a general viewpoint and specific tools designed for chaotic data analysis. We will see that tangible distinctions between randomness and chaos occur in state space, not in the time series.

3. Time Series Analysis Techniques for Chaos. Clearly the most convenient way to study a dynamical system is to collect measurements as it evolves in time. Accordingly most techniques developed to study chaotic systems are time series analysis methods. Correlation functions, power spectra, and other standard analytic tools, while useful in some problems, do not come to grips with the essential features of chaotic processes. These quantities are fundamentally unable to distinguish uncorrelated and independently distributed processes. This distinction may be important in nonlinear problems.

This section reviews some chaotic data analysis tools that have been developed in the last ten years or so. We begin by setting the stage. *State space* is the universal arena for chaotic systems analysis, both theoretical and experimental. We give some attention to the simplified state space called the *embedding space*, in which system evolution can be studied much more easily that in the full physical state space. This most important construct underlies all of the data analysis methods. We then briefly outline these techniques to inform the reader of what has been done and where to find details of specific algorithms.

3.1. Embedding. A convenient way to visualize the evolution of a physical system is to trace its path through state space (almost universally called *phase space*). The coordinates of this abstract space are a complete set of independent variables, such as particle positions and momenta. Any configuration of the system corresponds to a unique point of this space, and vice versa. The changing positions and velocities of the particles making up the system can be followed by visualizing a path through state space (Abraham and Shaw 1983).

In almost every scientific context it is impractical to measure the coordinates needed to specify points in state space. At best, one monitors a few variables chosen for ease of measurement, not dynamical completeness. At worst, one is presented with noisy, unrepeatable, incomplete sampling of a global quantity (e.g. temperature or brightness of a star) that is merely peripheral to the dynamics. Experimental data often merely constrain the dynamical variables, often in ways not fully understood and depending on imperfectly known parameters.

This gloomy picture is saved by a remarkable piece of mathematics. Measurements of just one variable can reveal the structure of the trajectories in the full multivariate state space. Under certain conditions on the dynamical system there is a simplified state space, the coordinates of which can be derived from a single observed variable (Packard, Crutchfield, Farmer, and Shaw 1980, Takens 1981). The trajectories in this space are simply related to the trajectories in the full physical state space. In particular, there is a smooth map from embedding to state space that preserves the topology of trajectories. Thus the essential features of the unobservable state space trajectories can be understood by studying the accessible trajectories in the embedding space.

There are many choices possible for the embedding space coordinates. The most common choice is the observed variable, evaluated at a set of lagged times:

$$(11) \qquad \mathbf{X} = (X_{n+m_1}, X_{n+m_2}, X_{n+m_3}, ..., X_{n+m_M}).$$

The integer M is the dimension of the embedding space, and the M lags m_i are arbitrary but distinct integers. Since all of the elements of $\mathbf{X}$ are measured quantities, it is straightforward to plot the data in the embedding space, to attempt to trace out the trajectories as the system evolves. Such a plot is called the phase portrait, and is the most important tool in the analysis of chaos.

As time goes by, points defined by Equation (11) follow embedding space trajectories topologically the same as the system's state space trajectories. Thus the phase portrait depicts the topology of the full state space dynamics. When the system satisfies the requisite mathematical conditions and the embedding space has high enough dimension, one says that a suitable embedding space has been found.

In its simplest form the phase portrait is just a plot of X_{n+1} against X_n. Figure 2 contains phase portraits for data from our example processes.

In practice, with real data, the situation is not so simple. It may be that the process must be embedded in a space of higher dimensions, using additional coordinates $X_{n+2}, X_{n+3}, \ldots$. The point is, one does not know a priori the dimension M of the embedding space – or the values of the lags m_i, for that matter. The theorems justifying the embedding procedure are for infinite, noise-free data streams. In this context the value of the lags do not matter and there is a simple prescription for the maximum value that M can assume. Unfortunately this theoretical value is usually much larger than values of M which provide quite satisfactory embeddings (see below). Moreover, observational errors always produce noise that somewhat blurs the phase portrait. However with a sufficiently large signal-to-noise ratio the phase portrait will reveal the nature of the dynamics in spite of the noise. Many workers are investigating these problems, to improve the effectiveness of the embedding procedure in practical data analysis.

It may be noted that embedding high-dimensional data in other spaces, e.g. of lower dimension, is not limited to dynamical systems. For example, *projection pursuit* (Friedman and Tukey 1974, Walden 1989) is a very similar way of looking at time series data, and may be effective on the sheet-like fractal structures which attractors often have.

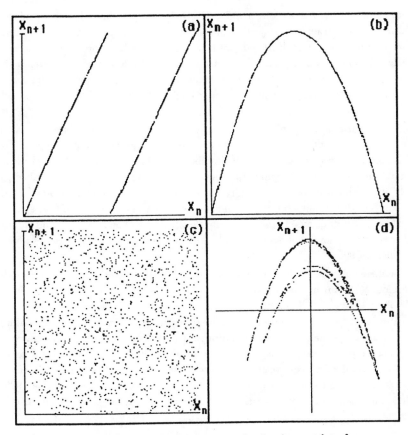

Fig. 2. Two-dimensional phase portraits for the raw data from (a) the one-sided Bernoulli shift, (b) the logistic map, (c) the cat map, and (d) the Hénon map.

3.2. Attractor Dimension. In most cases the long-time evolution of a physical system approaches a set of points in state space, called the *attractor*. Often the dimension of the set is much smaller than that of the state space. For periodic systems in discrete time the attractor is a finite set of points. In chaotic situations it can be strange indeed, for example a fractal. See references given earlier, especially (Takens 1981), for subtleties in the definitions of attractor, attracting set, strange attractor, etc.

Let D_A be the smallest integer not less than the dimension of this attractor. Imagine plotting the lagged time series data, *à la* Equation (11), in spaces of increasing dimension M. As long as $M < D_A$ the phase portrait will give a confused picture of the attractor – projection into a lower dimension flattens out its structure. On the other hand, if $M > D_A$, the phase portrait has more dimensions than needed to capture the attractor's structure, and is therefore confusing. Hence the least integer not less than the attractor dimension is the best embedding dimension. As stated above, theory provides an upper limit to the embedding dimension that is usually much larger than needed.

For its own interest, and because of this connection with embedding dimension, the determination of attractor dimension has received a great deal of attention. The most commonly used method, due to Grassberger and Procaccia (1983), implements the notion that the rate at which the volume contained in a hypersphere embedded in M-dimensional space grows with radius as R^M. Current research includes different kinds of dimension, statistics of dimension estimators, recognition that a single number does not always characterize fully the structure of an attractor, and various other generalizations – all described in hundreds of research papers. A recent review is (Theiler 1990).

A uncorrelated nonchaotic random process ("white noise") has an infinite dimension. (For an independently distributed process the number of points in an M-sphere grows as R^M no matter how large M is.) Some analysts have therefore assumed that the determination of a finite dimension, particularly a small noninteger value, proves that the underlying process is chaotic, not random. This is wrong, and some reported detections of chaos are almost certainly incorrect. Data generated by a random process with a nonwhite power spectrum has a finite dimension that has nothing to do with deterministic chaos (Osborne and Provenzale 1989). Thus dimension is a useful descriptor time series data if one knows that chaos is present, but it cannot be used to detect chaos.

3.3. Lyapunov Exponents. There are quantities that do indicate the presence of chaos. Perhaps the most well-studied are the *Lyapunov exponents*, which quantify sensitivity to initial conditions. Consider two close initial states $\mathbf{X}_1(0)$ and $\mathbf{X}_2(0)$, separated in state space by

$$(12) \qquad \delta\mathbf{X}(0) = \mathbf{X}_1(0) - X_2(0).$$

If the system is chaotic, the orbits starting from these initial states diverge exponentially:

$$(13) \qquad \delta\mathbf{X}(t) \sim \delta\mathbf{X}(0)e^{\lambda t},$$

where λ is the *Lyapunov exponent*. In an M-dimensional state space there are M exponents corresponding to the growth in different directions. Of most importance is the largest, λ_{max}: if it is > 0 the system is chaotic; if $\lambda_{max} \leq 0$ there is no exponential growth and therefore no sensitivity to initial conditions.

As with dimension, many algorithms have been proposed to estimate the Lyapunov exponents from time series data (e.g., Eckmann and Ruelle 1985, Wolf, Swift,

Swinney, and Vastano 1985). These methods rely on finding near state-space neighbors of a given point. How can a given point, say X', have near neighbors? The orbit starting at X' moves away and wanders around the state space, but because of ergodicity comes back; let X'' be the closest point to X' on a given revisit of the old neighborhood. A comparison of the evolution away from X' and X'' yields an estimate of the Lyapunov exponent.

There are several practical problems with this approach. One must first of all have enough data that many points in state space will indeed have near neighbors. For a space of high dimensions this leads to insatiable demands for data. Additionally, the noise cannot be very large, because an X'' closer to X' than the mean noise level is pretty useless. Another pitfall is that there are often places in state space where the trajectories arrive at critical points where some go left and others – a hair away – go right. If X' and X'' take different paths, the result is a catastrophically wrong contribution to the estimate of the Lyapunov exponent.

Thus, although the algorithms can provide accurate results (judged by comparison with exact values known for a few systems), their practical application is far from being trouble-free. Experimentalists (Frank, Lookman, Nerenberg and Essex 1990, to pick one recent example) are wrestling with these problems, and if their data are clean enough, they may get reasonable results.

3.4. Entropy, etc. Many other quantities of interest can be estimated from time series data. Examples include *Kolmogorov entropy*, the rate at which the dynamical system generates information, and system complexity – algorithmic or otherwise. The reader is referred to any of the general references listed above for more details on these interesting topics, which however are not relevant to the methods we describe in the next section.

3.5. Prediction and Noise Reduction. One topic that is relevant is prediction, as our deconvolution method is based on predictive deconvolution.

Prediction, of course, has two distinct applications. In real-time control, or economic time series analysis (e.g. Brock 1990), the importance of prediction obviously derives from interest in future values, literally, of the observed series. More generally, the ability of a model to predict X_n based on the past data $X_i, i < n$, is often taken as a measure of model quality. A good model that has captured the structure of the process underlying the data, it is argued, will be good (have a low RMS error) at prediction of the future. In *post facto* data analysis, which after all covers most scientific applications, this reasoning can be extended to include the "prediction" of X_n, based on future data, X_i for $i > n$, or based on both past and future data, $X_i, i \neq n$ (Scargle 1981).

For information on nonlinear prediction methods in the analysis of chaotic time series, consult, e.g., (Farmer and Sidorowich 1987, Casdagli 1990).

4. A New Tool: Linear Time Domain Models. Time domain models are useful for any dynamical process, random or chaotic. We first examine a theorem that guarantees the existence of a convenient, linear model for such processes as long as they are stationary. Then a model estimation procedure is outlined and demonstrated on synthetic data.

4.1. Existence Theory. Stationarity of a process implies that it has a remarkably simple linear representation as the convolution of a filter with a white noise process:

The Wold Decomposition Theorem: Any stationary process X can be written as

$$(14) \qquad X = R * C + D,$$

where D is a linearly deterministic process, C is a constant (linear) filter, and R is an uncorrelated ("white") process. In addition, C is causal and minimum delay, and R and D are not correlated with each other.

Since chaos is nonlinearly deterministic it resides in the convolutional term, not in the linearly deterministic part D. It is remarkable that the deterministic aspects (C and D) of any stationary process can be separated from the purely random component (R) in such an explicit and linear way. We remove any linearly deterministic part, e.g. a linear or periodic trend, so the Wold representation becomes:

$$(15) \qquad X = R * C,$$

or, explicitly,

$$(16) \qquad X_n = \sum_{k=1}^{\infty} C_k R_{n-k}.$$

This is a causal moving average (MA) model. It is sometimes necessary to generalize to acausal models, i.e. to allow negative k (Scargle 1981, Breidt, Davis, Lii and Rosenblatt 1990). Another useful time-domain representation is the autoregressive (AR) model,

$$(17) \qquad R = X * A,$$

or with a minor rearrangement

$$(18) \qquad X_n = -\sum_{\substack{-\infty \\ k \neq 0}}^{\infty} A_k X_{n-k} + R_n,$$

representing the process in terms of a linear memory. Note that as mentioned in §3.5 we have allowed "memory" of the future as well as the past; this is necessary because a given causal MA model is only guaranteed to have an inverse if acausality is permitted (Scargle 1981). Any process has both a MA and an equivalent AR representation, as long as one allows non-causal representations. The AR model is often easier to estimate, while the MA form has a more direct physical interpretation – random pulses, or shot noise.

4.2. Deconvolution as a Modeling Tool. We have just seen how a moving average model must exist for any stationary process, chaotic as defined above or random in the traditional sense. To make this useful we need a procedure to estimate the moving average model from time series data.

Of the many approaches known for this kind of problem, the one most suited to the current goal is to estimate the autoregressive model inverse to the moving average guaranteed by the Wold Decomposition Theorem. That is, given the data X, construct the convolution

$$(19) \qquad\qquad R'(A) = A * X,$$

meant as an estimate of the innovation of the process. In particular, if A were the correct convolutional inverse of the C in the Wold representation of X, then R' would be the actual innovation R in the Wold Decomposition. Thus, if we know that R has a certain property we can adjust the parameters of A to maximize some measure of this property. Since it is known that R is uncorrelated, A can be estimated by maximizing the whiteness of R'; indeed, this is the basis of least-squares ARMA modeling (Brockwell and Davis 1987). However, this approach is incapable of determining the correct C unless it is known (or assumed) that C is minimum-delay (Scargle 1981). In addition, optimizing a different property of the innovation turns out to be more appropriate for chaotic processes.

I have demonstrated empirically (Scargle 1990a) that a good procedure is to define a *cost function* for R' that is essentially the M-volume covered by the plot of the points

$$(20) \qquad\qquad \mathbf{R'} = (R'_{n+m_1}, R'_{n+m_2}, R'_{n+m_3}, ..., R'_{n+m_M}).$$

[*c.f.* Equ. (11)] in an embedding space for the innovation. In the computations, one constructs an algorithm for evaluating a *cost function* of R' that measures the embedding space volume of the points described above, given the data X and the filter A. It is then straightforward to minimize the cost function with respect to the parameters in A, thence determine estimates of C and R in the Wold Decomposition.

In addition to the usual problem of determining the order of the autoregressive model, one must find the dimension of the embedding space, as discussed earlier. I gave a description of how this can be done and other details of the method in (Scargle 1990a, 1990b). These papers present results of deconvolution experiments on simulated moving averages of known parameters and order. While far from a comprehensive study, they demonstrate how chaotic processes can be detected and distinguished from random processes, even in the presence of a background of randomly distributed observational errors. With good signal-to-noise the embedding dimension and the order of the moving average can be determined, in simple cases. These results suggest that a practical linear deconvolution procedure is possible and may provide insight into real-world dynamical systems.

4.3. Results. The rest of this section outlines some deconvolution results for noise-free data generated by the example chaotic systems considered earlier.

These do not represent realistic physical systems, but are meant to demonstrate the flavor of the results that are possible in cases simple enough that the uncertainties connected with experimental data do not obscure the basic phenomena.

Exact deconvolutions are presented for the processes generated by piecewise linear maps. The processes connected with quadratic maps are deconvolved numerically, so these results are not exact.

4.3.1. Deconvolution of the Bernoulli Process.

The Bernoulli process, derived from the piecewise linear doubling map, is so simple that it can be analyzed in complete detail (Scargle 1989a,1989b), yet contains all of the essential features of deterministic chaos.

As discussed in §2, with X_0 uniformly distributed on $[0,1]$, the map in Equations (3) or (4) generates a correlated, stationary stochastic process. Therefore the Wold theorem will yield a nontrivial decomposition. (White noise has a trivial representation as the convolution of a delta function with itself.) The Durbin-Levinson algorithm (Brockwell and Davis 1987, §5.2) gives the autoregressive coefficients exactly:

$$(21) \qquad A_1 = \frac{\rho(1)}{\rho(0)} = -\frac{1}{2}; \quad A_i = 0, \quad i = 2, 3, \ldots$$

That is, the filter that deconvolves the Bernoulli process best, in least-squares, is

$$(22) \qquad A = (1, -\frac{1}{2}),$$

where boldface indicates the origin of time. It is then straightforward, following the constructive proof of the Wold Theorem, to compute that the linearly deterministic part of Equation (14) is $D = 0$, and that the MA filter C is

$$(23) \qquad C_n = (\frac{1}{2})^n \quad all \ n \geq 0.$$

Of course A_n and C_n are 0 for all $n < 0$, because the Wold representation is causal, by assumption. The innovation R can also be computed explicitly:

$$(24) \qquad R_n = \frac{3}{2} X_{n-1} - \frac{1}{2} Sign(X_{n-1}).$$

In summary, the Wold decomposition of the Bernoulli process has no linearly deterministic part, and the MA filter is a simple exponential.

The minimum of the area of the 2-dimensional phase portrait can be found by inspection (or guessed at using numerical experimentation):

$$(25) \qquad A = (-\frac{1}{2}, 1),$$

i.e. an acausal, time-reversed version of the AR filter in the Wold representation. In this case, the innovation is simply

$$(26) \qquad R_n = \frac{1}{4} Sign(X_n).$$

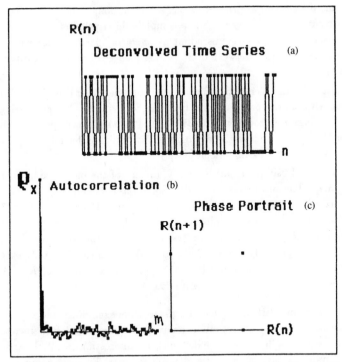

Fig. 3. Deconvolution of data generated by the one-sided Bernoulli shift (doubling map): (a) the deconvolved time series, (b) the autocorrelation function, and (c) the phase portrait.

Figure 3 shows the details of the resulting deconvolution. The innovation is discrete, with only two values (see below). Note that, even though we did not ask for an uncorrelated innovation, this one is; the deconvolution has removed the exponential correlation present in the original process [*c.f.* Equ. (5)]. In addition, the phase portrait has zero area, as it consists of four points.

Interestingly the innovation in Equation (26) coincides with the symbolic representation of X. In symbolic dynamics the range of the process is partitioned into a collection of cells and each cell is labeled with a different symbol. This "digital X-meter" reads out a sequence of symbols for any realization of the process. Even though this discretization process discards information, as does any digital meter with finite resolution, it may provide a complete description of the dynamical system. In particular, some systems admit a *generating partition*, for which there is a unique correspondence between orbits and sequences of symbols (Crutchfield and Packard 1983).

In most cases the identification of a generating partition is difficult. However it is well known that the partition into two cells, $< \frac{1}{2}$ and $\geq \frac{1}{2}$, is generating for the form of the Bernoulli process in Equation (3). These cells correspond precisely to 0 and 1 for the most significant binary digit of X_n; orbits and symbol sequences thus both correspond to real numbers on $[0, 1]$.

This extremely simple result is clearly a result of the piecewise linearity of the doubling map. The autoregressive model, Equations (17) and (18), represents linear dependence of successive values of X_n, so if the map, Equations (2) or (3), has a linear piece, an autoregressive filter A can exactly remove the effect of the map over this piece. In particular, one can see directly that the convolution of the filter in Equation (22) precisely undoes the map in Equation (3); The only exception is when the "mod" operates, producing an impulse in the output of $R = A * X$, as seen in the plot of this process in Figure 3a.

4.3.2. Deconvolution of the Logistic Process. Since the logistic map is not piecewise linear, it does not admit an exact deconvolution. But since the linear model of the Bernoulli process yielded such an interesting result, I carried out a numerical minimization of the phase-portrait volume. Figure 4 compares the raw and deconvolved data.

This is a two-dimensional problem, but we examine it in both two and three dimensions for comparison with the cat and Hénon maps, which are three dimensional. In two dimensions the phase portrait of the raw data is a parabola, as in Figure 2b. In three dimensions, i.e. a plot of X_{n+2} *vs.* X_{n+1} and X_n, the parabola splits into two branches, labeled **ABCD** and **DEFG**, forming a figure that projects into a simple parabola in views either along the X_n or the X_{n+2} axes, but into a quartic along the X_{n+1} axis.

The portraits of the deconvolved innovation are bent versions of those of the raw data. For example, in the three dimensional view the two branches **abcd** and **defg** have an apparent 90-degree bend at the points **c** and **e**. In two dimensions one can see that the effect of the bend is that along part of the trajectory the value of R_{n+1} is constant, independent of R_n. In this sense the innovation is a partially discretized, or symbolic dynamical, representation of the process.

4.3.3. Deconvolution of the Cat Map Process. Now turn to the first of the two dimensional cases. The toral automorphism is a three dimensional version of the doubling map. For the same reason as in that case – piecewise linearity of the map – the toral automorphisms can be exactly deconvolved into symbolic dynamic representations.

Because of the higher dimensionality of this case, we expect that the deconvolving filter will be one higher order than in the case of the doubling map. A

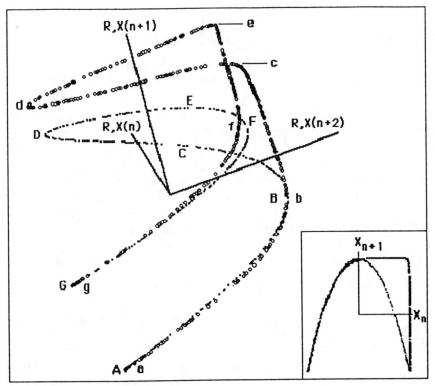

Fig. 4. Three dimensional state space plot of the deconvolution of data generated by the logistic map: raw data shown as
dots, deconvolved innovation as circles. The inset shows the
2-dimensional projection of the 3-dimensional state space.

straightforward evaluation of the convolution of a general second-order autoregressive filter with the system defined by Equations (7) and (8) shows that processes X
and Y are deconvolved by the same filter, namely

$$(27) \qquad A = (\mathbf{1}, -a - d, \pm 1).$$

Furthermore, the convolutions $A * X$ and $A * Y$ with this filter take on only a finite
set of discrete values.

For example, in the case of the Cat map Equation (27) gives

(28) $$A = (\mathbf{1}, -3, 1).$$

The innovation $A * X$ takes on only values from the set $\mathbf{S}$ consisting of the four values $\{-1, 0, 1, 2\}$. It is easy to prove from the invertibility of the filter A that this symbolic representation yields a generating partition of the 2-torus.

Figure 5 shows the phase portraits of the data generated by the cat map. In this case three is the smallest number of dimensions in which this process makes much sense.

Part a of the figure is the raw data X, which populates four sheets: the top and bottom fill out triangles, while the middle two consist of uniformly filled unit squares. The deconvolved innovation R occupies 42 of the possible 64 points, corresponding to the Cartesian product $\mathbf{S} \otimes \mathbf{S} \otimes \mathbf{S}$. The selection rules and transition probabilities connecting states of this innovation process can be understood by examining the conditions on the point (X_n, Y_n) under which the modulus operations are triggered in the transformations T and T^2 given by Equations (7) and (8).

4.3.4. Deconvolution of the Hénon Process. We now turn to the most complex case of the four, the Hénon map. Again the absence of piecewise linearity forces us to carry out a numerical deconvolution, and the dimensionality of the map forces us to display the data in three dimensions. Curiously, the numerical results for a cost-function defined in a three-dimensional embedding space were very close to those obtained in a two-dimensional embedding. The deconvolution depicted in Figure 6 is based on a numerical deconvolution, using the the two-dimensional area as the cost function.

The connection between the phase portraits of the derived innovation and the raw data is very similar to that found for the logistic map: the portrait of the deconvolved data can be obtained by a smooth transformation that is in the form of a bending of the sheet-like attractor of the raw data. Again the bend makes a right angle, and the result is a partial discretization of the innovation.

5. Summary. This paper outlines a scheme for estimating time-domain models for time series data generated by a measure-preserving dynamical system, which is a special kind of stationary stochastic processes. The Wold Decomposition serves as an existence theorem for moving average models. The embedding procedure, now pivotal in chaos studies, yields a cost function for a practical deconvolution method of model estimation. The four examples studied here demonstrate piecewise linear and quadratically nonlinear maps in one and two dimensions. Deconvolution of the Bernoulli shift yields its representation as the convolution of an exponential filter with an uncorrelated innovation that is equivalent to the symbolic dynamics of the shift. We conclude that any process generated by iteration of a piecewise linear map (e.g. the doubling and cat maps) is deconvolved exactly into a discrete representation related to the symbolic dynamics of the process. The quadratically nonlinear cases, the logistic and Hénon maps, have only approximate deconvolutions, which are in the form of a bending of the phase portrait. In these cases the bending produces a partial, not a complete, discretization of the process.

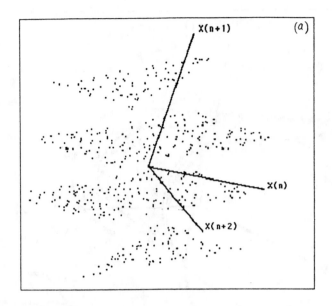

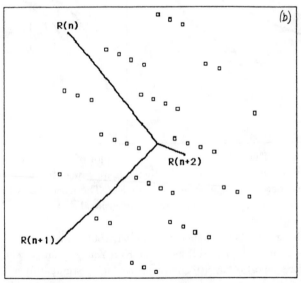

Fig. 5. Three dimensional state space plot of the deconvolution of data generated by the cat map: (a) raw data, (b) deconvolved innovation.

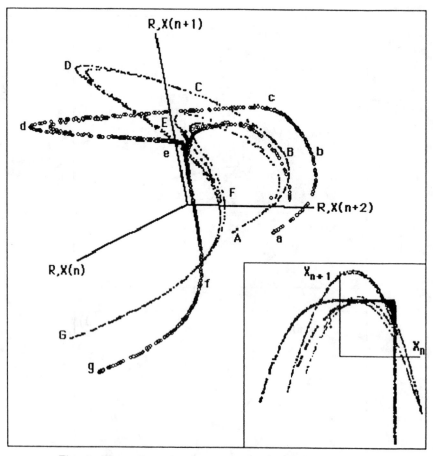

Fig. 6. Three dimensional state space plot of the deconvolution of data generated by the Hénon map: raw data shown as dots, deconvolved innovation as circles. The inset shows the 2-dimensional projection of the 3-dimensional state space.

Acknowledgements: I am indebted to many colleagues for useful suggestions, especially Jim Crutchfield, David Donoho and Karl Young. I am grateful to the IMA and its staff for support of my attendance of this highly stimulating and productive Workshop. Funding of various aspects of this research was provided by the NASA Astrophysics Software and Research Aids program (NRA 89-OSSA-8), the NASA-Ames Research Center Director's Discretionary Fund, and a special grant from Mike Smith of the NASA Office of Commercial Programs.

REFERENCES

ABRAHAM, R., SHAW, C., *Dynamics–The Geometry of Behavior, Part 2: Chaotic Behavior*, (Santa Cruz: Aerial Press), 1983.

ADLER, R.L., WEISS, B., *Similarity of Automorphisms of the Torus*, Memoirs Amer. Math. Soc, Vol. 98 (1970).

ALEKSEEV, V.M., YAKOBSON, M.V., *Symbolic Dynamics and Hyperbolic Dynamical Systems*, Phys. Reports, Vol. 75 (1981), pp. 287–325.

ARNOLD, V.I., AVEZ, A., *Ergodic Problems of Classical Mechanics*, Addison-Wesley Pub. Co., New York, 1989.

BAI-LIN, H., *Chaos*, World Scientific Pub. Co., Singapore, 1984.

BAI-LIN, H., *Directions in Chaos*, World Scientific Pub. Co., Singapore, 1988.

BAI-LIN, H., *Elementary Symbolic Dynamics and Chaos in Dissipative Systems*, World Scientific Pub. Co., Singapore, 1989.

BELLMAN, R., COOKE, K.L., *Differential-Difference Equations*, Academic Press, New York, 1963.

BERGE, P., POMEAU, Y., VIDAL, C., *Order within Chaos: Towards a Deterministic Approach to Turbulence*, New York: John Wiley & Sons, 1984.

BILLINGSLEY, P., *Ergodic Theory and Information*, John Wiley and Sons, New York, 1965.

BILLINGSLEY, P., *Probability and Measure*, John Wiley and Sons, New York, 1986.

BREIDT, F.J., DAVIS, R.A., LII, K-S., ROSENBLATT, M., *Maximum Likelihood Estimation for Noncausal Autoregressive Processes*, preprint (1990).

BROCK, W., *Causality, Chaos, Explanation and Prediction in Economics and Finance*, preprint, in *Beyond Belief: Randomness, Prediction, and Explanation in Science*, Casti and Karlqvist, eds. CRC Press: Boca Raton (to appear).

BROCKWELL, P., DAVIS, R., *Time Series: Theory and Methods*, New York: Springer, 1987.

CASDAGLI, M., *Nonlinear Prediction of Chaotic Time Series*, Physica, Vol. D35 (1989), pp. 335–356.

CORNFELD, I.P., FOMIN, S.V., SINAI, YA.G., *Ergodic Theory*, Springer-Verlag, New York, 1989.

CRUTCHFIELD, J., PACKARD, N., *Symbolic Dynamics of Noisy Chaos*, Physica, Vol. 7D, pp. 201-22.

CRUTCHFIELD, J., YOUNG, K., *Computation at the Onset of Chaos*, in *Entropy, Complexity, and the Physics of Information*, ed. W. Zurek, Addison-Wesley, New York, 1989.

CVITANOVIC, P., *Universality in Chaos*, Adam Hilger Ltd., Bristol, 1984.

DEVANEY, R.L., *An Introduction to Chaotic Dynamical Systems*, Benjamin, Cummings, Menlo Park, 1986.

DOOB, J., *Stochastic Processes*, John Wiley and Sons, New York, 1953.

ECKMANN, J.-P., RUELLE, D., *Ergodic Theory of Chaos and Strange Attractors*, Rev. Mod. Phys., 57, Part I, 617 (1985).

EUBANK, S., FARMER, D., *An Introduction to Chaos and Randomness 1989 Lectures in Complex Systems*, SFI Studies in the Sciences of Complexity, Lect. Vol. II, ed. Erica Jen, Addison-Wesley, New York, 1990.

FARMER, D., SIDOROWICH, J., *Exploiting Chaos to Predict the Future and Reduce Noise*, in *Evolution, Learning and Cognition*, ed. Y. Lee, World Scientific Pub. Co., Singapore, 1988.

FEDER, J., *Fractals*, Plenum Press, New York, 1988.

FEIGENBAUM, M.J., *Quantitative Universality for a Class of Nonlinear Transformations*, J. Stat. Phys., Vol. 19 (1978), pp. 25–52.

FRANK, G.W., LOOKMAN, T., NERENBERG, M.A.H., ESSEX, C., *Chaotic Time Series Analysis of Epileptic Seizures*, Physica, Vol. D46 (1990), pp. 427-438.

FRIEDMAN, J.H., TUKEY, J.W., *A Projection Pursuit Algorithm for Exploratory Data Analysis*, IEEE Trans. Computers, Vol. 23 (1974), pp. 881–890.

GRASSBERGER, P., PROCACCIA, I., *Characterization of Strange Attractors*, Phys. Rev. Letters, Vol. 50 (1983), p. 346.

GREBOGI, C., OTT, E., YORKE, J.A., *Chaotic Attractors in Crisis*, Phys. Rev. Letters, 48 (1982), pp. 1507–1510.

GUCKENHEIMER, J., HOLMES, P., *Nonlinear Oscillations, Dynamical Systems, and Bifurcations of Vector Fields*, Springer-Verlag, New York, 1983.

HÉNON, M., *A Two-Dimensional Mapping with a Strange Attractor*, Communications of Mathematical Physics, Vol. 50 (1976), p. 69.

JACKSON, E., *Perspectives of Nonlinear Dynamics, Vol. I*, Cambridge University Press, Cambridge, 1989.

LICHTENBERG, A.J., AND LIEBERMAN, M.A., *Regular and Stochastic Motion*, Springer-Verlag, New York, 1983.

MANDELBROT, B., *The Fractal Geometry of Nature*, W.H. Freeman and Company, New York, 1983.

MAY, R.M., *Simple Mathematical Models with Very Complicated Dynamics*, Nature, Vol. 261 (1976), pp. 459–467.

MOON, F., *Chaotic Vibrations: An Introduction for Applied Scientists and Engineers*, John Wiley & Sons, New York, 1987.

ORNSTEIN, D.S., *Ergodic Theory, Randomness, and 'Chaos'*, in Science, Vol. 24, 1989, pp. 182–186.

OSBORNE, A. AND PROVENZALE, A., *Finite Correlation Dimension for Stochastic Systems with Power-law Spectra*, Physica D35 (1989), p. 357.

PACKARD, N.H., CRUTCHFIELD, J.P., FARMER, J.D., AND SHAW, R.S., *Geometry from a Time Series*, Physical Review Letters, Vol. 45 (1980), p. 712.

PETERSEN, K., *Ergodic Theory*, Cambridge U. Press, Cambridge, 1983.

POINCARÉ , H., *Les Méthodes Nouvelles de la Méchanique Celeste*, Gauthier-Villars, Paris; in English (1967): *N.A.S.A. Translation TT F-450/452*. U.S. Federal Clearinghouse; Springfield, VA, USA, 1892.

QUINN, T., TREMAINE, S., *Roundoff error in long-term planetary orbit integrations*, preprint (1990).

RUELLE, D., *Chaotic Evolution and Strange Attractors: The Statistical Analysis of Time Series for Deterministic Nonlinear Systems*, Cambridge U. Press, 1989.

SCARGLE, J., *Studies in Astronomical Time Series Analysis. I. Modeling Random Processes in the Time Domain*, Astrophysical Journal Supplements, Vol. 45 (1981), pp. 1–71.

SCARGLE, J., *An Introduction to Chaotic and Random Time Series Analysis*, International Journal of Imaging Systems and Technology, Vol. 1 (1989a), pp. 243–253.

SCARGLE, J., *Random and Chaotic Time Series Analysis: Minimum Phase-Volume Deconvolution*, in Nonlinear Structures in Physical Systems, Proceedings of the Second Woodward Conference, eds. L. Lam and H. Morris, Springer-Verlag, New York, 1989b, pp. 131–134.

SCARGLE, J., *Studies in Astronomical Time Series Analysis. IV. Modeling Chaotic and Random Processes with Linear Filters*, Astrophysical Journal, 343 (1990a), pp. 469–482.

SCARGLE, J., *Astronomical Time Series Analysis: Modeling of Chaotic and Random Processes*, in Errors, Bias and Uncertainties in Astronomy, eds. C. Jaschek and F. Murtagh, Cambridge U. Press, Cambridge, 1990b, pp. 1–23.

SCHUSTER, H., *Deterministic Chaos*, VCH, New York, 1988.

SHANNON, C.E., WEAVER, W., *The Mathematical Theory of Communication*, U. of Illinois Press, Urbana, 1964.

SINAI, YA.G., *Introduction to Ergodic Theory*, Princeton University Press, Princeton, originally published in 1973 by Erevan University, 1977.

SMALE, S., *The Mathematics of Time: Essays on Dynamical Systems, Economic Processes and Related Topics*, Springer-Verlag, New York, 1980.

TAKENS, F., *Detecting Strange Attractors in Turbulence*, in Dynamical Systems and Turbulence, *Warwick, 1981*, Vol. 898 of *Lecture Notes in Mathematics* (Springer-Verlag, Berlin).

THEILER, J., *Estimating Fractal Dimension*, preprint, J. Opt. Soc. Am. A, Vol. 7 (1990) (to appear), pp. 1055–1073.

THOMPSON, J., STEWART, H., *Nonlinear Dynamics and Chaos*, New York: John Wiley, 1986.

TONG, H., *Non-Linear Time Series, A Dynamical System Approach*, Oxford University Press, Oxford, 1990.

WALDEN, A.T., *Clustering of Attributes by Projection Pursuit for Reservoir Characterization*, preprint, to appear, in Automated Pattern Recognition in Geophysical Exploration Society of Exploration Geophysicists, 1989.

WIGGINS, S., *Global Bifurcations and Chaos, Analytical Methods*, Springer-Verlag, New York, 1988.

WIGGINS, S., *Introduction to Applied Nonlinear Dynamical Systems and Chaos*, Springer-Verlag, New York, 1990.

WOLF, A., SWIFT, J.B., SWINNEY, H.L., VASTANO, J.A., *Determining Lyapunov Exponents from a Time Series*, Physica, Vol. 16 (1985), p. 285.

CONTRASTING ASPECTS OF NON-LINEAR TIME ANALYSIS

HOWELL TONG*

Abstract. The diversity of recent activities in non-linear time series analysis has given rise to many interesting contrasts. I comment on some of these in this paper. Some new results and some open problems are mentioned.

Key words. Additive noise, Bayesian, continuous state space, continuous time, deterministic, discrete state space, discrete time, drift criterion, empirical, ergodic, formal, frequency domain, frequentist, informed, long memory, multiplicative noise, non-ergodic, off-line, on-line, short memory, stability criterion, stochastic, substantive, time domain.

1. Introduction. Although it is still rather young, the subject of non-linear time series analysis is developing very rapidly, so much so that its diversity can be quite bewildering to a newcomer. A fairly comprehensive introduction to the basic theory and practice is now available in Tong (1990); earlier monographs include Granger and Andersen (1978), Tong (1983), Subba Rao and Gabr (1984) and Priestley (1988).

By focusing on the contrasting aspects of the different approaches and methodologies, I hope that the article would be found helpful in providing some informative perspectives. Comments of this nature cannot be wholly objective or exhaustive and this fact explains my choice of the pronouns, which is a deviation from my normal practice.

For convenience, the paper will be divided into three major themes: (I) Approaches, (II) Models, (III) Methodologies with the usual caveat that the division is fairly arbitrary and the allocation of subject matter to the themes is not always indisputable. Moreover, I shall, generally speaking, not amplify the interactions in the '3×2^k design', for lack of space and definitive results at present. In other words, these represent interesting areas for future development. Finally, I shall try to keep the technical detail to a minimum and only key references will be cited. Tong (1990) has a relatively comprehensive list of references.

2. Theme I: Approaches. This theme consists of 5 pairs: (i) frequency domain vs. time domain, (ii) deterministic vs. stochastic, (iii) continuous time vs. discrete time, (iv) continuous state space vs. discrete state space and (v) substantive vs. empirical.

2.1 Frequency Domain vs. Time Domain. There is a natural extension of the same pair in linear time series analysis. In the linear analysis, usually attention is confined to the second order moments, although third order moments are also used in the analysis of linear non-Gaussian time series. For example, Rosenblatt

---

*Institute of Mathematics & Statistics, University of Kent at Canterbury, Canterbury, Kent, CT2 7NF, United Kingdom

(1985) has discussed the used of the third order cumulant to deconvolute a linear non-Gaussian time series.

For nonlinear time series analysis moments higher than two are necessary. We can either base our analysis on the kth *order joint cumulant*, $\text{cum}(X_{t_1}, \ldots, X_{t_2})$, of a time series $\{X_t\}$, assumed k^{th} order stationary, or its $(k-1)$-fold Fourier transform, i.e. the *cumulant spectral density function of order* k. The former belongs to the time-domain approach and the latter the frequency-domain approach. Brillinger (1975), Subba Rao and Gabr (1984) and Rosenblatt (1985) are useful references for the latter. Unlike the usual spectral density function (which is identical to the cumulant spectral density function of order two), physical interpretation is not always obvious for cases with $k \geq 3$. The difficulties stem partly from the fact that the arguments of these functions do not admit any obvious physical interpretation. However, some successful applications are available in, e.g. Hasselman et al. (1963), Lii, Helland and Rosenblatt (1982) and other papers in this volume. Another source of difficulty is perhaps associated with the fact that the greater the variability of the estimate the greater is k. Thus, the most commonly used values of k are 2 and 3.

In the time-domain approach, the evaluation of kth order moments is not necessarily the only concern. Quite often, the conditional moments are of interest too. In fact, assuming strict stationarity, one is frequently also interested in the stationary marginal distributions and conditional distributions, mostly of low dimensions. Some numerical and analytic results are available (See, e.g. Tong, 1990). The time-domain approach is commonly facilitated by a finite parametric (non linear) model. Needless to say, this is not essential. For example, consider the problem of estimating the delay parameter, d, in the input/output relation

$$(2.1.1) \qquad X_t = f(Y_{t-d}) + \varepsilon_t, \quad t \in \mathbf{Z}, \quad d \in \mathbf{Z}^+$$

where $\{\varepsilon_t\}$ is the unobservable stationary (but not necessarily white) noise process independent of the observed stationary input signal $\{Y_t\}$, and $\{X_t\}$ is the observed stationary output signal. I propose here that by constructing kernel estimates of $\text{var}[X_t | Y_{t-s} = y]$ ($g_s(y)$, say) for $s = 0, 1, 2, \ldots$ (assumed to exist), one can obtain a good estimate of d without assuming any specific functional form of f. Note that $g_d(y)$ is constant whilst, for $s \neq d$, $g_s(y)$ is, in general, not a constant but a positive function of y bounded below by $g_d(y)$.

2.2 Deterministic vs. Stochastic. Recent development in deterministic dynamical systems, especially those which generate chaos, raises the issue of the deterministic approach vs. the stochastic approach. It might be tempting to entertain the thought that, with luck, one could explain real phenomena by purely deterministic nonlinear systems. Randomness is then associated with sensitivity to initial conditions. Thus, given real data, e.g. EEG, ECG, etc., attempts have been made to identify these with 'low-dimensional' attractors under certain conditions, (see, e.g., Babloyantz and Destexhe, 1986). That is to say, the deterministic approach tends to emphasize the 'regular component' of a time series, and to identify it as a low-dimensional attractor. To appreciate this point, one needs to make precise the notion of dimension, which can be done in a number of ways. (See, e.g., Farmer et al. 1983.) Consider the following discrete time dynamical system $f : R^k \to R^k$

$$(2.2.1) \qquad x_t = f(x_{t-1}), \quad t \in \{1, 2, \dots\}, \quad x_0 \in R^k .$$

First, think of an attractor roughly as a compact subset of R^k to which x_t tends as $t \to \infty$ if x_0 lies in the neighborhood of the subset. Suppose that f is differentiable and let $D_x f$ denote its derivative evaluated at x. Let $f^{(n)}$ denote the $n-th$ iterate of f. Let $a_i(n, x)$ denote the modulus of the ith eigenvalue of $D_x f^{(n)}$, ordered so that $a_1(n, x) \geq \cdots \geq a_k(n, x)$. Define the ith *Lyapunov exponent* $\lambda_i(x)$ as

$$(2.2.2) \qquad \lambda_i(x) = \lim_{n \to \infty} \frac{1}{n} \ln a_i(n, x), \qquad i = 1, \dots, k.$$

Suppose that these turn out to be independent of x. Denote by d_L the quantity $m - (\lambda_1 + \cdots + \lambda_m)/(\lambda_{m+1})$, where m is the largest integer for which $\lambda_1 + \lambda_2 + \cdots + \lambda_m \geq 0$. [If $\lambda_1 < 0$, define $d_L = 0$; if $\lambda_1 + \cdots + \lambda_k \geq 0$, define $d_L = k$.] This d_L is the so-called *Lyapunov dimension of the attractor*, and is one of an array of dimensions used in dynamical systems. Note that d_L may be non-integral. Another commonly used dimension is the so-called *Kolmogorov capacity*, d_C, of a compact set C. To describe this, cover C with squares of size ε. Let $N(\varepsilon)$ denote the number of such squares required. Then by definition

$$(2.2.3) \qquad d_C = \lim_{\varepsilon \to 0} \frac{\ln N(\varepsilon)}{\ln(1/\varepsilon)} .$$

It may shown (op.cit.) that in general

$$d_C \neq d_L ,$$

where C is the attractor of a dynamical system whose Lyapunov exponents define d_L.

By contrast, the tendency of the stochastic approach is to try to extract information about the 'irregular component' of a time series. In some, but not all, cases the regular component is considered a 'nuisance' to be removed, e.g., detrending, deseasonalising, etc. Often the irregular component tends to 'fill' the space in which it lives. More specifically, for every positive integer k, the trajectory of the k-vector $(X_{t-1}, X_{t-2} \dots, X_{t-k})$ constructed from the irregular component $\{X_t\}$ tends to fill R^k as t evolves over the parameter space. In this sense, the irregular component is said to be a high dimensional attractor.

My contention (elaborated in my discussion of Bartlett. 1990) is that it would be a mistake to hold a wholly deterministic view or a wholly stochastic view. Rather like the Yin and the Yang, the low dimensional attractor and the high dimensional attractor co-exist in any one real time series, i.e.
(2.2.4)
Real Time Series = Low Dimensional Attractor $\oplus$ High Dimensional Attractor.

The basic idea is not new in time series analysis but the emphasis seems worth repeating. Note that even a computer-generated chaotic time series cannot deviate from (2.2.4) because of the (deterministic) rounding errors being part of the driving noise. Now one tends to associate a regular component with a low dimensional attractor and an irregular component with a high dimensional attractor. Regular components are usually associated with slow variation and include, as special cases, constant levels, cycles or chaos, etc. Clearly, the concept of signal-to-noise ratio is still relevant to the 'detection of chaos'. Further, note that the decomposition need not be unique. One possible decomposition is to identify the high dimensional attractor with the innovation process. On adopting the Yin–Yang approach, the problem of detecting chaos in experimental data becomes a generalized signal extraction problem. Tong (1989) and Gerrard and Tong (1990) have discussed the effect on the Lyapunov exponents when a random driving term is added to the right hand side of (2.2.1).

It is particularly interesting to note that often problems which are found to be difficult in, say, the deterministic approach have their counterparts in the stochastic approach which are equally difficult. The converse is also true. Take the very fundamental problem of determination of the value k, the so-called *embedding dimension* in the deterministic approach, using experimental data. This problem is known to be difficult in the study of chaos. The analogous problem of determining the order, k, of a discrete time univariate nonlinear autoregressive model of the form

$$(2.2.5) \qquad X_t = f(X_{t-1}, X_{t-2}, \dots, X_{t-k}) + \varepsilon_t,$$

where f is *not* known, is equally difficult. In my discussion of Bartlett (1990)—see also Tong (1990), I have suggested three methods to determine k from the observations $(x_1, x_2, \dots, x_N)$. They are all based on nonparametric regression using the kernel method. It turns out that two of them are asymptotically equivalent and are non-parametric versions of Akaike's FPE. Specifically, let $\hat{x}_{t,p}$ denote the kernel estimate of $E[X_t|X_{t-1}, \dots, X_{t-p}]$, either by the FPE-like approach or the cross validation approach based on the observations $x_1, \dots, x_N$ omitting x_t. Cheng and Tong (1990) have shown that under general regularity conditions

$$CT(p) \stackrel{\Delta}{=} \frac{1}{N-r+1} \sum_{t=r}^{N} (x_t - \hat{x}_{t,p})^2 W(x_{t-1}, \dots, x_{t-p})$$

$$(2.2.6)$$

$$= \sigma^2 \left(\int W(x_1, \dots, x_p) dx_1 \dots dx_p \right) [1 + c(N\, B_N^p)^{-1} + o_R((N\, B_N^P)^{-1}],$$

where σ^2 is var(ε_t), $W(.)$ is a weight function, c is a constant dependent on the functional form of the kernel with bandwidth B_N, and o_R denotes the small 'o' in probability. For large N, $B_N < 1$. Then the term $c(N\, B_N^p)^{-1}$, which represents the penalty of the complexity (i.e. p) of the model, tends to infinity as p tends to infinity, in exactly the same spirit as Akaike's FPE but at an exponential rate dependent on B_N^{-1}.

The problem is much harder if (2.2.5) is generalized to the multiplicative noise case

$$(2.2.7) \qquad X_t = f(X_{t-1}, X_{t-2}, \ldots, X_{t-k}, \varepsilon_t),$$

with f remaining unknown.

2.3 Continuous Time vs. Discrete Time. As far as applications are concerned, it seems a matter of convenience as to which parameter space should be used. For example, if the sampliing is at regular intervals, then discrete time is quite natural. On the other hand, if (a) the recording is by an analogue device, or (b) the sampling is at irregular intervals (this includes the case of missing observations), or (c) the sampling rate is much greater than the 'natural frequency' of the system under study, then continuous time would be appropriate. Tong (1990, Ch. 5) has detail of the testing for linearity for case (b).

Now, it is well known that a continuous time linear autoregresive model may be obtained as the limit (in some sense) of a discrete time linear autoregressive model as the sampling interval decreases to zero and the autoregressive parameters tend to the boundary of the region of stationarity. Specifically consider a discrete time linear autoregressive model of order 1, namely

$$(2.3.1) \qquad X_n = \alpha X_{n-1} + \varepsilon_n, \quad |\alpha| < 1,$$

where ε_n are i.i.d $N(0,1)$ and $n \in \mathbf{Z}$. Let δ denote the sampling interval and let $t \in \mathbf{R}$ such that $n = \left[\dfrac{t}{\delta}\right]$. From (2.3.1),

$$X_n - X_{n-1} = -(1-\alpha)X_{n-1} + \varepsilon_n,$$

i.e.

$$(2.3.2) \qquad X_{\left[\frac{t}{\delta}\right]} - X_{\left[\frac{t}{\delta}-1\right]} = -\frac{(1-\alpha)}{\delta} X_{\left[\frac{t}{\delta}-1\right]}\delta + W_{\left[\frac{t}{\delta}\right]} - W_{\left[\frac{t}{\delta}\right]}$$

where W_n is defined to be $\Sigma^n \varepsilon_j$. Suppose that

$$1 - \alpha = a\delta.$$

Let $\delta \to 0$ (and thus $\alpha \to 1$). Equation (2.3.2) then tends (in the weak sense) to

$$(2.3.3) \qquad dX_t = aX_t dt + d\,W_t,$$

which is a continuous time linear autoregressive model of order 1. It would be interesting to see if a similar development could be applied to a non-linear autoregressive model. Consider, for example, a first order threshold autoregressive model

$$(2.3.4) \qquad X_n = \begin{cases} \alpha X_{n-1} + \varepsilon_n & \text{if } X_{n-1} \leq 0 \\ \beta X_{n-1} + \varepsilon_n & \text{if } X_{n-1} > 0 \end{cases}$$

where ε_n are i.i.d $N(0,1)$, $n \in \mathbf{Z}$ and $\alpha < 1$, $\beta < 1$, $\alpha\beta < 1$. Use the same set-up as previously but suppose now that

$$1 - \alpha = a\delta,$$

and

$$1 - \beta = b\delta.$$

Let $\delta \to 0$. It seems plausible that eqn. (2.3.4) tends (in the weak sense) to

$$(2.3.5) \qquad dX_t = \begin{cases} -aX_t dt + dW_t & \text{if } X_t \leq 0 \\ -bX_t dt + dW_t & \text{if } X_t > 0. \end{cases}$$

Detail will be reported elsewhere.*. Results of the above type are potentially useful to shed further light on the probabilistic structure of the discrete time model via that of its continuous time limit because the latter is often more amenable to analytical handling. A possible spin-off may be that the sampling properties of parameter estimates for the discrete time will become more forthcoming even under non-standard conditions, e.g., non-ergodic cases.

Conversely, starting with a continuous time autoregressive model, say

$$(2.3.6) \qquad dX_t = f(X_t)dt + d W_t,$$

where f is smooth and W_t is a Wiener process, one would be interested to know the result of discretization, i.e. sampling $\{X_t\}$ at regular intervals, say Δ. First, consider (2.3.6) without $d W_t$. In this case, on using Taylor's series, it holds that

$$(2.3.7) \qquad X_{t+\Delta} - X_t \doteq (e^{\Delta D} - 1)X_t,$$

where D is the differential operator, i.e. $D X_t = dX_t/dt$. Thus,

$$\frac{X_{t+\Delta} - X_t}{\Delta} \doteq \left(\frac{e^{\Delta D} - 1}{\Delta}\right) X_t$$
$$= \left(\frac{e^{\Delta D} - 1}{\Delta}\right) D^{-1} f(X_t)$$
$$(2.3.8) \qquad = \hat{f}(X_t), \quad \text{say}.$$

Now, Ozaki (1985) has proposed, as a discretization of (2.3.6), the discrete time autoregressive model

$$(2.3.9) \qquad \frac{X_{t+\Delta} - X_t}{\Delta} = \tilde{f}(X_t) + \frac{\varepsilon_{t+\Delta}}{\sqrt{\Delta}},$$

where $\varepsilon_t \sim N(0,1)$ and are i.i.d and

$$(2.3.10) \qquad \tilde{f}(x) = \begin{cases} \left(\frac{e^{\Delta J_x} - 1}{\Delta}\right) J_x^{-1} f(x), & \text{if } J_x \neq 0 \\ f(x) & \text{if } J_x = 0. \end{cases}$$

$$(2.3.11) \qquad J_x = D f(x).$$

*Note: This will be done in collaboration with Pham Dinh Tuan

Effectively, as far as the 'drift term' is concerned, he has replaced the operator D by a scalar multiplication. (Compare $\tilde{f}$ with $\hat{f}$.) This device was developed much earlier by the NASA engineers. See, e.g. Smith (§6.3, 1987), who referred to his own work in 1974, and R. Bowles' paper in 1973 under the name of *local linearization algorithm*. It was designed to simulate continuous-time nonlinear systems by *piecewise linear difference equations*. It is clear that Δ has to be sufficiently small. It is unclear to me how to rigorize Ozaki's heuristics for the complete model which includes the noise term. Brockwell et al. (1990), and Tong and Yeung (1990) have discussed the modelling of (2.3.6) using discrete time data where f is piecewise linear. At general level, it should be remarked that the solution of Ito's stochastic differential equation is based on the consideration of discretization. Results there are directly relevant beyond the Bowles–Smith–Ozaki quantization.*

One distinct advantage of continuous time models is the possibility of building on models suggested by subject matter considerations, e.g. from the background physics, biology, etc.

2.4 Continuous State Space vs. Discrete State Space. Most models in the non-linear time series literature of the narrowly defined sense are concerned with the continuous state space, typically $\boldsymbol{R}^k$. Of course, in a broader sense, discrete state space has a very long history at least in the theory of Markov chains in both discrete time and continuous time. Fairly recently, explicit discrete time series models with discrete state space have been constructed, which retain familiar second order structure, e.g. exponential damping. One such class is based on the ingenious device of *binomial thinning*, an idea which seems to originate in point processes. Specifically, let $\{B_i(\alpha)\}$ be a sequence of i.i.d. binary random variables such that

$$P[B_i(\alpha) = 1] = \alpha,$$

where $0 \le \alpha \le 1$. Define the *binomial thinning* operation, $*$, of a discrete random variable X by

$$(2.4.1) \qquad \alpha^* X = \sum_{i=1}^{X} B_i(\alpha).$$

As a typical example of such an operation, consider the first order autoregressive model in discrete time

$$(2.4.2) \qquad X_t = \alpha^* X_{t-1} + \varepsilon_t, \quad t \ge 1, \quad 0 < \alpha < 1,$$

where X_0 is Poisson with mean θ and ε_t is Poisson with mean $\theta(1 - \alpha)$ independent of X_{t-1}. It can be shown that X_t is stationary with a Poisson distribution (mean θ) and $\mathrm{corr}(X_0, X_k) = \alpha^k$. For detail see, e.g., McKenzie (1985).

---

*I owe this remark to Professor M. Rosenblatt.

2.5 Substantive vs. Empirical. Following Sir David Cox, by a substantive approach I mean one which builds on the subject matter under study. As an example of this approach, I would mention Rodriguez–Iturbe et al. (1986), who have modelled storm rainfall using specifically a physical model as the starting point, and Brillinger et al. (this volume). The complement of a substantive approach is called an empirical approach, sometimes also called a black-box approach. Models such as ARMA belong to this category.

If the background theory is sufficiently well developed (such as instances in the 'hard' sciences), then it would be quite natural to adopt a substantive approach. On the other hand, it is unclear to me if a substantive approach is necessarily profitable for the 'soft sciences'. Moreover, it seems to me that an empirical approach would always have a role to play even in the hard sciences when it comes to forecasting and control. The forecasting of sunspot numbers and the restoring of a human body to normal health are cases in point.

3. Theme II: Models. This theme consists of 4 pairs: (i) additive noise vs. multiplicative noise, (ii) on-line vs. off-line, (iii) long memory vs. short memory and (iv) univariate vs. multivariate.

3.1 Additive Noise vs. Multiplicative noise. Chapter 3 of Tong (1990) has listed no fewer than 12 classes of discrete time non-linear time series models, mostly of the empirical type. Let it be agreed that models of the form

$$(3.1.1) \qquad\qquad X_t = a(X_{t-1}) + \varepsilon_t, \qquad t \in Z,$$

where ε_t are i.i.d. and independent of $X_s, s < t$, are called models with additive noise, and those of the form,

$$X_t = m(X_{t-1}, \varepsilon_t)$$

or more generally

$$(3.1.2) \qquad\qquad X_t = m(X_{t-1}, \varepsilon_t, \varepsilon_{t-1}), \qquad t \in Z,$$

are called models with multiplicative noise. Extension to include higher lags is immediate but, for simplicity, only the simplest forms are illustrated. Before listing examples of $a(.)$ and $m(.)$, it might be instructive to think of a discrete time analogue of the diffusion process, namely

$$(3.1.3) \qquad\qquad X_t = E[X_t|X_{t-1}] + \{\mathrm{Var}[X_t|X_{t-1}]\}^{1/2}\varepsilon_t,$$

where the first summand is analogous to the 'drift-term' and the second to the 'diffusion term'. If the 'instantaneous variance' is a constant independent of states, then we have an additive noise model otherwise a multiplicative noise model. Examples of $a(.)$ include

(i)   *threshold autoregressive model with additive noise:*

$$a(x) = \begin{cases} \alpha + \beta x & \text{if } x \le r \\ \gamma + \delta x & \text{if } x > r \end{cases}$$

(ii)   *fractional autoregressive model:*

$$a(x) = \frac{\alpha + \beta x}{\gamma + \delta x}$$

(iii)   exponentially damped autoregressive model:

$$a(x) = (\alpha + \beta e^{-\gamma x^2})x \ .$$

Examples of $m(.)$ include

(i)   *bilinear model*

$$m(X_{t-1}, \varepsilon_t, \varepsilon_{t-1}) = \varepsilon_t + \alpha \varepsilon_{t-1} X_{t-1}$$

(ii)   *ARCH model:*

$$m(X_{t-1}, \varepsilon_t, \varepsilon_{t-1}) = \varepsilon_t \sqrt{\gamma + \beta X_{t-1}^2}$$

(iii)   non-linear moving average model:

$$m(X_{t-1}, \varepsilon_t, \varepsilon_{t-1}) = \varepsilon_t + \beta \varepsilon_t \varepsilon_{t-1}^2.$$

Extension of $m(.)$ to include η_t's, independent of ε_t's, yields more examples, e.g.

(iv)   *random coefficient autoregressive model*

$$m(X_{t-1}, \varepsilon_t, \eta_t) = (\alpha + \eta_t)X_{t-1} + \varepsilon_t$$

and

(v)   *newer exponential autoregressive model:*

$$m(X_{t-1}, X_{t-2}, \ldots, X_{t-p}, \varepsilon_t, \eta_t) = \beta^{(\eta_t)} X_{t-\eta_t} + \varepsilon_t,$$

where η_t takes values in $\{0, 1, \ldots, p\}$ and $\beta^{(j)} \in \boldsymbol{R}, j \in \{0, 1, \ldots, p\}$.

If one does not tie $a(.)$ and $m(.)$ down to any specific functional form, then one obtains quite general finite parametric models such as Priestley's state dependent models (1988) and Tjøstheim's doubly stochastic models (1986). I can see that the former provides a framework for a systematic development of an algorithmic technique and the latter permits some general discussion of non-linear modelling. It is, however, still unclear ot me what are the major advantages of a very general class of finite parametric models. Here, I am influenced by the fact that, in the usual non-linear regression modelling, one does not postulate a general class of models. If flexibility is desired, it would seem to me that a non-parametric approach might be quite practical, but then the sampel size required for modelling would be substantial. Recall the 'curse of dimensionality'. Casdagli (1989), An and Cheng (1990), Diebolt (1990) and Lewis and Stevens (1990) give important developments in this area, in which the idea of thresholds, i.e. dividing the state space into regimes, is exploited.

3.2 On-line vs. Off-line. The models listed in §3.1 are usually fitted off-line in the sense that the model fitting is not done recursively as new observations become available. Although off-line technology will probably remain useful in many applications, e.g. historical data, laboratory results already obtained, etc., there are situations where on-line technology will be indispensable, e.g. on-line process control. Kitagawa (1987) and West and Harrison (1989) are important recent developments in the on-line technology. Until quite recently, the emphasis in the on-line area seems to be linear (though possibly non-Gaussian) time series. It seems that the emphasis is being shifted to non-linear time series, see, e.g., Kitagawa (1989). Wei (1990) has discussed the connection between the residual sum of squares (RSS) of an on-line model with the RSS of its off-line counterpart.

3.3 Long Memory vs. Short Memory. Much attention has recently been given to long memory time series models. Cox (1984) is an excellent review of results up to then. A second order stationary time series is said to have *long memory* if its autocovariance function is not absolutely integrable (summable); otherwise it is said to have *short memory*. For linear time series in discrete time, the fractionally differenced ARMA model, independently proposed by Granger and Joyeux (1980) and Hosking (1981), seems to be the only example of long memory models. For non-linear time series in discrete time, I know of no examples which do not incorporate a fractional differencing or, more generally, a linear model with long memory.* It might also be relevant to consider more general definitions of long memory in the non linear case.

For continuous time, fractional Brownian motion proposed by Mandelbrot and van Ness (1968) has long memory.

3.4 Univariate vs. Multivariate. Multivariate non-linear time series models are only studied rather superficially, especially in terms of real applications. The special issue 'Time Series Analysis in Water Resources' (Ed. K.W. Hipel 1985) by the American Water Resources Association has articles in this area.

4. Theme III: Methodologies. This theme consists of 4 pairs: (i) informal vs. formal, (ii) drift criterion vs. stability criterion, (iii) Bayesian vs. frequentist, and (iv) ergodic vs. non-ergodic

4.1 Informal vs. Formal. By the informal methods is meant graphical methods. Tong (1990, Chapter 5) has a fairly comprehensive account, which includes time plots, histograms (univariate and bivariate), sample autocorrelation plots, sample partial autocorrelation plots, sample spectral density functions, sample bispectral density functions, reverse data plots, directed scatter plots, non-parametric lagged regressions, plots of diagonal elements of the hat matrices after fitting either a linear autoregressive model or a threshold autoregressive model, normal probability plots, half normal probability plots, skeleton plots, profile likelihood plots,

---

*Dr. Martin Casdagli has informed me at the IMA meeting of the paper by Bak, Tang and Weverfeld entitled 'Self organised criticality' in *Phys. Rev. Letters*, 1987, which might constitute an example

etc. These are useful for both exploratory analysis and diagnostics and most of these are readily available, even on a personal computer. For example, the PC package STAR3 accompanying Tong (1990) includes most of these. Formal methods usually consist of statistical tests, such as those for linearity, for whiteness, for normality and unimodality. Chapter 5 of Tong (1990) gives a fairly comprehensive account, which includes the bispectral test, Keenan's one-degree-of-freedom test, Tsay's test, CUSUM test, likelihood ratio test, Ljung-Box's test, Li-McLeod's test, Lin-Mudholkar's test, Silverman's test, etc. Like other areas of statistical modelling, the informal and the formal methods complement each other and should always be used together.

4.2 Drift Criterion vs. Stability Criterion. A standard method to prove ergodicity of a discrete time non-linear time series model which can be treated as a Markov chain on R^k say is to use the drift criterion, which originated in Foster (1950) and reached its maturity in Tweedie (1975) and Nummelin (1985). The basic ingredients are a non-negative measurable function g (rather like an "energy" function) and a special set, C, in R^k (called a *small set*, which is rather like the "centre" of the state space). Both g and C are problem specific. Then under quite general conditions on the Markov chain, ergodicity is assured if

(4.2.1)   (i) $$E[g(X_{t+1})|X_t = x] < B, \quad x \in C$$

and

(ii) $$E[r\, g(X_{t+1})|X_t = x] < g(x) - \gamma, \quad x \notin C ,$$

where $r > 1$, $\gamma > 0$ and $B > 0$ are real constants. Essentially, there is a 'drift' towards the 'centre C'. This criterion is purely probabilistic. Tjøsteim (1990) has extended the criterion by considering $E[g(X_{t+h})|X_t]$, where h is an integer greater than 1. Recalling the decomposition (2.2.4), it is tempting to enquire whether it is possible to establish ergodicity of the time series by focusing on the 'low dimensional attractor' i.e., the 'slowly varying component'. The advantage of doing so is the emergence of a method which is almost non-probabilistic. (It is not wholly non-probabilistic because some assumption is still necessary on the distribution of the driving (white) noise). Chan and I have shown that, under fairly general conditions, it is possible to deduce ergodicity from the stability (in an appropriate sense) of the slowly varying component (see, e.g., Tong, 1990, especially the Appendix by K.S. Chan).

In any given situation, if suitable g and C are readily available, then naturally the drift criterion should be adopted. Otherwise, the stability criterion may become useful either in its own right or as a vehicle leading to the identification of suitable g and C. In effect, one now has two stones to kill one bird. It should be emphasized that in the title of this subsection the word 'vs.' should really be replaced by 'and', in the spirit of father and son.

4.3 Bayesian vs. Frequentist. Up to now, most of the statistical tools in non-linear time series analysis belong to the frequentist school, with West and Harrison (1989) as a notable exception. The situation will almost certainly be different in the near future. For example, it is possible to obtain a Bayes factor for comparing a linear autoregression with a threshold autoregression, and to construct (Bayesian) confidence intervals for the threshold parameters of the latter. (See, e.g., Kheradmandnia and Tong, (1990.) These problems are far from being straightforward within the frequentist paradigm. Their non-standard nature has important implications too. (See Tong, 1990.) For example, if Akaike's information criterion is used for model selection, which is essentially a likelihood ratio minus twice the degrees of freedom, then the 'penalty' term may require modification. It is also quite possible that the Bayesian method will have an impact on the on-line modelling.

4.4 Ergodic vs. Non-ergodic. To date, almost all the statistical results in non-linear time series analysis assume ergodicity. Pham et al. (1989) is the only exception, to the best of my knowledge. They have obtained sufficient conditions for the strong consistency of the (conditional) least squares estimates of a non-ergodic threshold autoregressive model of order 1 with a known threshold. Much energy seems to have been expended in the study of *linear* non-ergodic time series, e.g. the study of the so-called 'unit root' problem. Is there any future for a "non-linear unit root"?

Acknowledgement. I wish to thank the French Government for awarding me a Visiting Professorship at the Fourier University of Grenoble and to Professors A. Le Breton and Pham Dinh Tuan for stimulating discussions during the preparation of this article. I also thank the IMA for its hospitality during my visit.

REFERENCES

AN, H.Z. AND CHENG, B. (1989), *A Kolmogorov–Smirnov type statistics with applications to test for nonlinearity in time series*, Tech. Rep., Institute of Appl. Maths., Academia Sinica, Beijing, China.

BABLOYANTZ, A. AND DESTEXHE, A. (1986), *Low dimensional chaos in an instance of epilepsy*, Proc. Math. Acad. Sci., U.S.A., 83, 3513–3517.

BARTLETT, M.S. (1990), *Chance or Chaos*, J.R. Statist. Soc. A. 153, 321–347.

BRILLINGER, D.R. (1975), *Time Series Data Analysis and Theory*, New York: Holt, Rinehardt and Winston.

BROCKWELL, P.J., R.J. HYNDMAN AND G.K. GRUNWALD (1990), *Continuous time threshold autoregressive models*, Tech. Rep. Colorado State University, U.S.A..

CASDAGLI, M. (1989), *Nonlinear prediction of chaotic time series*, Physica D, 35, 35, 335–356.

CHENG, B. AND TONG, H. (1990), *On order determination of unknown nonlinar autoregression*, Tech. Rep., July 1990, Institute of Mathematics and Statistics, University of Kent, Canterbury, Kent, U.K. (To be read Is the Royal Statistical Society, Oct., 1991).

COX, D.R. (1984), *Long range dependence: a review. Statisticss: An appraisal*, Ed. H.A. David and H.T. David, Iowa State Univ. Press. (Proceedings 50th Am. Conf. Iowa State Stat. Lab.).

DIEBOLT, J. (1990), *Testing the functions defining a non-linear autoregressive time series*, To appear in Stoch. Proc. Applic..

FARMER, J.D., OTT, E. AND YORKE, J.A. (1983), *The dimension of chaotic attractors*, Physica, 7d, 153–180.

FOSTER, F.G. (1953), *On the stochastic matrices associated with certain queuing processes*, Ann. Math. Stat., 24, 355–360.

GERRARD, R. AND TONG, H.(1990), *Noisy Chaos*, (Under preparation) first draft read to SERC Workshop on Nonlinear Time Series Analysis, Edinburgh, U.K. July 12–25 (1989).

GRANGER, C.W. J. and ANDERSEN, A.P. (1978), *An Introduction to Bilinear Time Series Models*, Gottingen: Vandehoek & Ruprecht.

GRANGER, C.W.J. AND JOYEUX, R. (1980), *An introduction to long-memory time series models and fractional differencing*, J. Time Series Anal., 1, 15–30.

HASSELMAN, K., MUNK, W. and MACDONALD, G. (1963), *Bispectrum of ocean waves*, Time Series Analysis (Ed. M. Rosenblatt), New York: J. Wiley, pp. 125–139.

HOSKING, J.R.M. (1981), *Fractional differencing*, Biometrika, 68, 165–176.

KHERADMANDNIA, M. AND TONG, H. (1990), *A Bayesian approach to threshold autoregressive modelling*, Tech. Rep., July 1990, Inst. of Maths & Stat., Univ of Kent, U.K.

KITAGAWA, G.(1987), *Non-Gaussian state space modelling of non-stationary time series*, J. Am. Stat. Assoc., 82, 1032–1063.

KITAGAWA, G. (1989), *Smoothing time series with non linear state space model*, Tech. Rep., Inst. Stat. Maths., Tokyo.

LEWIS, P.A.W. and STEVENS, J.G. (1990), *Nonlinear modelling of time series using multivariate adaptive regression splines (MARS)*, Rep. Naval Post. Sch. Monterey, California, U.S.A. (May 1990).

LII, K.S., HELLAND, K.N. AND ROSENBLATT, M. (1982), *Estimating three dimensional energy transfer in isotropic turbulence*, J. Time Series Anal., 3, 1–28.

MANDELBROT, B.B. and VAN NESS, J.W. (1968), *Fractional Brownian motions, fractional noise, and applications*, SIAM Rev., 10, 422–437.

McKENZIE, E. (1985), *Some simple models for discrete variate time series*, Time Series Analysis in Water Resources (Ed. K.W. Hipel) Am. Water. Res. Assoc. 656–50.

NUMMELIN (1984), *General Irreducible Markov Chains and Non-Negative Operators*, Cambridge: Cambridge University Press.

OZAKI, T. (1985), *Statistical identification of storage models with applications to stochastic hydrology*, In Time Series Analysis in Water Resources (Ed. K.W. Hipel) pp. 663–676. Am. Wate Res. Assoc.

PHAM, DIN THUAN, CHAN, K.S. AND TONG, H. (1989), *Strong consistency of the least squares estimator for a non-stationary threshold autoregressive model*, Contributed paper to the 47th Session of the International Statistical Institute, Paris.

PRIESTLEY, M.B. (1988), *Non Linear and Non Stationary Time Series Analysis*, London: Academic Press.

RODRIGUEZ-ITURBE, I., COX, D.R. and EAGLESON, P.S. (1986), *Spatial modelling of total storm rainfall*, Proc. R. Soc. London A., 403, 27–50.

ROSENBLATT, M. (1985), *Stationary Sequences and Random Fields*, Basel: Birkhäuser.

SMITH, J.M. (1987), *Mathematical modelling and digital simulation for engineers and scientists*, (Second ed., first edition in 1976.) New York: J. Wiley.

SUBBA RAO, T. AND GABR, M.M. (1984), *An introduction to bispectral analysis and bilinear time series models*, Lecture Notes in Statistics, No. 24, Springer-Verlag.

TJØSTHEIM, D. (1986), *Some doubly stochastic time series models*, J. Time Ser. Anal., 7, 51–72.

TJØSTHEIM, D. (1990), *Nonlinear time series and Markov chains*, Tech. Rep., Dept. Maths., Univ. Bergen, Norway.

TONG, H. (1983), *Threshold Models in Non-linear Time Series Analysis*, Lecture Notes in Statistics, No. 21, New York: Springer-Verlag.

TONG, H. (1989), *On the effect of noise on the Lyapunov exponents*, Tech. Rep, Institute of Mathematics and Statistics, University of Kent, U.K. July 1989.

TONG, H. (1990), *Non-linear Time Series: A Dynamical System Approach*, Oxford University Press.

TONG, H. AND YEUNG, I. (1991), *On threshold autoregressive modelling in continuous time*, To appear in Statistica Sinica, Vol 1 No. 2.

WEI, C.Z.(1990), *On predictive least squares principles*, Tech. Rep., Dept. of Maths. Univ. of Maryland.

WEST, M. AND HARRISON, P.J. (1989), *Bayesian Forecasting and Dynamic Models*, Heidelberg: Springer-Verlag.

A NONPARAMETRIC FRAMEWORK
FOR TIME SERIES ANALYSIS*

YOUNG K. TRUONG†

Abstract. Much of time series analysis deals with inference concerning the unknowns in the stochastic model for a random phenomena. In parametric approach the unknowns are a specific finite number of parameters while in the nonparametric approach they are smooth functions. This paper describes a nonparametric framework for examining the problem of estimating the conditional mean and conditional median function involving time series. Specifically, the effect of correlated structure on smoothing procedures such as kernel method based on local mean and local median will be examined, and recent results on selecting a sequence of estimators that achieves the optimal rates of convergence will also be addressed.

Key words. Nonparametric regression, local polynomial estimator, optimal rate of convergence, time series, autoregressive process.

AMS(MOS) subject classifications. Primary 62G05; Secondary 62E20.

1. Introduction. Let $(\mathbf{X}, Y)$ denote a pair of random variables such that $\mathbf{X}$ is $\mathbf{R}^d$ valued and Y is real-valued. Consider the situation in which it is desired to model the relationship between $\mathbf{X}$ and Y by the regression function $E(Y|\mathbf{X})$. For example, given an univariate time series $\{X_t : t = 0, \pm 1, \dots\}$, one is interested in the relationship between $Y = X_{t+1}$ and $\mathbf{X} = (X_{t-1}, \dots, X_{t-d})$ by examining the regression function

$$E(X_t | X_{t-1}, \dots, X_{t-d}).$$

In this context, $\mathbf{X}$ is referred to as the "past", and Y is the "present". In the bivariate case, one can explore the relation between the series $\{Z_t; t = 0, \pm 1, \dots\}$ and $\{X_t; t = 0, \pm 1, \dots\}$ via

$$E(Z_t | X_{t-1}, \dots, X_{t-d}).$$

More generally, one may also consider the adaptive model:

$$E(Z_t | Z_{t-1}, \dots, Z_{t-k}, X_{t-k-1}, \dots, X_{t-d})$$

with k being a positive integer equal to or less than d.

From the prediction point of view, the function $E(Y|\mathbf{X})$ (also called the L^2-predictor) arises naturally because it minimizes the *mean squared error*. Alternately, to address issues on robust time series analysis, especially when outliers may be present; one may choose to adopt the *mean absolute deviation* as a measure of accuracy in predicting Y and consider the conditional median function of Y on $\mathbf{X}$, med$(Y|\mathbf{X})$. See Bloomfield and Steiger (1983). This approach has provided regression analysis a very important alternative in modelling the relationship between $\mathbf{X}$

---

*This research was supported in part by the Institute for Mathematics and its Applications with funds provided by the National Science Foundation.

†Department of Biostatistics, University of North Carolina, Chapel Hill

and Y. In fact, for symmetric and light tailed conditional distributions of Y given $\mathbf{X}$, the conditional mean function would be a reasonable choice for exploring the association between the variables. However, the conditional median function would be more appropriate for skewed or heavy tailed conditional distributions.

Note that these functions depend on the unknown joint distribution of the $(\mathbf{X}, Y)$, hence one has to estimate them from a set of data or realization: $(\mathbf{X}_1, Y_1)$, $\ldots, (\mathbf{X}_n, Y_n)$. In the *parametric* approach, one starts with specific assumptions about the relationship between the response Y and the explanatory variable $\mathbf{X}$ and, about the variation in the response that may or may not be accounted for by the explanatory variables. For instance, the standard autoregressive method in time series starts with an a priori model for the regression function $\theta(\cdot)$ which, by assumption or prior knowledge, is a linear function that contains finitely many unknown parameters. Under the assumption that the joint distribution is Gaussian, it is an optimal prediction rule; if the distribution is non-Gaussian, it is not generally possible to determine the function $\theta(\cdot)$; so one might settle for the *best* linear predictor. By contrast, in the *nonparametric* approach, the regression function will be estimated directly without assuming such an a priori model for $\theta(\cdot)$. As pointed out in Stone (1985), this approach is more *flexible* than the standard regression method; *flexibility* means the ability of the model to provide accurate fits in a wide variety of realistic situations, inaccuracy here leading to *bias* in estimation. In recent years, nonparametric estimation has become an active area in time series analysis because of its flexibility in fitting data. See, for example, Bierens (1983), Collomb (1984), Collomb and Härdle (1986), Györfi et al. (1989), Robinson (1983), Truong (1988), Truong (1989), Truong and Stone (1988), Truong and Stone (1990) and Yakowitz (1985, 1987).

The purpose of the present paper is to give a general account for nonparametric estimation of conditional mean and median functions based on a sequence of mixing random variables. Its aim is to derive these nonparametric estimators heuristically and provide motivations for conditions under which results on the achievability of optimal rates of convergence can be established. Moreover, to stimulate interests in the development of nonparametric function estimation involving time series, there will be a discussion on some important inequalities and imbedding theorems that are useful for proving resutls for sequences of mixing random variables.

The rest of the paper is organized as follows. Section 2 derives a class of nonparametric estimators for the conditional mean and median functions. Section 3 describes optimal rates of convergence for iid sequence of random varaibles. Section 4 and its subsections discuss the conditions involved and various nonparametric function estimators for achieving the optimal rates of convergence in the time series context. These include local M-estimators, local polynomial estimators and some related open problems. Finally, numerical examples are treated in Section 5.

2. Nonparametric estimators. Let $\{(\mathbf{X}_i, Y_i), i = 0, \pm 1, \ldots\}$ be an $(d+1)$ vector-valued strictly stationary series and set $\theta(\mathbf{x}) = E(Y_0 | \mathbf{X}_0 = \mathbf{x}_0)$. A naive nonparametric estimator of $\theta(\cdot)$ can be derived as follows. Set

$$F(\mathbf{x}) = P(\mathbf{X}_0 \leq \mathbf{x}),$$
$$G(\mathbf{x}, y) = P(\mathbf{X}_0 \leq \mathbf{x}, Y_0 \leq y),$$
$$A = \text{nbhd}(\mathbf{x}_0; \delta_n),$$

where, for each $n \geq 1$, δ_n is a positive number. Let $F_n(\cdot)$, $G_n(\cdot, \cdot)$ denote the empirical cumulative distribution functions defined by $\mathbf{X}_1, \ldots, \mathbf{X}_n$ and $(\mathbf{X}_1, Y_1), \ldots,$ $(\mathbf{X}_n, Y_n)$, repectively. Then

$$\begin{aligned}
E(Y_0|\mathbf{X}_0 = \mathbf{x}_0) &\approx E(Y_0|\mathbf{X}_0 \in A) \\
&= \frac{\iint_{\mathbf{x} \in A} y \, G(d\mathbf{x}, dy)}{\int_{\mathbf{x} \in A} F(d\mathbf{x})} \\
&\approx \frac{\iint_{\mathbf{x} \in A} y \, G_n(d\mathbf{x}, dy)}{\int_{\mathbf{x} \in A} F_n(d\mathbf{x})} = \frac{\sum Y_i 1_{\{\mathbf{X}_i \in A\}}}{\sum 1_{\{\mathbf{X}_j \in A\}}},
\end{aligned}$$

provided δ_n is small.

Now set $I_n(\mathbf{x}) = \{i : 1 \leq i \leq n \text{ and } \|\mathbf{X}_i - \mathbf{x}\| \leq \delta_n\}$ and $N_n(\mathbf{x}) = \#I_n(\mathbf{x})$, where $\|\mathbf{x}\| = (x_1^2 + \cdots + x_d^2)^{1/2}$ for $\mathbf{x} = (x_1, \ldots, x_d) \in \mathbf{R}^d$. Let the local mean

$$\hat{\theta}_n(\mathbf{x}) = \frac{1}{N_n(\mathbf{x})} \sum_{I_n(\mathbf{x})} Y_i, \qquad \mathbf{x} \in \mathbf{R}^d$$

or, simply

$$\hat{\theta}_n(\mathbf{x}) = \text{ave}(Y_i : \mathbf{x} \in I_n(\mathbf{x}))$$

be a conditional mean function estimator. Similarly, let the local median

$$\hat{\theta}_n(\mathbf{x}) = \text{med}(Y_i : \mathbf{x} \in I_n(\mathbf{x})), \qquad \mathbf{x} \in \mathbf{R}^d$$

be a conditional median function estimator.

3. Optimal rates of convergence. In deriving the estimator of $\theta(\mathbf{x}) = E(Y|\mathbf{X} = \mathbf{x})$, we note that δ_n should be small. On the other hand, since the variance of the local mean estimator is proportional to $1/N_n(\mathbf{x}) \sim 1/n\delta_n^d$ (because $N_n(\mathbf{x})$ is the number of $\mathbf{X}_i$ falling into $\text{nbhd}(\mathbf{x}; \delta_n)$), one should also have $n\delta_n^d \to \infty$. Thus rates of convergence of the estimator is related to the question: "How fast should $\delta_n \to 0$?" For simplicity, let $(\mathbf{X}_1, Y_1), \ldots, (\mathbf{X}_n, Y_n)$ denote a random sample from the distribution of $(\mathbf{X}, Y)$ and set $\theta(\mathbf{x}) = E(Y|\mathbf{X} = \mathbf{x})$. Then

$$\hat{\theta}_n(\mathbf{x}) - \theta(\mathbf{x}) = N_n^{-1}(\mathbf{x}) \sum_{I_n(\mathbf{x})} \left(Y_i - \theta(\mathbf{X}_i) \right) + N_n^{-1}(\mathbf{x}) \sum_{I_n(\mathbf{x})} \left(\theta(\mathbf{X}_i) - \theta(\mathbf{x}) \right).$$

Conditional on $\mathbf{X}_1, \ldots, \mathbf{X}_n$, the first term on the right is the average of iid mean zero random variables, which can be considered as the *noise* or *variance* of $\hat{\theta}_n(\mathbf{x}) - \theta(\mathbf{x})$. Thus

$$E\left(\left(N_n^{-1}(\mathbf{x}) \sum_{I_n(\mathbf{x})} (Y_i - \theta(\mathbf{X}_i)) \right)^2 \middle| \mathbf{X}_1, \ldots, \mathbf{X}_n \right) = O\left(N_n^{-1}(\mathbf{x}) \right).$$

On the other hand, if $\theta(\cdot)$ has a bounded first derivative at $\mathbf{x}$, then

$$N_n^{-1}(\mathbf{x}) \sum_{I_n(\mathbf{x})} \left(\theta(\mathbf{X}_i) - \theta(\mathbf{x})\right) = O(\delta_n),$$

which, can be viewed as the *bias* of $\hat{\theta}_n(\mathbf{x}) - \theta(\mathbf{x})$. Now suppose the density of $\mathbf{X}$ is bounded from above and below at $\mathbf{x}$ so that $N_n(\mathbf{x}) = O_p(n\delta_n^d)$. Here, for sequence of random variables $\{V_n\}$ and sequence of numbers $\{b_n\}$, $V_n = O_p(b_n)$ means that

$$\lim_{c\to\infty} \overline{\lim_n} \, P(|V_n| > cb_n) = 0.$$

Then by the usual bias and variance decompostion,

$$E\left((\hat{\theta}_n(\mathbf{x}) - \theta(\mathbf{x}))^2 | \mathbf{X}_1, \dots, \mathbf{X}_n\right) = \text{bias}^2 + \text{variance} = O_p\left(\delta_n^2 + n^{-1}\delta_n^{-d}\right).$$

Minimize the right hand side with respect to δ_n, we obtain $\delta_n \sim n^{-1/(2+d)}$ and hence

$$E\left((\hat{\theta}_n(\mathbf{x}) - \theta(\mathbf{x}))^2 | \mathbf{X}_1, \dots, \mathbf{X}_n\right) = O_p\left(n^{-1/(2+d)}\right).$$

Let C be a fixed compact subset having a nonempty interior and let $g(\cdot)$ be a real-valued function on $\mathbf{R}^d$. Set

$$\|g(\cdot)\|_q = \left\{\int_C |g(\mathbf{x})|^q \, d\mathbf{x}\right\}^{\frac{1}{q}}, \quad (1 \le q < \infty); \qquad \|g(\cdot)\|_\infty = \sup_{\mathbf{x}\in C} |g(\mathbf{x})|.$$

Under appropriate conditions, Stone (1980, 1982) established the local rates of convergence

$$|\hat{\theta}_n(\mathbf{x}) - \theta(\mathbf{x})| = O_p\left(n^{-1/(2+d)}\right), \qquad \mathbf{x} \in C;$$

the L_q rates of convergence

$$\|\hat{\theta}_n(\cdot) - \theta(\cdot)\|_q = O_p\left(n^{-1/(2+d)}\right), \qquad 1 \le q < \infty,$$

and the L_∞ rate

$$\|\hat{\theta}_n(\cdot) - \theta(\cdot)\|_\infty = O_p\left((n^{-1}\log n)^{1/(2+d)}\right).$$

Moreover, these rates can not be improved. Hence they are called the optimal rates of convergence. (Stone's results are more general than the one presented here, we will address the general case later.)

The above results are the starting points for our investigation of function estimation in time series and that is the subject of the next section.

4. Nonparametric function estimation involving time series. In this section, we consider the problem of estimating the conditional mean and median functions $\theta(\cdot)$ based on a realization $(\mathbf{X}_1, Y_1), \ldots, (\mathbf{X}_n, Y_n)$ from the stationary series $\{(\mathbf{X}_t, Y_t) : t = 0, \pm 1, \ldots\}$. From the above heuristic arguments, the function $\theta(\cdot)$ should be continuous so that the bias can be estimated. It is also necessary to assume that the density of $\mathbf{X}$ be bounded from above and below in order to estimate the rate of growth of $N_n(\mathbf{x})$. Formally, these conditions are given as follows.

CONDITION A.

1. *The function $\theta(\cdot)$ has a bounded first derivative on C.*

2. *The density of $\mathbf{X}_0$ is bounded away from zero and infinity on C.*

3. *The conditional density of $\mathbf{X}_j$ ($j \geq 1$) given $\mathbf{X}_0$ is bounded away from zero and infinity on C.*

Condition A.3 is required for bounding the variance of various terms in establishing results on rates of convergence for mixing sequence of random variables.

The asymptotic properties of local mean or local median estimators depend on some sort of weak dependence condition (also called mixing) on the stationary sequence. This weak dependence can be formulated in a number of ways, among the weakest are α-mixing (Brillinger (1983) and Rosenblatt (1956, 1985). Let $\mathcal{F}_t$ and $\mathcal{F}^t$ denote the σ-fields generated respectively by $(\mathbf{X}_i, Y_i)$, $-\infty < i \leq t$, and $(\mathbf{X}_i, Y_i)$, $t \leq i < \infty$. Given a positive integer k set

$$\alpha(k) = \sup\{|P(A \cap B) - P(A)P(B)| : A \in \mathcal{F}_t \quad \text{and} \quad B \in \mathcal{F}^{t+k}\}.$$

The stationary sequence is said to be α-mixing or strongly mixing if $\alpha(k) \to 0$ as $k \to \infty$. Results on rates of convergence require (i), (ii), or (iii) of the following condition. (Note that (i), (ii) and (iii) are increasingly strong forms of α-mixing.)

4. (i) $\sum_{i \geq N} \alpha(i) = O(N^{-1})$ as $N \to \infty$.

    (ii) $\sum_{i \geq N} \alpha^{1-\frac{2}{\nu}}(i) = O(N^{-1})$ as $N \to \infty$, $(\nu > 2)$.

    (iii) $\alpha(N) = O(\rho^N)$ as $N \to \infty$ for some ρ with $0 < \rho < 1$.

Additional conditions related to the function $\theta(\cdot)$ will be given in the following sections. We consider the conditional mean function first.

4.1 Rates of convergence for local mean estimators

In this section, let $\theta(\cdot)$ denote the conditional mean function so that,

$$\theta(\mathbf{x}) = E(Y_0 | \mathbf{X}_0 = \mathbf{x}).$$

Also, denote its estimator by

$$\hat{\theta}_n(\mathbf{x}) = \text{ave}(Y_i : i \in I_n(\mathbf{x})).$$

To obtain the rates for various terms, the following condition is neccesary.

CONDITION B. *There is a positive constant $\nu > 2$ such that*

$$\sup_{\mathbf{x} \in U} E(|Y_0|^\nu | \mathbf{X}_0 = \mathbf{x}) < \infty.$$

THEOREM 1. *Suppose Conditions A and B hold. Then $n^{-1/(2+d)}$ is the optimal local and L_2 rates of convergence. Moreover, suppose that the series Y_i is bounded. Then $(n^{-1}\log n)^{1/(2+d)}$ is the optimal rates of convergence in the L_∞ norm.*

Note that the term "optimal" is included because a sequence of iid random variables is a special case of mixing stationary series. An interesting numerical example related to this result is given in Section 5

4.2 Rates of convergence for local median estimators. This section considers the problem of estimating the conditional median function for stationary sequence of random variables. Denote the conditional median function by $\theta(\mathbf{x}) = \mathrm{med}(Y_0|\mathbf{X}_0 = \mathbf{x})$ and its estimator by

$$\hat{\theta}_n(\mathbf{x}) = \mathrm{med}(Y_i : i \in I_n(\mathbf{x})).$$

The condition below is required to guarantee the uniqueness of the conditional median (uniqueness will ensure consistency) and also the achievability of the desired rate of conververgence. If the conditional density is not bounded away from zero around the median the desired rate of convergence will not be achievable. (The same condition is required in order to obtain the usual asymptotic result about the sample median in the univariate case.)

CONDITION C. *The conditional density of Y_0 given $\mathbf{X}_0 = \mathbf{x}$ is bounded away from zero and infinity over a neighborhood of the median.*

THEOREM 2. *Suppose that Conditions A and C hold. Then $n^{-1/(2+d)}$ is the optimal local and L_q ($1 \le q < \infty$) rates of convergence, and $(n^{-1}\log n)^{1/(2+d)}$ is the optimal L_∞ rates.*

4.3 Some useful inequalities

Although proofs of theorems in previous sections can be found in Truong and Stone (1988); however, we should mention the following important inequalities, not only they form the main crux for proving the above results, they are also useful for establishing asymptotic results for mixing sequence of random variables in general. The first is about tail probabilities approximation for dependent binomial random variables and its proof uses martingale inequalities.

PROPOSITION 1. *Let $\{\xi_j, j \ge 1\}$ be a strictly stationary sequence of real-valued random variables, centered at expectations and uniformly bounded by 1. Suppose that $\{\xi_j, j \ge 1\}$ is α-mixing and that $\sigma^2 = E\xi_1^2 + 2\sum_{j \ge 2} E\xi_1\xi_j < \infty$. Let c_1, c_2 and γ denote positive constants such that $0 < \gamma < \frac{1}{2}$. Then for any $R > 0$,*

$$P\left(|\textstyle\sum_{j \le n}\xi_j| > Rn^{1/2}\right)$$
$$\le \begin{cases} O\left(\exp\left(-c_1 R^2/\sigma^2\right) + n\alpha([n^\gamma])\left(\sigma^{-4} + R^{-2}\right)\right), & \text{if } R \le \sigma^2\sqrt{n}/n^\gamma; \\ O\left(\exp\left(-c_2 n\sigma^2/n^{2\gamma}\right) + n\alpha([n^\gamma])\left(\sigma^{-4} + R^{-2}\right)\right), & \text{if } R > \sigma^2\sqrt{n}/n^\gamma. \end{cases}$$

Proof. See Theorem 4 and Proposition 5.1 of Philipp (1982).

The following inequality is an extension of Whittle's inequalities to dependent variables, which is useful to establishing L^q rates of convergence for local median estimators. Suppose $0 < \epsilon < 1$ and let $\{\nu_n\}$ be a sequence of positive numbers so that $\nu_n \sim n^{-\epsilon}$.

PROPOSITION 2. *Let* $V_{n1}, \ldots, V_{nn}$ *be uniformly bounded random variables such that* V_{ni} *has mean zero and is a function of the α-mixing sequence* $\mathbf{X}_i$. *Suppose that* $E|V_{ni}| \leq \nu_n$ *and* $E|V_{ni}V_{nj}| \leq \nu_n^2$ *for* $1 \leq i < j \leq n$. *Suppose* $\alpha(N) = O(\rho^N)$, $N = 1, 2, \ldots$ *and let* k *be a positive integer. Then*

$$E(\textstyle\sum_i V_{ni})^k = O\left((n\nu_n)^{k/2}\right) \qquad \text{as } n \to \infty.$$

Proof. See Lemma 9 of Truong and Stone (1988).

These two inequalities are very useful for establishing results on rates of convergence for local median estimators. However, due to the boundedness conditions, they are not quite good enough for showing the achievability of optimal L^∞ rate of convergence for local mean estimators. The subsequent sections discuss various approaches to remedy this problem.

4.4 Rates of convergence for local M-estimators. Truong and Stone (1988) established optimal asymptotic properties for local mean and local median estimators under the α-mixing condition. In that approach, the L^∞ rate of convergence for local mean estimator is obtained by putting the boundedness condition on the response series Y_i. The current section addresses this problem by considering the local M-estimators.

Let $\psi(\cdot)$ denote a monotone function and let $\theta(\cdot) = \theta_\psi(\cdot)$ denote the function such that $E[\psi(Y_0 - \theta(\mathbf{X}_0))|\mathbf{X}_0] = 0$ almost surely. For example, $\theta(\mathbf{X}_0) = E(Y_0|\mathbf{X}_0)$ corresponds to $\psi(y) = y$, while $\psi(y) = \text{sign}(y)$ yields $\theta(\mathbf{X}_0) = \text{med}(Y_0|\mathbf{X}_0)$. The function $\theta(\cdot)$ defined through $\psi(\cdot)$ provides some interesting alternative to non-parametric regression function estimation. Indeed, Conditions B and C fail when the conditional distribution of Y_0 given $\mathbf{X}_0$ has heavy tails and its density is zero at the conditional median. To remedy this, it is necessary to take the function $\psi(y) = \max\{-k, \min\{y, k\}\}$ so that, $\theta(\mathbf{X}_0)$ is a form of trimmed mean of the conditional distribution of Y_0 given $\mathbf{X}_0$ (Huber, 1981). More generally, suppose the function $\psi(\cdot)$ is bounded and increasing with

$$E[\psi(Y - \theta(\mathbf{x}))|\mathbf{X} = \mathbf{x}] = 0, \qquad \mathbf{x} \in U.$$

Also,

$$|E[\psi(Y - \theta(\mathbf{x}) + t)|\mathbf{X} = \mathbf{x}]| > M_1|t| \qquad \text{for } |t| < M_2^{-1}, \mathbf{x} \in U,$$

$$|E[\psi(Y - \theta(\mathbf{x}) + t)|\mathbf{X} = \mathbf{x}]| > M_3 \qquad \text{for } |t| \geq M_2^{-1}, \mathbf{x} \in U,$$

and

$$\psi(y) = \begin{cases} -M_4 & \text{if } y \leq -M_5; \\ M_4 & \text{if } y \geq M_5; \end{cases}$$

where M_1–M_5 are positive constants such that $M_4 (\geq M_3)$ and $M_5 (\geq M_2^{-1})$. Given $\mathbf{x} \in \mathbf{R}^d$, define the estimator $\hat{\theta}_n(\mathbf{x})$ as the solution of the equation

$$\frac{1}{N_n(\mathbf{x})} \sum_{I_n(\mathbf{x})} \psi(Y_i - \hat{\theta}_n(\mathbf{x})) = 0.$$

THEOREM 3. *Suppose* $\delta_n \sim (n^{-1} \log n)^{1/(2+d)}$ *and that Condition A holds. Then there exists a positive constant c such that*

$$\lim_n P \left(\|\hat{\theta}_n(\cdot) - \theta(\cdot)\|_\infty \geq c(n^{-1} \log n)^r \right) = 0.$$

Moreover, suppose Condition A holds and that $\delta_n \sim n^{-1/(2+d)}$. *Then*

$$\|\hat{\theta}_n(\cdot) - \theta(\cdot)\|_q = O_p(n^{-r}), \qquad 1 \leq q < \infty.$$

This result indicates that the boundedness condition on the sequence Y_i can be relaxed by linearize the tail distribution of Y_i given $\mathbf{X}_i$ through the function $\psi(\cdot)$. For a proof of the above results, see Truong (1988).

Robust estimation was addressed by Collomb and Härdle (1986) on uniform consistency under ϕ-mixing, and by Boente and Fraiman (1989, 1990) under α-mixing conditions. The class of estimators considered there did not include local medians and the rates of convergence established by these authors are not optimal. Robinson (1984) established a central limit theorem for the local M-estimators under the α-mixing condition.

4.5. An embedding theorem. Another approach in removing the boundedness condition on the Y series is to use the truncation argument and an embedding theorem of Bradley (1983). The idea was first explored by Tran (1989).

PROPOSITION 3. *(Bradley, 1983) Let S denote a Borel space and let ξ and ζ be random variables taking values in S and $\mathbf{R}$, respectively. Suppose q and γ are positive constants such that $q \leq \|\zeta\|_\gamma = (E|\zeta|^\gamma)^{1/\gamma} < \infty$. Furthermore, suppose U is a uniform $[0,1]$ random variable independent of (ξ, ζ). Then there exists a real-valued random variable $\zeta^* = f(\xi, \zeta, U)$, where f is a measurable function from $S \times \mathbf{R} \times [0,1]$ into $\mathbf{R}$, such that*

(i) ζ^* *is independent of* ξ.

(ii) $\mathrm{dist}(\zeta^*) = \mathrm{dist}(\zeta)$

(iii) $P(|\zeta^* - \zeta| \geq q) \leq 18(\|\zeta\|_\gamma / q)^{\gamma/(2\gamma+1)} (\alpha(\xi, \zeta))^{2\gamma/(2\gamma+1)}$.

Let $\theta(\mathbf{x})$ denote the conditional mean function and $\hat{\theta}_n(\mathbf{x})$ be the local mean estimator.

THEOREM 4 (TRAN, 1989). *Suppose* $\delta_n \sim (n^{-1} \log n)^r$. *Also suppose Conditions A and B hold, and that $E|Y_0|^k < \infty$ for sufficiently large k. Then*

$$\lim_n P \left(\|\hat{\theta}_n(\cdot) - \theta(\cdot)\|_\infty \geq c(n^{-1} \log n)^r \right) = 0.$$

4.6 Semiparametric approaches. Let $\theta(\cdot)$ be a smooth real-valued function on $\mathbf{R}^d$ and consider the regression model:

$$Y_i = \theta(\mathbf{X}_i) + \sum_{u=-\infty}^{\infty} a_{i-u}\varepsilon_u,$$

where $\{a_u : u = 0, \pm 1, \pm 2, \dots\}$ is a sequence of unknown parameters satisfying $\sum_u |a_u| < \infty$, $\{\varepsilon_u : u = 0, \pm 1, \dots\}$ is a sequence of iid random variables with mean zero and variance σ^2. Furthermore, suppose the sequences $\{\mathbf{X}_i, i = 0, \pm 1, \pm 2, \dots\}$ and $\{\varepsilon_i, i = 0, \pm 1, \pm 2, \dots\}$ are independent, so that $\theta(\mathbf{X}_i) = E(Y_i|\mathbf{X}_i)$. Let $\phi(\cdot)$ denote the characteristic function of ε_0. Suppose the function $\log \phi(\cdot)$ has a Taylor expansion: (See Brillinger, 1981)

$$\sum_q C_q z^q/q! < \infty, \qquad z \text{ in some neighborhood of the origin,}$$

where $C_q = |\operatorname{cum}_q(\varepsilon_0)|$. Under these conditions and also suppose that Condition A holds, then $(n^{-1}\log n)^{1/(2+d)}$ is the optimal reates of convergence in the L_∞ norm.

Suppose now the function $\theta(\cdot)$ has bounded partial derivatives of order $(k+1)$ on C, where k is a nonnegative integer. Then the rates can be improved and that the above theorem becomes a special case of the following result. To achieve this, the local mean estimator will have to be replaced by the local polynomial estimator, which will now be introduced. Let $\hat{\theta}_n(\cdot; \mathbf{x})$ denote the polynomial of degree k that minimizes

$$\sum_{i \in I_n(\mathbf{x})} [Y_i - \hat{\theta}_n(\mathbf{X}_i; \mathbf{x})]^2.$$

Set $p = k + 1$ and $\gamma = p/(2p+d)$.

THEOREM 5. (Truong and Stone, 1990). *Suppose Conditions A.2–A.4 hold for the series $\mathbf{X}_i$ and that $\delta_n \sim (n^{-1}\log n)^{1/(2p+d)}$. Then there is a positive constant c such that*

$$\lim_n P\left(\|\hat{\theta}_n(\cdot) - \theta(\cdot)\|_\infty \geq c(n^{-1}\log n)^\gamma\right) = 0.$$

The above theorem can also be generalized to cover derivative estimation. For details, see Truong and Stone (1990).

Theorem 5 has an interesting application to parametric inference for the time series component. Specifically, let ν denote a positive interger and let the above regression model be given by

$$Y_i = \theta(\mathbf{X}_i) + Z_i, \quad i = 0, \pm 1, \dots,$$

with

$$Z_i = \beta_1 Z_{i-1} + \cdots + \beta_\nu Z_{i-\nu} + \varepsilon_i,$$

where the ν roots of the polynomial $1 - \beta_1 z - \cdots - \beta_\nu z^\nu$ are all greater than 1 in the absolute value, and ε_i, $i = 0, \pm 1, \dots$ are independent and identically distributed with mean zero and finite variance σ^2. Then $\{Z_i\}$ is an autoregressive process of order ν. By Theorem 3.1.1 of Brockwell and Davis (1987), there is a sequence a_u, $u = 0, \pm 1, \pm 2, \ldots$ of real numbers such that $\sum |a_u| < \infty$ and $Z_i = \sum_u a_{i-u}\varepsilon_u$ for $i = 0, \pm 1, \ldots$. We also assume that $\mathbf{X}_i$, $i = 0, \pm 1, \ldots$ is a C-valued stationary time series that is independent of ε_i, $i = 0, \pm 1, \pm 2, \ldots$, here C is a compact subset of $\mathbf{R}^d$ having a nonempty interior (for example, a d-dimensional rectangle).

Let $\hat{\theta}_n(\cdot; \mathbf{x})$ denote the local polynomial estimator of degree k with δ_n to be chosen according to the theorem below. Set $\hat{Z}_i = Y_i - \hat{\theta}_n(\mathbf{X}_i)$, $i = 1, \ldots, n$. Let $\hat{\beta}_n$ denote the parametric estimator of β obtained by regressing $\hat{Z}_i$ on $\hat{Z}_{i-1}, \ldots, \hat{Z}_{i-\nu}$.

THEOREM 6. Suppose $p > d/2$ and that $n^{-1/2d}(\log n)^{1/d} \ll \delta_n \ll n^{-1/4p}$. Then $\mathrm{dist}(\sqrt{n}(\hat{\beta}_n - \beta)) \to N(0, \sigma^2 \Gamma^{-1})$. Here Γ is the $\nu \times \nu$ given by $\Gamma = (\mathrm{cov}(Z_i, Z_j))_{i,j=1}^{\nu}$.

When there is only one covariate, i.e. $d = 1$, the above result on root-n parametric inference is valid by simply using local mean estimators. However, when there are two or three covariates, i.e. $d = 2$ or 3, it is necessary to use local polynomials of degree one for constructing the residual series $\hat{Z}_i = Y_i - \hat{\theta}_n(\mathbf{X}_i)$. In general, kth-degree local polynomials would be required to draw the usual parametric inference for the time series components when there are d covariates. The above result also indicates that the parametric estimates are insentitive to the δ_n. See Section 5 for a numerical example. Proof of this theorem can be found in Truong and Stone (1990).

4.7 Open problems. There are several interesting open questions related to the above developments.

QUESTION 1. Based the nonparametric setup, can the embedding theorem of Bradley be applied to establish optimal rates of convergence for local polynomial estimators?

QUESTION 2. What is the optimal rate of convergence for local polynomial estimators if the *least absolute deviation* is adopted as a measure of accuracy?

QUESTION 3. In the semiparametric approach, is Theorem 6 still valid if C is not bounded?

QUESTION 4. How would the order ν be consistently or efficiently selected?

QUESTION 5. Let $d \geq 2$ and suppose the function $\theta(\cdot)$ is additive:

$$\theta(\mathbf{x}) = \theta_1(x_1) + \cdots + \theta_d(x_d).$$

Is $\{n^{p/(2p+1)}\}$ still an achievable rate of convergence?

5. Numerical examples

EXAMPLE 1. An univariate time series of 200 observations is generated from the model

$$X_t = W(X_{t-1}) + \epsilon_t, \quad \epsilon_t \sim_{\mathrm{iid}} N(0, 0.5^2), \quad t = 1, \ldots, 200;$$

where

$$
W(x) = \begin{cases}
-x - 2, & \text{if } x < -1, \\
2x + 1, & \text{if } -1 \le x < 0, \\
-2x + 1, & \text{if } 0 \le x < 1, \\
x - 2, & \text{if } 1 \le x.
\end{cases}
$$

The series is plotted in Figure 1. The scatter plot of X_{t+1} vs X_t, $t = 1, \dots, 199$, is given in Figure 2, which displays some nonlinearity. The structure becomes transparent (Figure 3) when the scatter plot is smoothed by a local mean estimator described in Section 2. This smooth estimate is obtained by setting $\delta_n = 0.25$, which is chosen from a sequence of δ_n to provide the most reasonable fit. Figure 4 compares the estimate and the function $W(\cdot)$.

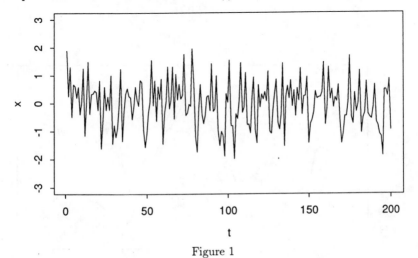

Figure 1

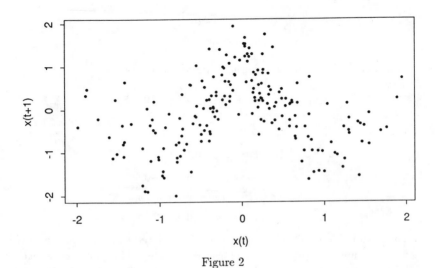

Figure 2

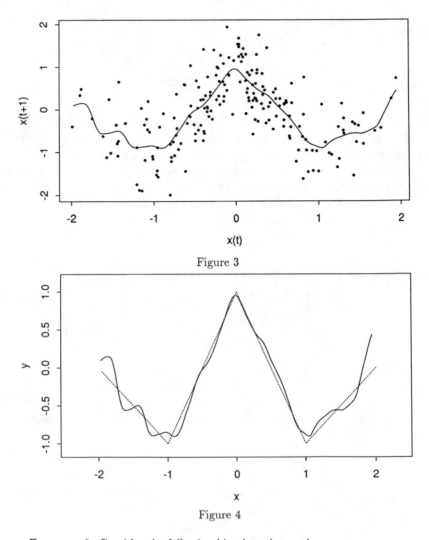

Figure 3

Figure 4

EXAMPLE 2. Consider the following bivariate time series

$$x_i = 0.5 \cdot x_{i-1} + \epsilon_i, \quad \epsilon_i \sim \text{iid } N(0, 0.5),$$

$$y_i = \theta(x_i) + z_i,$$

$$z_i = 0.5 \cdot z_{i-1} + \xi_i, \quad \xi_i \sim \text{iid } N(0, 0.5), \qquad i = 1, \ldots, 200,$$

$$\theta(x) = \begin{cases} -x, & \text{if } x < -0.5; \\ x^2 + 0.25, & \text{if } |x| \leq 0.5; \\ x, & \text{if } x > 0.5. \end{cases}$$

The X and Y series are plotted in Figure 5 and 6, respectively. The scatterplot is given in Figure 7 which shows a nonlinear relationship. This plot is smoothed by a local polynomial estimate with degree 1 and $\delta_n = 0.20$ (Figure 8). This is done by

383

trying several δ_n and 0.20 was chosen to provide the best fit. Figure 9 compares $\theta(\cdot)$ and $\hat{\theta}_n(\cdot)$, which indicates that the estimate is reasonable. However, it is not clear the linear curvatures were really there on the intervals $(-\infty, -0.5)$ and $(0.5, \infty)$. To see this feature, a derivative estimate is provided. Figure 10 shows the estimate and the actual derivative, which indicates that the estimate is reasonable in the middle, but not so well toward the ends where data are sparse.

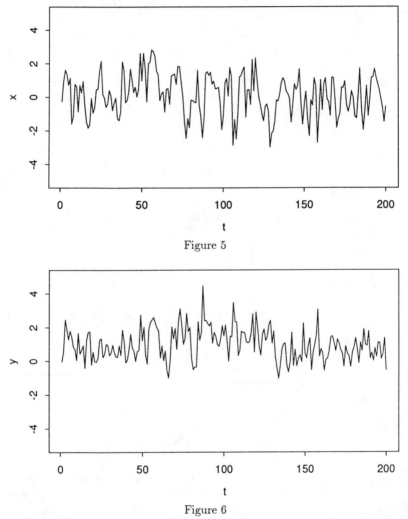

Figure 5

Figure 6

To see the effect of fitting parametric time series Z_t, the residuals series is obtained with $\delta_n = .20$. The estimates are obtained by solving the Yule-Walker equations, and the results are summerized in Table 1. Notice that the model selection was done by either AIC or BIC, and they are quite consistent. It is also important to point out in fitting the error process, the parametric estimates and the model selection rules are insentitive to δ_n, which has been indicated in Theorem 6.

384

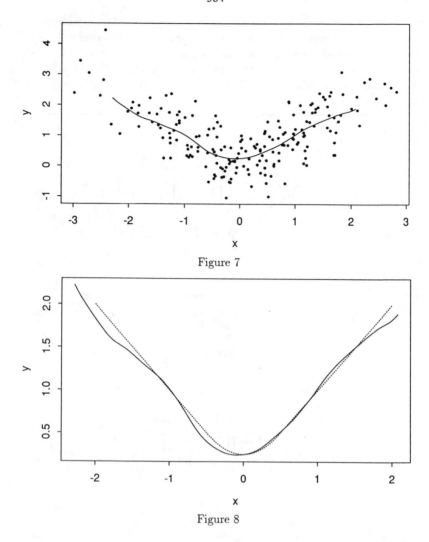

Figure 7

Figure 8

Table 1. Fitting parametric errors

p	$\hat{\phi}$	Resid Var	$AIC(p)$	$BIC(p)$
1	$\hat{\phi}_1 = 0.46672$	0.46601	-148.71	-142.11
2	$\hat{\phi}_1 = 0.48672, \hat{\phi}_2 = 0.04285$	0.46515	-147.08	-137.18
3	$\hat{\phi}_1 = 0.49138, \hat{\phi}_2 = 0.09580$			
	$\hat{\phi}_3 = 0.10879$	0.45965	-147.46	-134.27
4	$\hat{\phi}_1 = 0.49415, \hat{\phi}_2 = 0.09824$			
	$\hat{\phi}_3 = 0.12133, \hat{\phi}_4 = 0.25517$	0.45935	-145.59	-129.10

Acknowledgment. The author wishes to thank Chuck Stone for many helpful discussions.

385

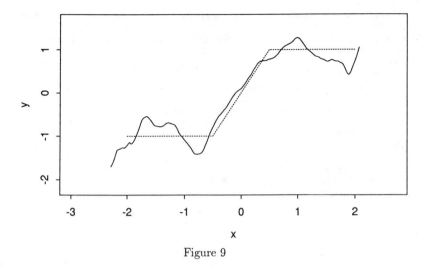

Figure 9

REFERENCES

BIERENS, H. J., *Uniform consistency of kernel estimators of a regression function under generalized conditions*, J. Amer. Statist. Assoc. 78 (1983), pp. 699–707.

BLOOMFIELD, P. AND STEIGER, W. L., *Least Absolute Deviations*, Birkhauser, Boston, 1983.

BOENTE, G. AND FRAIMAN, R., *Robust nonparametric regression estimation for dependent variables*, Ann. Statist. 17 (1989), pp. 1242–1256.

BOENTE, G. AND FRAIMAN, R., *Asymptotic distribution of robust estimators for nonparametric models from mixing processes*, Ann. Statist. 18 (1990) (to appear).

BRADLEY, R. C., *Approximation Theorems for strongly mixing random variables*, Michigan Math. J. 30 (1983), pp. 69–81.

BRILLINGER, D. R., *Time series: Data analysis and theory*, Holden-Day, San Francisco, 1981.

BRILLINGER, D. R., *The finite Fourier transform of stationary process*, in D. R. Brillinger and P. R. Krishnaiah, eds., *Handbook of Statistics, Vol 3.*, Elsevier Science Publishers B.V., 1983, pp. 21–37.

COLLOMB, B. G., *Propriétés de convergence presque complète du prédicteur à noyau*, Z. Wahrsch. verw. Gebiete 66 (1984), pp. 441–460.

COLLOMB, B. G. AND HÄRDLE, W., *Strong uniform convergence rates in robust nonparametric time series analysis and prediction: kernel regression estimation from dependent observations*, Stochastic Processes and their Applications 23 (1986), pp. 77–89.

GYÖRFI, L., W. HÄRDLE, SARDA AND P. VIEU, *Nonparametric estimation from time series*, Springer Lecture Notes, 1989.

HALL, P. AND HEYDE, C. C., *Martingale limit theory and its applications*, Academic Press, New York, 1980.

HUBER, P., *Robust Statistics*, John Wiley & Sons, Inc., New York, 1981.

PHILIPP, W., *Invariance principles for sums of mixing random elements and the multivariate empirical process*, Colloquia Mathematica Societatis Janos Bolyai. 36. Limit Theorems in Probability and Statistics (1982), pp. 843–873.

ROBINSON, P. M., *Nonparametric estimators for time series*, J. Time Series Anal. 4 (1983), pp. 185–207.

ROBINSON, P. M., *Robust nonparametric regression*, Robust and nonlinear time series analysis Franke, J., Härdle, W. and Martin, D., (eds.) Springer Lecture Notes, New York (1984).

ROSENBLATT, M., *A central limit theorem and a strong mixing condition*, Proc. Nat. Acad. Sci. U.S.A. 42 (1956), pp. 43–47.

ROSENBLATT, M., *Staionary Sequences and Random Fields*, Birkhäuser, Boston, 1985.

STONE, C. J., *Optimal rates of convergence for nonparametric estimators*, Ann. Statist. 8 (1980), pp. 1348–1360.

STONE, C. J., *Optimal global rates of convergence for nonparametric regression*, Ann. Statist. 10 (1982), pp. 1040–1053.

STONE, C. J., *Additive regression and other nonparametric models*, Ann. Statist. 13 (1985), pp. 689–705.

TRAN, L. T., *Nonparametric function estimation for time series by local average and local median estimators*, Manuscript (1989).

TRUONG, Y. K., *Nonparametric functional estimation from dependent observations*, Manuscript (1988).

TRUONG, Y. K., *Nonparametric curve estimation with time series errors*, Manuscript (1989).

TRUONG, Y. K. AND STONE, C. J., *Nonparametric function estimation involving time series*, Manuscript (1988).

TRUONG, Y. K. AND STONE, C. J., *Semiparametric time series regression*, Manuscript (1990).

YAKOWITZ, S., *Nonparametric density estimation, prediction, and regression for Markov Sequence*, J. Amer. Statist. Assoc. 80 (1985), pp. 215–221.

YAKOWITZ, S., *Nearest-neighbour methods for time series analysis*, J. Time Series Anal. 8 (1987), pp. 235–247.

REFLECTIONS*

JOHN TUKEY†

Perhaps the title, picked by Manny, is intended to hint that the content is "something glimpsed as in a mirror darkly"! Clearly I am supposed to look back to the past, and forward to the future.

ORIGIN (FOR JWT)

How did my own work in time series start? Very much from contact with real data and real applications! To illustrate this, let me summarize/sketch a half dozen early stimuli; two military and four geophysical.

1. W.T. Budenbom wanted to talk about the spectrum of tracking error for a new radar. Bud wanted estimates from data, so we calculated covariances and took a crude Fourier transform. R.W. Hanning noticed that smoothing with weights $1/4, 1/2, 1/4$ improved the picture greatly. Dick and I then spent a few months finding out why.

2. The Cornell Aeronautical Laboratory was studying turbulence in the free air by flying a highly-instrumental airplane (over a Great Lake) for a few hundred miles. (The answers were to be relevant to naval anti-aircraft fire.). Here we learned that visual "averaging" in reading strip charts of instrument indication can play merry hades with complex cross-spectral analyses.

3. Hans Panofsky was using John von Neumann's new computer to analyze, in spectrum terms, wind velocity vectors at various heights on the Brookhaven tower. The meaning of the cospectrum, as the frequency analysis of the Reynolds stress, was easily accepted. The meaningfulness of the quadrature spectrum required selling (zero-lift balloon). A result was the discovery of "eddies" rolling along the ground and steadily increasing in size from early morning to late afternoon.

4. Bill Pierson and his concern with (one- and two-dimensional) spectra of the sea surface was another stimulus.

5. Walter Munk and I interacted, over a long time span, in connection with a wide variety of oceanographic spectrum analyses.

6. Mike Healy came to Murray Hill with the intention of learning how statistics was used in an industrial context. Thanks to international Nuclear Test Ban negotiations, however, he, Bruce Bogert, and I worked on seismology with emphasis on discriminating between explosions and earthquakes. Out of this came, among other things, cepstra and pseudo-autocorrelations.

*Presented at Institute for Mathematics and its Applications Summer Program: New Directions in Time Series Analysis, University of Minnesota, July 16, 1990. Prepared in connection with research at Princeton University, sponsored by the Army Research Office (Durham), DAAL03-86-K-0073

†Princeton University, Fine Hall, Washington Road, Princeton, NJ 08544–1000

STATUS

My feeling – not necessarily correct, but $\cdots$ – is that our current frequency/time techniques are quite well developed (at least so far as the present cycle goes), so that the most difficult questions are not "how to *solve* it" but rather either "how to *formulate* it", or "how do we extend applicability to *less comfortable* conditions". Notice that either effective formulation, or effective broadening, almost certainly require deeper and more extensive interaction with data.

If we can find enough interesting and useful problems to try to formulate in somewhat familiar terms, we may continue to make some progress in our familiar framework, at least for a time. But it may well be desirable to go out in search of other kinds of problems:

- Those whose foundations seem to escape formulation in our familiar terms.
- Those whose behavior is too hard for the conventional techniques to handle.

Indeed, it is probably only through such routes that we can star a new cycle, using new techniques.

UNIFICATION

This morning, Manny Parzen began his account of the "soft mathematics" that gives "the big picture". We badly need to be deeply and carefully *ambivalent* about most unifications. On the one hand, unification can often produce a *big picture* (not *the* big picture), and thereby let us structure more effectively and efficiently whatever concepts and techniques are currently available.

But any such big picture has to be a *limited* picture. If we fail to be ambivalent – if we fail to respond to the Scots proverb that one may have a very fine house, but must "sit loose" to it – we can let each new Big Picture keep us from going further, even when some further progress becomes important or necessary.

AN EXAMPLE

In inference, for instance, we have learned – as some weeks that were spent here last summer discussing "robustness" show – to recognize that there are aspects of adequately broadly formulated problems – aspects such as distribution shape – which it is better not to try to estimate – either explicitly or implicitly. So far as I can see, this breaks the mold of any presently available "big picture". Yet the need to extend robust techniques across almost all areas of statistics is manifestly vital. We dare not hold to current "big pictures", if we are to be able to do what needs to be done.

BACK TO THE GENERAL CASE

We need then, to use big pictures to compact what we know. We need to do this without allowing these big pictures to keep us from breaking new grounds. (This needed new ground is rarely broken by something growing out of the big picture as a whole – rather some restricted part of the big picture gets generalized, or deeply modified.) In my experience we will need to study new real problems, keeping our ideas and approaches flexible, in order to break out into new ground as we should.

If we do not compact into big pictures, our attention will be absorbed by details. If we do not sit loose to our big pictures, we will fail to learn from new problems – fail to develop the new techniques that would open new doors, as well as providing the raw material needed as the main source of seeds for the new cycle, the next compaction, the next big picture. The aim of a big picture is not to learn the same amount in less time, but rather to save time for learning new things, overall learning *more* in the *same* amount of time.

Whenever we are not willing to be iconoclastic, to break the mold of the current "big picture", we shall be likely to miss major opportunities.

Those who have read Thomas Kuhn will realize that this view makes statisticians resemble physicists, working for a significant period within a single framework, but breaking out at intervals.

It is not enough, when we try to evaluate our position in the very large, to look only at the big picture – we must, at least occasionally – look at the *bigger* picture.